Descriptive Inorganic Chemistry

Descriptive Inorganic Chemistry

Geoff Rayner-Canham

W. H. Freeman and Company
New York

Library of Congress Cataloging-in-Publication Data

Rayner-Canham, Geoff.
Descriptive inorganic chemistry / by Geoff Rayner-Canham.
p. cm.
Includes index.
ISBN 0-7167-2819-2
1. Chemistry, Inorganic. I. Title.
QD151.5.R39 1995 95-11084
546—dc20 CIP

Printed in the United States of America

First printing 1995, HAM

Overview

Contents

Preface

The elements of the periodic table make up everything in the universe, and inorganic chemistry is the study of those elements and the compounds they form. In this book, I focus on the properties of the more interesting, important, and unusual elements and compounds rather than providing an exhaustive analysis of all the elements, which would resemble a chemical dictionary. However, inorganic chemistry is not just the study of properties and reactions; it includes explanations. To explain "why," it is necessary to look at the principles of chemistry, such as atomic structure, bonding, intermolecular forces, thermodynamics, and acid-base behavior. In this text, only those principles that are needed to explain chemical behavior are introduced.

The text begins by developing each important theoretical concept. Such topics are extensions of those introduced in general chemistry courses, although I have attempted to overlap the new and the old material rather than leave students with gaps in their knowledge of chemical principles. Throughout the text, the periodic table is used as a theme, although, as I repeatedly show, there are many aspects of chemical behavior that do not fit the simple division into groups and periods.

Inorganic chemistry goes beyond academic interest; it is an important part of our lives. Every chapter opens with some interesting human aspect of the subject. In the descriptive chapters, trends and patterns in the group are discussed next, and then elements and compounds that are interesting as a result of their uses or their unique chemistry are described. Wherever possible, I try to explain chemical behavior in terms of fundamental chemical principles. For the many compounds that play a major role in the economy, I describe the industrial synthesis and uses. Finally, every descriptive chapter ends with a section on biochemical relevance, for life would be impossible without the presence of a wide range of chemical elements.

This book has been written as an attempt to pass on my enthusiasm for descriptive inorganic chemistry to another generation. Thus the comments of the readers, both students and instructors, will be sincerely appreciated. Please send them to the author, care of W. H. Freeman and Company, 41 Madison Avenue, New York, NY 10010.

Many thanks must go to the people at W. H. Freeman who have been involved in bringing this project from concept to fruition. In particular, I am indebted to my editor, Deborah Allen, who guided the early stages, and Mary Louise Byrd, who assumed control in the later stages. The clarity of the text owes much to Jodi Simpson, the copy editor, who combed through the manuscript.

I wish to acknowledge the following reviewers, whose criticism and comments were much appreciated: E. Joseph Billo, Boston College; David

Finster, Wittenberg University; Stephen J. Hawkes, Oregon State University; Martin Hocking, University of Victoria; Vaké Marganian, Bridgewater State College; Edward Mottel, Rose-Hulman Institute of Technology; and Alex Whitla, Mount Allison University.

This book would never have been written but for the enthusiasm for inorganic chemistry that I acquired over the years from my teachers and mentors, three of whom deserve special recognition: Brian Bourne, Harvey Grammar School; Margaret Goodgame, Imperial College of Science and Technology; and Derek Sutton, Simon Fraser University. My interest has been sustained over the years by fruitful interactions with my many colleagues in the chemistry department at Sir Wilfred Grenfell College and with Dr. Howard Clase of Memorial University, St. John's. The late Ferris Hodgett, former vice-principal of Sir Wilfred Grenfell College, is remembered for his unwavering support over the years. Some of the work on this book was accomplished during a summer at the New College of the University of South Florida, and I thank Dr. Jane Stephens for making my stay possible. The project was completed while I was on sabbatical leave at the University of York, courtesy of Professor David Waddington and the Science Education Group.

Finally, I must express eternal gratitude for the support and encouragement of my spouse, Marelene F. Rayner-Canham.

CHAPTER

1

The Electronic Structure of the Atom: A Review

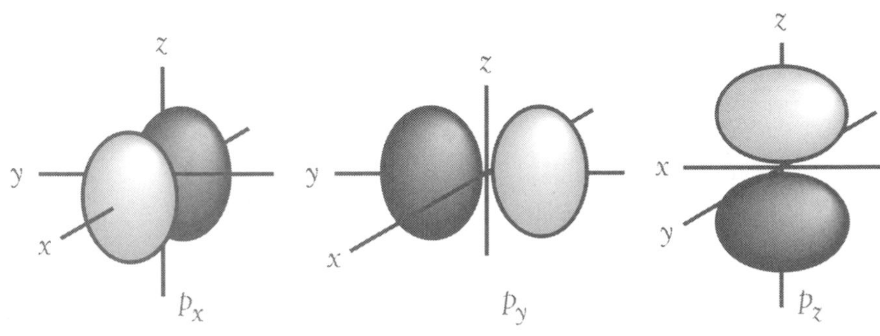

In the first quarter of the twentieth century, scientists realized that the behavior of electrons in atoms can only be described in terms of a probability model. To describe and predict the bonding in inorganic substances, it is essential to use orbital concepts derived from this probability model of electronic behavior.

It is amazing that Isaac Newton discovered anything at all, for he was the original model for the absent-minded professor. Supposedly, he always timed the boiled egg he ate at breakfast; one morning, his maid found him standing by the pot of boiling water, holding an egg in his hand and gazing intently at the watch in the bottom of the pot! Nevertheless, he initiated the study of the electronic structure of the atom in about 1700 when he noticed that the passage of sunlight through a prism produced a continuous visible spectrum (Figure 1.1). Much later, in 1860, Robert Bunsen (of burner fame) investigated the light emissions from flames and gases. Bunsen observed that the emission spectra, rather than being continuous, were series of colored lines (line spectra). He noted that each chemical element produced a unique and characteristic spectrum (Figure 1.2). Other investigators subsequently

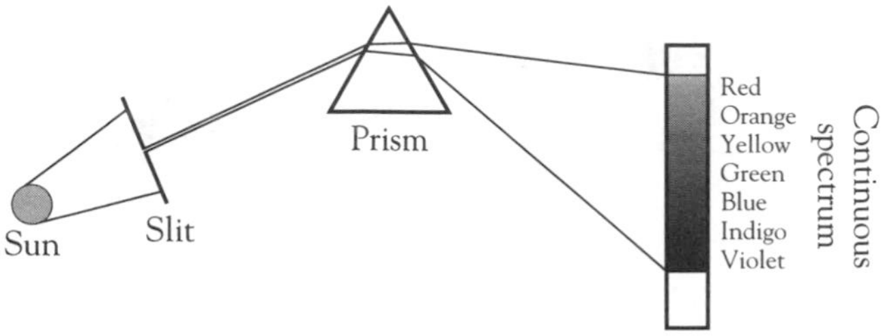

Sun
Slit
Prism
Red
Orange
Yellow
Green
Blue
Indigo
Violet
Continuous spectrum

Figure 1.1 A prism splits white light into the wavelengths of the visible spectrum.

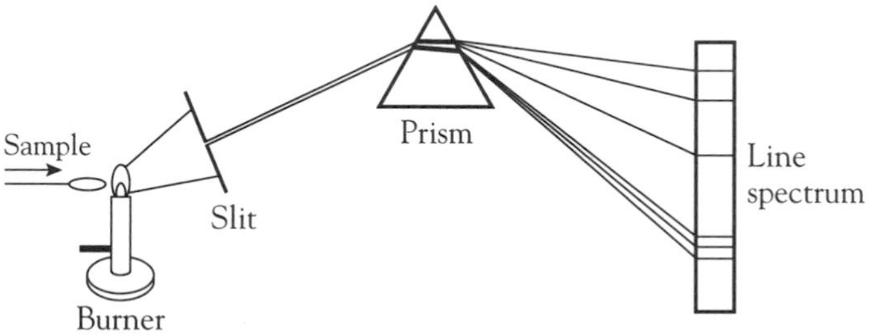

Sample
Slit
Prism
Burner
Line spectrum

Figure 1.2 A line spectrum is produced when an element is heated in a flame.

showed that there were, in fact, several sets of spectral lines for the hydrogen atom: one set in the ultraviolet region, one in the visible region, and a number of sets in the infrared part of the electromagnetic spectrum (Figure 1.3).

The explanation of the spectral lines was one of the triumphs of the *Bohr model* of the atom. In 1913 Niels Bohr proposed that the electrons could occupy only certain energy levels that corresponded to various circular orbits around the nucleus. He identified each of these levels with an integer, which he called a *quantum number.* The value of this parameter could range from 1 to ∞. He argued that, as energy was absorbed by an atom from a flame or electrical discharge, electrons moved from one quantum level to one or more higher energy levels. The electrons sooner or later returned to lower quantum levels, closer to the nucleus; and when they did, light was emitted. The wavelength of the emitted light directly corresponded to the energy of separation of the initial and final quantum levels. When the electrons occupied the lowest possible energy level, they were said to be in the *ground state.* If one or more electrons absorbed enough energy to move away from the nucleus, then they were said to be in an *excited state.*

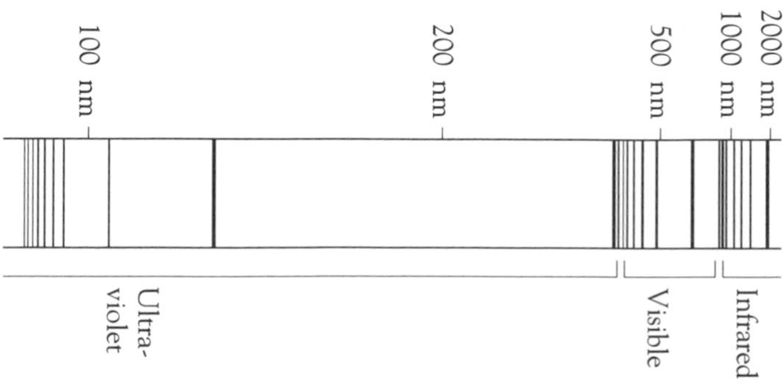

Figure 1.3 Emission spectrum of hydrogen.

Atomic Absorption Spectroscopy

A glowing body, such as the sun, is expected to emit a continuous spectrum of electromagnetic radiation. However, in the early nineteenth century, a German scientist, Josef von Fraunhofer, noticed that the visible spectrum from the sun actually contained a number of dark bands. Later investigators realized that the bands were the result of the absorption of particular wavelengths by cooler atoms in the "atmosphere" above the surface of the sun. The electrons of these atoms were in the ground state, and they were absorbing radiation at wavelengths corresponding to the energies needed to excite them to higher energy states. A study of these "negative" spectra led to the discovery of helium. Such spectral studies are still of great importance in cosmochemistry—the study of the chemical composition of stars.

In 1955 two groups of scientists, one in Australia and the other in Holland, finally realized that the absorption method could be used to detect the presence of elements at very low concentrations. Each element has a particular absorption spectrum corresponding to the various separations of (differences between) the energy levels in its atoms. When light from a powerful source is passed through a vaporized sample of an element, the particular wavelengths corresponding to the various energy separations will be absorbed. The higher the concentration of the atoms, the greater the proportion of the light that will be absorbed. The sensitivity of this method is extremely high, and concentrations of parts per million are easy to determine; some elements can be detected at the parts per billion level. Atomic absorption spectroscopy has now become a routine analytical tool in chemistry, metallurgy, geology, medicine, forensic science, and many other fields of science—and it simply requires the movement of electrons from one energy level to another.

However, the Bohr model had a number of flaws. For example, the spectra of multielectron atoms had far more lines than the simple Bohr model predicted. Nor could the Bohr model explain the splitting of the spectral lines in a magnetic field (a phenomenon known as the *Zeeman effect*). Within a short time, a radically different model, the quantum mechanical model, was proposed to account for these observations.

The Schrödinger Wave Equation and Its Significance

The more sophisticated quantum mechanical model of atomic structure was derived from the work of Louis de Broglie. De Broglie showed that, just as electromagnetic waves could be treated as streams of particles (photons), moving particles could exhibit wavelike properties. Thus it was equally valid to picture electrons either as particles or as waves. Using this wave-particle duality, Erwin Schrödinger developed a partial differential equation to represent the behavior of an electron around an atomic nucleus. This equation, given here for a one-electron atom, shows the relationship between the wave function of the electron, ψ, and E and V, the total and potential energies of the system. The second differential terms represent the wave function along each of the Cartesian coordinates x, y, and z; m is the mass of an electron, and h is the Planck constant.

$$\frac{\partial^2\psi}{\partial x^2} + \frac{\partial^2\psi}{\partial y^2} + \frac{\partial^2\psi}{\partial z^2} + \frac{8\pi^2 m}{h^2}(E - V)\psi = 0$$

The derivation of this equation and the method of solving it are in the realm of physics and physical chemistry. But the solution itself is of great importance to inorganic chemists. We should always keep in mind, however, that the wave equation is simply a mathematical formula. We attach meanings to the solution simply because most people need concrete images in order to think about subatomic phenomena. Fortunately, these images do not have to have any basis in reality to be useful.

Schrödinger argued that the real meaning of the equation could be found from the square of the wave function, ψ^2, which represents the probability of finding the electron at any point in the region surrounding the nucleus. There are a number of solutions to a wave equation. Each solution describes a different orbital and, hence, a different probability distribution for an electron in that orbital. Each of these orbitals is uniquely defined by a set of three integers: n, l, and m_l. Like the integers in the Bohr model, these integers are also called quantum numbers.

In addition to the three quantum numbers derived from the original theory, a fourth quantum number had to be defined to explain the results of a later experiment. In this experiment, it was found that passing a beam of hydrogen atoms through a magnetic field caused about half the atoms to be deflected in one direction and the other half in the opposite direction. Other investigators proposed that the observation was the result of two different electronic spin orientations. The atoms possessing an electron with one spin were deflected one way, and the atoms whose electron had the opposite spin were deflected in the converse direction. This spin quantum number was assigned the symbol m_s.

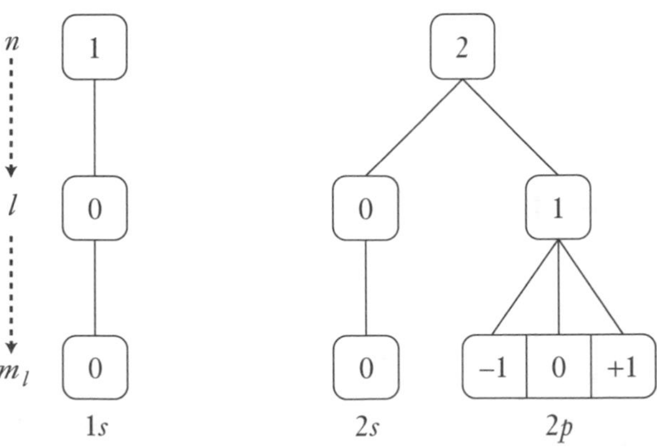

Figure 1.4 The possible sets of quantum numbers for $n = 1$ and $n = 2$.

Table 1.1 Correspondence between angular momentum number l and orbital designation

l value	Orbital designation
0	*s*
1	*p*
2	*d*
3	*f*

The possible values of the quantum numbers are defined as follows:

n, the *principal quantum number*, can have all positive integer values from 1 to ∞.

l, the *angular momentum quantum number*, can have all integer values from $n - 1$ to 0.

m_l, the *magnetic quantum number*, can have all integer values from $+l$ through 0 to $-l$.

m_s, the *spin quantum number*, can have values of $+\frac{1}{2}$ and $-\frac{1}{2}$.

For a principal quantum number of 1, there is only one possible set of quantum numbers n, l, and m_l (1, 0, 0); whereas for a principal quantum number of 2, there are four sets of quantum numbers (2, 0, 0; 2, 1, −1; 2, 1, 0; 2, 1, +1). This situation is shown diagrammatically in Figure 1.4. To identify the electron orbital that corresponds to each set of quantum numbers, we use the value of the principal quantum number n, followed by a letter for the angular momentum quantum number l. Thus, when $n = 1$, there is only the 1*s* orbital; when $n = 2$, there is one 2*s* orbital and three 2*p* orbitals (corresponding to the m_l values of +1, 0, and −1). The letters *s*, *p*, *d*, and *f* are derived from categories of the spectral lines: sharp, principal, diffuse, and fundamental. The correspondences are shown in Table 1.1.

When the principal quantum number $n = 3$, there are nine sets of quantum numbers (Figure 1.5). These sets correspond to one 3*s*, three 3*p*, and five

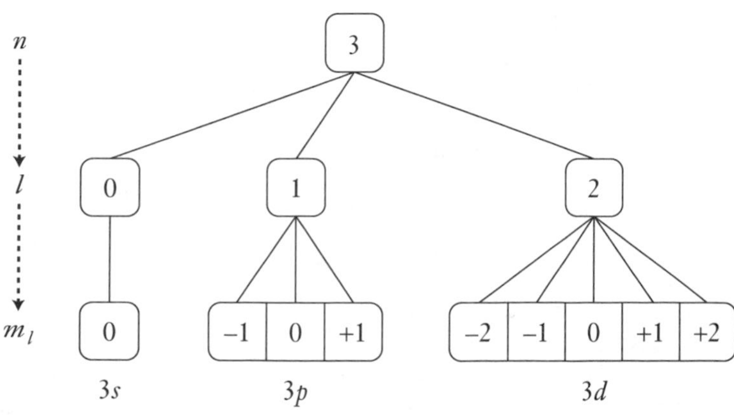

Figure 1.5 The possible sets of quantum numbers for $n = 3$.

Table 1.2 Correspondence between angular momentum number l and number of orbitals

l value	Number of orbitals
0	1
1	3
2	5
3	7

3*d* orbitals. A similar diagram for the principal quantum number $n = 4$ would show 16 sets of quantum numbers, corresponding to one 4*s*, three 4*p*, five 4*d*, and seven 4*f* orbitals (Table 1.2). Theoretically, we can go on and on, but as we shall see, the *f* orbitals represent the limit of orbital types among the elements of the periodic table.

Shapes of the Atomic Orbitals

Representing the solutions to a wave equation on paper is not an easy task. In fact, we would need four-dimensional graph paper (if it existed) to display the complete solution for each orbital. Thus our representations here will be somewhat simplified.

Each of the three quantum numbers derived from the wave equation represents a different aspect of the orbital:

The principal quantum number n indicates the size of the orbital.

The angular momentum quantum number l represents the shape of the orbital.

The magnetic quantum number m_l represents the spatial direction of the orbital.

The spin quantum number m_s has little physical meaning; it merely allows two electrons to occupy the same orbital.

An orbital diagram is used to indicate the probability of finding an electron at any instant at any location. An alternative viewpoint is to consider the locations of an electron over a lengthy period of time. We define a location where an electron seems to spend most of its time as an area of high *electron density*. Conversely, locations rarely visited by an electron are called areas of low electron density.

The s Orbitals

The *s* orbitals are spherically symmetric about the atomic nucleus. As the principal quantum number increases, the electron tends to be found farther from the nucleus. To express this idea in a different way, we say that, as the principal quantum number increases, the orbital becomes more diffuse.

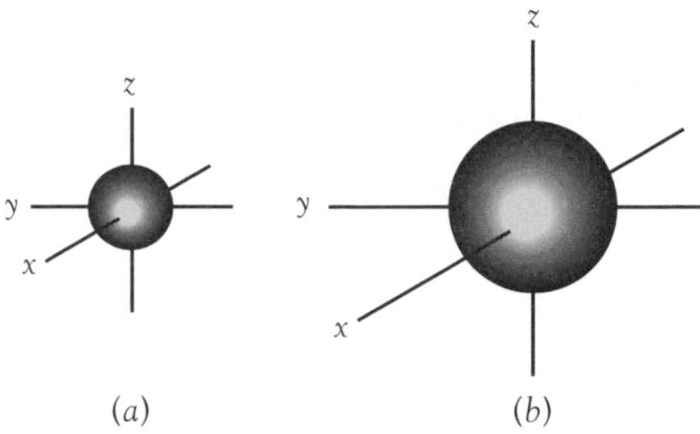

Figure 1.6 Representations of the shapes of the 1*s* and 2*s* orbitals. (Adapted from D. A. McQuarrie and P. A. Rock, *General Chemistry*, 2nd ed. [New York: W. H. Freeman, 1991], p. 322.)

Same-scale representations of the shapes of the 1*s* and 2*s* orbitals of an atom are compared in Figure 1.6. The volume of a 2*s* orbital is about four times greater than that of a 1*s* orbital. In both cases, the nucleus occupies a tiny volume at the center of the spheres. These spheres represent the 99 percent probability of finding an electron occupying that particular orbital. The total probability cannot be represented, for the probability of finding an electron drops to zero only at an infinite distance from the nucleus.

In addition to the enormous difference in size between the 1*s* and 2*s* orbitals, the 2*s* orbital has, at a certain distance from the nucleus, a spherical surface on which the electron density is zero. A surface on which the probability of finding an electron is zero is called a *nodal surface.* When the principal quantum number increases by 1, the number of nodal surfaces also increases by 1. We can visualize nodal surfaces more clearly by plotting a graph of the radial density distribution function as a function of distance from the nucleus for any direction. Figure 1.7 shows plots for the 1*s*, 2*s*, and 3*s* orbitals. These plots show that the electron tends to be farther from the nucleus as the principal quantum number increases. The areas under all three curves are the same.

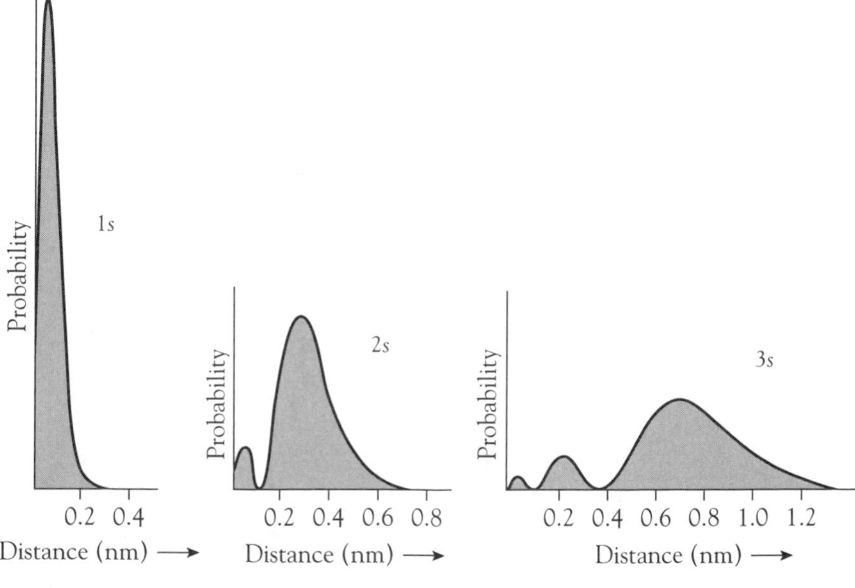

Figure 1.7 The variation of the radial density distribution function with distance from the nucleus for electrons in the 1*s*, 2*s*, and 3*s* orbitals of a hydrogen atom.

Electrons in an *s* orbital are different from those in *p*, *d*, or *f* orbitals in two significant ways. First, only the *s* orbital has an electron density that varies in the same way in every direction from the atomic nucleus out. Second, there is a finite probability that an electron in an *s* orbital is at the nucleus of the atom. Every other orbital has a node at the nucleus.

The *p* Orbitals

Unlike the *s* orbitals, the *p* orbitals are not spherically symmetric. In fact, the *p* orbitals consist of two separate volumes of space (lobes), with the nucleus located between the two lobes. Because there are three *p* orbitals, we assign each orbital a direction according to Cartesian coordinates: we have p_x, p_y, and p_z. Figure 1.8 shows representations of the three 2*p* orbitals. At right angles to the axis of higher probability, there is a nodal plane through the nucleus. For example, the $2p_z$ orbital has a nodal surface in the *xy* plane.

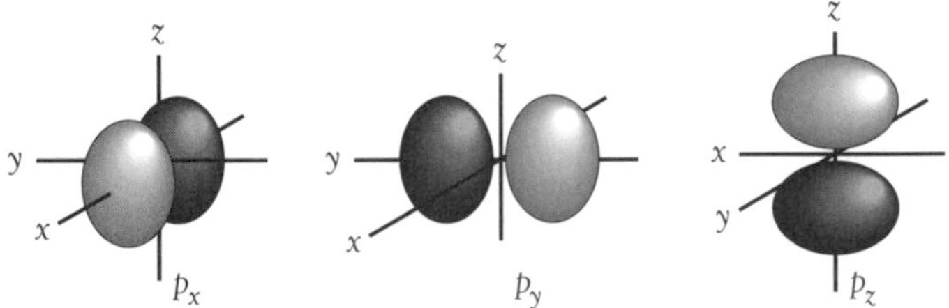

Figure 1.8 Representations of the shapes of the $2p_x$, $2p_y$, and $2p_z$ orbitals. (Adapted from D. F. Shriver, P. Atkins, and C. H. Langford, *Inorganic Chemistry*, 2nd ed. [New York: W. H. Freeman, 1994], p. 24.)

If we compare graphs of electron density as a function of atomic radius for the 2*s* orbital and a 2*p* orbital (the latter plotted along the axis of higher probability), we find that the 2*s* orbital has a much greater electron density close to the nucleus than does the 2*p* orbital (Figure 1.9). Conversely, the second maximum of the 2*s* orbital is farther out than the single maximum of the 2*p* orbital. However, the mean distance of maximum probability is the same for both orbitals.

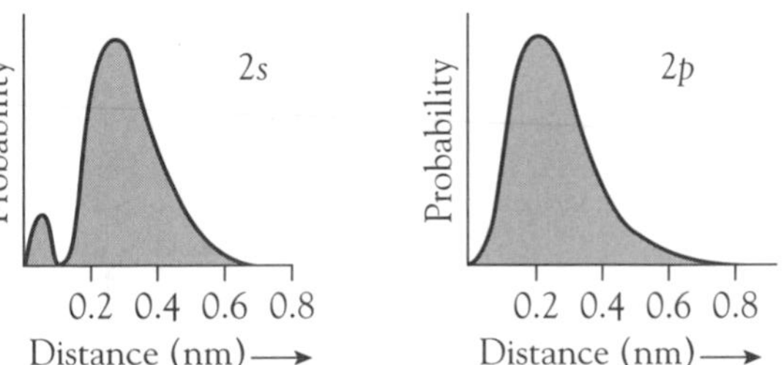

Figure 1.9 The variation of the radial density distribution function with distance from the nucleus for electrons in the 2*s* and 2*p* orbitals of a hydrogen atom.

Like the *s* orbitals, the *p* orbitals develop additional nodal surfaces within the orbital structure as the principal quantum number increases. Thus a 3*p* orbital does not look exactly like a 2*p* orbital. However, the detailed differences in orbital shapes for a particular angular momentum quantum number are of little relevance in the context of basic inorganic chemistry.

The *d* Orbitals

The five *d* orbitals have more complex shapes. Three of them are located between the Cartesian axes, and the other two are oriented along the axes. In all cases, the nucleus is located at the intersection of the axes. Three orbitals each have four lobes that are located between pairs of axes (Figure 1.10). These orbitals are identified as d_{xy}, d_{xz}, and d_{yz}. The other two *d* orbitals, d_{z^2} and $d_{x^2-y^2}$, are shown in Figure 1.11. The d_{z^2} orbital looks somewhat similar to a p_z orbital (see Figure 1.8), except it has an additional donut-shaped ring of high electron density in the *xy* plane. The $d_{x^2-y^2}$ orbital is identical to the d_{xy} orbital but has been rotated through 45°.

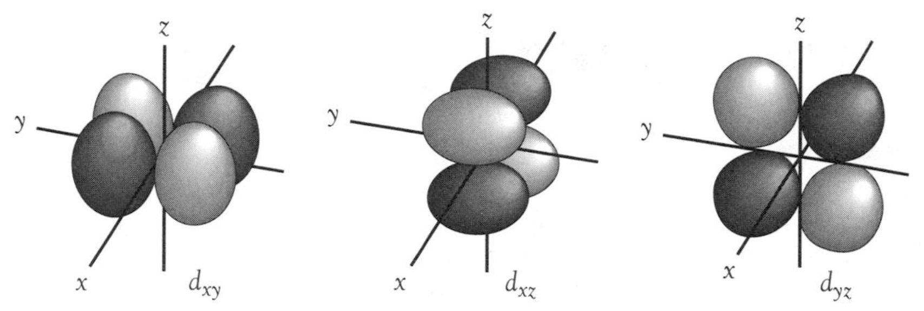

Figure 1.10 Representations of the shapes of the $3d_{xy}$, $3d_{xz}$, and $3d_{yz}$ orbitals. (Adapted from D. F. Shriver, P. Atkins, and C. H. Langford, *Inorganic Chemistry*, 2nd ed. [New York: W. H. Freeman, 1994], p. 25.)

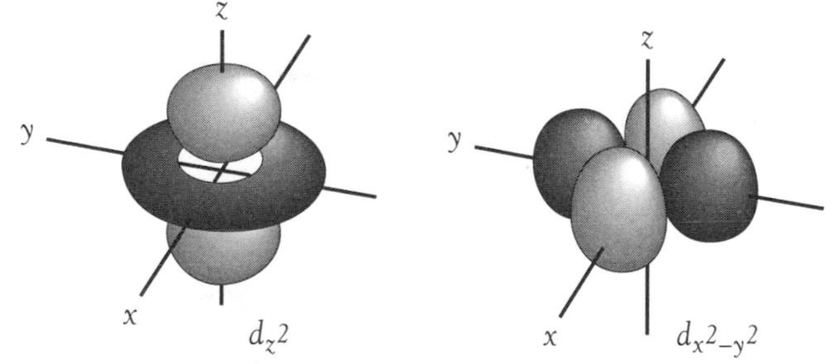

Figure 1.11 Representations of the shapes of the $3d_{z^2}$ and $3d_{x^2-y^2}$ orbitals. (Adapted from D. F. Shriver, P. Atkins, and C. H. Langford, *Inorganic Chemistry*, 2nd ed. [New York: W. H. Freeman, 1994], p. 25.)

The *f* Orbitals

The *f* orbitals are even more complex than the *d* orbitals. There are seven *f* orbitals, four of which have eight lobes. The other three look like the d_{z^2} orbital but have two donut-shaped rings instead of one. These orbitals are rarely involved in bonding, so we do not need to consider them in any detail.

The Polyelectronic Atom

In our model of the polyelectronic atom, the electrons are distributed among the orbitals of the atom according to the *Aufbau* (building-up) *principle.* This simple idea proposes that, when the electrons of an atom are all in the ground state, they occupy the orbitals of lowest energy, thereby minimizing the atom's total electronic energy. Thus the configuration of an atom can be

The "Other" Wave Equation

The Schrödinger wave equation is usually presented as the definitive representation of the electrons of an atom. But it is not. The equation fails to take into account the fact that the electrons are moving at relativistic speeds. In other words, Einstein's special theory of relativity has to be considered for a more precise determination of the electron energies. The Schrödinger equation works well for low atomic number atoms, but for high atomic number atoms, this relativistic effect becomes very significant for the electrons in orbitals with lobes close to the nucleus (the *s* orbitals).

It is possible to modify the Schrödinger equation for relativistic effects, but a better approach is to use the Dirac wave equation, derived by the English physicist P. A. M. Dirac in 1928. The principal quantum number has the same significance in the Schrödinger and Dirac equations, but the other three quantum numbers have different meanings. In addition, the shapes of Dirac orbitals are quite different from those of Schrödinger orbitals. The real test of the Dirac equations occurs when we consider the chemistry of the heavier elements. Once we progress into Periods 6 and 7 of the periodic table (see Chapter 2), some of the element properties, such as the low melting point and low chemical reactivity of mercury, can be understood only by considering relativistic effects. The Dirac equation is even more complex than the Schrödinger equation but, because this is a descriptive chemistry text, you do not have to worry about the details of either!

Figure 1.12 Electron configuration of a hydrogen atom.

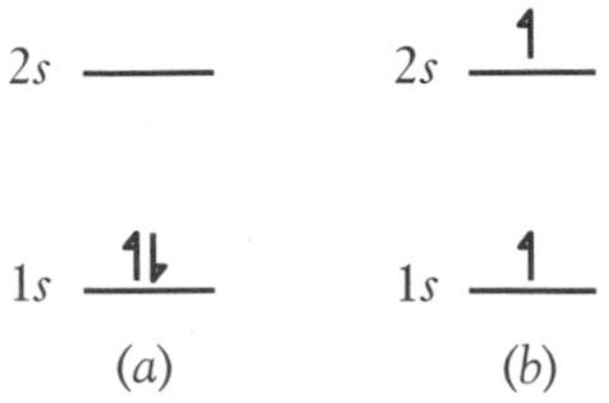

2s 2s

1s 1s

(a) (b)

Figure 1.13 Two possible electron configurations for helium.

described simply by adding electrons one by one until the total number required for the element has been reached.

Before starting to construct electron configurations, there is a second rule to be considered: the *Pauli exclusion principle*. According to this rule, no two electrons in an atom may possess identical sets of the four quantum numbers. Thus there can be only one orbital of each three-number quantum set per atom and each orbital can hold only two electrons, one with $m_s = +\frac{1}{2}$ and the other with $m_s = -\frac{1}{2}$.

The simplest configuration is that of the hydrogen atom. According to the Aufbau principle, the single electron will be located in the 1*s* orbital. This configuration is the ground state of the hydrogen atom. Adding energy would raise the electron to one of the many higher energy states. These configurations are referred to as excited states. In the diagram of the ground state of the hydrogen atom (Figure 1.12), a half-headed arrow is used to indicate the direction of electron spin. The electron configuration is written as $1s^1$, the superscript 1 indicating the number of electrons in that orbital.

With a two-electron atom (helium), there is a choice: The second electron could go in the 1*s* orbital (Figure 1.13a) or the next higher energy orbital, the 2*s* orbital (Figure 1.13b). Although it might seem obvious that the second electron would enter the 1*s* orbital, it is not so simple. If the second electron entered the 1*s* orbital, it would be occupying the same volume of space as the electron already in that orbital. The very strong electrostatic repulsions would discourage the occupancy of the same orbital. For helium, the energy separation between the 1*s* and 2*s* orbitals is about 4 $MJ \cdot mol^{-1}$, whereas the *pairing energy*, the energy needed to overcome the interelectronic repulsive forces, is about 3 $MJ \cdot mol^{-1}$. Hence the actual configuration will be $1s^2$, although it must be emphasized that electrons pair up in the same orbital *only* when pairing is the lower energy option.

In the lithium atom the 1*s* orbital is filled by two electrons, and the third electron must be in the next higher energy orbital, the 2*s* orbital. Thus lithium has the configuration of $1s^2 2s^1$. Because the energy separation of an *s* and its corresponding *p* orbitals is always greater than the pairing energy in a polyelectronic atom, the electron configuration of beryllium will be $1s^2 2s^2$ rather than $1s^2 2s^1 2p^1$.

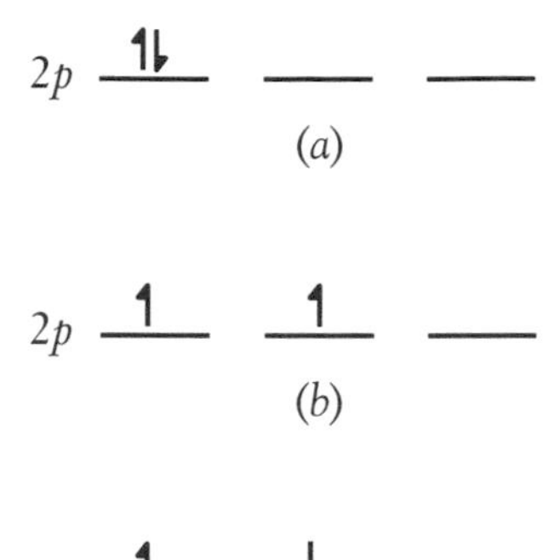

Figure 1.14 Possible 2*p* electron configurations for carbon.

Boron marks the beginning of the filling of the 2*p* orbitals. A boron atom has an electron configuration of $1s^2 2s^2 2p^1$. Because the *p* orbitals are *degenerate* (that is, they all have the same energy), it is impossible to decide which one of the three orbitals contains the electron.

Carbon is the second ground-state atom with electrons in the *p* orbitals. Its electron configuration provides another challenge. There are three possible arrangements of the two 2*p* electrons (Figure 1.14): both electrons in one orbital; two electrons with parallel spins in different orbitals; and two electrons with opposed spins in different orbitals. On the basis of electron repulsions, the first possibility (a) can be rejected immediately. The decision between the other two possibilities is less obvious and requires a deeper knowledge of quantum theory. In fact, if the two electrons have parallel spins, there is a zero probability of them occupying the same space. However, if the spins are opposed, there is a finite possibility that the two electrons will occupy the same region in space, thereby resulting in some repulsion and a higher energy state. Hence the parallel spin situation (b) will have the lowest energy. This preference for unpaired electrons with parallel spins has been formalized in the *Hund rule*: When filling a set of degenerate orbitals, the number of unpaired electrons will be maximized and these electrons will have parallel spins.

After the completion of the 2*p* electron set at neon ($1s^2 2s^2 2p^6$), the 3*s* and 3*p* orbitals start to fill. Rather than write the full electron configurations, a shortened form can be used. In this notation, the inner electrons are represented by the noble gas symbol having that configuration. Thus magnesium, whose full electron configuration would be written as $1s^2 2s^2 2p^6 3s^2$, can be represented as having a neon noble gas core, and its configuration is written as $[Ne]3s^2$. At this point, we have finished our analysis of the electron configuration of the two short *periods* (rows) of the periodic table (Figure 1.15): Period 2, lithium to neon, and Period 3, sodium to argon.

After completely filling the 3*p* orbitals in argon, the 3*d* and 4*s* orbitals start to fill. It is here that the simple orbital energy level concept breaks down, because the energy levels of the 4*s* and 3*d* orbitals are very close. What becomes most important is not the minimum energy for a single electron but the configuration that results in the least number of interelectron repulsions for all of the electrons. For potassium, this is $[Ar]4s^1$, and for calcium, $[Ar]4s^2$. To illustrate how this delicate balance changes with increasing numbers of

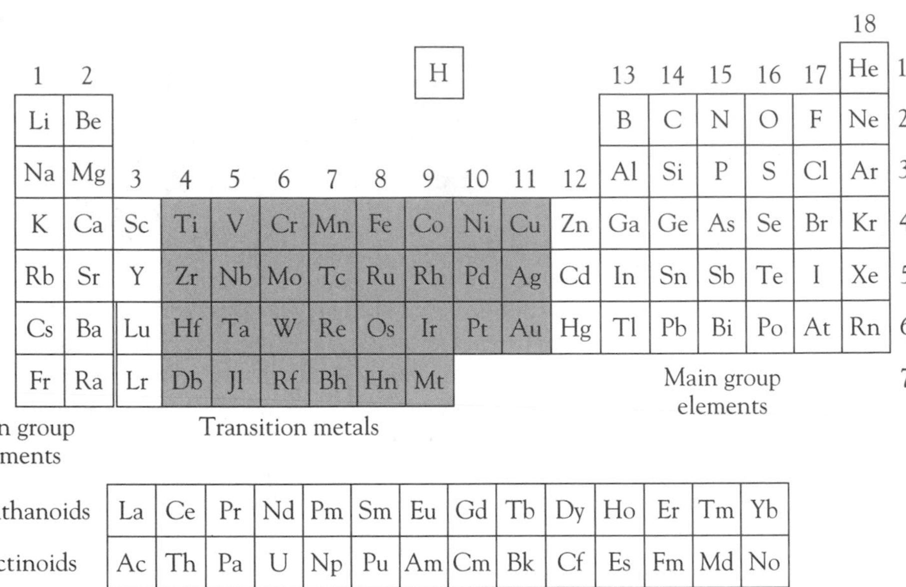

Figure 1.15 Essential features of the periodic table. The numbers across the top designate groups (columns) of elements. The numbers down the right-hand side designate periods (rows).

protons and electrons, the outer electrons in each of the three series of transition elements (including Group 12 elements) are listed below.

Atom	Configuration	Atom	Configuration	Atom	Configuration
Sc	$4s^2 3d^1$	Y	$5s^2 4d^1$	Lu	$6s^2 5d^1$
Ti	$4s^2 3d^2$	Zr	$5s^2 4d^2$	Hf	$6s^2 5d^2$
V	$4s^2 3d^3$	Nb	$5s^1 4d^4$	Ta	$6s^2 5d^3$
Cr	$4s^1 3d^5$	Mo	$5s^1 4d^5$	W	$6s^2 5d^4$
Mn	$4s^2 3d^5$	Tc	$5s^2 4d^5$	Re	$6s^2 5d^5$
Fe	$4s^2 3d^6$	Ru	$5s^1 4d^7$	Os	$6s^2 5d^6$
Co	$4s^2 3d^7$	Rh	$5s^1 4d^8$	Ir	$6s^2 5d^7$
Ni	$4s^2 3d^8$	Pd	$5s^0 4d^{10}$	Pt	$6s^1 5d^9$
Cu	$4s^1 3d^{10}$	Ag	$5s^1 4d^{10}$	Au	$6s^1 5d^{10}$
Zn	$4s^2 3d^{10}$	Cd	$5s^2 4d^{10}$	Hg	$6s^2 5d^{10}$

In general, the lowest overall energy for each transition metal is obtained by filling the *s* orbitals first; the remaining electrons then occupy the *d* orbitals. However, for certain elements, the lowest energy is obtained by shifting one or both of the *s* electrons to *d* orbitals. Looking at the first series in isolation would lead to the conclusion that there is some preference for a half-full or full set of *d* orbitals by chromium and copper. Although palladium and silver in the second transition series both favor the $4d^{10}$ configuration, it is more accurate to say that the interelectron repulsion between the two *s* electrons is sufficient in several cases to result in an s^1 configuration.

For the elements from lanthanum (La) to ytterbium (Yb), the situation is even more fluid because the 6*s*, 5*d*, and 4*f* orbitals all have similar energies.

For example, lanthanum has a configuration of $[Xe]6s^2 5d^1$, whereas the next element, cerium, has a configuration of $[Xe]6s^2 4f^2$. The most interesting electron configuration in this row is that of gadolinium, $[Xe]6s^2 5d^1 4f^7$, rather than the predicted $[Xe]6s^2 4f^8$. This configuration provides more evidence of the importance of interelectron repulsion in the determination of electron configuration when adjacent orbitals have similar energies. Similar complexities occur among the elements from actinium (Ac) to nobelium (No), in which the 7*s*, 6*d*, and 5*f* orbitals have similar energies.

In spite of these minor fluctuations in configurations throughout the *d*-block and *f*-block elements, the order of filling is surprisingly systematic. Thus the orbitals fill in the order

1*s* 2*s* 2*p* 3*s* 3*p* 4*s* 3*d* 4*p* 5*s* 4*d* 5*p* 6*s* 4*f* 5*d* 6*p* 7*s* 5*f* 6*d* 7*p*

This order is also shown in Figure 1.16. The orbitals fill in this order because the energy differences between the *s*, *p*, *d*, and *f* orbitals of the same angular momentum quantum number become greater than the energy differences between the orbitals of different principal quantum numbers. It is important

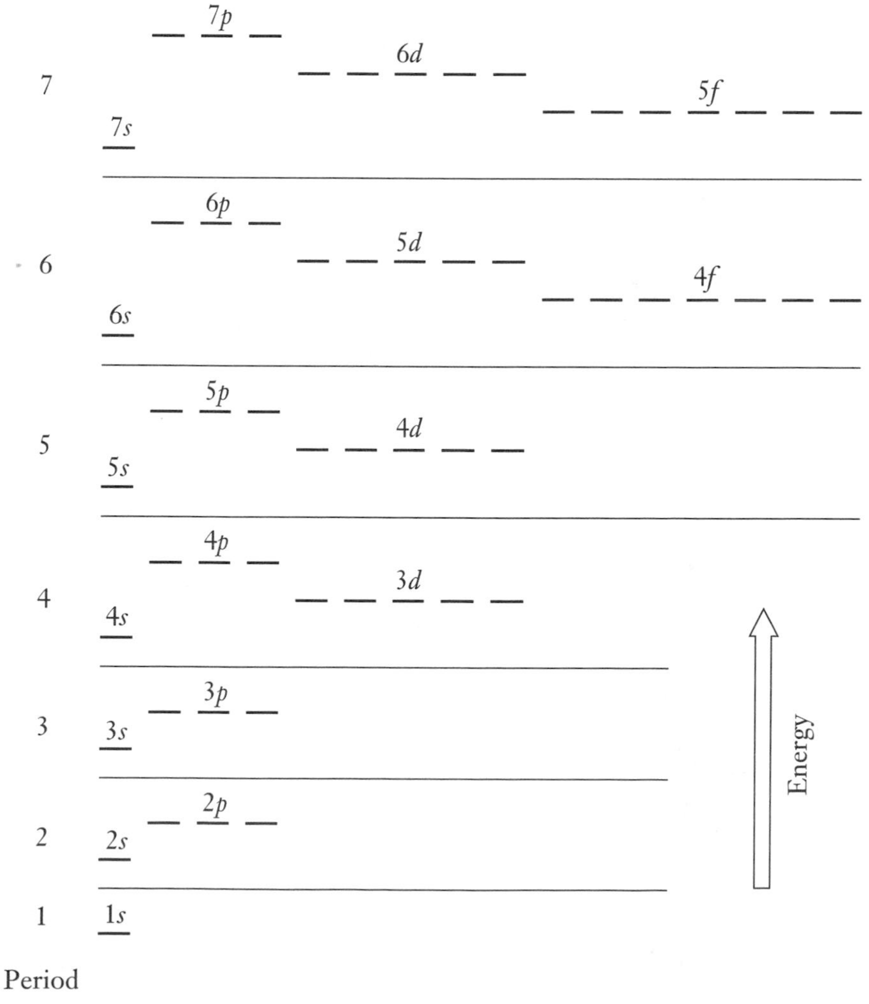

Figure 1.16 Representation of the comparative energies of the atomic orbitals for filling order purposes.

to note that Figure 1.16 shows the filling order, not the order for any particular element. For example, for elements beyond Period 4, the $3d$ orbitals are far lower in energy than the $4s$ orbitals are. However, at this point, the $3d$ orbitals have become "inner" orbitals and have no role in chemical bonding. Hence their precise ordering is unimportant.

Ion Electron Configurations

For the early main group elements, the common ion electron configurations can be predicted quite readily. Thus metals tend to lose all of the electrons in the outer orbital set. This situation is illustrated for sodium, magnesium, and aluminum:

Atom	Electron configuration	Ion	Electron configuration
Na	[Ne]$3s^1$	Na^+	[Ne]
Mg	[Ne]$3s^2$	Mg^{2+}	[Ne]
Al	[Ne]$3s^23p^1$	Al^{3+}	[Ne]

Ions possessing the same electron configuration are said to be *isoelectronic*.

Nonmetals gain electrons to complete the outer orbital set. This situation is shown for nitrogen, oxygen, and fluorine:

Atom	Electron configuration	Ion	Electron configuration
N	[He]$2s^22p^3$	N^{3-}	[Ne]
O	[He]$2s^22p^4$	O^{2-}	[Ne]
F	[He]$2s^22p^5$	F^-	[Ne]

Some of the later main group metals form two ions with different charges. For example, lead forms Pb^{2+} and (rarely) Pb^{4+}. The 2+ charge can be explained by the loss of the p electrons only, whereas the 4+ charge results from loss of both s and p electrons:

Atom	Electron configuration	Ion	Electron configuration
Pb	[Xe]$6s^25d^{10}6p^2$	Pb^{2+}	[Xe]$6s^25d^{10}$
		Pb^{4+}	[Xe]$5d^{10}$

Notice that the electrons of the higher principal quantum number are lost first. This principle is found to be true for all the elements. For the transition metals, the s electrons are always lost first when a metal cation is formed. In other words, for the transition metal cations, the $3d$ orbitals are always lower in energy than the $4s$ orbitals; and a charge of 2+, representing the loss of the two s electrons, is common for the transition metals and the Group 12 metals. For example, zinc always forms an ion of 2+ charge:

Atom	Electron configuration	Ion	Electron configuration
Zn	[Ar]$4s^23d^{10}$	Zn^{2+}	[Ar]$3d^{10}$

Iron forms ions with charges of 2+ and 3+ and, as shown here, it is tempting to ascribe the formation of the 3+ ion to a process in which interelectron repulsion "forces out" the only paired *d* electron:

Atom	Electron configuration	Ion	Electron configuration
Fe	$[Ar]4s^23d^6$	Fe^{2+}	$[Ar]3d^6$
		Fe^{3+}	$[Ar]3d^5$

It is dangerous, however, to read too much into the electron configurations of atoms as a means of predicting the ion charges. The series of nickel, palladium, and platinum illustrate this point: They have different configurations as atoms, yet their ionic charges and corresponding ion electron configurations are similar:

Atom	Electron configuration	Ion	Electron configuration
Ni	$[Ar]4s^23d^8$	Ni^{2+}	$[Ar]3d^8$
Pd	$[Kr]5s^04d^{10}$	Pd^{2+}, Pd^{4+}	$[Kr]4d^8$, $[Kr]4d^6$
Pt	$[Xe]6s^15d^9$	Pt^{2+}, Pt^{4+}	$[Xe]5d^8$, $[Xe]5d^6$

Magnetic Properties of Atoms

In the discussions of electron configuration, we saw that some atoms possess unpaired electrons. The presence of unpaired electrons in the atoms of an element can be determined easily from the element's magnetic properties. If atoms containing only spin-paired electrons are placed in a magnetic field, they are weakly repelled by the field. This phenomenon is called *diamagnetism.* Conversely, atoms containing one or more unpaired electrons are attracted by the magnetic field. This behavior of unpaired electrons is named *paramagnetism.* The attraction of each unpaired electron is many times stronger than the repulsion of all of the spin-paired electrons in that atom.

To explain paramagnetism in simple terms, we can visualize the electron as a particle spinning on its axis and generating a magnetic moment, just as an electric current flowing through a wire does. This permanent magnetic moment results in an attraction into the stronger part of the field. When electrons have their spins paired, the magnetic moments cancel each other. As a result, the paired electrons are weakly repelled by the lines of force of the magnetic field. We will encounter this phenomenon again in our discussions of covalent bonding and the bonding in transition metal compounds.

Exercises

1.1. Define the following terms: (a) nodal surface; (b) Pauli exclusion principle; (c) paramagnetic.

1.2. Define the following terms: (a) orbital; (b) degenerate; (c) Hund rule.

1.3. Construct a quantum number tree for the principal quantum number $n = 4$ similar to that depicted for $n = 3$ in Figure 1.5.

1.4. Determine the lowest value of n for which m_l can (theoretically) have a value of +4.

1.5. Identify the orbital that has $n = 5$ and $l = 1$.

1.6. Identify the orbital that has $n = 6$ and $l = 0$.

1.7. How does the quantum number n relate to the properties of an orbital?

1.8. How does the quantum number l relate to the properties of an orbital?

1.9. Explain concisely why carbon has two electrons in different p orbitals with parallel spins rather than the other possible arrangements.

1.10. Explain concisely why beryllium has a ground-state electron configuration of $1s^2 2s^2$ rather than $1s^2 2s^1 2p^1$.

1.11. Write noble gas core ground-state electron configurations for atoms of (a) sodium; (b) nickel; (c) copper.

1.12. Write noble gas core ground-state electron configurations for atoms of (a) calcium; (b) chromium; (c) lead.

1.13. Write noble gas core ground-state electron configurations for ions of (a) potassium; (b) scandium 3+; (c) copper 2+.

1.14. Write noble gas core ground-state electron configurations for ions of (a) chlorine; (b) cobalt 2+; (c) manganese 4+.

1.15. Predict the common charges of the ions of thallium. Explain your reasoning in terms of electron configurations.

1.16. Predict the common charges of the ions of tin. Explain your reasoning in terms of electron configurations.

1.17. Predict the common charge of the silver ion. Explain your reasoning in terms of electron configurations.

1.18. Predict the highest possible charge of a zirconium ion. Explain your reasoning in terms of electron configurations.

1.19. Use diagrams similar to Figure 1.14 to determine the number of unpaired electrons in atoms of (a) oxygen; (b) magnesium; (c) chromium.

1.20. Use diagrams similar to Figure 1.14 to determine the number of unpaired electrons in atoms of (a) nitrogen; (b) silicon; (c) iron.

1.21. The next set of orbitals after the f orbitals are the g orbitals. How many g orbitals would there be? What would be the lowest principal quantum number n that would possess g orbitals? Deduce the atomic number of the first element at which g orbitals would begin to be filled on the basis of the patterns of the d and f orbitals.

CHAPTER

2

An Overview of the Periodic Table

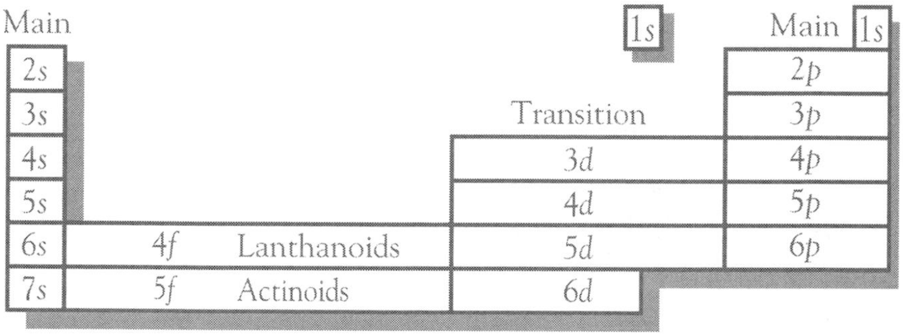

Main | 1s | Main | 1s
2s | 2p
3s | Transition | 3p
4s | 3d | 4p
5s | 4d | 5p
6s | 4f Lanthanoids | 5d | 6p
7s | 5f Actinoids | 6d

The periodic table is the framework upon which much of our understanding of inorganic chemistry is based. In this chapter, we provide the basic information that you will need for the more detailed discussions of the individual groups in later chapters.

The search for patterns among the chemical elements really started with the work of the German chemist Johann Döbereiner in 1817. He noticed that there were similarities in properties among various groups of three elements, such as calcium, strontium, and barium. He named these groups "triads." Almost 50 years later, John Newlands, a British sugar refiner, realized that, when the elements were placed in order of increasing atomic weights, a cycle of properties repeated with every eight elements. Newlands called this pattern the law of octaves. At the time, scientists had started to look for a unity of the physical laws that would explain everything, so to correlate element organization with the musical scale seemed natural. Unfortunately, his proposal was laughed at by most of the chemists of the time. Only a few years later, the Russian chemist Dmitri Mendeleev independently devised the same concept (without linking it to music). It attracted little attention until Lothar

I	II	III	IV	V	VI	VII	VIII
H							
Li	Be	B	C	N	O	F	
Na	Mg	Al	Si	P	S	Cl	
K	Ca		Ti	V	Cr	Mn	Fe Ni Co Cu
Cu?	Zn			As	Se	Br	
Rb	Sr	Yt	Zr	Nb	Mo		Ru Pd Rh Ag
Ag?	Cd	In	Sn	Sb	Te	I	
Cs	Ba		Ce				
		Er	La	Ta	W		Os Pt Ir Au
Au?	Hg	Tl	Pb	Bi			
			Th		U		

Figure 2.1 The organization of one of Mendeleev's designs for the periodic table.

Meyer, a German chemist, published his own report on the periodic relationship. Meyer did acknowledge that Mendeleev had had the same idea first.

Mendeleev and Meyer would hardly recognize our contemporary design. In Mendeleev's proposal, the elements known at the time were organized in an eight-column format in order of increasing atomic mass. He claimed that each eighth element had similar properties. Groups I to VII each contained two subgroups, and Group VIII contained four subgroups. The organization of one of his designs is shown in Figure 2.1. To ensure that the patterns in the properties of elements fitted the table, it was necessary to leave spaces. Mendeleev assumed that these spaces corresponded to unknown elements. He argued that the properties of the missing elements could be predicted on the basis of the chemistry of its neighbors in the same group. For example, the missing element between silicon and tin, called eka-silicon (Es) by Mendeleev, should have properties intermediate between those of silicon and tin. Table 2.1 compares Mendeleev's predictions with the properties of germanium, discovered 15 years later.

However, the Mendeleev periodic table had three major problems:

1. If the order of increasing atomic mass was consistently followed, elements did not always fit in the group that had the matching properties. Thus the order of nickel and cobalt had to be reversed, as did that of iodine and tellurium.

2. Elements were being discovered, such as holmium and samarium, for which no space could be found. This difficulty was a particular embarrassment.

Table 2.1 Comparison of Mendeleev's predictions for eka-silicon and the actual properties of germanium

Element	Atomic weight	Density ($g \cdot cm^{-3}$)	Oxide formula	Chloride formula
Eka-silicon	72	5.5	EsO_2	$EsCl_4$
Germanium	72.3	5.47	GeO_2	$GeCl_4$

3. Elements in the same group were sometimes quite different in their chemical reactivity. This discrepancy was particularly true of the first group, which contained the very reactive alkali metals and the very unreactive coinage metals (copper, silver, and gold).

As we now know, there was another flaw: To establish a group of elements, at least one element has to be known already. Because none of the noble gases was known at that time, no space was left for them. Conversely, some spaces in Mendeleev's table were completely erroneous. This was because he tried to fit the elements into repeating rows (periods) of eight. Now, of course, we know that the periods are not consistently eight members long but, instead, increase regularly; successive rows have 2, 8, 8, 18, 18, 32, and 32 elements.

The crucial transition to our modern ideas was provided by Henry Moseley, a British physicist. It had been discovered that directing a beam of X-rays onto an element caused its atoms to emit X-rays of wavelengths unique to that element. Moseley showed that the wavelengths of the emitted X-rays fitted a formula that yielded a unique integer for each element. This integer, he argued, matched the number of positive charges in the atom of the element. Ordering the elements according to this number (the atomic number) removed the irregularities of the table that was based on atomic masses, and it defined exactly the spaces in the table where elements still needed to be found.

Shortly after his discovery, Moseley was drafted to fight in the First World War. He was killed in the Gallipoli campaign of 1915. He was only 27 at the time of his death, and many chemists believe that, had he survived the war, he would have become one of the greatest scientists of this century. Unfortunately, Nobel prizes are never awarded posthumously, so Moseley's crucial role will never receive that ultimate recognition. Nor, surprisingly, has an element been named after him, even though many others have received that honor.

Organization of the Modern Periodic Table

In the modern periodic table, the elements are placed in order of increasing atomic number (the number of protons). There have been numerous designs of the table over the years, but the two most common are the long form and the short form. The long form (Figure 2.2) shows all of the elements in numerical order.

The start of a new period always corresponds to the introduction of the first electron into the *s* orbital of a new principal quantum number. The number of elements in each period corresponds to the number of electrons required to fill those orbitals (Figure 2.3). In a particular period, the principal quantum number of the *p* orbitals is the same as that of the *s* orbitals, whereas the *d* orbitals are one less and the *f* orbitals are two less. The *main groups* correspond to elements in which the *s* and *p* orbitals are being filled; the transition metals are elements in which the *d* orbitals are being filled.

Each group contains elements of similar electron configuration. For example, all Group 1 elements have an outer electron that is ns^1, where n is the principal quantum number. Although elements in a group have similar

																						H									He
Li	Be																									B	C	N	O	F	Ne
Na	Mg																									Al	Si	P	S	Cl	Ar
K	Ca															Sc	Ti	V	Cr	Mn	Fe	Co	Ni	Cu	Zn	Ga	Ge	As	Se	Br	Kr
Rb	Sr															Y	Zr	Nb	Mo	Tc	Ru	Rh	Pd	Ag	Cd	In	Sn	Sb	Te	I	Xe
Cs	Ba	La	Ce	Pr	Nd	Pm	Sm	Eu	Gd	Tb	Dy	Ho	Er	Tm	Yb	Lu	Hf	Ta	W	Re	Os	Ir	Pt	Au	Hg	Tl	Pb	Bi	Po	At	Rn
Fr	Ra	Ac	Th	Pa	U	Np	Pu	Am	Cm	Bk	Cf	Es	Fm	Md	No	Lr	Db	Jl	Rf	Bh	Hn	Mt									

Figure 2.2 Long form of the periodic table.

properties, it is important to realize that every element has its own unique character. Thus, although nitrogen and phosphorus are sequential elements in the same group, nitrogen gas is very unreactive and phosphorus is so reactive that it spontaneously reacts with the oxygen in the air.

Because the long form of the periodic table is a very elongated diagram and because the elements from lanthanum to ytterbium and from actinium to nobelium show similar chemical behavior, the short form displays these two sets of elements in rows beneath the remainder of the table and the resulting space is closed up. Figure 2.4 shows this more compact, short form of the table.

According to the recommendations of the International Union of Pure and Applied Chemistry (IUPAC), the groups of elements are numbered from 1 to 18. This system replaces the old system of using a mixture of Roman numerals and letters, a notation that caused confusion because of differences in numbering between North America and the rest of the world. For example, in North America IIIB referred to the group containing scandium, whereas in the rest of the world this designation was used for the group starting with boron. Numerical designations are not used for the series of elements from lanthanum (La) to ytterbium (Yb) and from actinium (Ac) to nobelium (No), because there is much more resemblance in properties within each of those rows of elements than vertically in groups.

Groups 1 and 2 and 13 through 18 represent the main group elements; Groups 4 through 11 are classified as the transition metals. In the transition metals, the *d* orbitals are being filled. The discussion of the main groups will

Main				Main
			1*s*	1*s*
2*s*				2*p*
3*s*			Transition	3*p*
4*s*			3*d*	4*p*
5*s*			4*d*	5*p*
6*s*	4*f*	Lanthanoids	5*d*	6*p*
7*s*	5*f*	Actinoids	6*d*	

Figure 2.3 Electron orbital filling sequence in the periodic table.

1	2	3	4	5	6	7	8	9	10	11	12	13	14	15	16	17	18
								H									He
Li	Be											B	C	N	O	F	Ne
Na	Mg											Al	Si	P	S	Cl	Ar
K	Ca	Sc	Ti	V	Cr	Mn	Fe	Co	Ni	Cu	Zn	Ga	Ge	As	Se	Br	Kr
Rb	Sr	Y	Zr	Nb	Mo	Tc	Ru	Rh	Pd	Ag	Cd	In	Sn	Sb	Te	I	Xe
Cs	Ba	* Lu	Hf	Ta	W	Re	Os	Ir	Pt	Au	Hg	Tl	Pb	Bi	Po	At	Rn
Fr	Ra	** Lr	Db	Jl	Rf	Bh	Hn	Mt									

Lanthanoids*	La	Ce	Pr	Nd	Pm	Sm	Eu	Gd	Tb	Dy	Ho	Er	Tm	Yb
Actinoids **	Ac	Th	Pa	U	Np	Pu	Am	Cm	Bk	Cf	Es	Fm	Md	No

Figure 2.4 Short form of the periodic table displaying the group numbers. Main group elements are shaded.

take up the majority of space in this text because it is these elements that cover the widest range of chemical and physical properties. The elements of Group 12, although sometimes included among the transition metals, have a very different chemistry from that series; hence Group 12 will be considered separately. The elements of Group 3 closely resemble the lanthanoids, and they will all be studied together in this text. Several of the main groups have been given specific names: *alkali metals* (Group 1); *alkaline earth metals* (Group 2); *chalcogens* (a little-used term for Group 16); *halogens* (Group 17); and *noble gases* (Group 18). The elements in Group 11 are sometimes called the *coinage metals.*

Although the elements in the periodic table are arranged according to electron structure, we make an exception for helium ($1s^2$). Rather than placing it with the other ns^2 configuration elements, the alkaline earth metals, it is placed with the other noble gases (of configuration ns^2np^2) because of chemical similarities (see Figure 2.3). Hydrogen is even more of a problem. Although some versions of the periodic table show it as a member of Group 1 or Group 17, or both, its chemistry is unlike either the alkali metals or the halogens. For this reason, it is placed on its own in the tables in this text, to indicate its uniqueness.

The elements corresponding to the filling of the $4f$ orbitals are called the *lanthanoids* and those corresponding to the filling of the $5f$ orbitals are called the *actinoids*. They used to be named the lanthanides and actinides, but the *-ide* ending more correctly means a negative ion, such as oxide or chloride. For a few years, IUPAC was suggesting the names of lanthanons and actinons, but because the ending *-on* is preferred for nonmetals (and the lanthanoids and actinoids are all metallic elements), the *-oid* ending is now recommended. The elements scandium (Sc), yttrium (Y), lutetium (Lu), and the lanthanoid elements lanthanum to ytterbium (La to Yb) are collectively known as the *rare earth elements* because the chemistry of scandium, yttrium, and lutetium more closely resembles that of the lanthanoids than that of the transition metals. We discuss this point in Chapter 21.

Existence of the Elements

To understand why there are so many elements and to explain the pattern of the abundances of the elements, we must look at the most widely accepted theory of the origin of the universe. This is the big bang theory, which assumes that the universe started from a single point. About one second after the universe came into existence, the temperature had dropped to about 10^{10} K, at which point protons and neutrons could exist. During the next three minutes, hydrogen-1, hydrogen-2, helium-3, helium-4, beryllium-7, and lithium-7 nuclei formed (the number following the hyphen represents the *mass number*, the sum of protons and neutrons, of that isotope). After these first few minutes, the universe had expanded and cooled to the point where nuclear fusion reactions could no longer occur. At this point, as is still true today, most of the universe consisted of hydrogen-1 and some helium-4.

Through gravitational effects, the atoms became concentrated in small volumes of space; indeed, the compression was great enough to cause exothermic nuclear reactions. These volumes of space we call stars. In the stars, hydrogen nuclei are fused to give more helium-4 nuclei. About 10 percent of the helium in the present universe has come from hydrogen fusion within stars. As the larger stars become older, buildup of helium-4 and additional gravitational collapse cause the helium nuclei to combine to form beryllium-8, carbon-12, and oxygen-16. At the same time, the fragile helium-3, beryllium-7, and lithium-7 are destroyed. For most stars, oxygen-16 and traces of neon-20 are the largest (highest atomic number) elements produced. However, the temperature of the very massive stars increases to a maximum as high as 10^9 K, and their density increases to about 10^6 g·cm^{-3}. Under these conditions, the tremendous repulsion between the high positive charges of carbon and oxygen nuclei can be overcome, a condition leading to the formation of all the elements up to iron. Iron is the limit, however, because, beyond iron, synthesis (fusion) is endothermic rather than exothermic.

When the more massive elements have accumulated in the core of the star and the energy from nuclear syntheses is no longer balancing the enormous gravitational forces, a catastrophic collapse occurs. This can happen in as short a time as a few seconds. It is during the brief time of this explosion, what we see as a supernova, that there is sufficient free energy to cause the formation of large atomic nuclei (greater than 26 protons) in endothermic nuclear reactions. All the elements from the supernovas that happened early in the history of the universe have spread throughout the universe. It is these elements that make up our solar system and, indeed, ourselves. So it is really true when songwriters and poets say that we are "stardust."

Stability of the Elements and Their Isotopes

In the universe, there are only 81 stable elements (Figure 2.5). For these elements, one or more isotopes do not undergo spontaneous radioactive decay. No stable isotope exists for any element above bismuth, and two elements in the earlier part of the table, technetium and promethium, exist only as radioactive isotopes. Two elements, uranium and thorium, for which only

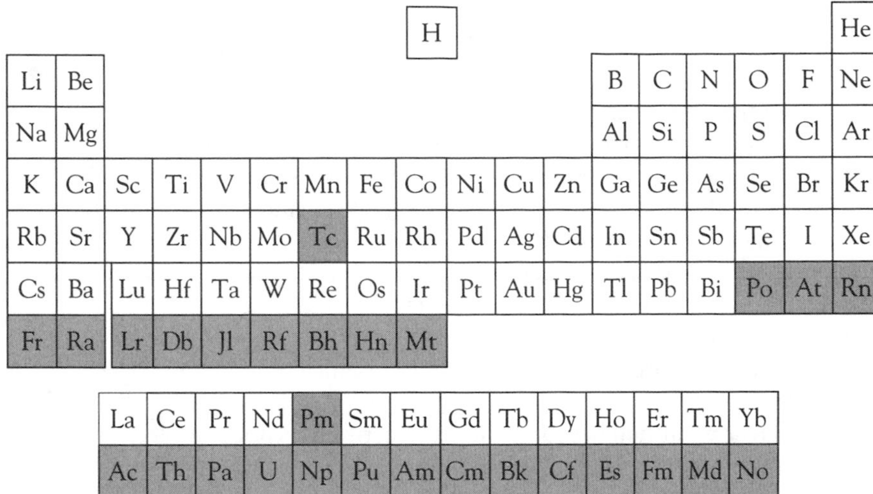

Figure 2.5 Elements that have only radioactive isotopes (shaded).

radioactive isotopes exist, are found quite abundantly on Earth because the half-lives of some of their isotopes—10^8 to 10^9 years—are almost as great as the age of Earth itself.

The fact that the number of stable elements is limited can be explained by recalling that the nucleus contains positively charged protons. Repulsive forces exist between the protons, just like the repulsive forces between electrons discussed in Chapter 1. We can visualize the neutrons simply as material separating the positive charges. Table 2.2 shows that as the number of protons increases, the number of neutrons in the most common isotope of each element increases at a faster rate. Beyond bismuth, the number of positive charges in the nucleus becomes too large to maintain nuclear stability, and the repulsive forces prevail.

To gain a better understanding of the nucleus, we can devise a quantum (or shell) model of the nucleus. Just as Bohr visualized electrons as existing in

Table 2.2 Neutron-proton ratios for common isotopes

Element	No. of protons	No. of neutrons	Neutron-proton ratio
Hydrogen	1	0	0.0
Helium	2	2	1.0
Carbon	6	6	1.0
Iron	26	30	1.2
Iodine	53	74	1.4
Lead	82	126	1.5
Bismuth	83	126	1.5
Uranium	92	146	1.6

The Origin of the Shell Model of the Nucleus

The proposal that the nucleus might have a structure came much later than Bohr's work on electron energy levels. Of the contributors to the discovery, probably the most crucial work was accomplished by Maria Goeppert Mayer. In 1946, Mayer was studying the abundances of the different elements in the universe, and she noticed that certain nuclei were far more abundant than those of their neighbors. The higher abundances had to reflect a greater stability of those particular nuclei. She realized that the stability could be explained by considering that the protons and neutrons were not just a solid core but were themselves organized in energy levels just like the electrons.

Mayer published her ideas, but the picture was not complete. She could not understand why the numbers of nucleons to complete each energy level were 2, 8, 20, 28, 50, 82, and 126. After working on the problem for three years, the flash of inspiration came one evening and she was able to derive theoretically the quantum levels and sublevels. Another physicist, Hans Jensen, read her ideas on the shell model of the nucleus and, in the same year as Mayer, independently came up with the same theoretical results. Mayer and Jensen met and collaborated to write the definitive book on the nuclear structure of the atom. Mayer and Jensen, who became good friends, shared the 1963 Nobel prize in physics for their discovery of the structure of the nucleus.

quantum levels, so we can visualize layers of protons and neutrons (together called the *nucleons*). Thus, within the nucleus, protons and neutrons will independently fill energy levels corresponding to the principal quantum number n. However, the angular momentum quantum number l is not limited as it is for electrons. In fact, for nucleons, the filling order starts with 1s, 1p, 2s, 1d. . . . Each nuclear energy level is controlled by the same magnetic quantum number rules as electrons, so there are one s level, three p levels, and five d levels. Both nucleons have spin quantum numbers that can be $+\frac{1}{2}$ or $-\frac{1}{2}$.

Using these rules, we find that, for nuclei, completed quantum levels contain 2, 8, 20, 28, 50, 82, and 126 nucleons of one kind (compared with 2, 10, 18, 36, 54, and 86 for electrons). Thus the first completed quantum level corresponds to the $1s^2$ configuration, the next with the $1s^21p^6$ configuration, and the following one with $1s^21p^62s^21d^{10}$. These levels are filled independently for protons and for neutrons. We find that, just like the quantum levels of electrons, completed nucleon levels confer a particular stability on a nucleus. For example, the decay of all the radioactive elements beyond lead results in the formation of lead isotopes, all of which have 82 protons.

The influence of the filled energy levels is apparent in the patterns among stable isotopes. Thus tin, with 50 protons, has the largest number of stable isotopes (10). Similarly, there are seven different elements with isotopes containing 82 neutrons and six different elements with isotopes containing 50 neutrons.

If the possession of a completed quantum level of one nucleon confers a stability to the nucleus, then we might expect that nuclei with filled levels for both nucleons would be even more favored. This is indeed the case. In particular, helium-4 with $1s^2$ configurations of both protons and neutrons is the second most common isotope in the universe, and the helium-4 nucleus (the α particle) is ejected in many nuclear reactions. Similarly, we find that it is the next doubly completed nucleus, oxygen-16, that makes up 99.8 percent of oxygen on this planet. Calcium follows the trend with 97 percent of the element being calcium-40. As we saw in Table 2.2, the number of neutrons increases more rapidly than those of protons. Thus the next doubly stable isotope is lead-208 (82 protons and 126 neutrons). This is the most massive stable isotope of lead, and the most common in nature.

Different from electron behavior, spin pairing is an important factor for nucleons. In fact, of the 273 stable nuclei, only four have odd numbers of both protons and neutrons. Elements with even numbers of protons tend to have large numbers of stable isotopes, whereas those with odd numbers of protons tend to have one or, at the most, two stable isotopes. For example, cesium (55 protons) has just one stable isotope, whereas barium (56 protons) has seven stable isotopes. Technetium and promethium, the only elements before bismuth to exist only as radioactive isotopes, both have odd numbers of protons.

The greater stability of even numbers of protons in nuclei can be related to the abundance of elements on Earth. As well as the decrease of abundance with increasing atomic number, we see that elements with odd numbers of protons have an abundance about one-tenth that of their even-numbered neighbors (Figure 2.6).

Figure 2.6 Solar system abundances of the elements as percentages on a logarithmic scale. (Adapted from P. A. Cox, *The Elements* [Oxford: Oxford University Press, 1989], p. 17.)

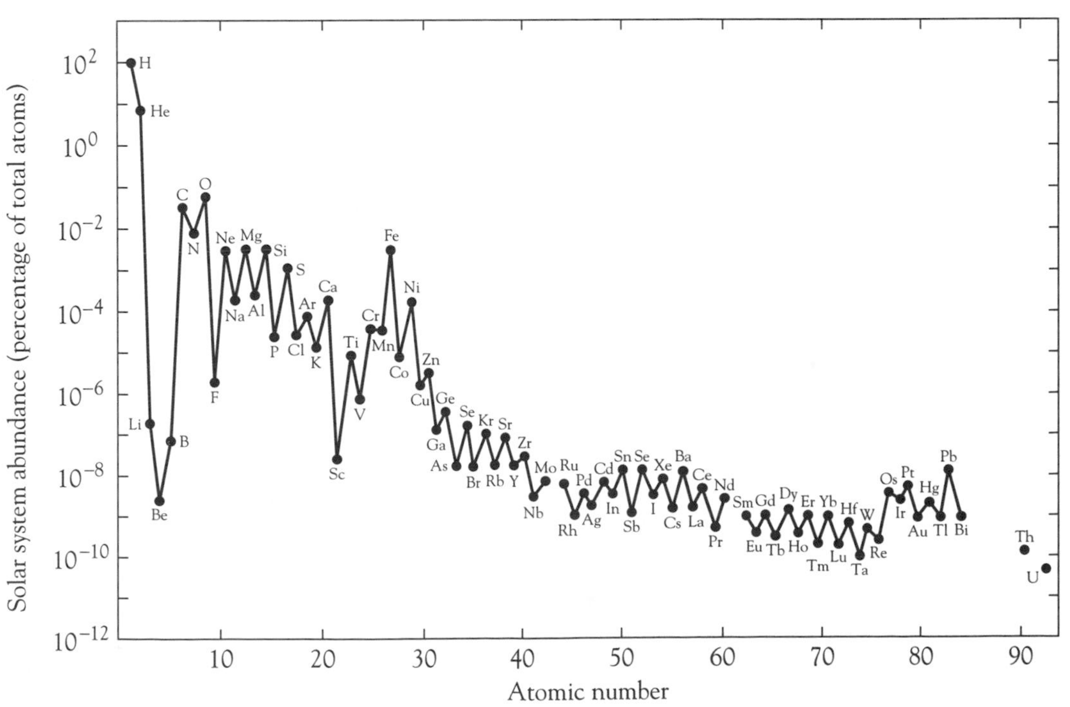

Classifications of the Elements

There are numerous ways in which the elements can be classified. The most obvious is by phase at *standard ambient temperature* (25°C) *and pressure* (100 kPa), conditions that are referred to as SATP (not to be confused with the old standard, STP, of 0°C and 101 kPa pressure). Among all the elements, there are only two liquids and eleven gases (Figure 2.7) at SATP. It is important to define the temperature precisely, because two metals have melting points just slightly above the standard temperatures: cesium, m.p. 29°C, and gallium, m.p. 30°C.

Another very common classification scheme has two categories, metals and nonmetals. But what is meant by a metal? A lustrous surface is not a good criterion, for several elements that are regarded as nonmetals—silicon and iodine are two—have very shiny surfaces. Even a few compounds, such as the mineral pyrite, FeS_2 (also known as fool's gold), look metallic. Density is not a good guide either, because lithium has a density one-half that of water, and osmium has a density 40 times that of lithium. Hardness is an equally poor guide, because the alkali metals are very soft. The ability of the element to be flattened into sheets (*malleability*) or to be pulled into wires (*ductility*) is sometimes cited as a common property of metals, but some of the transition metals are quite brittle. High thermal conductivity is common among the elements we call metals, but diamond, a nonmetal, has one of the highest thermal conductivities of any element. So that classification is not valid.

High three-dimensional electrical conductivity is the best criterion of a metal. We have to stipulate three rather than two dimensions because graphite, an allotrope of carbon, has high electrical conductivity in two dimensions. There is a difference of 10^2 in conductivity between the best electrical conducting metal (silver) and the worst (plutonium). But even plutonium has an electrical conductivity about 10^5 times better than the best conducting nonmetallic element. But, to be precise, the SATP conditions of 100 kPa pressure and 25°C have to be stipulated, because below 18°C, the stable allotrope of tin is nonelectrically conducting. Furthermore, under

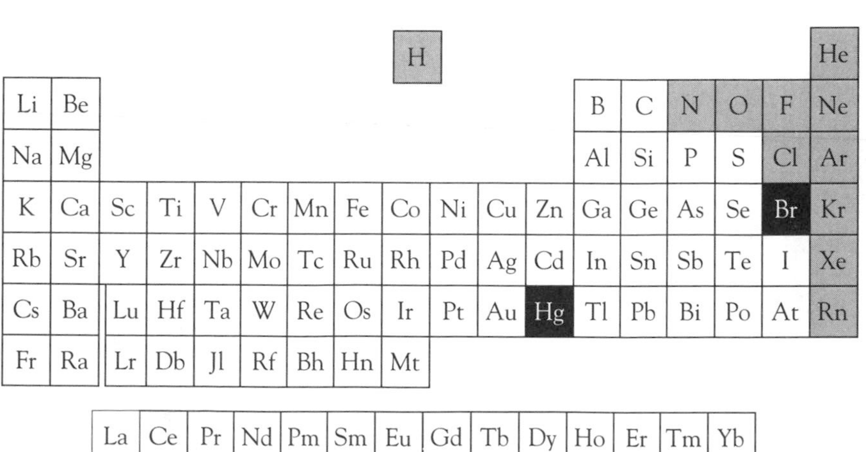

Figure 2.7 Classification of the elements into gases (shaded), liquids (black), and solids (white).

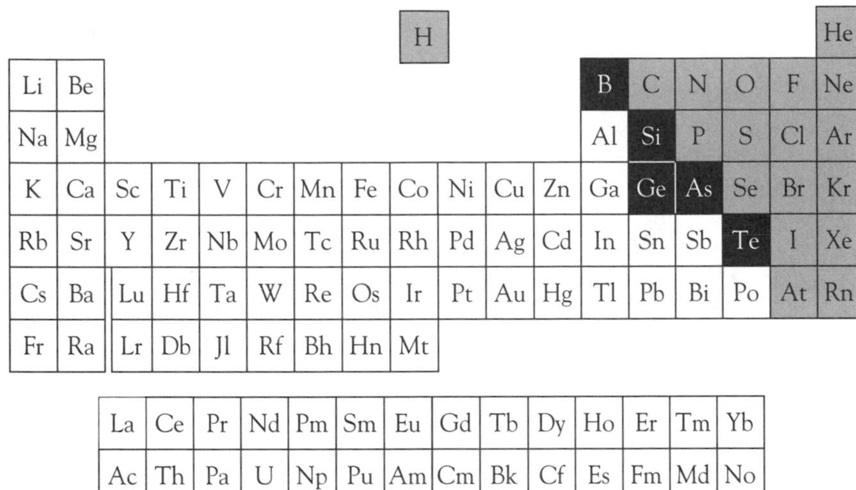

Figure 2.8 Classification of the elements into metals (white), semimetals (black), and nonmetals (shaded).

readily obtainable pressures, iodine becomes electrically conducting. A more specific physical criterion is the temperature dependence of the electrical conductivity, for the conductivity of metals decreases with increasing temperature whereas that of nonmetals increases.

For chemists, however, the most important feature of an element is its pattern of chemical behavior, in particular, its tendency toward covalent bond formation or its preference for cation formation. And no matter which criteria are used, some elements always fall in the border region of the metal-nonmetal divide. As a result, most inorganic chemists agree that boron, silicon, germanium, arsenic, and tellurium can be assigned an ambiguous status as *semimetals* (Figure 2.8), formerly called metalloids.

Even then, the division of elements into three categories is a simplification. There is a subgroup of the metals, the ones closest to the borderline, that exhibit some chemical behavior that is more typical of the semimetals. These nine "weak metals" are beryllium, aluminum, zinc, gallium, tin, lead, antimony, bismuth, and polonium.

Periodic Properties: Atomic Radius

One of the most systematic periodic properties is atomic radius. What is the meaning of atomic size? Because the electrons can be defined only in terms of probability, there is no real boundary to an atom. Nevertheless, there are two common ways in which we can define *atomic radius.* We can define it as the half-distance between the nuclei of two atoms joined in a covalent bond, called the *covalent radius,* r_{cov}. Or we can define it as the half-distance between the nuclei of two atoms of neighboring molecules, called the *van der Waals radius,* r_{vdw} (Figure 2.9). Furthermore, for the metallic elements, it is possible to measure a *metallic radius:* the half-distance between the nuclei of two neighboring atoms in the solid metal.

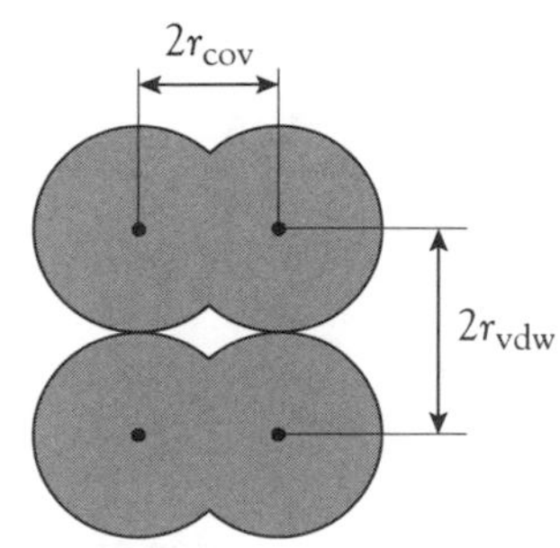

Figure 2.9 Comparison of the covalent radius, r_{cov}, and the van der Waals radius, r_{vdw}.

Fairly reliable values of covalent radii are known for almost all of the elements; thus we will use these values in our comparisons. However, covalent radii are experimental values, so there are slight variations in the values

Li 134	Be 91	B 82	C 77	N 74	O 70	F 68
Na 154						
K 196						
Rb 216						
Cs 235						

Figure 2.10 Atomic radii (pm) of a typical group and short period.

derived from different sets of measurements. A typical set of values for radii (in picometers, 10^{-12} m) of the Group 1 and the Period 2 elements is shown in Figure 2.10.

To explain these trends, we must examine the model of the atom. Let us start with lithium. A lithium atom contains three protons, and its electron configuration is $1s^22s^1$. The apparent size of the atom is determined by the size of the outermost filled orbital, in this case, the $2s$ orbital. The electron in the $2s$ orbital is shielded from the full attraction of the protons by the electrons in the $1s$ orbital (Figure 2.11). Hence the *effective nuclear charge*, Z_{eff}, felt by the $2s$ electron will be much less than three and closer to one. The electrons in the inner orbitals will not completely shield the $2s$ electron, however, because the volumes of the $2s$ and $1s$ orbitals overlap, so the Z_{eff} will be slightly greater than one. In fact, its value can be estimated as 1.3 charge units.

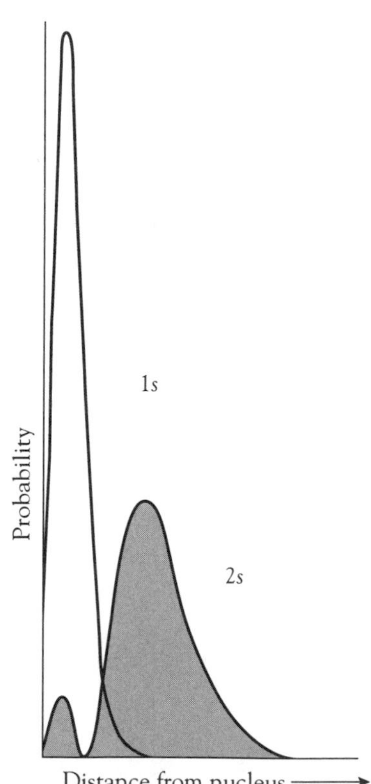

Figure 2.11 The variation of electron probability with distance from the nucleus for electrons in $1s$ and $2s$ orbitals.

A beryllium nucleus has four protons, and the electron configuration of beryllium is $1s^22s^2$. Each $2s$ electron will offer little shielding to the other because they are occupying the same volume of space. The stronger nuclear attraction (higher Z_{eff}) will cause a contraction of the $2s$ orbitals, and the atom will appear to be smaller. Proceeding across the period, the contraction continues: The Z_{eff} increases as a result of the increasing nuclear charge and has a greater and greater effect on the electrons that are being added to orbitals (s and p of the same principal quantum number) that overlap substantially. Thus, as atomic number increases in the period, the atoms become smaller and smaller. In other words, it is the value of Z_{eff} for the outer electrons that determines the apparent outer orbital size and hence the radii of the atoms across a period.

Descending a group, the atoms become larger (see Figure 2.10). This trend also is explainable in terms of the increasing size of the orbitals and the influence of the shielding effect. Let us compare a lithium atom (3 protons) with the larger sodium atom (11 protons). Because of the greater number of protons, one might expect the greater nuclear charge to cause sodium to have the smaller atomic radius. However, sodium has 10 "inner" electrons, $1s^22s^22p^6$, shielding the electron in the $3s^1$ orbital. As a result, the $3s$ electron will feel a nuclear attraction only a little greater than would an electron in the $3s$ orbital of a hydrogen atom. Thus the apparent covalent radius will be quite large.

There are a few minor variations in the smooth trend. For example, gallium has the same covalent radius (126 pm) as aluminum, the element above it. If we compare the electron configurations—aluminum is [Ne]$3s^2 3p^1$ and gallium is [Ar]$4s^2 3d^{10} 4p^1$—we see that gallium has 10 additional protons in its nucleus; these protons correspond to the electrons in the $3d$ orbitals. However, the $3d$ orbitals do not shield outer orbitals very well. Thus, the $4p$ electrons are exposed to a higher Z_{eff} than expected. As a result, the radius is reduced to a value similar to that of the preceding member of the group.

Before leaving this topic, it is worth recalling that the nucleus makes up a very small part of the atom. For example, the covalent radius of an oxygen atom is 70 pm, but its nucleus has a radius of only 0.0015 pm. In terms of volume, the nucleus represents only about a 10^{-11}th part of the atom.

Periodic Properties: Ionization Energy

One trend that relates very closely to electron configuration is that of ionization energy. Usually we are interested in the *first ionization energy*, that is, the energy needed to remove one electron from the outermost occupied orbital of a free atom X:

$$X(g) \rightarrow X^+(g) + e^-$$

Whereas the values of covalent radii depend on which molecules are studied and the errors in measurement, ionization energies can be measured with great precision. Figure 2.12 shows the first ionization energies for the first two periods.

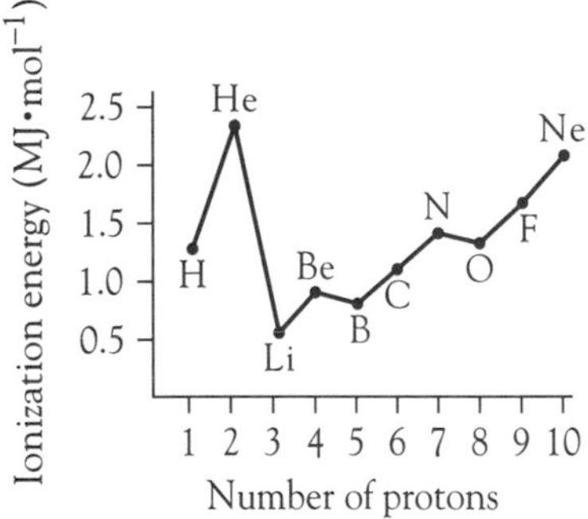

Figure 2.12 First ionization energies for the elements in Periods 1 and 2.

The explanation for the substantial increase from hydrogen to helium involves the second proton in the nucleus. Each electron in the $1s$ orbital of helium is only slightly shielded by the other. Thus an ionizing electron of helium would have to overcome an attraction almost twice that of the electron lost from the hydrogen atom.

In the lithium atom, the ionizing $2s$ electron is shielded from the nuclear attraction by the two electrons in the $1s$ orbital. With a weaker attraction to overcome, the energy needed should be much less; and that is what we find experimentally. The first ionization energy of beryllium is higher than that of lithium; and again we use the concept of little shielding between electrons in the same orbital set—in this case, the $2s$ orbital—to explain this result.

The slight drop in ionization energy for boron shows a phenomenon that is not apparent from a comparison of covalent radii; it is an indication that the s orbitals do partially shield the corresponding p orbitals. This effect is not unexpected, for we showed in Chapter 1 that the s orbitals penetrate closer to the nucleus than do the matching p orbitals. Following boron, the trend of increasing ionization energy resumes as the Z_{eff} increases and the additional electrons are placed in the same p orbital set.

The final deviation from the trend comes with oxygen. The drop in first ionization energy here can only be explained in terms of interelectron repulsions. That is, the one paired electron can be lost more readily than would otherwise be the case, leaving the oxygen ion with an electron configuration of $1s^2 2s^2 2p^3$. Beyond oxygen, the steady rise in first ionization energy continues to the completion of Period 2. Again, this pattern is expected as a result of the increase in Z_{eff}.

Li 0.52										F 1.68
Na 0.50										Cl 1.26
K 0.42	Ca 0.59	Sc 0.63	. . .	Cu 0.75	Zn 0.91	Ga 0.58	Ge 0.76	As 0.95	Se 0.94	Br 1.14
Rb 0.40										I 1.01
Cs 0.38										At 0.91

Figure 2.13 Ionization energies ($MJ \cdot mol^{-1}$) of two typical groups and part of a long period.

Proceeding down a group, the first ionization energy generally decreases. Using the same argument as that for the increase in atomic radius, we conclude that the inner orbitals shield the electrons in the outer orbitals and the successive outer orbitals themselves are larger. For example, compare lithium and sodium again. Although the number of protons has increased from 3 in lithium to 11 in sodium, sodium has 10 shielding inner electrons. Thus the Z_{eff} for the outermost electron of each atom will be essentially the same. At the same time, the volume occupied by the electron in the 3*s* orbital of sodium will be significantly larger than that occupied by the electron in the 2*s* orbital of lithium. Hence the 3*s* electron of sodium will require less energy to ionize than the 2*s* electron of lithium. A similar trend is apparent among the halogen atoms, although the values themselves are much higher than those of the alkali metals (Figure 2.13).

Looking at a long period, such as that from potassium to bromine, we find that the ionization energy increases substantially from left to right in the transition metals. This trend is accounted for by the poor shielding of the 4*s* electrons by the 3*d* electrons. Hence Z_{eff} becomes larger as the number of protons increases, even though the number of electrons increases in the same proportion. Once the filling of the 4*p* orbitals commences (at Ga), the 3*d* orbitals are inner electrons and become more effective at shielding. Hence the ionization energy for Ga is similar to that of Ca. The important point to note is that the later transition groups, particularly Zn and the Group 12 elements, have first ionization energies as high as the nonmetals in Group 17. Thus cation formation will not be strongly favored for these metals.

We can also gain information from looking at successive ionizations of an element. For example, the second ionization energy corresponds to the process

$$X^{+}(g) \rightarrow X^{2+}(g) + e^{-}$$

Lithium provides a simple example of such trends, with a first ionization energy of 0.52 $MJ \cdot mol^{-1}$, a second ionization energy of 7.4 $MJ \cdot mol^{-1}$, and a third ionization energy of 11.8 $MJ \cdot mol^{-1}$. The second electron, being one of the 1*s* electrons, requires greater than 10 times more energy to remove than it takes to remove the 2*s* electron. To remove the third and last electron takes even more energy. This trend shows that even within the same orbital, one electron does partially shield the other electron.

Periodic Properties: Electron Affinity

Just as ionization energy represents the loss of an electron by an atom, so electron affinity represents the gain of an electron. *Electron affinity* is defined as the energy needed to add an electron to the lowest energy unoccupied orbital of a free atom:

$$X(g) + e^- \rightarrow X^-(g)$$

There are conflicting sets of values for experimental electron affinities, but the trends are always consistent, and it is the trends that are important to inorganic chemists. One source of confusion is that electron affinity is often defined as the energy *released* when an electron is added to an atom. This definition yields signs that are the opposite of those on the values discussed here. To identify which sign convention is being used, recall that halogen atoms become halide ions exothermically (that is, the electron affinities for this group should have a negative sign). A typical data set is shown in Figure 2.14.

Note that addition of an electron to an alkali metal is an exothermic process. Because losing an electron by ionization is endothermic (requires energy) and gaining an electron is exothermic (releases energy), for the alkali metals, forming a negative ion is energetically preferred to forming a positive ion! This statement contradicts the dogma often taught in introductory chemistry (the point is relevant to the discussion of ionic bonding in Chapter 5). However, we must not forget that ion formation involves competition between the elements for the electrons. Because the formation of an anion by a nonmetal is more exothermic (releases more energy) than that for a metal, it is the nonmetals that gain an electron rather than the metals.

To explain the low electron affinity for beryllium, we have to assume that the electrons in the $2s$ orbital shield any electron added to the $2p$ orbital. Thus the attraction of a $2p$ electron to the nucleus is close to zero. The high electron affinity of carbon indicates that addition of an electron to give the $1s^22s^22p^3$ half-filled p orbital set of the C^- ion does provide some energy advantage. The near-zero value for nitrogen suggests that the added inter-electron repulsion when a $2p^3$ configuration is changed to that of $2p^4$ is a very significant factor. The high values for oxygen and fluorine can be related to the high Z_{eff} of these two atoms.

Li −60	Be 0	B −26	C −154	N −7	O −141	F −328	Ne 0
Na −53							
K −48							
Rb −47							
Cs −46							

Figure 2.14 Electron affinities ($kJ \cdot mol^{-1}$) of a typical group and period.

Finally, just as there are sequential ionization energies, there are sequential electron affinities. These values, too, have their surprises. Let us look at the first and second electron affinities for oxygen:

$$O(g) + e^- \rightarrow O^-(g) \qquad -141\ kJ{\cdot}mol^{-1}$$
$$O^-(g) + e^- \rightarrow O^{2-}(g) \qquad +744\ kJ{\cdot}mol^{-1}$$

Thus the addition of a second electron is an endothermic process. This energetically unfavorable process is not surprising from the point of view of adding an electron to a species that is already negatively charged, but then it becomes surprising that the oxide ion exists. In fact, as we will see in Chapter 5, the oxide ion can only exist where there is some other driving force, such as the formation of a crystal lattice.

Biochemistry of the Elements

Bioinorganic chemistry, the study of the elements in the context of living organisms, is one of the most rapidly growing areas of chemistry. The details of the role of the elements in biological systems will be discussed in the framework of each group, but an overview of the essential elements for life is provided here.

An element is considered essential when a lack of that element produces an impairment of function and addition of the element restores the organism to a healthy state. Fourteen chemical elements are required in considerable quantities (Figure 2.15).

The essentiality of these bulk-requirement elements is easy to determine, but it is very challenging to identify elements that organisms need in tiny quantities—the ultratrace elements. Because we need so little of them, it is almost impossible to eliminate them from a normal diet in order to examine the effects of any deficiency. Up to now, 16 additional elements seem to be needed for a healthy life. It is amazing that our bodies require almost one-

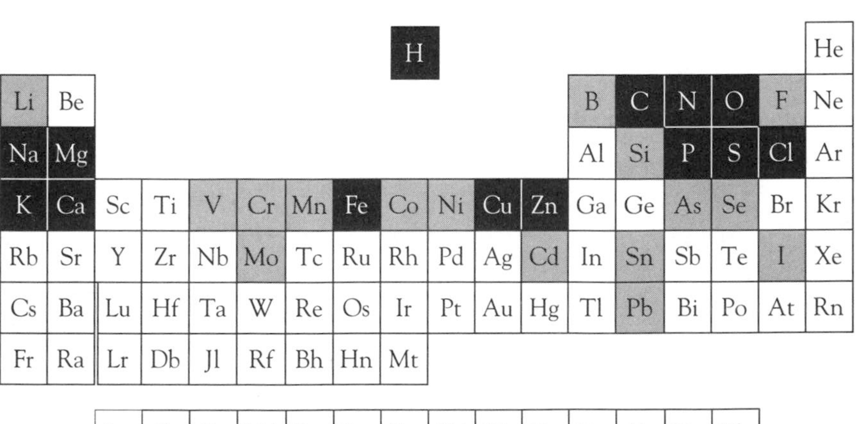

Figure 2.15 The elements necessary for life in large quantities (black) and in ultratrace amounts (shaded).

third of the stable elements for healthy functioning. The precise functions of some of these ultratrace elements are still unknown. As biochemical techniques become more sophisticated, it is possible that more elements will be added to the list of those required.

For almost all the essential elements, there is a range of intake that is optimum, whereas below and above that range some harmful effects are experienced. This principle is known as the *Bertrand rule* (Figure 2.16). Many people are aware of the Bertrand rule in the context of iron intake, where too little can cause anemia but children have died after consuming large quantities of iron supplement pills. The range of optimum intake varies tremendously from element to element. One of the narrow ranges is that of selenium, for which the optimum intake is between 50 μg·day^{-1} and 200 μg·day^{-1}. Less than 10 μg·day^{-1} will cause severe health problems, whereas death ensues from intake levels above 1 mg·day^{-1}. Fortunately, most people, through their normal food intake, ingest levels of selenium in the required range.

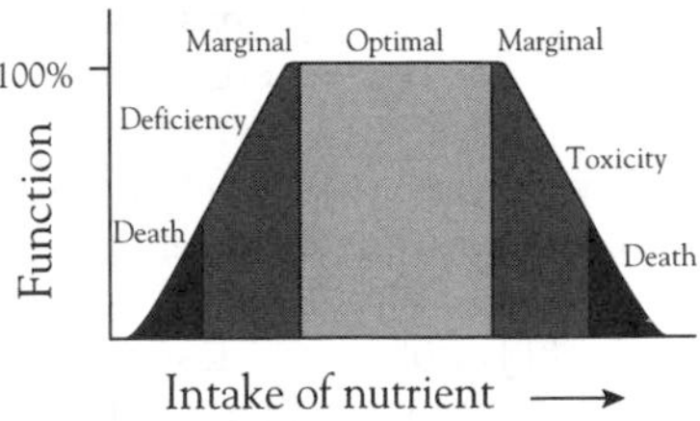

Figure 2.16 Variation of response with intake dose (Bertrand rule).

Exercises

2.1. Define the following terms: (a) rare earth metals; (b) van der Waals radius; (c) effective nuclear charge.

2.2. Define the following terms: (a) second ionization energy; (b) electron affinity; (c) Bertrand rule.

2.3. Explain the two reasons why the discovery of argon posed problems for the original Mendeleev periodic table.

2.4. Explain why the atomic mass of cobalt is greater than that of nickel even though the atomic number of cobalt is less than that of nickel.

2.5. What are one advantage and one disadvantage of the long form of the periodic table?

2.6. Suggest why the Group 11 elements are sometimes called the coinage metals.

2.7. Suggest why it would be more logical to call element 2 helon rather than helium. Why is the *-ium* ending inappropriate?

2.8. Why were the names lanthanides and actinides inappropriate for those series of elements?

2.9. Why is iron the highest atomic number element formed in stellar processes?

2.10. Why must the heavy elements on this planet have been formed from the very early supernovas that exploded?

2.11. Identify

(a) the highest atomic number element for which stable isotopes exist

(b) the only transition metal for which no stable isotopes are known

(c) the only liquid nonmetal at SATP

2.12. Identify the only two radioactive elements to exist in significant quantities on Earth. Explain why they are still present.

2.13. Which element, sodium or magnesium, is likely to have only one stable isotope? Explain your reasoning.

2.14. Suggest the number of neutrons in the most common isotope of calcium.

2.15. Neon-20 and iron-56 are the most common isotopes of those two elements. On the basis of the shell model, suggest a possible explanation for this fact.

2.16. Suggest why polonium-210 and astatine-211 are the isotopes of those elements with the longest half-lives.

2.17. In the classification of elements into metals and nonmetals,

(a) Why is a metallic luster a poor guide?

(b) Why can't thermal conductivity be used?

(c) Why is it important to define electrical conductivity in three dimensions as the best criteria for metallic behavior?

2.18. On what basis are elements classified as semimetals?

2.19. Which atom should have the larger covalent radius, potassium or calcium? Give your reasoning.

2.20. Which atom should have the larger covalent radius, fluorine or chlorine? Give your reasoning.

2.21. Suggest a reason why the covalent radius of germanium (122 pm) is almost the same as that of silicon (117 pm), even though germanium has 18 more electrons than silicon.

2.22. Suggest a reason why the covalent radius of hafnium (144 pm) is less than that of zirconium (145 pm), the element above it in the periodic table.

2.23. Which element should have the higher ionization energy, silicon or phosphorus? Give your reasoning.

2.24. Which element should have the higher ionization energy, arsenic or phosphorus? Give your reasoning.

2.25. An element has the following first through fourth ionization energies in MJ·mol^{-1}: 0.7, 1.5, 7.7, 10.5. Deduce to which group in the periodic table it probably belongs. Give your reasoning.

2.26. Contrary to the general trend, the first ionization energy of lead (715 kJ·mol^{-1}) is higher than that of tin (708 kJ·mol^{-1}). Suggest a reason for this.

2.27. Which element, sodium or magnesium, should have an electron affinity closer to zero? Give your reasoning.

2.28. Would you expect the electron affinity of helium to be positive or negative in sign? Explain your reasoning.

2.29. Why is it wise for your food to come from a number of different geographical locations?

2.30. What part of the periodic table contains the elements that we need in large quantities? How does this correspond to the element abundances?

CHAPTER

3

Covalent Bonding

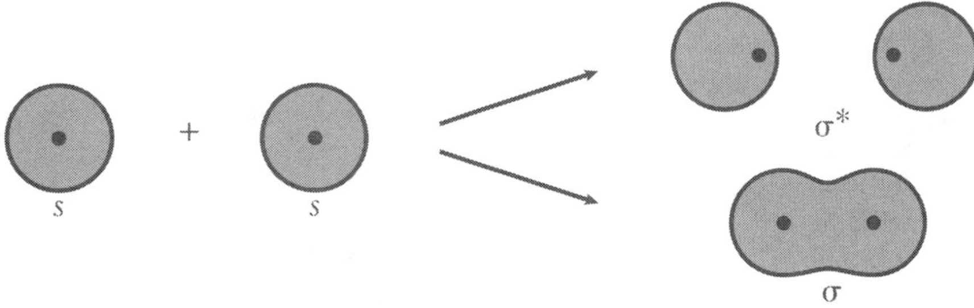

The covalent bond, one of the most crucial concepts in chemistry, is explained best by molecular orbital theory. For complex molecules, more simplistic theories can be used to predict their shapes. Intermolecular forces account for the values of the melting points and boiling points of small covalently bonded molecules.

One of the tantalizing questions raised at the beginning of the twentieth century was, How do atoms combine to form molecules? A great pioneer in the study of bonding was Gilbert N. Lewis, who was raised on a small farm in Nebraska. He suggested in 1916 that the outer (valence) electrons could be visualized as sitting at the corners of an imaginary cube around the nucleus. An atom that was deficient in the number of electrons needed to fill eight corners of the cube could share edges with another atom to complete its octet (Figure 3.1).

As with most revolutionary ideas, many of the chemists of the time rejected the proposal. The well-known chemist Kasimir Fajans commented:

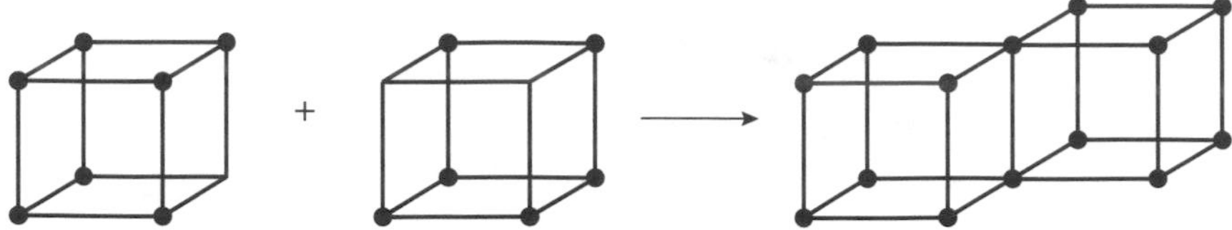

Figure 3.1 The Lewis cube model of the bonding of two halogen atoms.

> Saying that each of two atoms can attain closed electron shells by sharing a pair of electrons is equivalent to a husband and wife, by having a total of two dollars in a joint bank account and each having six dollars in individual bank accounts, have got eight dollars apiece.

Despite the initial criticism, the Lewis concept of shared electron pairs was generally accepted, although the cube diagrams lost favor.

The classical view of bonding was soon overtaken by the rise of quantum mechanics. In 1937 Linus Pauling devised a model that involved the overlapping of atomic orbitals. Pauling was awarded the Nobel prize in chemistry in 1954 for his work on the nature of the chemical bond. He was later awarded a second Nobel prize—the 1962 peace prize—for his work to end the testing of nuclear weapons. As a result of his campaigns against nuclear weapons testing in the atmosphere during the 1950s, he was considered a threat to U.S. national security, and his passport was withdrawn.

Introduction to Molecular Orbital Theory

Molecular orbitals provide the most sophisticated and useful method of explaining how atoms combine to form covalent molecules. For this reason, we will look at this approach first. However, when applied to molecules containing more than two atoms, molecular orbital theory becomes very complex. For these cases, we will use more simplistic and older bonding theories.

When two atoms approach each other, their atomic orbitals mix. The electrons no longer belong to one atom but to the molecule as a whole. To represent this process, we can combine the two atomic wave functions to give two molecular orbitals. This realistic representation of the bonding in covalent compounds involves the linear combination of atomic orbitals and is called *LCAO theory*.

If it is *s* orbitals that mix, then the molecular orbitals formed are given the representation of σ and σ^*. Figure 3.2 shows simplified electron density plots for the atomic orbitals and the resulting molecular orbitals.

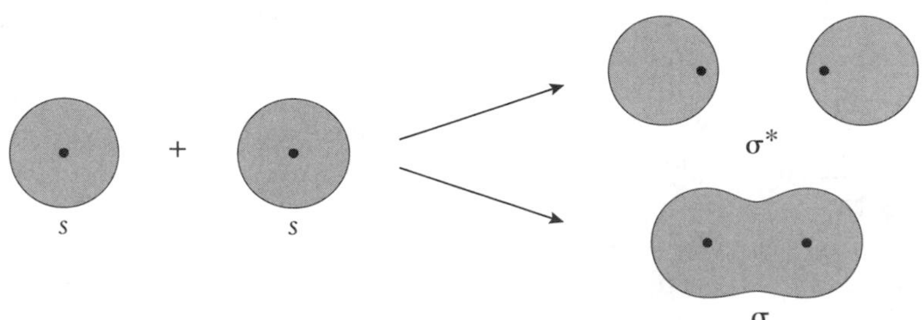

Figure 3.2 The combination of two *s* atomic orbitals to form σ and σ^* molecular orbitals.

For the σ orbital, the electron density between the two nuclei is increased relative to that between two independent atoms. There is an electrostatic attraction between the positive nuclei and this area of higher electron density, and the orbital is called a *bonding orbital*. Conversely, for the σ* orbital, the electron density between the nuclei is decreased, and the partially exposed nuclei cause an electrostatic repulsion between the two atoms. Thus the σ* orbital is an *antibonding orbital*. Figure 3.3 illustrates the variation in the energies of these two molecular orbitals as the atoms are brought together.

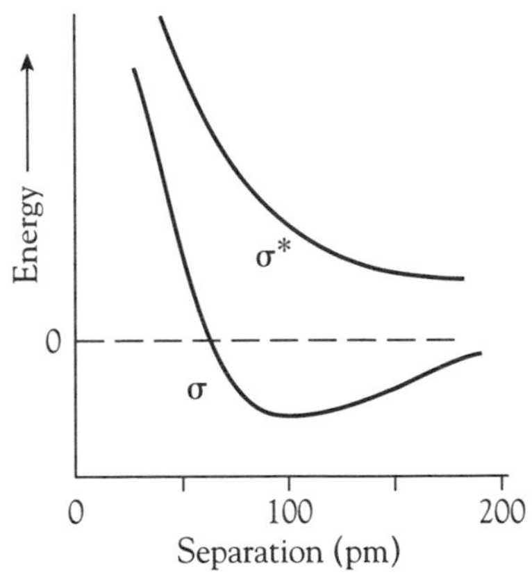

Figure 3.3 Molecular orbital energies as a function of atom separation for two hydrogenlike atoms.

When the atoms are an infinite distance apart, there is no attraction or repulsion; thus under those conditions they can be considered as having a zero energy state. Bringing together two atoms with electrons in bonding orbitals results in a decrease in energy as a result of electrostatic attraction. Figure 3.3 shows that the energy of the bonding orbital reaches a minimum at a certain internuclear separation. This point represents the normal bond length in the molecule. At that separation, the attractive force between the electron of one atom and the nucleus of the other atom is just balanced by the repulsions between the two nuclei. When the atoms are brought closer together, the repulsive force between the nuclei becomes greater, and the energy of the bonding orbital starts to rise.

For electrons in the antibonding orbital, there is no energy minimum. Electrostatic repulsion increases continuously as the partially exposed nuclei come closer and closer.

Several general statements can be made about molecular orbitals:

1. Whenever two atomic orbitals mix, two molecular orbitals are formed, one of which is bonding and the other antibonding. The bonding orbital is always lower in energy than the antibonding orbital.
2. For significant mixing to occur, the atomic orbitals must be of similar energy.
3. Each orbital can hold a maximum of two electrons, one with spin $+\frac{1}{2}$, the other $-\frac{1}{2}$.
4. The electron configuration of a molecule can be constructed by using the Aufbau principle by filling the lowest energy molecular orbitals in sequence.
5. When electrons are placed in different molecular orbitals of equal energy, the parallel arrangement (Hund rule) will have the lowest energy.
6. The bond order in a molecule is defined as the number of bonding electron pairs minus the number of antibonding pairs.

In the next section, we will first see how the molecular orbital theory can be used to explain the properties of diatomic molecules of Period 1 and then look at the slightly more complex cases of Period 2 elements.

Molecular Orbitals for Period 1 Diatomic Molecules

The simplest diatomic molecule is that formed between a hydrogen atom and a hydrogen ion. Figure 3.4 is an energy level diagram that depicts the occupancy of the atomic orbitals and the resulting molecular orbitals. Subscripts are used to indicate from which atomic orbitals the molecular orbitals are derived. Hence the σ orbital arising from the mixing of two 1*s* atomic orbitals is labeled as σ_{1s}. Notice that the energy of the electron is lower in the σ_{1s} molecular orbital than it was in the 1*s* atomic orbital. It is the net reduction in total electron energy that is the driving force in covalent bond formation.

The electron configuration of the dihydrogen ion is written as $(\sigma_{1s})^1$. A

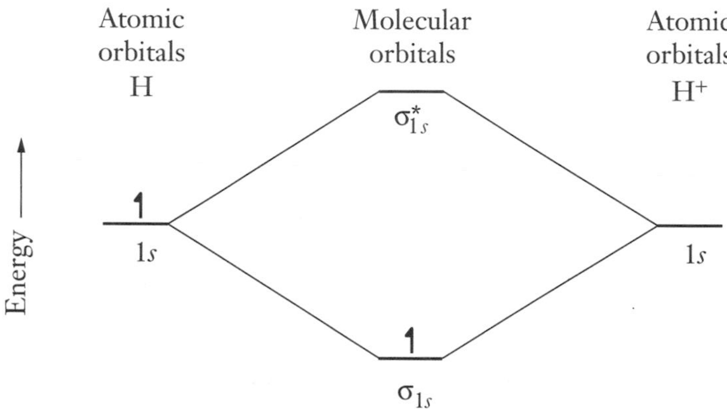

Figure 3.4 Molecular orbital diagram for the H_2^+ molecular ion.

"normal" covalent bond consists of one pair of electrons. Because there is only one electron in the dihydrogen ion bonding orbital, the bond order is $\frac{1}{2}$. Experimental studies of this ion show that it has a bond length of 106 pm and a bond strength of 255 kJ·mol^{-1}.

The energy level diagram for the hydrogen molecule is shown in Figure 3.5. In this case, the bond order is one. The greater the bond order, the greater the strength of the bond and the shorter the bond length. This correlation matches our experimental findings of a shorter bond length (74 pm) and a much stronger bond (436 kJ·mol^{-1}) than that in the dihydrogen ion. The electron configuration is written as $(\sigma_{1s})^2$.

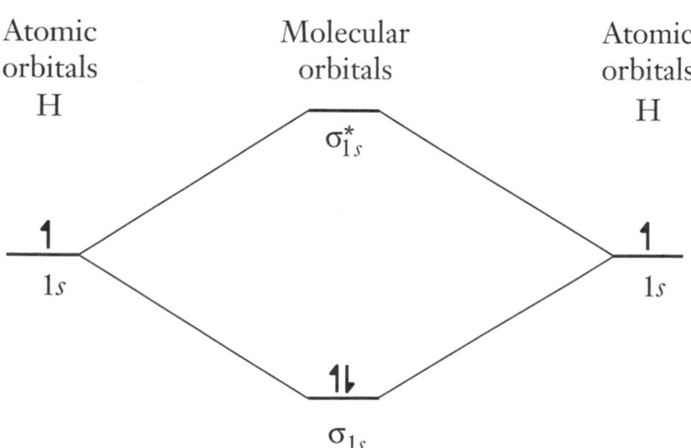

Figure 3.5 Molecular orbital diagram for the H_2 molecule.

It is possible under extreme conditions to combine a helium atom and a helium ion to give the He_2^+ molecular ion. In this species, the third electron will have to occupy the σ^* orbital (Figure 3.6). The molecular ion has an electron configuration of $(\sigma_{1s})^2(\sigma^*_{1s})^1$ and the bond order is $(1-\frac{1}{2})$ or $\frac{1}{2}$. The existence of a weaker bond is confirmed by the bond length (108 pm) and bond energy (251 kJ·mol^{-1})—values about the same as those of the dihydrogen ion.

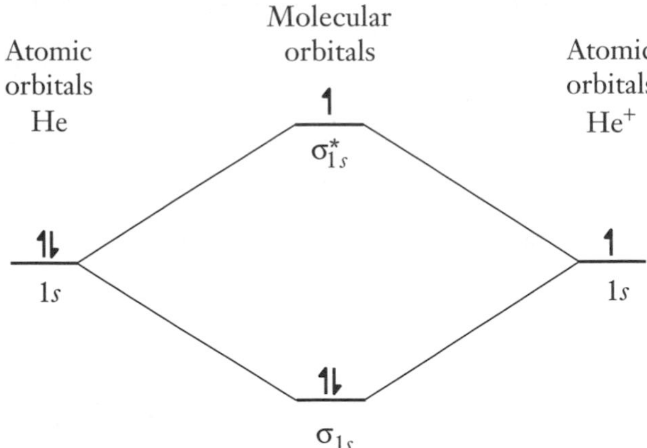

Figure 3.6 Molecular orbital diagram for the He_2^+ ion.

We can make up a molecular orbital diagram for the He_2 molecule (Figure 3.7). Two electrons decrease in energy upon formation of the molecular orbitals while two electrons increase in energy by the same quantity. Thus there is no net decrease in energy by bond formation. An alternative way of expressing the same point is that the net bond order will be zero. Thus no covalent bonding would be expected to occur, and, indeed, helium is a monatomic gas.

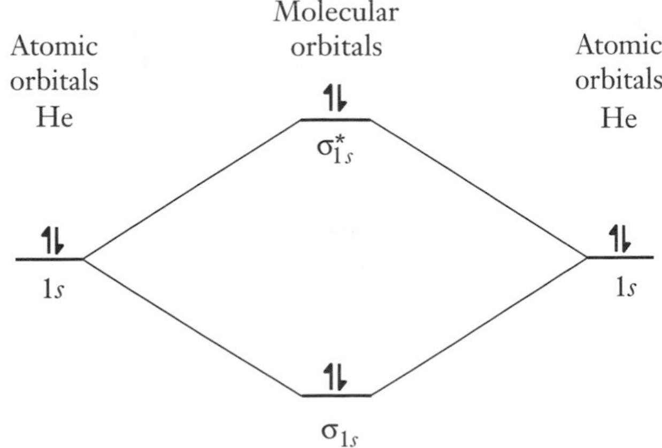

Figure 3.7 Molecular orbital diagram for the 1*s* atomic orbitals of the (theoretical) He_2 molecule.

Molecular Orbitals for Period 2 Diatomic Molecules

With greater effective nuclear charge—in all of the Period 2 elements (and those of later periods)—the 1*s* atomic orbitals contract closer to the nucleus. As a result, they are not involved in the bonding process. Hence, for the first two elements of Period 2, we need only construct a molecular orbital energy diagram corresponding to the 2*s* atomic orbitals. These outermost orbitals, at the "edge" of the molecule, are often called the *frontier orbitals*, and they are always the crucial orbitals for bonding.

Lithium is the simplest of the Period 2 elements. In both solid and liquid phases, the bonding is metallic, a topic that we discuss in the next chapter. In the gas phase, however, there is evidence for the existence of diatomic molecules. The two electrons from the 2*s* atomic orbitals occupy the σ_{2s} molecular orbital, thereby producing a bond order of 1 (Figure 3.8). Both the measured bond length and bond energy are consistent with this value for the bond order. To represent the complete electron configuration, we use the symbol KK to represent the inner orbitals of the lithium atoms; so the complete electron configuration of lithium is written as $KK(\sigma_{2s})^2$.

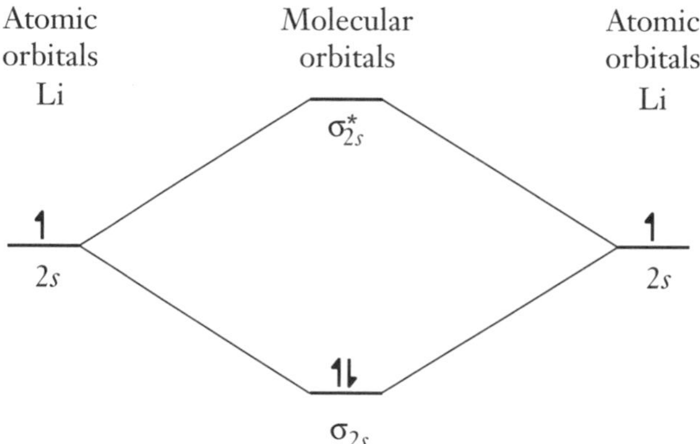

Figure 3.8 Molecular orbital diagram for the 2*s* atomic orbitals of the Li_2 (gas phase) molecule.

Before we consider the heavier Period 2 elements, we must examine the formation of molecular orbitals from 2*p* atomic orbitals. These orbitals can mix in two ways. First, they can mix end to end. When this orientation occurs, a pair of bonding and antibonding orbitals is formed and resembles those of the σ_{1s} orbitals. These orbitals are designated the σ_{2p} and σ^*_{2p} molecular orbitals (Figure 3.9).

Alternatively, the 2*p* atomic orbitals can mix side to side. The bonding and antibonding molecular orbitals formed in this way are designated π

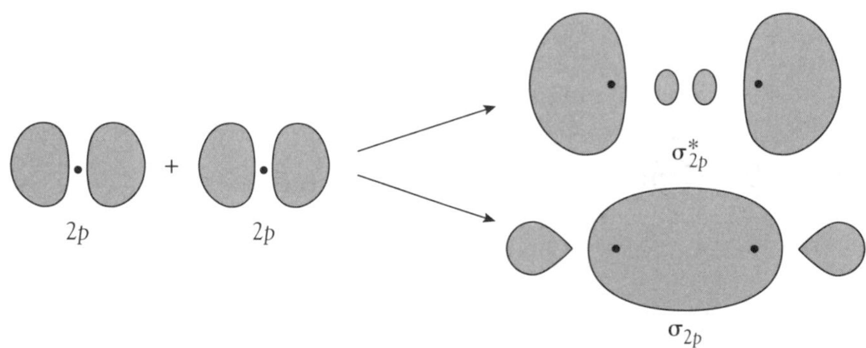

Figure 3.9 The combination of two 2*p* atomic orbitals end to end to form σ_{2p} and σ^*_{2p} molecular orbitals.

orbitals (Figure 3.10). For π orbitals, the increased electron density in the bonding orbital is not between the two nuclei, but above and below a plane containing the nuclei. Because every atom has three $2p$ atomic orbitals when two such atoms combine, there will be three bonding and three antibonding molecular orbitals formed. If we assume the bonding direction to be along the z-axis, the orbitals formed in that direction will be σ_{2p_z} and $\sigma^*_{2p_z}$. At right angles, the other two $2p$ atomic orbitals form π_{2p_x}, $\pi^*_{2p_x}$, π_{2p_y}, and $\pi^*_{2p_y}$ molecular orbitals.

It must be emphasized that bonding models are developed to explain experimental observations. The shorter the bond length and the higher the

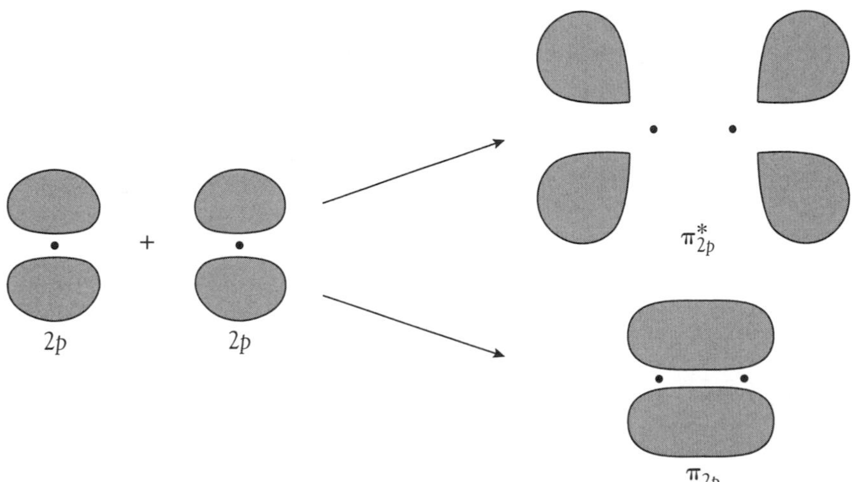

Figure 3.10 The combination of two $2p$ atomic orbitals side to side to form π_{2p} and π^*_{2p} molecular orbitals.

bond energy, the stronger the bond. For the Period 2 elements, bonds with energies of 200–300 $kJ{\cdot}mol^{-1}$ are typical for single bonds; those with energies of 500–600 $kJ{\cdot}mol^{-1}$ are defined as double bonds; and those with energies of 900–1000 $kJ{\cdot}mol^{-1}$ are defined as triple bonds. Thus, for dinitrogen, dioxygen, and difluorine, the molecular orbital model must conform to the bond orders deduced from the measured bond information shown in Table 3.1.

For all three of these diatomic molecules, N_2, O_2, and F_2, the bonding and antibonding orbitals formed from both $1s$ and $2s$ atomic orbitals are filled; thus there will be no net bonding contribution from these orbitals.

Table 3.1 Bond order information for the heavier Period 2 elements

Molecule	Bond length (pm)	Bond energy ($kJ{\cdot}mol^{-1}$)	Assigned bond order
N_2	110	942	3
O_2	121	494	2
F_2	142	155	1

Hence we need only consider the filling of the molecular orbitals derived from the 2*p* atomic orbitals.

The bonding in the dioxygen molecule, O_2, is of particular interest. In 1845 Michael Faraday showed that oxygen gas was the only common gas to be attracted into a magnetic field; that is, dioxygen is paramagnetic, so it must possess unpaired electrons. Later, bond strength studies showed that the dioxygen molecule has a double bond and two unpaired electrons.

For the Period 2 elements beyond dinitrogen, the σ_{2p} orbital is the lowest in energy, followed in order of increasing energy by π_{2p}, π^*_{2p}, and σ^*_{2p}. When the molecular orbital diagram is completed for the dioxygen molecule (Figure 3.11), we see that, according to the Hund rule, there are indeed two unpaired electrons; this diagram conforms with experimental measurements. Furthermore, the bond order of 2 $[3 - (2 \times \frac{1}{2})]$ is consistent with bond length and bond energy measurements. Thus the molecular orbital model explains our experimental observations perfectly.

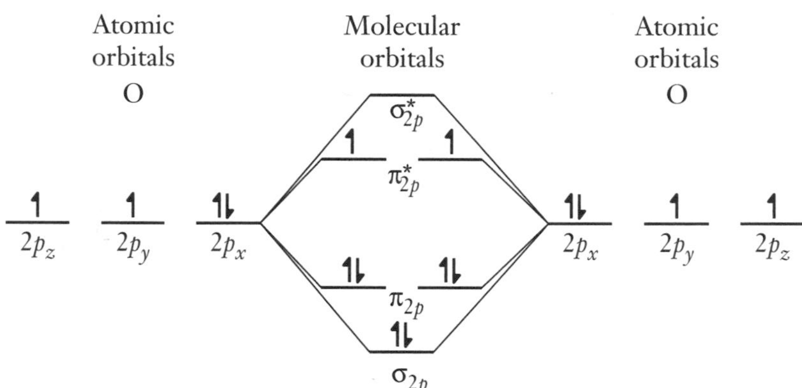

Figure 3.11 Molecular orbital diagram for the 2*p* atomic orbitals of the O_2 molecule.

In difluorine, two more electrons are placed in the antibonding orbital (Figure 3.12). Hence the bond order of 1 represents the net bonding arising from three filled bonding orbitals and two filled antibonding orbitals. The complete electron configuration is represented as $KK(\sigma_{2s})^2(\sigma^*_{2s})^2(\sigma_{2p})^2(\pi_{2p})^4(\pi^*_{2p})^4$.

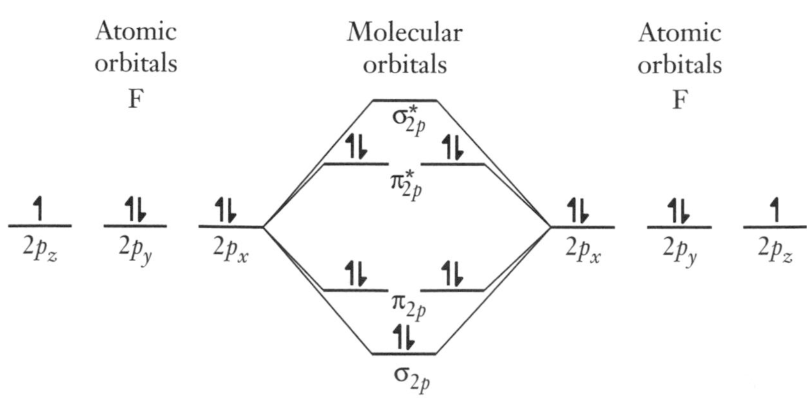

Figure 3.12 Molecular orbital diagram for the 2*p* atomic orbitals of the F_2 molecule.

Neon is the last (heaviest) element in Period 2. If a molecular orbital diagram is constructed for the theoretical Ne_2 molecule, all the bonding and antibonding orbitals derived from the *2p* atomic orbitals are filled; as a result, the net bond order is 0. This prediction is consistent with the observation that neon exists as a monatomic gas.

Up to now, we have avoided discussion of the elements lying in the middle part of Period 2, particularly dinitrogen. The reason concerns the relative energies of the *2s* and *2p* orbitals. For fluorine, with a high Z_{eff}, the *2p* atomic energy level is about 2.5 $MJ \cdot mol^{-1}$ higher in energy than that of the *2s* level. However, at the beginning of the period, the levels differ in energy by only about 0.2 $MJ \cdot mol^{-1}$. In these circumstances, the wave functions for the *2s* and *2p* orbitals become mixed. One result of the mixing is an increase in energy of the σ_{2p} molecular orbital to the point where it has greater energy than the π_{2p} orbital. This ordering of orbitals applies to dinitrogen and the preceding elements in Period 2. When we use this modified molecular orbital diagram to fill the molecular orbitals from the *2p* atomic orbitals for the dinitrogen molecule, we get a configuration with a bonding order of 3 (Figure 3.13). This calculation corresponds with the strong bond known to exist in the molecule. The complete electron configuration of dinitrogen is $KK(\sigma_{2s})^2(\sigma^*_{2s})^2(\pi_{2p})^4(\sigma_{2p})^2$.

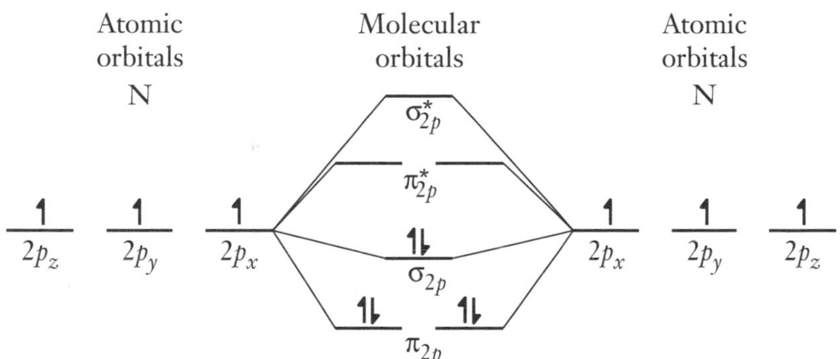

Figure 3.13 Molecular orbital diagram for the *2p* atomic orbitals of the N_2 molecule.

How are we so sure that the molecular orbital energies are actually what we predict? The orbital energies can be measured by a technique known as ultraviolet photoelectron spectroscopy (UV-PES). High-frequency ultraviolet radiation is directed at the molecule, causing an electron from one of the outer orbitals to be ejected. When an electron is lost from a dinitrogen molecule, a dinitrogen cation remains:

$$N_2(g) \xrightarrow{\text{UV}} N_2^+(g) + e^-$$

The various electrons removed have specific energies, and we can correlate those energies with the different molecular orbitals. This technique is illustrated for dinitrogen in Figure 3.14, where the three highest occupied molecular orbitals are matched in energy to the observed UV-PES spectrum (the several lines for the electrons ejected from π_{2p} orbitals result from molecular vibrations).

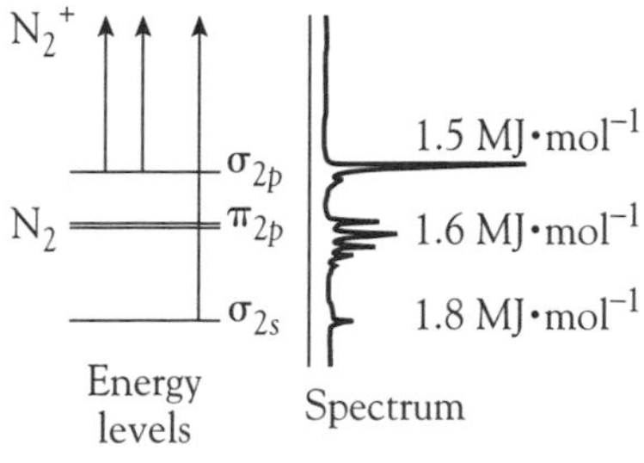

Figure 3.14 Correspondence between the three highest occupied molecular orbitals of the dinitrogen molecule and its photoelectron spectrum.

Molecular Orbitals for Heteronuclear Diatomic Molecules

When we combine atomic orbitals from different elements, we have to consider that the atomic orbitals will have different energies. For elements from the same period, the higher the atomic number, the higher the Z_{eff}, and hence the lower the orbital energies. We can use molecular orbital theory to visualize the bonding of carbon monoxide. A simplified diagram of the molecular orbitals derived from the 2*s* and 2*p* atomic orbitals is shown in Figure 3.15. The oxygen atomic orbitals are lower in energy than those of carbon, but they are close enough in energy that we can construct a molecular orbital diagram similar to that of the homonuclear diatomic molecules. A major difference between homonuclear and heteronuclear diatomic molecules is that the molecular orbitals derived primarily from the 2*s* atomic orbitals of one element overlap significantly in energy with those derived from the 2*p* atomic orbitals of the other element. Thus we must consider molecular orbitals derived from both these atomic orbitals in our diagram. Furthermore, because of the asymmetry of the orbital energies, the bonding molecular orbitals are derived mostly from the oxygen atomic orbitals, whereas the antibonding molecular orbitals are derived mostly from the carbon atomic orbitals. Finally, there are two molecular orbitals whose energies are between those of the contributing atomic orbitals. These orbitals, σ^{NB}, are defined as *nonbonding*, that is, they do not contribute significantly to the bonding. To determine the bond order of carbon monoxide, the number of antibonding pairs (0) is subtracted from the number of bonding pairs (3), a calculation leading to the prediction of a triple bond. The very high bond energy of 1072 kJ·mol^{-1} supports this prediction.

The molecular orbital approach can be applied to diatomic molecules containing atoms of different periods. However, it is then necessary to iden-

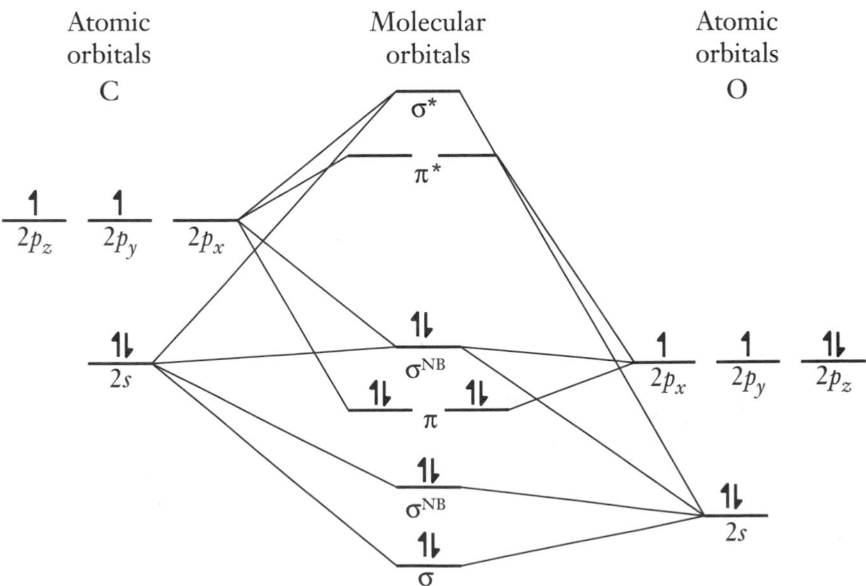

Figure 3.15 Simplified molecular orbital diagram for the 2*s* and 2*p* atomic orbitals of the CO molecule.

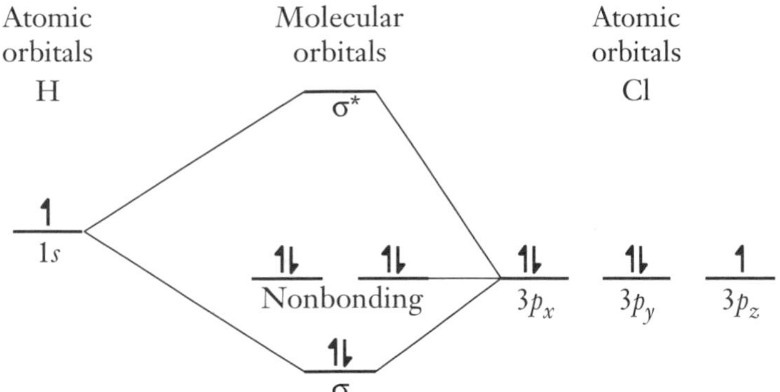

Figure 3.16 Molecular orbital diagram for the 1*s* atomic orbital of hydrogen and the 3*p* atomic orbitals of chlorine in the HCl molecule.

tify the orbitals of similar energies on the two atoms, a task well beyond the coverage of this descriptive inorganic chemistry course. It is instructive, however, to do one example. Consider the hydrogen chloride molecule (Figure 3.16). Calculations show that the 3*p* orbitals of chlorine have slightly lower energy than the 1*s* orbital of hydrogen does. Each 3*p* orbital is oriented in a different direction; but *s* orbitals can only form σ bonds, which can form only with the *p* orbital that is aligned along the bonding axis (traditionally, the p_z orbital). Hence, we conclude that a pair of σ bonding and σ antibonding orbitals will be formed between the 1*s* and 3*p* orbitals. With each atom contributing one electron, the bonding molecular orbital will be filled. This configuration yields a single bond. The two other 3*p* orbitals are oriented in such a way that no net overlap (hence, no mixing) with the 1*s* orbital of hydrogen can occur. As a result, the electron pairs in these orbitals are considered to be nonbonding orbitals. That is, they have the same energy in the molecule as they did in the independent chlorine atom.

Molecular orbital theory can also be used to develop bonding schemes for molecules containing more than two atoms. However, the energy diagrams and the orbital shapes become more and more complex. In later chapters, we will look specifically at the π molecular orbital energy diagrams of some triatomic molecules in order to explain the bond orders that we find from experiment. Nevertheless, for most of the polyatomic molecules, we are more interested in the prediction of the shapes of molecules rather than the finer points of orbital energy levels.

Lewis Theory

To account for the molecular shape of complex molecules, we will use the bonding approach of G. N. Lewis rather than molecular orbital theory, which would carry us beyond the scope of this text. The Lewis, or electron-dot, approach to covalent bond formation is covered extensively in high school and freshman chemistry; hence only a brief review is provided here. Lewis theory explains the driving force of bond formation as being the desire of each atom in the molecule to attain an octet of electrons in its outer (valence) energy level (except hydrogen, where a duet is required). Completion of the octet is accomplished by a sharing of electron pairs between bonded atoms. For example, the nitrogen atom in nitrogen trifluoride has five outer

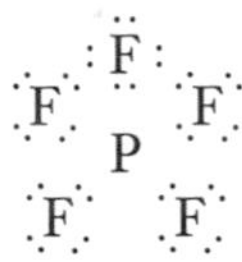

Figure 3.17 Electron-dot diagram for nitrogen trifluoride.

electrons of its own and shares three others, one from each fluorine atom, to give a total of eight [5 + (3 × 1)]. Each fluorine atom possesses seven outer electrons and shares one of the nitrogen's electrons for its total of eight (Figure 3.17).

There are a few anomalous molecules in which the central atom has fewer than eight electrons. There also are a substantial number of molecules in which the central atom has shares in more than eight bonding electrons. Lewis did not realize that the maximum of eight was only generally applicable to the Period 2 elements, for which the sum of the *s* and *p* electrons could not exceed eight. For elements in Period 3 and higher periods, *d* orbitals can be used in bonding to give a theoretical maximum of 18 bonding electrons. In reality, compounds of some of the higher period elements almost always have central atoms with 8, 10, or 12 bonding electrons. For example, phosphorus pentafluoride "pairs up" its five outer electrons with one electron of each of the fluorine atoms to attain an outer set of 10 electrons (Figure 3.18).

Figure 3.18 Electron-dot diagram for phosphorus pentafluoride.

One of the shortcomings of Lewis theory is its inability to account for the bonding in the dioxygen molecule. As mentioned during our earlier discussion of molecular orbitals, the dioxygen molecule is known to have a double bond and two unpaired electrons. When we attempt to draw a normal electron-dot (Lewis) diagram for dioxygen, we arrive at the structure shown in Figure 3.19. This molecule has a pair of electrons shared by each atom to give a double bond. Each atom "possesses" eight electrons.

Figure 3.19 Electron-dot diagram for dioxygen with a double bond.

However, the diagram does not show any unpaired electrons. Alternatively, we can draw a structure that has two unpaired electrons (Figure 3.20). But this diagram shows a molecule with a single, not a double, bond. Furthermore, each oxygen atom only "possesses" seven electrons—a situation that is unknown for any other simple molecule. Hence the molecular orbital theory provides a satisfactory representation of bonding in the dioxygen molecule, whereas the Lewis approach fails.

Figure 3.20 Electron-dot diagram for dioxygen with two unpaired electrons.

Partial Bond Order

In some cases the only structure that can be drawn does not correlate with our measured bond information. The nitrate ion illustrates this situation. A conventional electron-dot diagram for the nitrate ion is shown in Figure 3.21. In this structure, one nitrogen-oxygen bond is a double bond, whereas the other two nitrogen-oxygen bonds are single bonds. However, it has been shown that the nitrogen-oxygen bond lengths are all the same at 122 pm. This length is significantly less than the "true" (theoretical) nitrogen-oxygen single bond length of 141 pm. We explain this discrepancy by arguing that the double bond is shared between the three nitrogen-oxygen bond locations—a concept called *resonance*. The three alternatives could be represented by three electron-dot diagrams for the nitrate ion, each with the double bond in a different location (Figure 3.22).

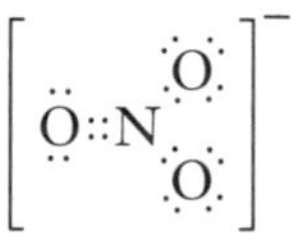

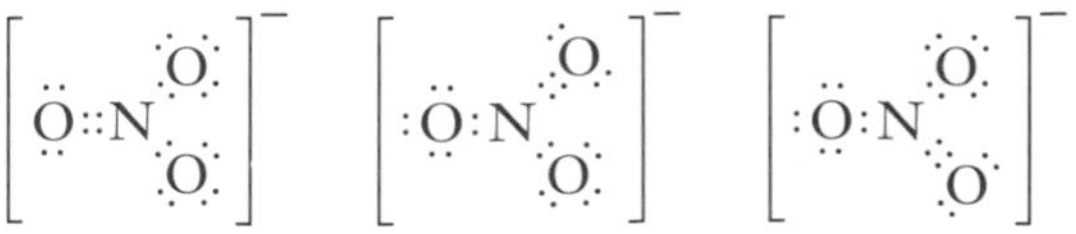

Figure 3.21 Electron-dot diagram for the nitrate ion.

Figure 3.22 The three resonance structures of the nitrate ion.

A third approach is to use a structural formula with broken lines to represent a fractional bond order (Figure 3.23). In this case, because the double bond character is shared by three bonds, the average bond order would be $1\frac{1}{3}$.

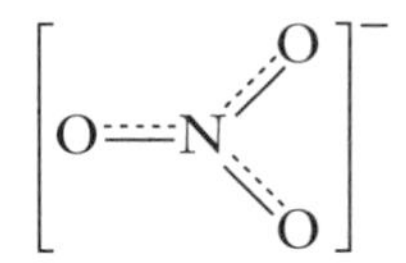

$[O \cdots N(\cdots O)(\cdots O)]^-$

Figure 3.23 Representation of the partial multiple bond character of the nitrate ion.

Formal Charge

In some cases, such as dinitrogen oxide, we can draw more than one feasible electron-dot diagram. It is known that N_2O is an asymmetrical linear molecule with a central nitrogen atom, but there are a number of possible electron-dot diagrams, three of which are shown in Figure 3.24.

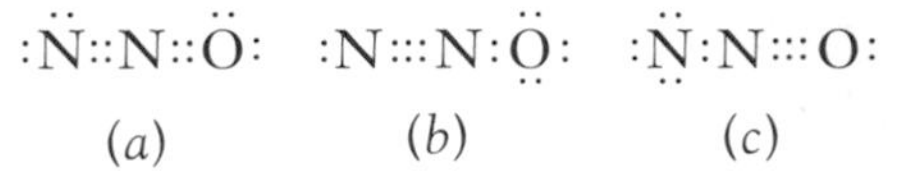

:N̈::N::Ö: :N:::N:Ö: :N̈:N:::O:

(*a*)   (*b*)   (*c*)

Figure 3.24 Three possible electron-dot diagrams for the dinitrogen oxide molecule.

To help decide which possibilities are unrealistic, we can use the concept of formal charge. To find the *formal charge*, we divide the bonding electrons equally among the constituent atoms and compare the number of assigned electrons for each atom with its original number of valence electrons. Any difference is identified by a charge sign (Figure 3.25). For example, in structure *a*, the left-hand nitrogen atom is assigned six electrons; the free atom has five. Hence its formal charge is (5 – 6), or –1. The central nitrogen atom has four assigned electrons and a formal charge of (5 – 4), or +1; the oxygen atom has the same number of electrons as a free atom (6 – 6), so its formal charge is 0.

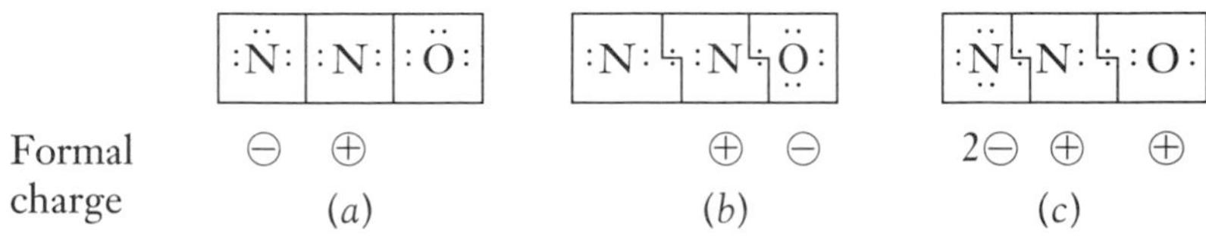

Figure 3.25 Assignment of formal charges to three electron-dot diagrams for the dinitrogen oxide molecule.

According to the concept of formal charge, the lowest energy structure will be the one with the smallest formal charges on the atoms. In the case of dinitrogen oxide, structure *c* is eliminated, but both *a* and *b* have equal but different formal charge arrangements. The best representation, then, is likely to be a resonance mixture of these two possibilities. This resonance form can best be represented by partial bonds, as shown in Figure 3.26. If these two resonance forms contributed equally, there would be an N–N bond order of $2\frac{1}{2}$ and an N–O bond order of $1\frac{1}{2}$, which is close to that estimated from measurement of bond lengths.

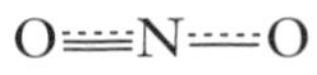

O≡N⋯O

Figure 3.26 Representation of the partial bond order in the dinitrogen oxide molecule.

Valence Shell Electron Pair Repulsion Theory

Electron-dot diagrams can be used to derive the probable molecular shape. To accomplish this, we can use a very simplistic theory that tells us nothing about the bonding but is surprisingly effective at predicting molecular shapes—the valence shell electron pair repulsion theory, known as VSEPR theory. In this theory, it is assumed that repulsions between electron pairs in the outermost occupied energy levels on a central atom will cause those electron pairs to be located as far from one another as is geometrically possible. For the purposes of this theory, we must ignore the differences between the energies of the *s*, *p*, and *d* orbitals and simply regard them as degenerate. It is these outer electrons that are traditionally called the valence electrons.

VSEPR theory is concerned with electron groupings around the central atom. An electron grouping can be an electron pair of a single bond, the two electron pairs in a double bond, the three electron pairs in a triple bond, a lone pair of electrons, or the rare case of a single electron. For simplicity, lone pairs are shown only for the central atom in diagrams of molecular geometry. In the following sections, we will look at each of the common configurations in turn.

Linear Geometry

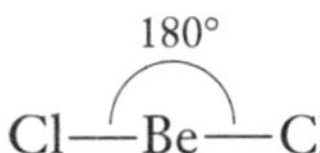

Figure 3.27 Predicted and actual geometry for the gaseous beryllium chloride molecule.

All diatomic molecules and ions are, by definition, linear. However, our main interest is the few common examples of this simplest geometry with triatomic molecules and ions. The most often used example is beryllium chloride. This compound has a complex structure in the room-temperature solid phase, but when it is heated above its boiling point of 820°C, it forms simple triatomic molecules. According to Lewis theory, the two outer electrons of beryllium pair with one electron of each chlorine atom and form two electron pairs around the central beryllium atom. Because there are only two electron groupings around the central atom, the bonds will be furthest apart when the angle between them is 180°. Hence the molecule should be linear, and that is indeed what we find (Figure 3.27).

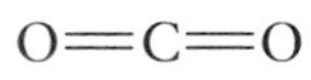

O=C=O

Figure 3.28 Predicted and actual geometry for the carbon dioxide molecule.

Another example of a molecule with two electron groupings is carbon dioxide. Although both carbon-oxygen bonds involve two electron pairs, each double bond represents only one electron grouping (Figure 3.28). Hence the carbon dioxide molecule is linear.

Trigonal Planar Geometry

Boron trifluoride is the common example of trigonal planar geometry. The three outer electrons of the boron atom pair with one electron of each of the fluorine atoms to produce three electron pairs. The maximum separation of three electron pairs requires an angle of 120° between each pair, as shown in Figure 3.29.

F 120°
B—F
F

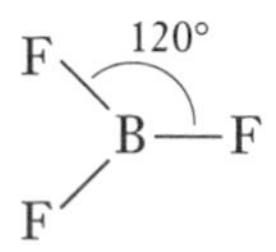

Figure 3.29 Predicted and actual geometry for the boron trifluoride molecule.

The nitrite ion is a good example of a species containing a lone pair on the central atom. The eight electrons on the central nitrogen are assembled as follows: The nitrogen atom itself has five, to which the one electron producing the negative charge is added. This gives six. One oxygen atom forms a double bond with the nitrogen atom, providing two electrons to make eight for the nitrogen atom. Finally, the second oxygen atom is attached to a pair of electrons provided by the nitrogen atom (a coordinate covalent bond).

The electron pair arrangement around the nitrogen atom is trigonal planar (Figure 3.30). However, we cannot detect lone pairs experimentally. The molecular shape we actually observe is V-shaped (also called angular or bent). The lone pair must occupy that third site, however; otherwise the molecule would be linear.

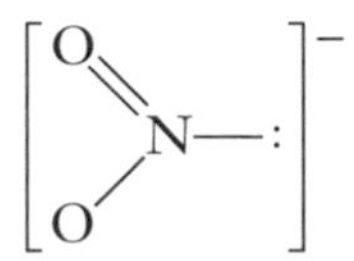

Figure 3.30 Predicted and actual geometry for the nitrite ion.

The nitrite ion illustrates a phenomenon that is generally true for all molecules and ions containing lone pairs. The molecular angles deviate from those of the theoretical geometric figure. For example, the O–N–O bond angle is "squashed" down to 115° from the anticipated 120° value. One suggested explanation is that lone pairs of electrons occupy a greater volume of space than bonding pairs do. We can use a series of ions and molecules to illustrate this concept. The nitryl ion, NO_2^+, with only two electron groupings, is linear; the neutral nitrogen dioxide molecule, NO_2, with three electron groupings (one an "odd" electron), has an O–N–O bond angle of 134°; the nitrite ion, which has a lone pair rather than a single electron, has an observed bond angle of 115° (Figure 3.31). Thus, even though we cannot experimentally "see" lone pairs, they must play a major role in determining molecular shape. The names of the shapes for central atoms that have three electron groupings are given in Table 3.2.

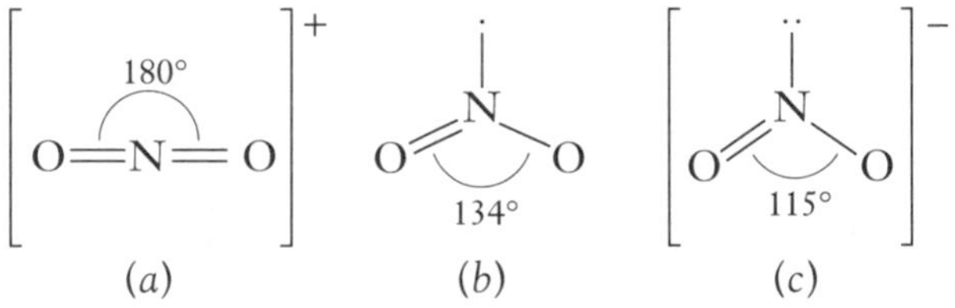

180° 134° 115°

(*a*) (*b*) (*c*)

Figure 3.31 Actual geometries for the nitryl ion (NO_2^+), the nitrogen dioxide molecule (NO_2), and the nitrite ion (NO_2^-).

Table 3.2 Molecules and ions with trigonal planar geometry

Bonding pairs	Lone pairs	Shape
3	0	Trigonal planar
2	1	V

Tetrahedral Geometry

The most common of all molecular geometries is that of the tetrahedron. To place four electron pairs as far apart as possible, molecules adopt this particular three-dimensional geometry in which the bond angles are $109\frac{1}{2}$. The simplest example is the organic compound methane (CH_4), shown in Figure 3.32. To represent the three-dimensional shape on two-dimensional paper, it is conventional to use a solid wedge to indicate a bond directed above the plane of the paper and a broken line to denote a bond angled below the plane of the paper.

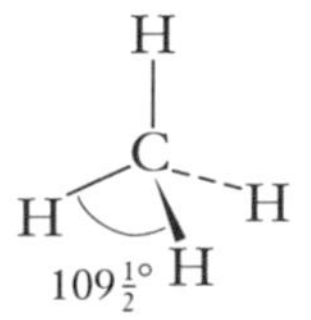

$109\frac{1}{2}°$

Figure 3.32 Predicted and actual geometry for the methane molecule.

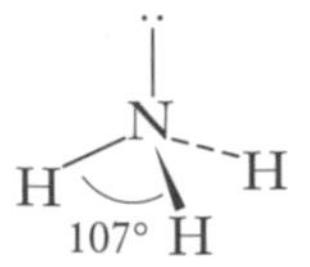

Figure 3.33 Actual geometry for the ammonia molecule.

Ammonia provides the simplest example of a molecule that has tetrahedral geometry and one lone pair. The resulting molecular shape is trigonal pyramidal (Figure 3.33). Like the earlier example of the nitrite ion, the H–N–H bond angle of 107° is slightly less than the expected $109\frac{1}{2}$°. The most familiar molecule with two lone pairs is water (Figure 3.34). The H–O–H bond angle in this V-shaped molecule is reduced from the expected $109\frac{1}{2}$° to 104°. The names of the shapes for central atoms that have four electron groupings are given in Table 3.3.

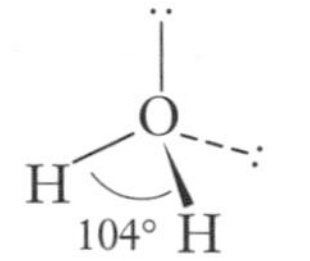

Figure 3.34 Actual geometry for the water molecule.

Table 3.3 Molecules and ions with tetrahedral geometry

Bonding pairs	Lone pairs	Shape
4	0	Tetrahedral
3	1	Trigonal pyramidal
2	2	V

Trigonal Bipyramidal Geometry

As mentioned earlier, atoms beyond Period 2 can possess more than four electron pairs when occupying the central position in a molecule. An example of five electron pairs around the central atom is provided by phosphorus pentafluoride in the gas phase (Figure 3.35). This is the only common molecular geometry in which the angles are not equal. Thus three (equatorial) bonds lie in a single plane and are separated by angles of 120°; the other two (axial) bonds extend above and below the plane and make an angle of 90° with it.

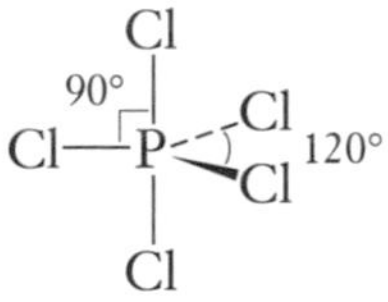

Figure 3.35 Predicted and actual geometry for the gaseous phosphorus pentafluoride molecule.

Sulfur tetrafluoride provides an example of a molecule that has trigonal bipyramidal geometry and one lone pair. There are two possible locations for the lone pair: one of the two axial positions (Figure 3.36a) or one of the three equatorial locations (Figure 3.36b). Two general rules apply to lone pairs: Lone pairs are located so that, first, they are as far from one another as possible and, second, they are as far from the bonding pairs as possible. Sulfur tetrafluoride possesses one lone pair, so only the second rule is applicable. If the lone pair were in an axial position, there would be three bonding pairs at 90° and one at 180°. However, if the lone pair were in an equatorial position, there would be only two bonding pairs at an angle of 90° and the other two at 120°. It is the second possibility, the seesaw shape, that provides the opti-

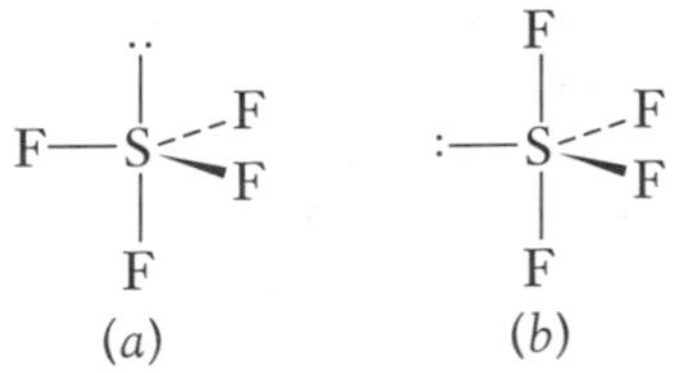

(*a*) (*b*)

Figure 3.36 Possible geometries for the sulfur tetrafluoride molecule (*a*) with the lone pair in the axial position and (*b*) with the lone pair in the equatorial position.

mum situation. This arrangement has been confirmed by bond angle measurements. In the measured angles, the axial fluorine atoms are bent away from the lone pair by $93\frac{1}{2}°$ rather than by 90°. Much more striking is the compression of the F–S–F equatorial angle from 120° to 103°, presumably as a result of the influence of the lone pair (Figure 3.37).

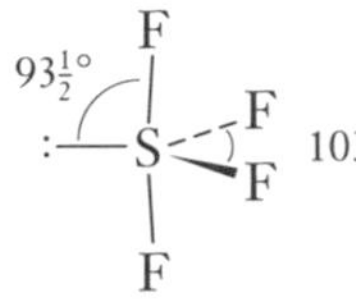

An example of trigonal bipyramidal geometry with two lone pairs is the bromine trifluoride molecule (Figure 3.38). The minimum electron repulsions occur with both lone pairs in the equatorial plane. Hence the molecule is essentially T-shaped, but the axial fluorine atoms are bent away from the vertical to form a F_{axial}–Br–$F_{equatorial}$ angle of only 86°.

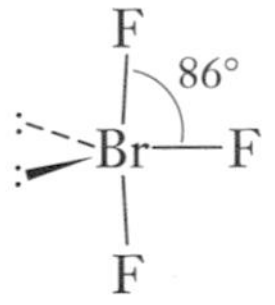

There are a number of examples of molecules with trigonal bipyramidal geometry and three lone pairs. One of these is the xenon difluoride molecule (Figure 3.39). The third lone pair occupies the equatorial position as well. Hence the observed molecular shape is linear. The names for the shapes of molecules and ions that have a trigonal bipyramidal geometry are given in Table 3.4.

Figure 3.37 Actual geometry for the sulfur tetrafluoride molecule.

Figure 3.38 Actual geometry for the bromine trifluoride molecule.

Table 3.4 Molecules and ions with trigonal bipyramidal geometry

Bonding pairs	Lone pairs	Shape
5	0	Trigonal bipyramidal
4	1	Seesaw
3	2	T
2	3	Linear

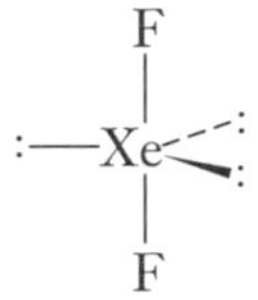

Figure 3.39 Predicted and actual geometry for the xenon difluoride molecule.

Octahedral Geometry

The common example of a molecule with six electron groupings is sulfur hexafluoride. The most widely spaced possibility arises from bonds at equal angles of 90°, the octahedral arrangement (Figure 3.40).

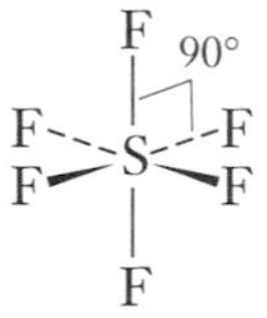

Iodine pentafluoride provides an example of a molecule with five bonding electron pairs and one lone pair around the central atom. Because theoretically all the angles are equal, the lone pair can occupy any site (Figure 3.41), thus producing an apparent square-based pyramidal shape. However, experimental measurements show that the four equatorial fluorine atoms are slightly above the horizontal plane, thus giving a F_{axial}–I–$F_{equatorial}$ angle of only 82°. Once again, this result indicates that the lone pair occupies a greater volume than do the bonding pairs.

Finally, xenon tetrafluoride proves to be an example of a molecule that has four bonding pairs and two lone pairs around the central xenon atom. The lone pairs occupy opposite sides of the molecule, thereby producing a square planar arrangement of fluorine atoms (Figure 3.42).

The names for the shapes of molecules and ions that have an octahedral geometry are given in Table 3.5.

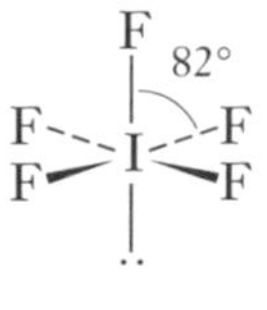

Figure 3.40 Predicted and actual geometry for the sulfur hexafluoride molecule.

Figure 3.41 Actual geometry for the iodine pentafluoride molecule.

Figure 3.42 Predicted and actual geometry for the xenon tetrafluoride molecule.

Table 3.5 Molecules and ions with octahedral geometry

Bonding pairs	Lone pairs	Shape
6	0	Octahedral
5	1	Square-based pyramidal
4	2	Square planar

(*a*)

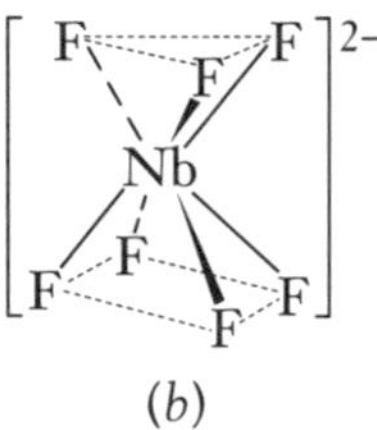

(*b*)

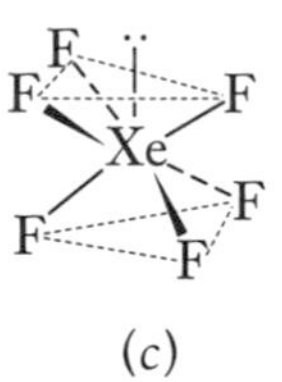

(c)

Figure 3.43 (*a*) The pentagonal bipyramidal structure of uranium(V) fluoride; (*b*) the capped trigonal prismatic structure of niobium(V) fluoride; (*c*) the probable capped octahedral structure of xenon hexafluoride.

Greater Than Six Bonding Directions

There are a few examples of molecules and ions in which the central atom is bonded to more than six neighbors. To accommodate seven or eight atoms around a central atom, the central atom itself has to be quite large and the surrounding atoms and ions quite small. Thus heavier elements from the lower portion of the periodic table combined with the small fluoride ion provide examples of these structures. The MX_7 species are particularly interesting because they can assume three possible geometries: pentagonal bipyramid, capped trigonal prism, and capped octahedron. The pentagonal bipyramid resembles the trigonal bipyramid and octahedron, except it has five rather than three and four bonds, respectively, in the equatorial plane. The capped trigonal prism has three atoms in a triangular arrangement above the central atom and four atoms in a square plane below the central atom. The capped octahedron is simply an octahedral arrangement in which three of the bonds are opened up from the 90° angle and a seventh bond inserted between. These three structures must be almost equally favored in terms of relative energy and atom spacing because all are found: The uranium(V) fluoride ion, UF_7^{2-}, adopts the pentagonal bipyramidal arrangement, whereas the niobium(V) fluoride ion, NbF_7^{2-}, adopts the capped trigonal prismatic structure, and it is believed that xenon hexafluoride, XeF_6, adopts the capped octahedral structure in the gas phase (Figure 3.43).

Orbital Hybridization

Both the Lewis structures and VSEPR theory assume that the outer electrons are of equivalent energies even though we know that the *s*, *p*, *d*, and *f* orbitals of the same principal quantum number have different energies. However, we can provide a theoretical foundation for both of these approaches by using the concept of orbital hybridization. Hybridization is a useful concept because it helps us visualize the bonding process, even though it may have little basis in reality.

The concept of hybridization was developed by chemists who were trying to explain in terms of atomic orbitals how molecules had the shapes that they did. For example, how could the bond angles in methane be $109\frac{1}{2}°$ when the *p* orbitals on the carbon atom were at 90° to one another? One would expect the electron on one hydrogen atom to pair with the *s* electron on the carbon atom and the electrons on the other hydrogen atoms to pair with the three *p* electrons on the carbon. Thus three bonds should be 90° apart, and the bond to the spherical *s* orbital could have any direction.

Table 3.6 Numbers of hybrid orbitals and type of hybridization for various molecular geometries

Orbitals			Type of hybridization	Number of hybrid orbitals	Resulting molecular geometry
s	*p*	*d*			
1	1	0	*sp*	2	Linear
1	2	0	sp^2	3	Trigonal planar
1	3	0	sp^3	4	Tetrahedral
1	3	1	sp^3d	5	Trigonal bipyramidal
1	3	2	sp^3d^2	6	Octahedral

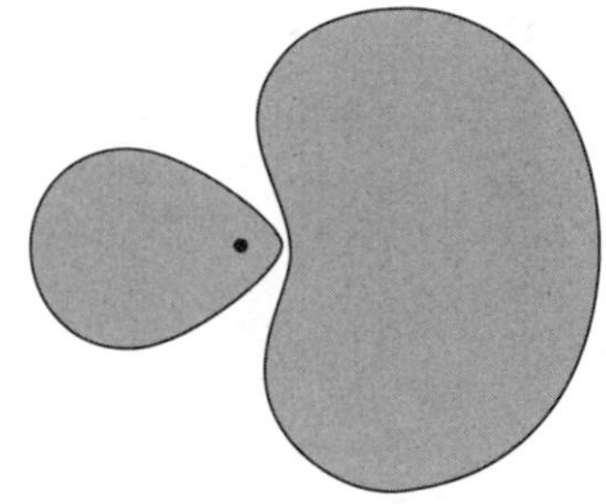

Figure 3.44 The 90 percent probability surface of a hybrid orbital involving combinations of *s* and *p* orbitals. The black dot in the smaller lobe identifies the location of the nucleus.

The *orbital hybridization* concept asserts that the wave functions of atomic orbitals of an atom (usually the central atom of a molecule) can mix together during bond formation to form hybrid atomic orbitals. According to this theory, electrons in these hybrid orbitals are still the property of the donor atom. If the wave functions of an *s* orbital and one or more *p* orbitals are combined, the hybrid orbitals produced are like those shown in Figure 3.44. These orbitals are oriented in a particular direction and should overlap more with the orbitals of another atom than do those of a spherical *s* orbital or of a two-lobed *p* orbital. A greater overlap means that the wave functions will mix better and form a stronger covalent bond.

The number of hybrid orbitals formed will equal the sum of the number of atomic orbitals that are involved in the mixing of wave functions. Like *s* and *p* orbitals, *d* orbitals also can be mixed in. The number of atomic orbitals used, the symbol for the hybrid orbital, and the geometry of the resulting molecule are all listed in Table 3.6.

Boron trifluoride can be used to illustrate hybridization. The boron atom has an electron configuration of [He]$2s^22p^1$ (Figure 3.45a). Suppose that one of the 2*s* electrons moves to a 2*p* orbital (Figure 3.45b). The wave functions of the three orbitals each containing a single electron then mix to provide three equivalent and lower energy sp^2 orbitals (Figure 3.45c). These orbitals, oriented at 120° to one another, overlap with the singly occupied 2*p* orbital on each fluorine atom to give three σ covalent bonds (Figure 3.45d). This explanation matches our experimental findings of equivalent boron-fluorine bonds, each forming 120° angles with the other two—the trigonal planar geometry.

Figure 3.45 The concept of hybrid orbital formation applied to boron trifluoride. (*a*) The electron configuration of the free atom. (*b*) The shift of an electron from the 2*s* orbital to the 2*p* orbital. (*c*) The formation of three sp^2 hybrid orbitals. (*d*) Pairing of boron electrons with three electrons (open half-arrows) of the fluorine atoms.

(*a*) 2*p* ↿ __ __ ; 2*s* ↿⇂

(*b*) 2*p* ↿ ↿ __ ; 2*s* ↿

(*c*) 2*p* __ ; sp^2 ↿ ↿ ↿

(*d*) 2*p* __ ; sp^2 ↿⇂ ↿⇂ ↿⇂

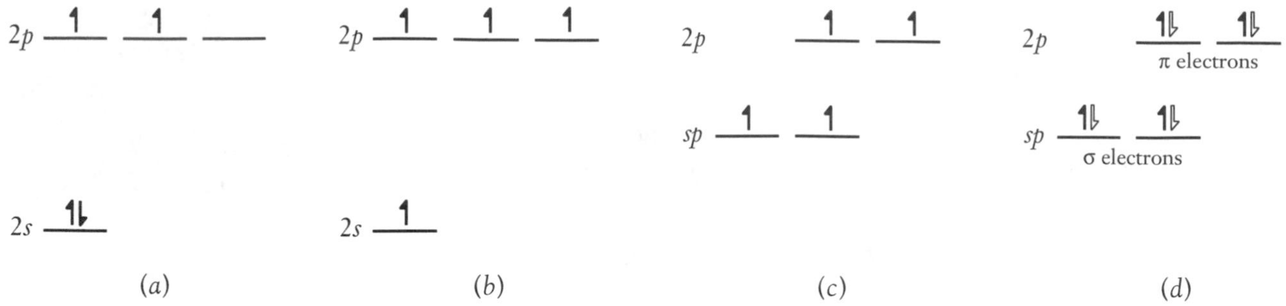

Figure 3.46 The concept of hybrid orbital formation applied to carbon dioxide. (*a*) The electron configuration of the free carbon atom. (*b*) The shift of an electron from the $2s$ orbital to the $2p$ orbital. (*c*) The formation of two sp hybrid orbitals. (*d*) Pairing of the carbon electrons with four oxygen electrons (open half-arrows).

Carbon dioxide provides an example of a molecule in which not all of the occupied orbitals are hybridized. We assume that the $[He]2s^22p^2$ configuration of the carbon atom (Figure 3.46a) is altered to $[He]2s^12p^3$ (Figure 3.46b). The s orbital and one of the p orbitals hybridize (Figure 3.46c). The resulting sp hybrid orbitals are 180° apart, and they overlap with one $2p$ orbital on each oxygen atom to provide a single σ bond and a linear structure. This leaves single electrons in the other two $2p$ orbitals of the carbon atom. Each p orbital overlaps side to side with a singly occupied $2p$ orbital on an oxygen atom to form a π bond with each of the two oxygen atoms (Figure 3.46d). Thus the concept of hybridization can be used to explain the linear nature of the carbon dioxide molecule and the presence of two carbon-oxygen double bonds.

To review, the formation of hybrid orbitals can be used successfully to account for a particular molecular shape. However, hybridization is simply a mathematical manipulation of wave functions, and we have no evidence that it actually happens. Furthermore, the hybridization concept is not a predictive tool; we can only use it once the molecular structure has actually been established. Molecular orbital theory provides a much better representation. Molecular orbital theory is predictive; it pictures the electrons as the property of the molecule as a whole. The most realistic image of bonding proceeds from a calculation of the lowest possible electron energy situation for the molecule as a whole. The problem, of course, with molecular orbital theory is the complexity of the calculations that are needed to deduce molecular shape.

Bonding in the Water Molecule

Earlier in this chapter, we saw that the properties of the dioxygen molecule conflicted with the simple electron-dot representation. For the molecule, molecular orbital theory finally provided an agreement between theory and experimental data. Similarly, another simple molecule, water, causes us to believe that hybridization theory has to be used with great caution.

According to hybridization theory, the outer electrons of oxygen should all be in sp^3 hybrid orbitals, although it is probable that those in the lone pairs would have slightly different energies than those in the bonding pairs. Therefore, the photoelectron spectrum of water would be expected to show two peaks, one corresponding to the ionization of lone pair electrons and the

The False Claim of Molecular Twins

From time to time, scientific claims are made and later shown to be false. One of the longer lasting cases in recent years has been the claim that there can exist pairs of molecules that differ only in the length of their covalent bonds—so-called bond stretch isomers. This proposal surfaced in 1971 when a team of British chemists claimed to have prepared two forms of the same compound, one blue and one green, that differed only in the length of the molybdenum-oxygen covalent bond in the compound. Little notice was taken of this claim, but the report of a second pair of compounds found by German chemists prompted Roald Hoffmann, an American Nobel prize–winning chemist, to propose a theory as to how pairs of compounds with different bond lengths could exist. With such an influential name involved in the discussions and a theory to explain the findings, bond stretch isomerism became an accepted phenomenon—until 1991, when the pairs of compounds were restudied. It was found that in one pair of compounds some of the molecules contained a chlorine atom instead of an oxygen atom. The larger size of the chlorine atom made the bond length appear longer. The chlorine-containing compound was yellow; thus the green color was resulting simply from a mixture of the real (blue) compound and a little yellow impurity. In addition, several people have argued that the theoretical calculations are flawed and that differences in bond lengths cannot be justified by theory either. There are still chemists who believe bond stretch isomers are possible, but, for most chemists, this concept has been disproved. Still, the theory survived for 20 years!

other the ionization of bonding electrons. However, the photoelectron spectrum of water actually shows four, not two, energy levels! If we derive the molecular orbital diagram for water (Figure 3.47), we see that four levels are indeed expected: a weakly bonding orbital resulting from the mixing of the 2*s* orbital of oxygen with the 1*s* orbitals of hydrogen (σ_{2s}); a strongly bonding orbital resulting from the mixing of a 2*p* orbital of oxygen with 1*s* orbitals of hydrogen (σ_{2p}); an essentially nonbonding orbital resulting from the mixing of 2*s* and 2*p* orbitals on the oxygen with 1*s* orbitals of hydrogen (σ^{NB}); a nonbonding orbital resulting from a 2*p* orbital of the oxygen ($2p^{NB}$); and two empty antibonding orbitals (σ^*).

The predominant contribution of oxygen to the bonding comes from the σ_{2p} molecular orbital, with a lesser contribution from the σ_{2s} orbital. This representation is not hugely different from the hybridization representation (except that the four molecular orbitals have different energies). The major difference is in our interpretation of the role of the nonbonding electrons. The molecular orbital description matches our findings from photoelectron spectroscopy and shows that the two electron pairs are in very different environments; one is the conventional "lone pair" (σ^{NB}), whereas the other is in the 2*p* orbital of the oxygen at right angles to the H–O–H plane ($2p^{NB}$). These nonbonding orbitals are shown in Figure 3.48.

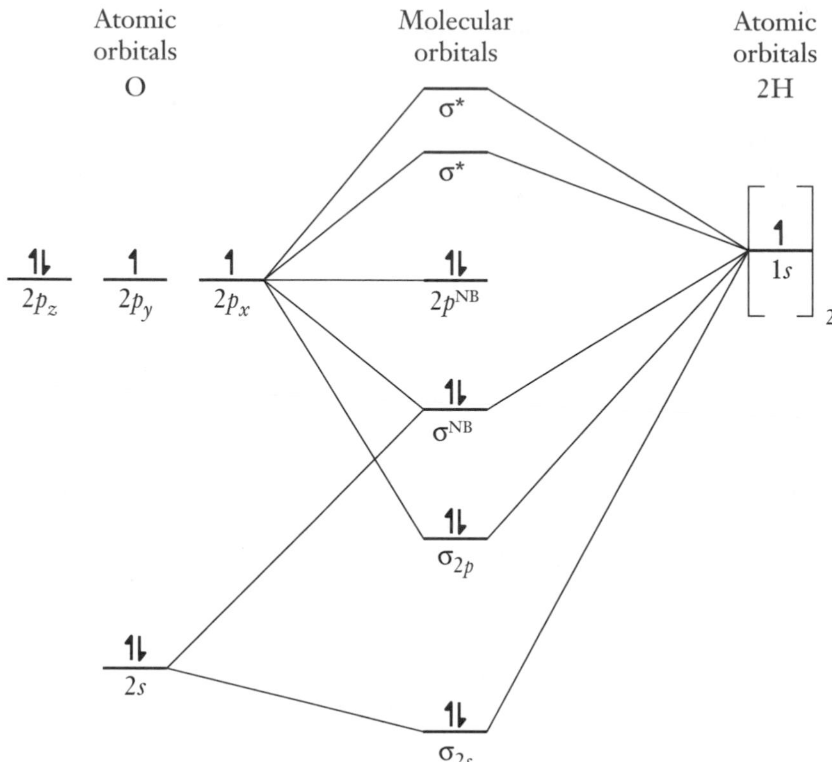

Figure 3.47 Molecular orbital diagram for the water molecule.

The relative energies of molecular orbitals change as bond angles change. The actual bond angle of $104\frac{1}{2}°$ represents the maximum bonding situation for the molecule. Although we know the orbital energies and we can match them to the photoelectron spectrum, there is disagreement about which energy levels are bonding and which are nonbonding. Thus, an alternative interpretation is that the lowest of the molecular orbitals, σ_{2s}, is essentially nonbonding (Figure 3.49). In this view, the mixed $2s/2p$ orbital, σ^{NB}, is actually a bonding orbital.

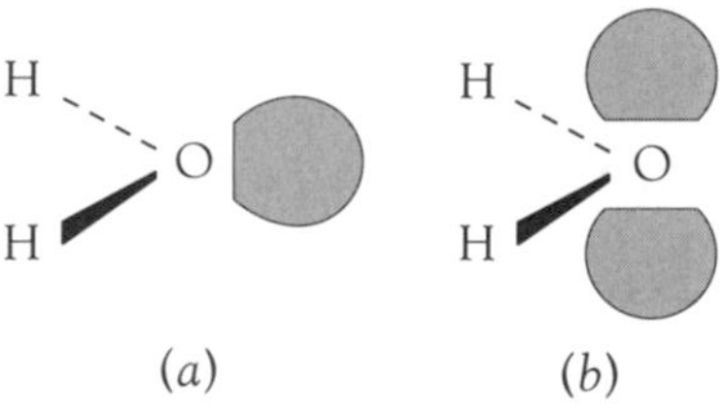

Figure 3.48 Depictions of the nonbonding orbitals of the water molecule: (*a*) the σ^{NB} orbital and (*b*) the $2p^{NB}$ orbital.

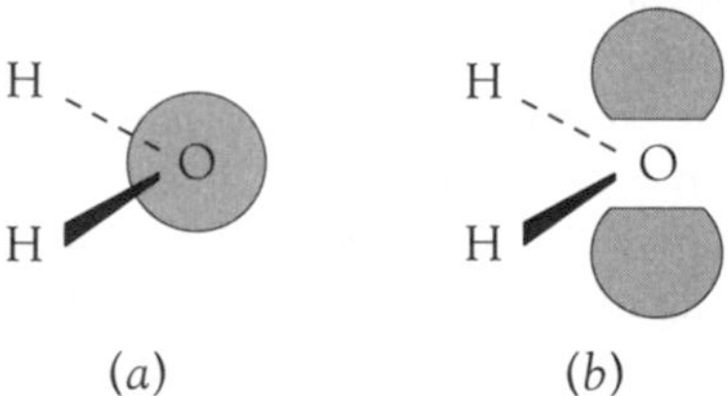

Figure 3.49 Alternative depictions of the nonbinding orbitals of the water molecule: (*a*) the σ_{2s} orbital and (*b*) the $2p^{NB}$ orbital.

Network Covalent Substances

In a diamond or a crystal of quartz, all of the atoms are held together by covalent bonds. Diamond is a form of carbon in which each and every carbon atom is bonded in a tetrahedral arrangement to all its neighbors (Figure 3.50). Such linking by covalent bonds throughout a substance is known as network covalent bonding. The whole crystal is one giant molecule. The second common example of network covalent bonding is quartz, the crystalline form of silicon dioxide, SiO_2. In this compound, each silicon atom is surrounded by a tetrahedron of oxygen atoms, and each oxygen atom is bonded to two silicon atoms.

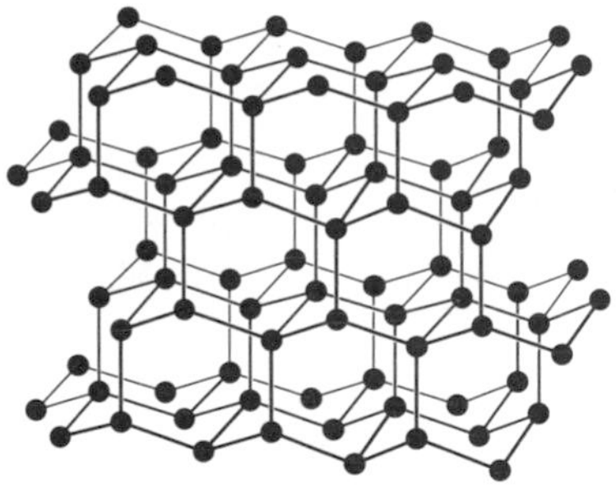

Figure 3.50 Arrangement of carbon atoms in a diamond.

To melt a substance that contains network covalent bonds, one must break the covalent bonds. But covalent bonds have energies in the range of hundreds of kilojoules per mole, so very high temperatures are needed to accomplish this cleavage. Thus diamond sublimes at about 4000°C, and silicon dioxide melts at 2000°C. For the same reason, network covalent substances are extremely hard; diamond is the hardest naturally occurring substance known. Furthermore, such substances are insoluble in all solvents.

Intermolecular Forces

Almost all covalently bonded substances consist of independent molecular units. If there were only intramolecular forces (the covalent bonds), there would be no attractions between neighboring molecules, and, consequently, all covalently bonded substances would be gases at all temperatures. We know this is not the case. Thus there must be forces between molecules, or intermolecular forces. Indeed, there is one intermolecular force that operates between all molecules: induced dipole attractions, also called dispersion forces or London forces (after the scientist Fritz London, not the British capital). The other types of forces—dipole-dipole, ion-dipole, and hydrogen bonding—only occur in specific circumstances, which we discuss later in this section.

Dispersion (London) Forces

In the orbital representation of atoms and molecules, the probability distribution of the electrons (electron density) is a time-averaged value. It is the oscillations from this time-averaged value that lead to the attractions between neighboring molecules. The noble gas atoms provide the simplest example. On average, the electron density should be spherically symmetric around the atomic nucleus (Figure 3.51a). However, most of the time, the electrons are asymmetrically distributed; consequently, one part of the atom has a higher electron density and another part has a lower electron density (Figure 3.51b). The end at which the nucleus is partially exposed will be slightly more positive (δ+), and the end to which the electron density had shifted will be partially negative (δ−). This separation of charge is called a temporary dipole. The partially exposed nucleus of one atom will attract electron density from a neighboring atom (Figure 3.52a), and it is this induced dipole between

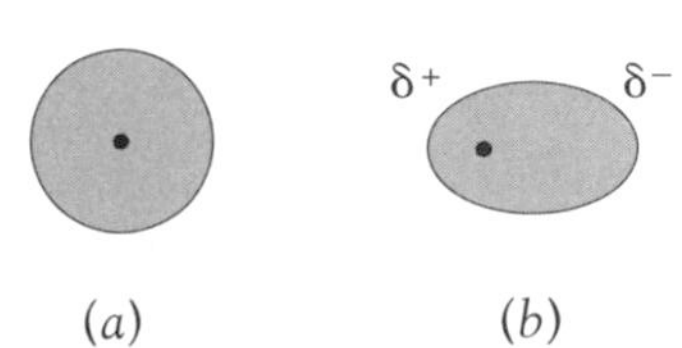

Figure 3.51 (*a*) Average electron density for an atom. (*b*) Instantaneous electron density producing a temporary dipole.

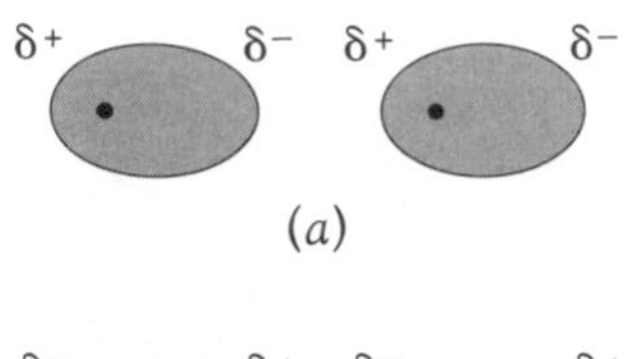

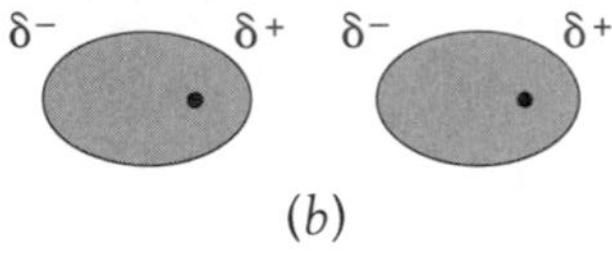

Figure 3.52 (*a*) The instantaneous attraction between neighboring molecules. (*b*) The reversal of polarity in the next instant.

molecules that represents the dispersion force between atoms and molecules. However, an instant later, the electron density will have shifted, and the partial charges involved in the attraction will be reversed (Figure 3.52b).

The strength of the dispersion force mainly depends on the number of electrons in the atom or molecule. In turn, it is the strength of these intermolecular forces that determines the melting point and boiling point of the substance—the stronger the intermolecular forces, the higher will be both the melting and the boiling points. This relationship is illustrated by the chart in Figure 3.53, which shows the dependence of the boiling points of the Group 14 hydrides on the number of electrons in the molecule.

Molecular shape is a secondary factor affecting the strength of dispersion forces. A compact molecule will allow only a small separation of charge, whereas an elongated molecule can allow a much greater charge separation. A good comparison is provided by sulfur hexafluoride, SF_6, and decane, $CH_3CH_2CH_2CH_2CH_2CH_2CH_2CH_2CH_2CH_3$. The former has 70 electrons and a melting point of −51°C, whereas the latter has 72 electrons and a melting point of −30°C. Hence, the dispersion forces are greater between the long decane molecules than between the near-spherical sulfur hexafluoride molecules.

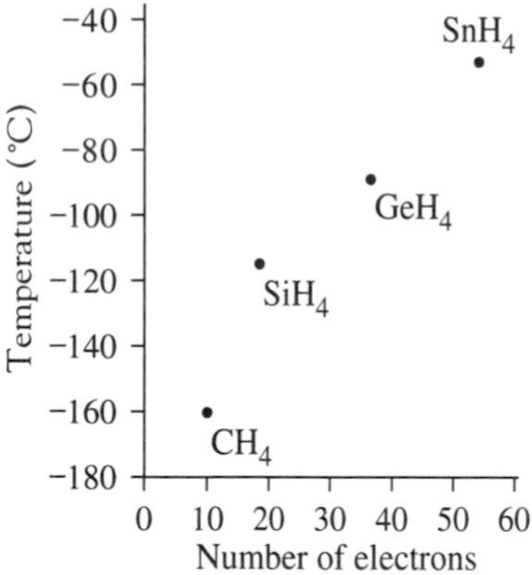

Figure 3.53 Dependence of the boiling points of the Group 14 hydrides on the number of electrons.

Electronegativity

There is a very simple experiment that shows the existence of two types of molecules. In this experiment, a positively charged rod is held near a stream of liquid. Many liquids (for example, carbon tetrachloride) are unaffected by the charged rod, whereas others (for example, water) are attracted by the rod. If the positively charged rod is replaced by a negatively charged rod, those unaffected by the positive charge are also unaffected by the negative charge, whereas those attracted by the positive charge are also attracted by the negative charge. To explain these observations, we infer that the deflected liquids consist of molecules in which there is a permanent charge separation (a permanent dipole). Thus the partially negative ends of the molecules are attracted toward the positively charged rod, and the partially positive ends are attracted toward the negatively charged rod. But why should some molecules have a permanent charge separation? For an explanation, we need to look at another concept of Linus Pauling's—electronegativity.

Pauling defined electronegativity as the power of an atom in a molecule to attract electrons to itself. This relative attraction for bonding electron pairs really reflects the comparative Z_{eff} of the two atoms on the shared electrons. Thus the values increase from left to right across a period and decrease down a group in the same way as ionization energies do. Electronegativity is a relative concept, not a measurable function. The Pauling electronegativity scale is an arbitrary one, with the value for fluorine defined as 4.0 (some useful values are shown in Figure 3.54).

Thus, in a molecule such as hydrogen chloride, the bonding electrons will not be shared equally between the two atoms. Instead, the higher Z_{eff} of chlorine will cause the bonding pair to be more closely associated with the chlorine atom than with the hydrogen atom. As a result, there will be a per-

			H 2.2			
Li 1.0	Be 1.6	B 2.0	C 2.5	N 3.0	O 3.4	F 4.0
			Si 1.9	P 2.2	S 2.6	Cl 3.2
						Br 3.0
						I 2.7

Figure 3.54 Pauling electronegativity values of various main group elements.

manent dipole in the molecule. This dipole is depicted in Figure 3.55, using the δ sign to indicate a partial charge and an arrow to indicate the dipole direction.

Individual bond dipoles can act to "cancel" each other. A simple example is provided by carbon dioxide, where the bond dipoles are acting in opposite directions. Hence the molecule does not possess a net dipole; in other words, the molecule is nonpolar (Figure 3.56).

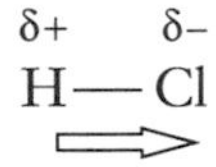

$$\overset{\delta+}{H}—\overset{\delta-}{Cl}$$
$$\Longrightarrow$$

Figure 3.55 Permanent dipole of the hydrogen chloride molecule.

$$\overset{\delta-}{O}{=}\overset{\delta+}{C}{=}\overset{\delta-}{O}$$
$$\Longleftarrow \quad \Longrightarrow$$

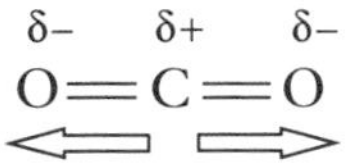

Figure 3.56 Because it has opposing bond dipoles, the carbon dioxide molecule is nonpolar.

Dipole-Dipole Forces

A permanent dipole results in an enhancement of the intermolecular forces. For example, carbon monoxide has higher melting and boiling points (68 K and 82 K, respectively) than dinitrogen does (63 K and 77 K), even though the two compounds are isoelectronic (have the same number of electrons).

It is important to realize that dipole-dipole attractions are a secondary effect in addition to the induced dipole effect. This point is illustrated by comparing hydrogen chloride with hydrogen bromide. The electronegativity difference between hydrogen and chlorine is 1.0 for hydrogen chloride and is 0.8 between the atoms in hydrogen bromide, so the dipole-dipole attractions between neighboring hydrogen chloride molecules will be stronger than those between neighboring hydrogen bromide molecules. Yet the boiling point of hydrogen bromide (206 K) is higher than that of hydrogen chloride (188 K). Thus induced dipole (dispersion) forces, which will be higher for the hydrogen bromide (36 electrons in HBr and 18 in HCl), must be the predominant factor. In fact, complex calculations show that dispersion forces account for 83 percent of the attraction between neighboring hydrogen chloride molecules and 96 percent of the attraction between neighboring hydrogen bromide molecules.

Hydrogen Bonding

If we look at the trend in boiling points of the Group 17 hydrides (Figure 3.57), we see that hydrogen fluoride has an anomalously high value. Similar

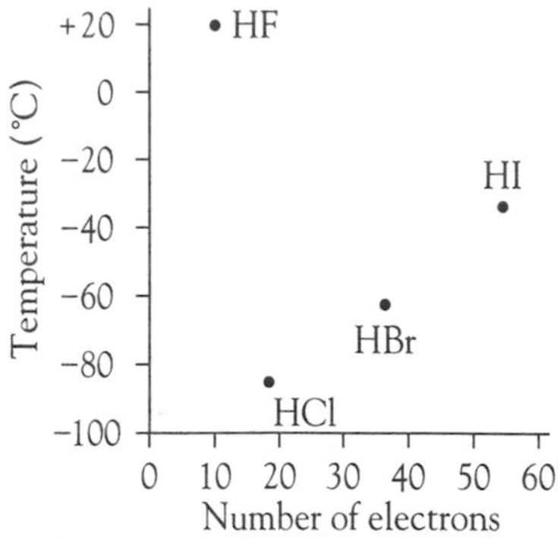

Figure 3.57 Boiling points of the Group 17 hydrides.

plots for the Groups 15 and 16 hydrides show that the boiling points of ammonia and water also are anomalous. The elements involved have high electronegativities; thus it is argued that the much stronger intermolecular forces are a result of exceptionally strong dipole-dipole forces. These forces are given the special name of hydrogen bonds.

Hydrogen bonding, then, is by far the strongest intermolecular force; indeed, it can represent 5 to 20 percent of the strength of a covalent bond. The strength of the hydrogen bond between molecules does depend on the identity of the nonhydrogen element. Thus hydrogen bond strength decreases in the order H–F > H–O > H–N, and this order parallels the decrease in electronegativity differences. However, this factor cannot be the whole answer, because the H–Cl bond is more polar than the H–N bond is, yet hydrogen chloride molecules do not exhibit very strong intermolecular attractions.

Because the distances between two molecules sharing a hydrogen bond are significantly less than the sum of the van der Waals radii, it is argued that electron density is shared across a hydrogen bond. In this approach, a hydrogen bond is less of an intermolecular force and more of a weak covalent bond. We return to this point and discuss it in more detail in Chapter 7.

Covalent Bonding and the Periodic Table

In this chapter, we have emphasized the existence of covalent bonds in compounds containing nonmetals and semimetals. The chemistry of nonmetals is dominated by the covalent bond. Yet as we will see throughout this text, covalent bonding is also very important for compounds containing metals. Metal-containing polyatomic ions provide examples of these. The permanganate ion, MnO_4^-, contains manganese covalently bonded to four oxygen atoms. We will see in Chapter 5 that the bonding of many metallic compounds is more easily described as covalent rather than ionic. However, it is better to discuss this topic in the context of the chapter on ionic bonds, because we need to discuss the nature of the ionic bond before we can appreciate why a compound adopts one or the other of the bonding types.

Exercises

3.1. Define the following terms: (a) LCAO theory; (b) σ orbital; (c) VSEPR theory; (d) hybridization.

3.2. Define the following terms: (a) network covalent molecules; (b) intramolecular forces; (c) electronegativity; (d) hydrogen bonding.

3.3. Use a molecular orbital diagram to determine the bond order of the H_2^- ion. Would the ion be diamagnetic or paramagnetic?

3.4. Would you expect Be_2 to exist? Use a molecular orbital energy diagram to explain your reasoning.

3.5. Use a molecular orbital diagram to determine the bond order in the N_2^+ ion. Write an electron configuration $[KK(\sigma_{2s})^2, \ldots]$ for this ion.

3.6. Use a molecular orbital diagram to determine the bond order in the O_2^+ ion. Write an electron configuration $[KK(\sigma_{2s})^2, \ldots]$ for this ion.

3.7. Assuming that it has molecular orbital energies similar to those of carbon monoxide, deduce the bond order of the NO^+ ion.

3.8. Assuming that it has molecular orbital energies similar to those of carbon monoxide, deduce the bond order of the NO^- ion.

3.9. Construct a molecular orbital diagram for diboron, B_2. What would you predict for the bond order? Construct a similar diagram for diboron, using the ordering for the heavier Period 2 elements, and compare the two results. What experimental property could be used to confirm this different ordering?

3.10. Construct a molecular orbital diagram and write the electron configuration of the dicarbon anion and cation, C_2^- and C_2^+. Determine the bond order in each of these ions.

3.11. Construct electron-dot diagrams for (a) oxygen difluoride; (b) phosphorus trichloride; (c) xenon difluoride; (c) the iodine tetrachloride ion, ICl_4^-.
3.12. Construct electron-dot diagrams for (a) the ammonium ion; (b) carbon tetrachloride; (c) the silicon hexafluoride ion, SiF_6^{2-}; (d) the sulfur pentafluoride ion, SF_5^-.
3.13. Construct an electron-dot diagram for the nitrite ion. Draw the structural formulas of the two resonance possibilities for the ion and estimate the average nitrogen-oxygen bond order. Draw a partial bond representation of the ion.
3.14. Construct an electron-dot diagram for the carbonate ion. Draw the structural formulas of the three resonance possibilities for the ion and estimate the average carbon-oxygen bond order. Draw a partial bond representation of the ion.
3.15. The thiocyanate ion, NCS^-, is linear, with a central carbon atom. Construct all feasible electron-dot diagrams for this ion, then use the concept of formal charges to identify the most probable contributing structures. Display the result by using a partial bond representation.
3.16. The boron trifluoride molecule is depicted as having three single bonds and an electron-deficient central boron atom. Use the concept of formal charge to suggest why a structure involving a double bond to one fluorine, which would provide an octet to the boron, is not favored.
3.17. For each of the molecules and polyatomic ions in Exercise 11, determine the electron pair arrangement and the molecular shape according to VSEPR theory.
3.18. For each of the molecules and polyatomic ions in Exercise 12, determine the electron pair arrangement and the molecular shape according to VSEPR theory.
3.19. For each of the molecules and polyatomic ions in Exercise 11, identify in which cases distortion from the regular geometric angles will occur as a result of the presence of one or more lone pairs.
3.20. For each of the molecules and polyatomic ions in Exercise 12, identify in which cases distortion from the regular geometric angles will occur as a result of the presence of one or more lone pairs.
3.21. For each of the electron pair arrangements determined in Exercise 11, identify the hybridization of the electrons on the central atom.
3.22. Using Figure 3.45 as a model, show how the concept of hybrid orbitals can be used to explain the bonding in the gaseous beryllium chloride molecule.
3.23. Using Figure 3.45 as a model, show how the concept of hybrid orbitals can be used to explain the bonding in the methane molecule.
3.24. For each of the electron pair arrangements determined in Exercise 12, identify the hybridization of the electrons on the central atom.
3.25. Which would you expect to have the higher boiling point, hydrogen sulfide, H_2S, or hydrogen selenide, H_2Se? Explain your reasoning clearly.
3.26. Which would you expect to have the higher melting point, dibromine, Br_2, or iodine monochloride, ICl? Explain your reasoning clearly.
3.27. For each of the molecules and polyatomic ions in Exercise 11, determine whether they are polar or nonpolar.
3.28. For each of the molecules and polyatomic ions in Exercise 12, determine whether they are polar or nonpolar.
3.29. Which would you expect to have the higher boiling point, ammonia, NH_3, or phosphine, PH_3? Explain your reasoning clearly.
3.30. Which would you expect to have the higher boiling point, phosphine, PH_3, or arsine, AsH_3? Explain your reasoning clearly.

CHAPTER

4

Metallic Bonding

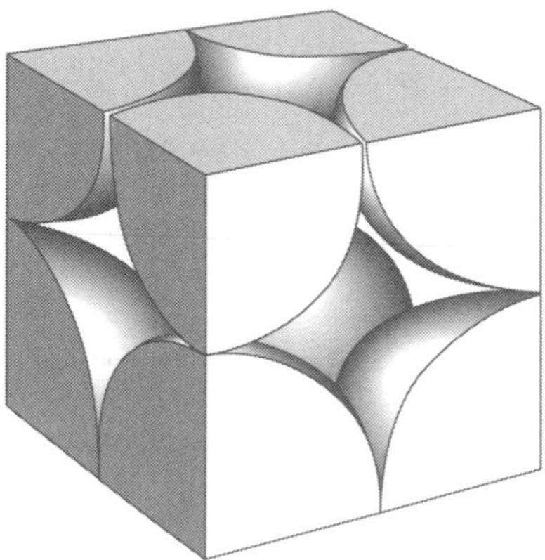

The bonding in metals is explained best in terms of the molecular orbital theory that we have already discussed in the context of covalent bonding. The arrangement of atoms in a metal crystal can be interpreted in terms of the packing of hard spheres. The packing arrangements are common to both metals and ionic compounds. Thus a study of metallic bonding provides a link between covalent and ionic bonding.

The extraction of metals from their ores coincided with the rise of civilization. Bronze, an alloy of copper and tin, was the first metallic material to be widely used. As smelting techniques became more sophisticated, iron became the preferred metal because it is a harder material and more suitable than bronze for swords and ploughs. For decorative use, gold and silver were easy to use because they are very malleable metals (that is, they can be deformed easily).

Over the ensuing centuries, the number of metals known climbed to its present large number, the great majority of the elements in the periodic table. Yet in the contemporary world, it is still a small number of metals that dominate our lives, particularly iron, copper, aluminum, and zinc. The metals that we choose must suit the purpose for which we need them, yet the availability of an ore and the cost of extraction are often the main reasons why one metal is chosen over another.

Bonding Models

Any theory of metallic bonding must account for the key properties of metals, the most important feature of which is the high electrical conductivity (see Chapter 2). Furthermore, any model should account for the high thermal conductivity and the high reflectivity (metallic luster) of metals.

The simplest metallic bonding model is the electron-sea (or electron-gas) model. In this model, the valence electrons are free to move through the bulk metal structure (hence the term *electron sea*) and even leave the metal, thereby producing positive ions. It is valence electrons, then, that convey electric current, and it is the motion of the valence electrons that transfers heat through a metal. However, this model does not explain the high reflectivity of metals, and it can be used for few quantitative calculations of metal properties.

Molecular orbital theory provides a more comprehensive model of metallic bonding. This extension of molecular orbital theory is sometimes called *band theory*, which we will illustrate by looking at the orbitals of lithium. In Chapter 3, we saw that two lithium atoms combined in the gas phase to form the dilithium molecule. The molecular orbital diagram showing the mixing of two $2s$ atomic orbitals is given in Figure 4.1 (both sets of atomic orbitals are shown on the left). Now suppose that the atomic orbitals of four lithium atoms are mixed. Again, there must be the same number of σ_{2s} molecular orbitals as $2s$ atomic orbitals, half of which are bonding and the other half antibonding. To avoid violating the quantum rules, the energies of the orbitals cannot be degenerate. That is, one σ_{2s} cannot have exactly the same energy as the other σ_{2s} orbital. Figure 4.2 shows the resulting orbital arrangement.

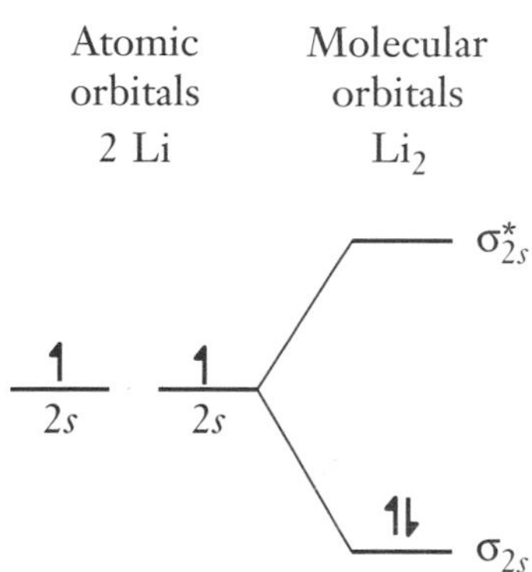

Figure 4.1 Molecular orbital diagram for the dilithium (gas phase) molecule.

In a large metal crystal, the orbitals of n atoms, where n is some enormous number, are mixed. These orbitals interact throughout the three dimensions of the metal crystal, yet the same principles of bonding apply.

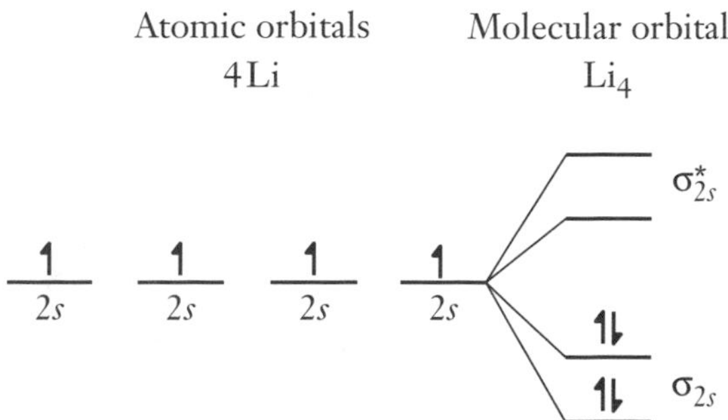

Figure 4.2 Molecular orbital diagram for the combination of four lithium atoms.

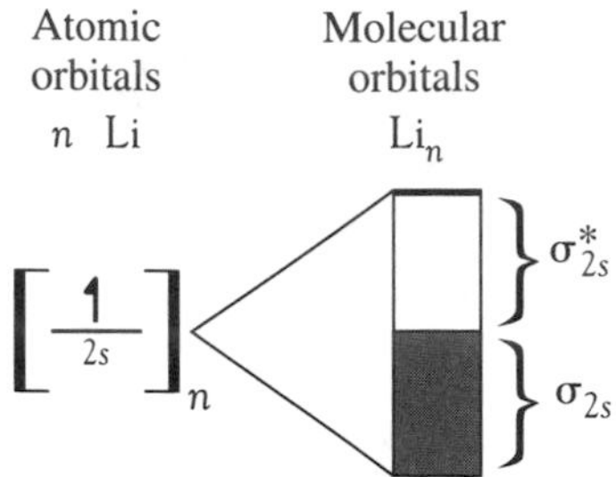

Figure 4.3 Band derived from the 2*s* atomic orbitals by the combination of *n* lithium atoms.

There will be $\frac{1}{2}n$ σ_{2s} (bonding) molecular orbitals and $\frac{1}{2}n$ σ^*_{2s} (antibonding) molecular orbitals. With such a large number of energy levels, the spacing of levels becomes so close that they essentially constitute a continuum. This continuum is referred to as a band. For lithium, the band derived from the 2*s* atomic orbitals will be half-filled. That is, the σ_{2s} part of the band will be filled and the σ^*_{2s} part will be empty (Figure 4.3).

We can visualize electrical conductivity simplistically as the gain by an electron of the infinitesimally small quantity of energy needed to raise it into the empty antibonding orbitals. It can then move freely through the metal structure, as electric current. Similarly, the high thermal conductivity of metals can be visualized as "free" electrons transporting translational energy throughout the metal structure. It is important to remember, however, that the "real" explanation of these phenomena requires a more thorough study of band theory.

In Chapter 1, we saw that light is absorbed and emitted when electrons move from one energy level to another. The light emissions are observed as a line spectrum. With the multitudinous energy levels in a metal, there is an almost infinite number of possible energy level transitions. As a result, the atoms on a metal surface can absorb any wavelength and then reemit light at that same wavelength as the electrons release that same energy when returning to the ground state. Hence band theory accounts for the reflectivity of metals.

Beryllium also fits the band model. With an atomic configuration of [He]2s^2, both the σ_{2s} and σ^*_{2s} molecular orbitals will be fully occupied. That is, the band derived from overlap of the 2*s* atomic orbitals will be completely filled. At first, the conclusion would be that beryllium could not exhibit metallic properties because there is no space in the band in which the electrons can "wander." However, the empty 2*p* band overlaps with the 2*s* band, thus enabling the electrons to "roam" through the metal structure (Figure 4.4).

We can use band theory to explain why some substances are electrical conductors, some are not, and some are semiconductors. In the metals, bands overlap and allow a free movement of electrons. In nonmetals, bands are widely separated, so no electron movement can occur (Figure 4.5a). These elements are called insulators. In a few elements, the bands are close enough to allow only a small amount of electron excitation in an upper unoccupied band (Figure 4.5b). These elements are known as intrinsic *semiconductors*. Our

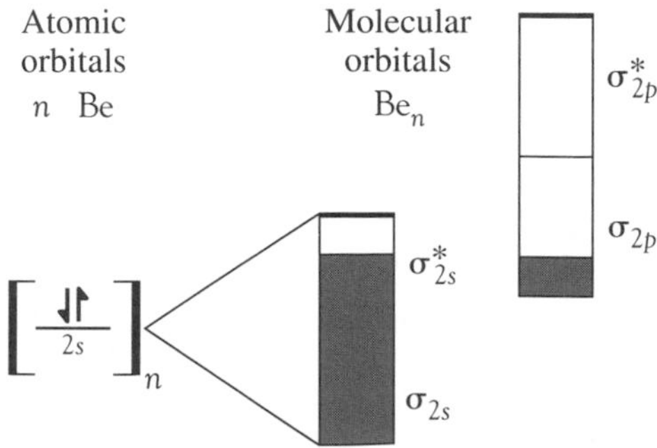

Figure 4.4 Bands derived from the frontier orbitals (2*s* and 2*p*) of beryllium.

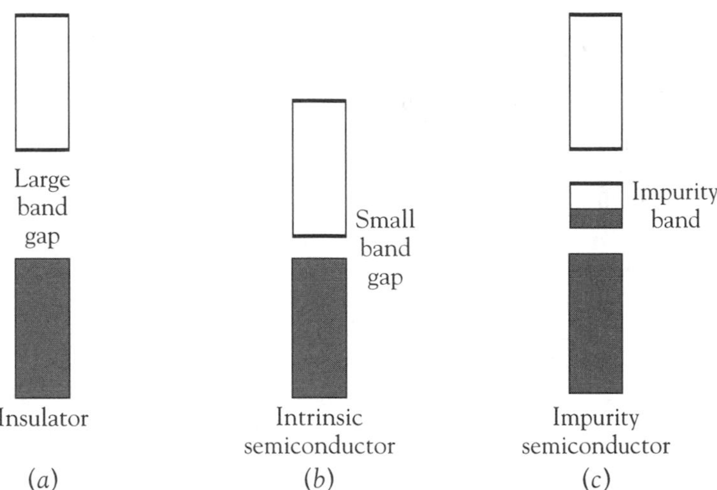

Figure 4.5 Schematic of the band structure of (*a*) a nonmetal, (*b*) an intrinsic semiconductor, and (*c*) an impurity semiconductor.

modern technology depends on the use of semiconducting materials, and it has become necessary to synthesize semiconductors with very specific properties. This can be done by taking an element with a wide band gap and "doping" it with some other element, that is, adding a trace impurity. The added element has an energy level between that of the filled and empty energy levels of the main component (Figure 4.5c). This impurity band can be accessed by the electrons in the filled band, enabling some conductivity to occur. By this means, the electrical properties of semiconductors can be adjusted to meet any requirements.

Finally, we should mention the phases in which metallic behavior can occur. Even in the liquid phase, there is enough orbital mixing for metallic properties to be maintained, but in the gas phase, metallic behavior cannot exist. Thus the boiling point of a metal represents the temperature at which the metallic bond ceases to exist. In fact, one of the most consistent characteristics of metals is a high boiling point, a property indicative of the strength of the metallic bond. For example, sodium melts at 98°C, but it does not boil until 890°C. Gaseous metallic elements like Li_2 exist as pairs or like Be as individual atoms. In any event, they do not look metallic. For example, vaporizing potassium produces a bright green, transparent gas.

Structure of Metals

The way in which metal atoms pack together in a crystal is interesting in itself, but, of equal importance to an inorganic chemist, it also provides a basis from which to discuss the ion packing in an ionic compound (which we do in the next chapter). The concept of crystal packing assumes that the atoms are hard spheres. In a metal crystal, the atoms are arranged in a repeating array that is called a *crystal lattice*. The packing of metal atoms is really a problem in geometry. That is, we are concerned with the different ways in which spheres of equal size can be arranged.

It is easiest to picture the atomic arrays by arranging one layer and then placing successive layers over it. The simplest possible arrangement is that in

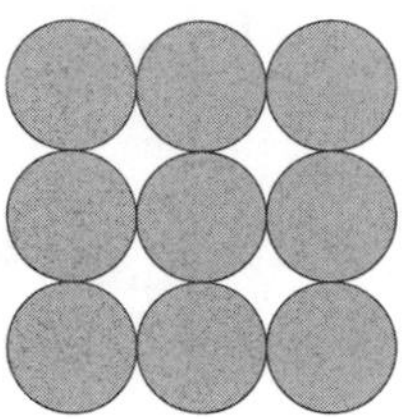

Figure 4.6 Simple cubic packing. Successive layers are superimposed over the first array.

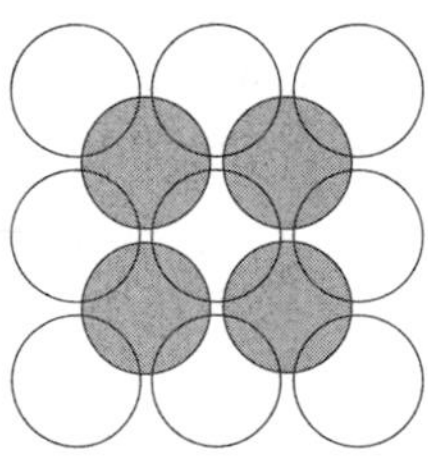

Figure 4.7 Body-centered cubic packing. The second layer (shaded) is placed over holes in the first layer and the third over the holes in the second.

which the atoms in the base are packed side by side. The successive layers of atoms are then placed directly over the layer below. This is known as *simple cubic packing* (Figure 4.6). Each atom is touched by four other atoms in its own plane plus one atom above and one below, a total of six neighboring atoms. As a result, it is said that each atom has a *coordination number* of six.

The simple cubic arrangement is not very compact and is unknown in metal structures, although as we will see in Chapter 5, it is found in some ionic compounds. An alternative cubic packing arrangement is to place the second layer of atoms over the holes in the first layer. The third layer then fits over the holes in the second layer—which happens to be exactly over the first layer. This more compact arrangement is called *body-centered cubic* (Figure 4.7). Each atom is touched by four atoms above and four atoms below its plane. Thus body-centered cubic results in a coordination number of eight.

The other two possibilities are based on a hexagon arrangement for each layer; that is, each atom is surrounded by six neighbors in the plane. Notice that in the hexagonal arrangement the holes between the atoms are much closer together than in the cubic arrangement (Figure 4.8). When the second hexagonal layer is placed over the first, it is physically impossible to place atoms over all the holes in the first layer. In fact, only half the holes can be covered. If the third layer is placed over holes in the second layer so that it is superimposed over the first layer, then a *hexagonal close-packed arrangement* is obtained. The fourth layer is then superimposed over the second layer. Hence it is also known as an *abab* packing arrangement (Figure 4.9).

An alternative hexagonal packing arrangement involves placement of the third layer over the holes in the first and second layers. It is the fourth layer, then, that repeats the alignment of the first layer. This *abcabc* packing arrangement is known as *cubic close-packed* (Figure 4.10). Both packings based on the hexagonal arrangement are 12-coordinate.

The types of packing, the coordination numbers, and the percentage of occupancy (filling) of the total volume are shown in Table 4.1. An occupancy of 60 percent means that 60 percent of the crystal volume is occupied by atoms and the spaces between atoms accounts for 40 percent. Hence the higher the percentage of occupancy, the more closely packed the atoms.

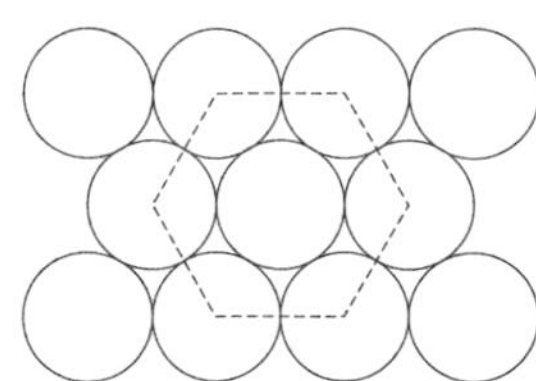

Figure 4.8 The first layer of the hexagonal arrangement.

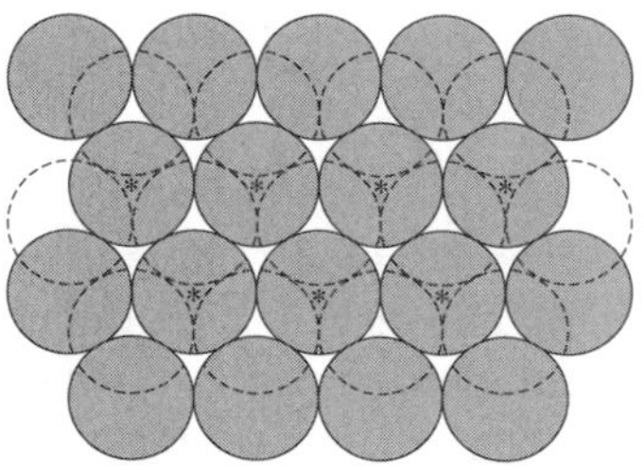

Figure 4.9 In hexagonal packing, the second layer (shaded) fits over alternate holes in the first layer. The hexagonal close-packed arrangement involves placing the third layer (locations marked with asterisks) over the top of the first layer.

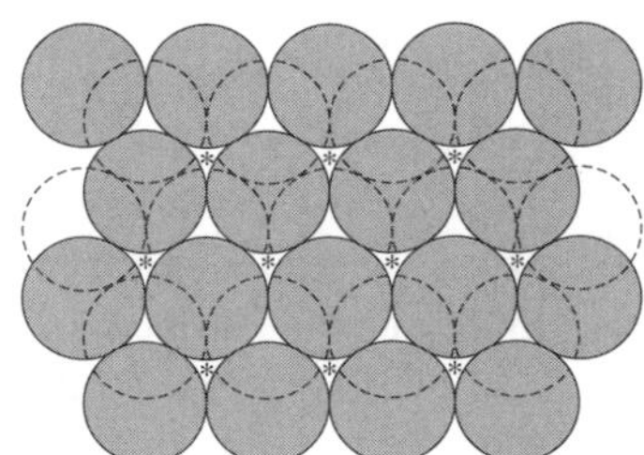

Figure 4.10 The cubic close-packed arrangement has the third layer (locations marked with asterisks) placed over voids in both the first and second layers.

Table 4.1 Properties of the different packing types

Packing type	Coordination number	Occupancy (%)
Simple cubic (sc)	6	52
Body-centered cubic (bcc)	8	68
Hexagonal close-packed (hcp)	12	74
Cubic close-packed (ccp)	12	74

Metals adopt all three of the more compact arrangements (bcc, hcp, ccp). The hard sphere model of packing does not enable us to predict which arrangement a particular metal will adopt. However, there seems to be a general rule that as the number of outer electrons increases, the preferred packing arrangement changes from bcc to hcp and finally to ccp. Thus the alkali metals all adopt the bcc packing, as do most of the metals in Groups 2 to 6. Metals in Groups 7, 8, and 12 are hcp, whereas those of Groups 9 to 11 are ccp packed. This is only a general trend, and there are some exceptions. For example, magnesium has hcp packing, and strontium is known to crystallize with each of these three arrangements, depending on the conditions.

Unit Cells

The simplest arrangement of spheres, which, when repeated, will reproduce the whole crystal structure is called a *unit cell.* The cell is easiest to see in the simple cubic case (Figure 4.11). In the unit cell, we have cut a cube from the center of eight atoms. Inside the unit cell itself, there are eight pieces, each one-eighth of an atom. Because 8(1/8) = 1, we can say that each unit cell contains one atom.

To obtain a unit cell for the body-centered cubic, we must take a larger cluster, one that shows the repeating three-layer structure. Cutting out a cube provides one central atom with eight one-eighth atoms at the corners. Hence the unit cell contains 1 + [8(1/8)] or two atoms (Figure 4.12).

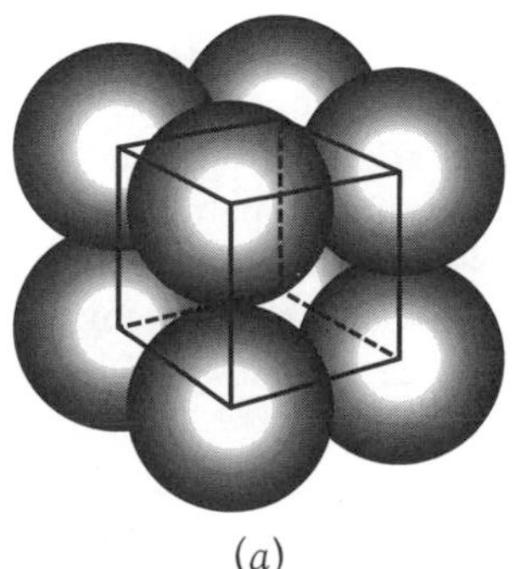

(*a*)

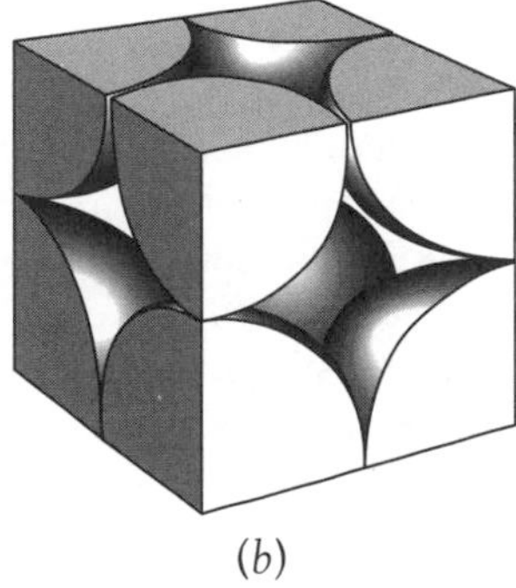

(*b*)

Figure 4.11 (*a*) Eight atoms in the simple cubic array. (*b*) The resulting unit cell. (From G. Rayner-Canham et al., *Chemistry: A Second Course* [Don Mills, ON: Addison-Wesley, 1989], p. 64.)

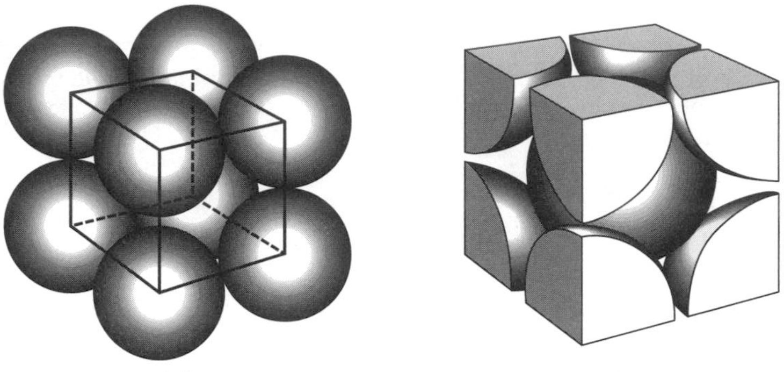

Figure 4.12 (*a*) Nine atoms in the body-centered cubic array. (*b*) The resulting unit cell. (From G. Rayner-Canham et al., *Chemistry: A Second Course* [Don Mills, ON: Addison-Wesley, 1989], p. 64.)

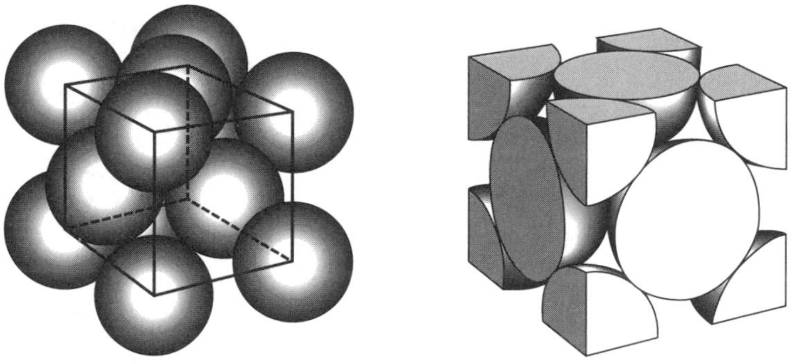

(*a*) (*b*)

Figure 4.13 (*a*) Fourteen atoms in the face-centered cubic array. (*b*) The resulting unit cell. (From G. Rayner-Canham et al., *Chemistry: A Second Course* [Don Mills, ON: Addison-Wesley, 1989], p. 65.)

(*a*) (*b*)

At first inspection, the hexagonal close-packed arrangement does not provide a simple unit cell. However, if a slice is taken through the corner of the face-centered cubic array, we can construct a cube in which there is an atom at each corner and an atom in the middle of each face (Figure 4.13). When we do this, the cube contains $6(1/2) + 8(1/8)$ or four atoms.

Once we know the crystal packing arrangement and the density of a metal, we can calculate the metallic radius of the element.

Alloys

A combination of two or more solid metals is called an *alloy*. The number of possible alloys is enormous. Alloys play a vital role in our lives, yet chemists rarely mention them. The atoms in alloys are held together by metallic bonds just like the component metallic elements. This bonding parallels the covalent bonding of nonmetals. Covalent bonds hold together molecules formed from pairs of different nonmetallic elements as well as pairs of identical nonmetallic elements. Similarly, the metallic bonds of an alloy hold together atoms of different metallic elements.

There are two types of alloys: the solid solutions and the alloy compounds. In the former, the molten metals blend to form a homogeneous mix-

Mercury Amalgam in Teeth

Most of us have mercury in our mouths in the form of dental fillings. The fillings consist of an amalgam—a homogeneous mixture of a liquid metal (mercury) and a number of solid metals. Typically, the dental amalgam has compositions in the following range: mercury (50–55 percent); silver (23–35 percent); tin (1–15 percent); zinc (1–20 percent); and copper (5–20 percent). The soft mixture is placed in the excavated tooth cavity while it is still a suspension of particles of the solid metals in mercury. In the cavity, the mercury atoms infiltrate the metal structure to give a solid amalgam (the equivalent of an alloy). As reaction occurs, there is a slight expansion that holds the filling in place.

Mercury is a very toxic element. However, its amalgamation with solid metals decreases its vapor pressure, so it does not present the same degree of hazard as pure liquid mercury. It is the verdict of the American Dental Association that mercury fillings are quite safe, but there are some who argue that even at very low levels the mercury released from fillings presents a hazard. The real problem is that we have no acceptable substitute at this time. Investigators are currently trying to synthesize a material that will chemically bond to the tooth surface and be strong enough to withstand the immense pressures that we place on our back (grinding) teeth.

As cremation becomes more common, particularly in heavily populated countries, we have to recognize the potential for mercury pollution. During the incineration of the bodies, the mercury amalgam decomposes, releasing mercury vapor into the atmosphere. Thus environmental controls of crematoria emissions are a new concern.

ture. To form a solid solution, the atoms of the two metals have to be about the same size, and the two metallic crystals must have the same structure. In addition, the metals must have similar chemical properties. Gold and copper, for example, form a single phase all the way from 100 percent gold to 100 percent copper. These two metals have similar metallic radii (144 pm for gold and 128 pm for copper), and they both adopt cubic close-packed structures. Lead and tin, however, have very similar metallic radii (175 pm and 162 pm, respectively), but lead forms a face-centered cubic structure and tin has a complex packing arrangement. Very little of one metal will crystallize with the other. Plumber's solder contains 30 percent tin and 70 percent lead, and it is this very immiscibility that enables it to function. During the cooling of the liquid, there is a range of about 100°C over which crystallization occurs. A solid solution cannot exist with more than 20 percent tin. As a result, the crystals are richer in the higher melting lead and the remaining solution has a lower solidification temperature. This "slushy" condition enables plumbers to work with the solder.

Table 4.2 Common alloys

Name	Composition (%)	Properties	Uses
Brass	Cu 70–85, Zn 15–30	Harder than pure Cu	Plumbing
Gold, 18-carat	Au 75, Ag 10–20, Cu 5–15	Harder than pure (24-carat) gold	Jewelry
Stainless steel	Fe 65–85, Cr 12–20, Ni 2–15, Mn 1–2, C 0.1–1, Si 0.5–1	Corrosion resistance	Tools, chemical equipment

In some cases in which the crystal structures of the components are different, mixing molten metals results in the formation of precise stoichiometric phases. For example, copper and zinc form three "compounds": CuZn, Cu_5Zn_8, and $CuZn_3$. Table 4.2 lists some of the common alloys, their compositions, and uses.

Exercises

4.1. Define the following terms: (a) electron-sea model of bonding; (b) unit cell; (c) alloy.
4.2. Define the following terms: (a) crystal lattice; (b) coordination number; (c) amalgam.
4.3. What are the three major characteristics of a metal?
4.4. What are the four most widely used metals?
4.5. Using a band diagram, explain how magnesium can exhibit metallic behavior when its 3*s* band is completely full.
4.6. Construct a band diagram for aluminum.
4.7. Explain why metallic behavior does not occur in the gas phase.
4.8. Explain how the addition of a trace impurity can turn an insulator into a semiconductor.
4.9. What are the two types of layer arrangements in metals? Which has the closer packing?
4.10. What is the difference in layer structure between cubic close-packed and hexagonal close-packed arrangements?
4.11. Draw the simple cubic unit cell and show how the number of atoms per unit cell is derived.
4.12. Draw the body-centered cubic unit cell and show how the number of atoms per unit cell is derived.
4.13. What conditions are necessary for the formation of a solid solution alloy?
4.14. Suggest two reasons why zinc and potassium are unlikely to form a solid solution alloy.

CHAPTER

Ionic Bonding

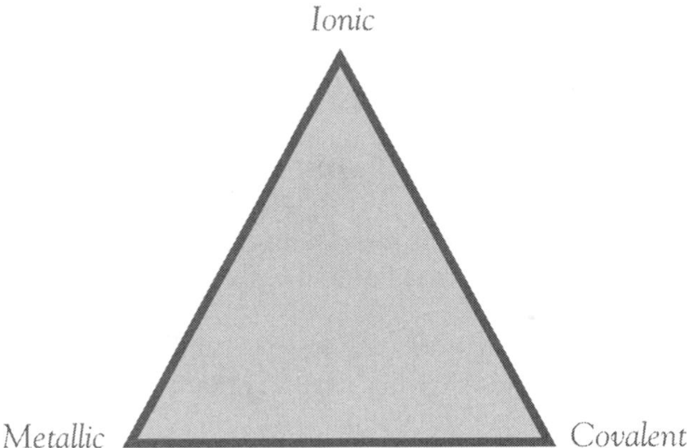

Chemical bonding can also occur through transfer of electrons and the subsequent electrostatic attraction between the charged particles. Elementary chemistry courses separate ionic and covalent bonding. In fact, there are few cases of "pure" ionic compounds.

One of the simplest chemical experiments is to take a beaker of deionized water and insert a light bulb conductivity tester. The bulb does not light. Then some salt is stirred into the water and the bulb lights. This experiment was crucial in the history of chemistry. In 1884 Svante Arrhenius proposed the modern explanation of this experiment. At the time, hardly anyone accepted his theory of electrolytic dissociation. In fact, his doctoral dissertation on the subject was given a low grade in view of the unacceptability of his conclusions. Not until 1891 was there general support for his proposal that the particles in salt solutions dissociated into ions. In 1903, when the significance of his work was finally realized, Arrhenius's name was put forward to share in both the chemistry and physics Nobel prizes. The physicists balked at the proposal and, as a result, Arrhenius was the recipient of the 1903 Nobel prize in chemistry for this work.

Although we ridicule those who opposed Arrhenius, at the time the opposition was quite understandable. The scientific community was divided between those who believed in atoms (the atomists) and those who did not. The atomists were convinced of the indivisibility of atoms. Enter Arrhenius, who argued against both sides: He asserted that sodium chloride broke down into sodium *ions* and chloride *ions* in solution but that these ions were not the same as sodium *atoms* and chlorine *atoms*. That is, the sodium was no longer reactive and metallic, nor was the chlorine green and toxic. No wonder his ideas were rejected until the era of J. J. Thomson and the discovery of the electron.

Characteristics of Ionic Compounds

Whereas covalent substances at room temperature can be solids, liquids, or gases, all ionic compounds are solids and have the following properties:

1. Crystals of ionic compounds are hard and brittle.
2. Ionic compounds have high melting points.
3. When heated to the molten state (if they do not decompose), ionic compounds conduct electricity.
4. Many ionic compounds dissolve in high-polarity solvents (such as water), and, when they do, the solutions are electrically conducting.

The Ionic Model and the Size of Ions

According to the Pauling concept of electronegativity, as the electronegativity difference between two covalently bonded atoms increases, the bond becomes increasingly polar. Eventually, the difference becomes so large that any "sharing" of electrons is negligible, and we define the bond as ionic. An ionic bond is simply the electrostatic attraction between a positive ion (cation) and a negative ion (anion).

Figure 5.1 shows a relationship devised by Pauling between electronegativity difference and the ionic character. Notice that the relationship is a continuum and that there is no actual dividing line between covalent and ionic behaviors. In fact, there are many compounds for which the bonding seems to fall into the intermediate range, where it could equally well be considered as very polar covalent bonding or partially covalent ionic bonding.

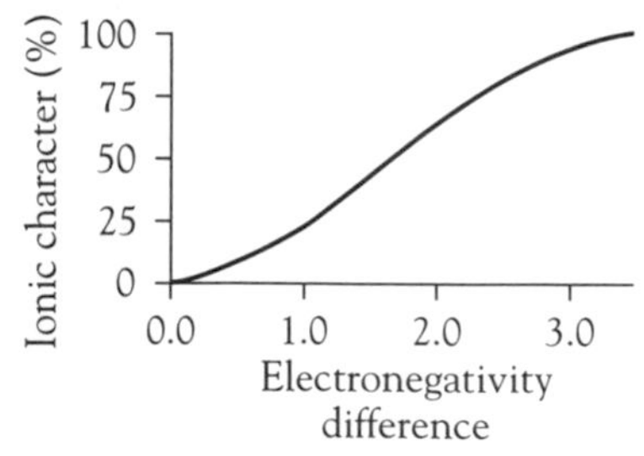

Figure 5.1 Relationship between electronegativity and ionic character.

Because metals have low electronegativities and nonmetals high electronegativities, the combinations from these two classes generally fit the category of the ionic bond. According to the "pure" ionic model, some of the outermost electrons have been completely transferred from the element of lower electronegativity to the element of higher electronegativity. This model is surprisingly useful, even though Pauling's plot indicates that there is always some small degree of covalency, even when the electronegativity difference is very large. As we study the chemistry of the different groups, we will see many examples of covalent character in supposedly ionic compounds.

In Chapter 2, we saw that the size of atoms decreases from left to right in a period as a result of an increase in Z_{eff}. However, the conversion of atoms to ions causes a major change in comparative size. The formation of metal ions from metal atoms usually involves the removal of all the outer (valence) electrons. The cation that remains possesses only the core electrons. Thus the cation will be much smaller than the parent atom. For example, the metallic radius of sodium is 186 pm, whereas its ionic radius is only 116 pm. In fact, the decrease in size is really more dramatic. The volume of a sphere is given by the formula $V = (4/3)\pi r^3$. Hence, the reduction of the radius of sodium upon ionization actually means that the ion is one-fourth the size of the atom!

Trends in Ionic Radii

The cation radii become even smaller if the ions have a multiple charge. We can see this from the set of isoelectronic ions in Table 5.1. Each of the ions has a total of 10 electrons ($1s^2 2s^2 2p^6$). The only difference is the number of protons in the nucleus; the larger the proton number, the higher the effective nuclear charge, Z_{eff}, and hence the stronger the attraction between electrons and nucleus and the smaller the ion.

For anions, the reverse situation is true: The negative ion is larger than the corresponding atom. For example, the covalent radius of the oxygen atom is 74 pm, whereas the radius of the oxide ion is 124 pm; there is a fivefold increase in volume. It can be argued that, with added electrons, the Z_{eff} on each individual outer electron will be less. As a result of the weaker nuclear attraction, the anion will be larger than the atom. For example, the oxygen atom has a single-bond covalent radius of 74 pm, whereas the oxide ion has an ionic radius of 124 pm. Table 5.2 shows that, for an isoelectronic series, the smaller the nuclear charge, the larger the anion. These anions are isoelectronic with the cations in Table 5.1 and illustrate how much larger anions

Table 5.1 Selected isoelectronic cation radii

Ion	Radius (pm)
Na^+	116
Mg^{2+}	86
Al^{3+}	68

Table 5.2 Selected isoelectronic anion radii

Ion	Radius (pm)
N^{3-}	132
O^{2-}	124
F^-	117

Table 5.3 Radii of the Group 17 anions

Ion	Radius (pm)
F^-	117
Cl^-	167
Br^-	182
I^-	206

are than cations. It is *generally* true, then, that the metal cations are smaller than the nonmetal anions.

Proceeding down a group, as the atoms become larger, so do the ions, both anions and cations. Values for the Group 17 anions are given in Table 5.3. Finally, it should be noted that ionic radii cannot be measured directly and are thus subject to error. For example, we can measure precisely the distance between the centers of a pair of sodium and chloride ions in a salt crystal, but this gives the sum of the two radii. The choice of how to apportion the distance between the two ions relies on an empirical formula rather than a definitive measurement. In this text we will use the Shannon-Prewitt values of ionic radii, which are more accurate than the values usually used.

Trends in Melting Points

The ionic bond, then, is a result of the attraction of one ion by the ions of opposite charge that surround it in the crystal lattice. The melting process involves partially overcoming the ionic attractions and allowing the free movement of the ions in the liquid phase. Ionic compounds have high melting points, so the ionic bond must be quite strong. The smaller the ion and hence the more concentrated the charge, the higher the melting point. This trend can be illustrated by the melting points of the potassium halides (Table 5.4), for which there is a strong inverse relationship between anionic radius and melting point. A much more dramatic difference in ionic melting points occurs when the charges themselves are higher. Thus magnesium oxide ($Mg^{2+}O^{2-}$) has a melting point of 2800°C, whereas that of sodium chloride (Na^+Cl^-) is only 801°C.

Table 5.4 Melting points of the potassium halides

Compound	Melting point (°C)
KF	857
KCl	772
KBr	735
KI	685

Polarization and Covalency

Even though most combinations of metals and nonmetals have the characteristics of ionic compounds, there are a variety of exceptions. These exceptions arise when the outermost electrons of the anion are so strongly attracted to the cation that a significant degree of covalency is generated in the bond; that is, the electron density of the anion is distorted toward the cation. This distortion from the spherical shape of the ideal anion is referred to as *polarization*.

The chemist Kasimir Fajans developed the following rules summarizing the factors favoring polarization of an ionic bond and hence the increase in covalency:

1. A cation will be more polarizing if it is small and highly positively charged.
2. An anion will be more easily polarized if it is large and highly negatively charged.
3. Polarization is favored by cations that do not have a noble gas configuration.

A measure of the polarizing power of a cation is its *charge density*. The charge density is the ion charge (number of charge units times the proton charge in coulombs) divided by the ion volume. For example, the sodium ion has a charge of 1+ and an ionic radius of 116 pm (we use 1.16×10^{-7} mm to give an exponent-free charge density value). Hence

$$\text{charge density} = \frac{1 \times 1.60 \times 10^{-19}\text{C}}{(4/3) \times \pi(1.16 \times 10^{-7}\ \text{mm})^3} = 24\ \text{C·mm}^{-3}$$

Similarly, the charge density of the aluminum ion can be calculated as 364 C·mm^{-3}. With a much greater charge density, the aluminum ion is much more polarizing than the sodium ion and hence more likely to favor covalency in its bonding.

One of the most obvious ways of distinguishing ionic behavior from covalent behavior is by observing melting points: Those of ionic compounds (and network covalent compounds) tend to be high; those of small-molecule covalent compounds, low. To illustrate the effects of anion size, we can compare aluminum fluoride (m.p. 1290°C) and aluminum iodide (m.p. 190°C). The fluoride ion, with an ionic radius of 117 pm, is much smaller than the iodide ion of radius 206 pm. In fact, the iodide ion has a volume more than five times greater than that of the fluoride ion. The fluoride ion cannot be polarized significantly by the aluminum ion. Hence the bonding is essentially ionic. The electron density of the iodide ion, however, is distorted toward the high charge density aluminum ion to such an extent that covalently bonded aluminum triiodide molecules are formed.

Because the ionic radius is itself dependent on ion charge, we find that the value of the cation charge is often a good guide in determining the degree of covalency in a simple metal compound. With a cation charge of 1+ or 2+, ionic behavior will usually predominate. With a cation charge of 3+, only

Table 5.5 Melting points of the silver halides

Compound	Melting point (°C)
AgF	435
AgCl	455
AgBr	430
AgI	558

compounds with poorly polarizable anions, such as fluoride, are likely to be ionic. Cations that theoretically have charges of 4+ or above do not actually exist as ions, and their compounds can always be considered to have a predominantly covalent character. This principle is illustrated by a comparison of two of the manganese oxides. Manganese(II) oxide, MnO, has a melting point of 1785°C, whereas manganese(VII) oxide, Mn_2O_7, is a liquid at room temperature. Studies have confirmed that manganese(II) oxide forms an ionic crystal lattice, whereas manganese(VII) oxide consists of covalently bonded Mn_2O_7 molecules. Comparing charge densities, we find that the manganese(II) ion has a charge density of 84 C·mm^{-3}, whereas that of the manganese(VII) ion, if it existed, would be 1240 C·mm^{-3}. The latter figure is so high that the manganese(VII) ion is capable of polarizing all anions and exclusively forming covalent bonds.

The third Fajans rule relates to cations that do not have a noble gas electron configuration. Most common cations, such as calcium, have an electron configuration that is the same as that of the preceding noble gas (for calcium, [Ar]). However, some do not. The silver ion (Ag^+), with an electron configuration of [Kr]$4d^{10}$, is a good example (among the others are Cu^+, Sn^{2+}, and Pb^{2+}). Table 5.5 shows the melting points of the silver halides. There are two major differences from the trend shown for the potassium halides (Table 5.4). First, the values themselves are much lower—by about 300°C. Second, the steadily decreasing trend down the potassium halide series is replaced by a scattering of values for the silver halides.

In the solid phase, the silver ions and halide ions are arranged in a crystal lattice, like any other "ionic" compound. However, the overlap of electron density between each anion and cation is sufficiently high, it is argued, that we can consider the melting process to involve the formation of actual silver halide molecules. Apparently, the energy needed to change from a partial ionic solid to covalently bonded molecules is less than that needed for the normal melting process of an ionic compound.

Another indication of a difference in the bonding behaviors of the potassium ion and the silver ion is their different aqueous solubilities. All of the potassium halides are highly water-soluble, whereas the silver chloride, bromide, and iodide are essentially insoluble in water. The solution process, as we will see later, involves the interaction of polar water molecules with the charged ions. If the ionic charge is decreased by partial electron sharing (covalent bonding) between the anion and cation, then the ion-water interaction will be weaker and the tendency to dissolve will be less. Unlike the other silver halides, silver fluoride is soluble in water. This observation is consistent with the Fajans rules which predict that silver fluoride should have the weakest polarization and the most ionic bonding of all the silver halides.

Often in chemistry there is more than one way to explain an observed phenomenon. This is certainly true for the properties of ionic compounds. To illustrate this point, we can compare the oxides and sulfides of sodium and of copper(I). Both these cations have about the same radius, yet sodium oxide and sodium sulfide behave as typical ionic compounds, reacting with water, whereas copper(I) oxide and copper(I) sulfide are almost completely insoluble in water. We can explain this in terms of the third rule of Fajans, that is, the non-noble gas configuration cation has a greater tendency toward covalency. Alternatively, we can use the Pauling concept of electronegativity and say that the electronegativity difference for, say, sodium oxide of 2.5 would indicate predominantly ionic bonding, whereas that of copper(I) oxide, 1.5, would indicate a major covalent character to the bonding.

Hydration of Ions

If the electrostatic attractions between the ions provide the ionic bond, what is the driving force that allows many ionic compounds to dissolve in water? It is the formation of ion-dipole interactions with the water molecules. The water molecules are polar (as we discussed in Chapter 3)—the oxygen atoms being partially negative and the hydrogen atoms partially positive. In the dissolving process, we picture the oxygen ends of water molecules surrounding the cations and the hydrogen ends of other water molecules surrounding the anions. If the ion-dipole interactions are stronger than the sum of the ionic attractions and the intermolecular forces of the water molecules, then solution will occur. The dissolution process is illustrated in Figure 5.2.

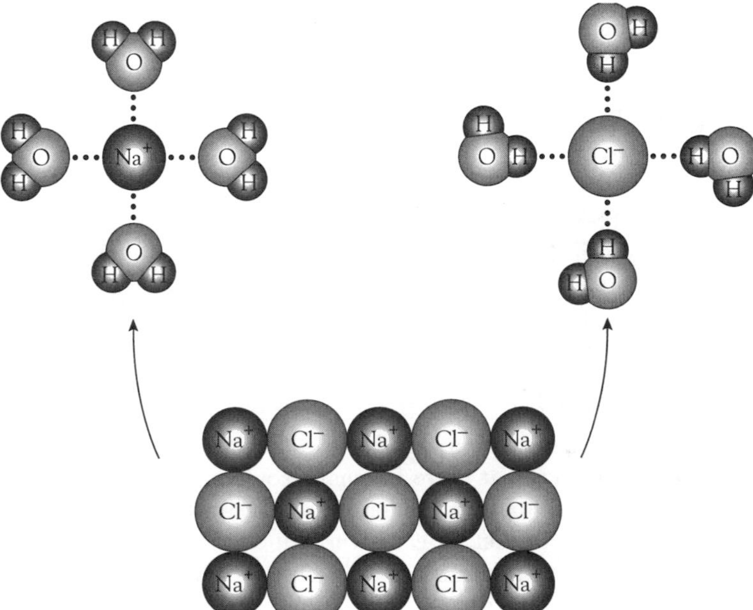

Figure 5.2 Representation of the dissolving process for sodium chloride in water. (From G. Raynor-Canham et al., *Chemistry: A Second Course* [Don Mills, ON: Addison-Wesley, 1989], p. 350.)

A more quantitative discussion of the process is given in the next chapter, but it is important to realize that ionic compounds will only dissolve when the ion-dipole interactions formed with the solvent are very strong compared with the ionic bond. Thus the solvent itself must be very polar. Among the common liquids, only water is polar enough to dissolve ionic compounds.

When an ionic compound crystallizes from an aqueous solution, water molecules often become incorporated in the solid crystal. These water-containing ionic compounds are known as hydrates. In some hydrates, the water molecules are simply sitting in holes in the crystal lattice, but in the majority of hydrates, the water molecules are associated closely with either the anion or the cation, usually the cation. For example, aluminum chloride crystallizes as aluminum chloride hexahydrate, $AlCl_3{\cdot}6H_2O$. In fact, the six water molecules are organized in an octahedral arrangement around the aluminum ion, with the oxygen atoms oriented toward the aluminum ion. Thus the solid compound is more accurately represented as $[Al(OH_2)_6]^{3+}{\cdot}3Cl^-$, hexaaqua-aluminum chloride (the water molecule is written reversed to indicate that it is the oxygen that forms the ion-dipole interaction with the aluminum ion). In the crystal of hydrated aluminum chloride, there are alternating arrays of hexaaquaaluminum cations and chloride anions.

The extent of hydration of ions in the solid phase usually correlates with the ion charge and size. Thus the simple binary alkali metal salts, such as sodium chloride, are anhydrous because both ions have low charge. Crystallization of an ion with a 3+ charge from aqueous solution always results in a hexahydrated ion in the crystal lattice. That is, the small, highly charged cation causes the ion-dipole interaction to be particularly strong. The extent of hydration of anions also depends on charge and sign. The more highly charged oxyanions are almost always hydrated, although not to the extent of the cations. For example, zinc sulfate, $ZnSO_4$, is a heptahydrate: Six of the water molecules are associated with the zinc ion and a seventh is associated with the sulfate ion. Thus the compound is more accurately represented as $[Zn(OH_2)_6]^{2+}[SO_4(H_2O)]^{2-}$. Many other dipositive metal sulfates form heptahydrates with the same structure as that of the zinc compound.

The Ionic Lattice

In the previous chapter we showed four different packing arrangements for metal atoms. The same packing arrangements are common among ionic compounds as well. Generally, the anions are much larger than the cations; thus it is the anions that form the array and the smaller cations fit in holes (called *interstices*) between the anions. Before discussing the particular types of packing, however, we should consider general principles that apply to ionic lattices.

1. Ions are assumed to be charged, incompressible, nonpolarizable spheres. We have seen that there is usually some degree of covalency in all ionic compounds, yet the hard sphere model seems to work quite well for most of the compounds that we classify as ionic.

2. Ions try to surround themselves with as many ions of the opposite charge as possible and as closely as possible. This principle is of particular importance for the cation. Usually, in the packing arrangement adopted, the cation is just large enough to allow the anions to surround it without touching one another.

3. The cation to anion ratio must reflect the chemical composition of the compound. For example, the crystal structure of calcium chloride, $CaCl_2$, must consist of an array of chloride anions with only half that number of calcium cations in the crystal lattice.

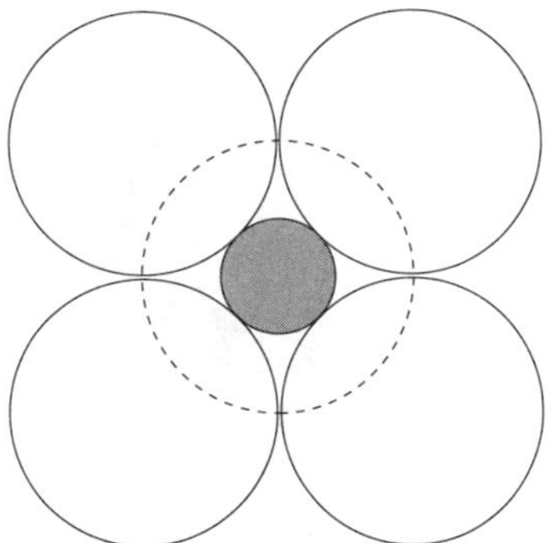

Figure 5.3 Representation of six anions surrounding a cation (shaded).

As mentioned in point 2, the packing arrangement adopted by an ionic compound is usually determined by the comparative sizes of the ions. Figure 5.3 shows four solid circles representing the anions of part of a cubic layer and a dashed circle representing the anions below and above the plane. To fit exactly in the space between these six anions, the cation has to be the size shown by the shaded circle. By using the Pythagorean theorem, we can calculate that the optimum ratio of cation radius to anion radius is 0.414. The numerical value, r_+/r_-, is called the *radius ratio*.

If the cation is larger than one giving the optimum 0.414 ratio, then the anions will be forced apart. In fact, this happens in most cases, and the increased anion-anion distance decreases the anion-anion electrostatic repulsion. However, when the radius ratio reaches 0.732, it becomes possible for eight anions to fit around the cation. Conversely, if the radius ratio is less than 0.414, the anions will be in close contact and the cations will be "rattling around" in the central cavity. Rather than allowing this to happen, the anions rearrange to give smaller cavities surrounded by only four anions. A summary of the radius ratios and packing arrangements is given in Table 5.6.

The Cubic Case

The best way to picture the ionic lattice is to consider the anion arrangement first and then look at the coordination number of the interstices in the anion arrays. The only packing arrangement that has holes with eight anions around them is the simple cubic (see Figure 4.6). The classic example is cesium chloride, and this compound gives its name to the lattice arrangement. The chloride anions adopt a simple cubic packing arrangement, with each cation sitting at the center of a cube. The radius ratio of 0.934 indicates that the cations are sufficiently large to prevent the anions from contacting one another. The unit cell of this crystal is shown in Figure 5.4; it contains

Table 5.6 The range of radius ratios corresponding to different ion arrangements

r_+/r_- values	Coordination number preferred	Name
0.732 up	8	Cubic
0.414 to 0.732	6	Octahedral
0.225 to 0.414	4	Tetrahedral

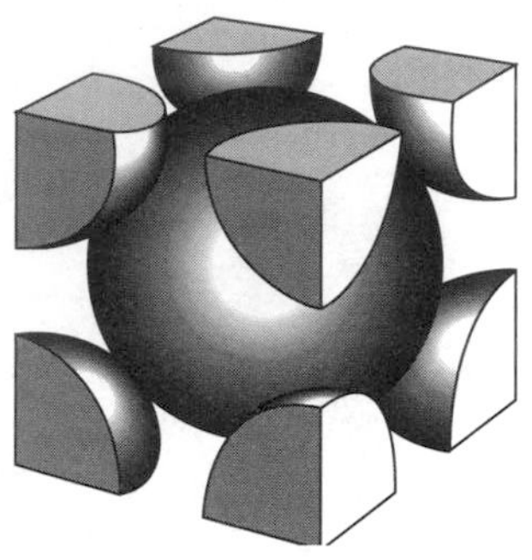

Figure 5.4 Unit cell of cesium chloride. (Adapted from G. Rayner-Canham et al., *Chemistry: A Second Course* [Don Mills, ON: Addison-Wesley, 1989], p. 72.)

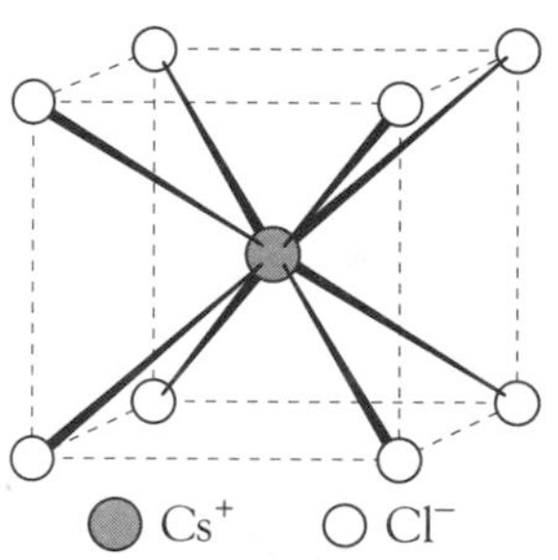

Figure 5.5 Ionic lattice diagram of cesium chloride. (Adapted from A. F. Wells, *Structural Inorganic Chemistry*, 5th ed. [New York: Oxford University Press, 1984], p. 246.)

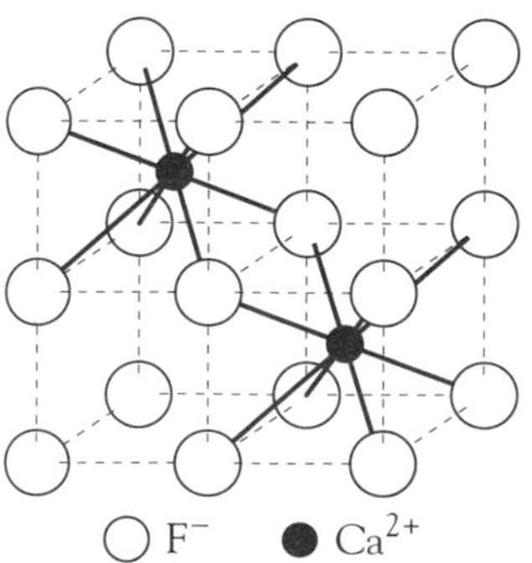

Figure 5.6 Partial ionic lattice diagram of calcium fluoride. (Adapted from A. F. Wells, *Structural Inorganic Chemistry*, 5th ed. [New York: Oxford University Press, 1984], p. 256.)

one cesium ion and 8(1/8) chloride ions. Hence each unit cell contains, in total, one formula unit. The cesium cation separates the chloride anions, so the ions only make contact along a diagonal line that runs from one corner through the center of the unit cell to the opposite corner. This diagonal has a length equal to the sum of two anion radii and two cation radii. To enhance our visualization of the various ion arrangements, we will display most of the ionic structures as *ionic lattice diagrams*. In these diagrams, the ionic spheres have been shrunk in size and solid lines have been inserted to represent points of ionic contact. The ionic lattice diagram for cesium chloride is shown in Figure 5.5.

If the stoichiometry of cation to anion is not 1:1, then the less common ion occupies a certain proportion of the spaces. A good example is calcium fluoride, CaF_2 (mineral name, fluorite), in which the cation to anion ratio is 1:2. Each calcium ion is surrounded by eight chloride ions, similar to the cesium chloride structure. However, each alternate space (that is, every other space) in the lattice is empty, thus preserving the 1:2 cation to anion ratio (Figure 5.6).

It is also possible to have cation to anion ratios of 2:1, as seen in lithium oxide. The structure is again based on the cesium chloride lattice, but this time every alternate anion site is empty. Because the unoccupied lattice spaces in the lithium oxide structure are the opposite of those left unoccupied in the calcium fluoride (fluorite) structure, the name given to this arrangement is the antifluorite structure.

The Octahedral Case

When the radius ratio falls below 0.732, the anions in the cesium chloride structure are no longer held apart by the cations. The potential repulsions between the anions cause the octahedral geometry to become the preferred arrangement. For this smaller radius ratio, six anions can fit around a cation without touching one another (see Figure 5.3). The actual anion arrangement is based on the cubic close-packed array in which there are octahedral holes and tetrahedral holes. Figure 5.7 shows the array with an asterisk (*) marking the location of the octahedral holes in which the cations can fit.

In the octahedral packing, all the octahedral holes are filled with cations and all of the tetrahedral holes are empty. Sodium chloride adopts this particular packing arrangement, and it gives its name to the structure. In the unit cell—the smallest repeating unit of the structure—the chloride anions form a face-centered cubic arrangement. Between each pair of anions, a cation is located. Because the cations are acting as separators of the anions, alternating cations and anions touch along the edge of the cube (Figure 5.8). The unit cell contains one central sodium ion plus 12(1/4) sodium ions along the edges. The centers of the faces hold 6(1/2) chloride ions, and the corners of the cube hold 8(1/8) more. As a result, the sodium chloride unit cell contains four formula units. The length of the side of the cube is the sum of two anion radii and two cation radii. The ionic lattice diagram shows that each sodium ion has six nearest-neighbor chloride ions and each chloride anion is surrounded by six sodium ions (Figure 5.9). Notice that the ionic lattice diagram does not show the large difference in size between the anion and the cation.

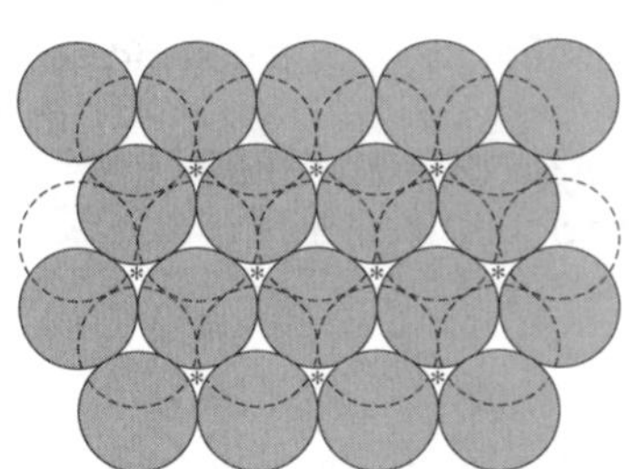

Figure 5.7 The first two layers of the cubic close-packed anion array, showing the octahedral holes (*) in which cations can fit.

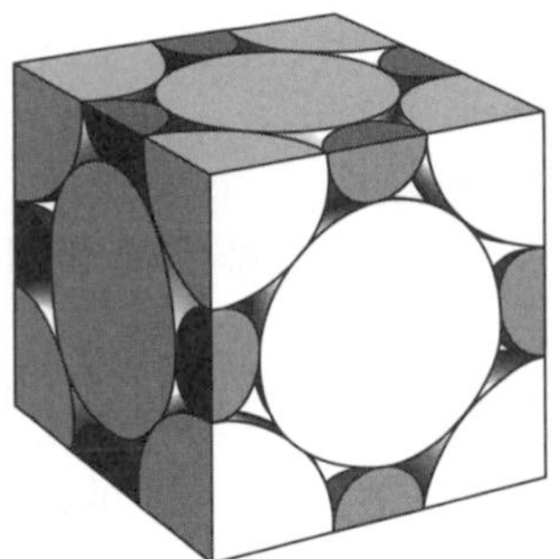

Figure 5.8 Unit cell of sodium chloride. (From G. Rayner-Canham et al., *Chemistry: A Second Course* [Don Mills, ON: Addison-Wesley, 1989], p. 71.)

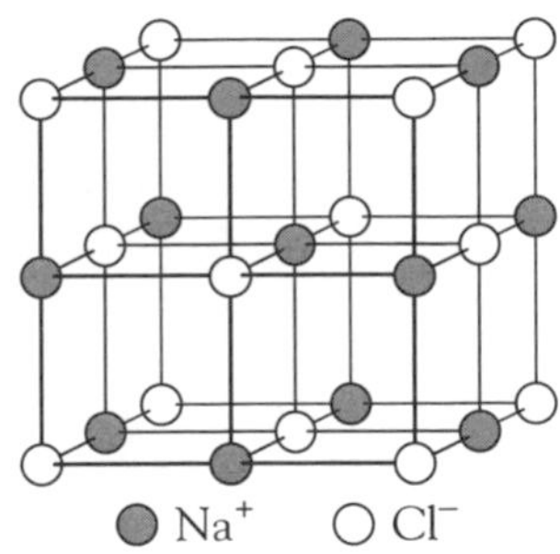

Figure 5.9 Ionic lattice diagram of sodium chloride. (Adapted from A. F. Wells, *Structural Inorganic Chemistry*, 5th ed. [New York: Oxford University Press, 1984], p. 239.)

It is also possible to have octahedral packing for compounds with stoichiometries other than 1:1. The classic example is that of titanium(IV) oxide, TiO_2 (mineral name, rutile). For the crystal, it is easiest to picture the titanium(IV) ions as forming a distorted body-centered array (even though they are much smaller than the oxide anions), with the oxide ions fitting in between (Figure 5.10).

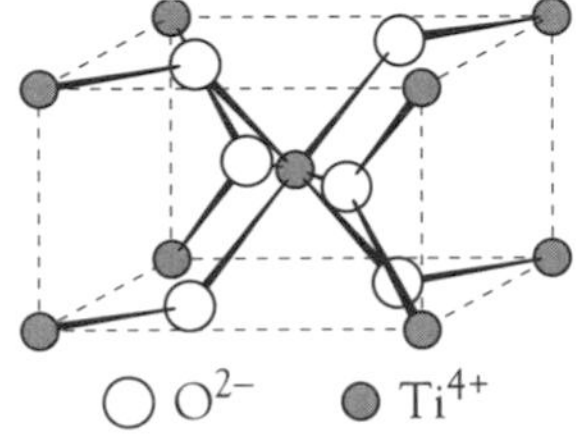

Figure 5.10 Ionic lattice diagram of titanium(IV) oxide. (Adapted from A. F. Wells, *Structural Inorganic Chemistry*, 5th ed. [New York: Oxford University Press, 1984], p. 247.)

The Tetrahedral Case

Ionic compounds in which the cations are very much smaller than the anions can be visualized as close-packed arrays of anions, with the cations fitting into the tetrahedral holes. Both hexagonal close-packed (hcp) and cubic close-packed (ccp) arrangements are possible, and usually a compound will adopt one or the other, although the reasons for particular preferences are not well understood. Figure 5.11 shows the cubic close-packed array; an asterisk (*) marks the location of the tetrahedral holes.

The prototype of this class is zinc sulfide, ZnS, which exists in nature in two crystal forms: the common mineral zinc blende, in which the sulfide ions form a cubic close-packed array; and wurtzite, in which the anion array is hexagonal close-packed (see Chapter 4). Both structures have twice as many tetrahedral holes as cations, so only alternate cation sites are filled (Figure 5.12).

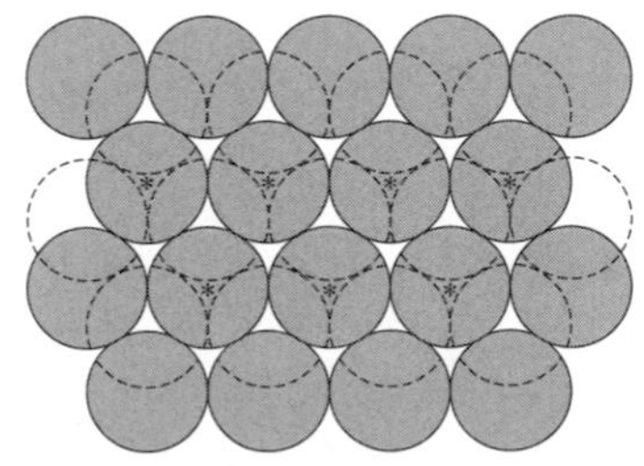

Figure 5.11 The first two layers of the cubic close-packed anion array, showing the tetrahedral holes (*) in which cations can fit.

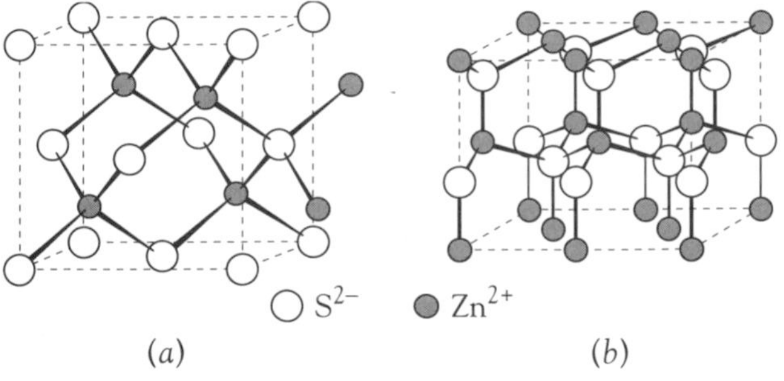

Figure 5.12 Two ionic lattice diagrams of zinc sulfide: (*a*) zinc blende; (*b*) wurtzite. (Adapted from A. F. Wells, *Structural Inorganic Chemistry*, 5th ed. [New York: Oxford University Press, 1984], p. 121.)

Exceptions to the Packing Rules

Up to now, we have discussed the different packing arrangements and their relationship to the radius ratio. However, the radius ratio is only a guide, and although many ionic compounds do adopt the predicted packing arrangement, there are many exceptions. In fact, the packing rules appear to predict the correct arrangement in about two-thirds of the cases. Table 5.7 shows some of the most extreme exceptions.

Chemistry is not a simplistic subject, and to reduce the reasons for a particular packing arrangement to one particular criterion, the radius ratio, is to disregard many factors. In particular, we discussed earlier that there is an appreciable degree of covalent bonding in most ionic compounds. Thus the hard sphere model of ions is not considered valid for many compounds. For example, mercury(II) sulfide is likely to have such a high degree of covalency in its bonding that the compound might equally well be regarded as a network covalent substance, like diamond or silicon dioxide (see Chapter 3). A high degree of covalency would specifically explain the preference of mercury(II) sulfide for the tetrahedral coordination of the ZnS structure, because in its covalent compounds, mercury(II) often forms four covalent bonds arranged at the tetrahedral angles.

Partial covalent behavior is also observed in lithium iodide. On the basis of standard values for ionic radii, its adoption of the octahedral coordination of the sodium chloride lattice makes no sense. The iodide anions would be in contact with one another, and the tiny lithium ions would "rattle around" in the octahedral holes. However, the bonding in this compound is believed to be about 30 percent covalent, and crystal structure studies show that the electron density of the lithium is not spherical but stretched out toward each of the six surrounding anions. Thus lithium iodide, too, cannot be considered as a "true" ionic compound.

Furthermore, there is evidence that the energy differences between the different packing arrangements are often quite small. For example, rubidium chloride normally adopts the unexpected sodium chloride structure (Table 5.7), but crystallization under pressure results in the cesium chloride structure. Thus the energy difference in this case between the two packing arrangements must be very small.

Finally, we must keep in mind that the values of the ionic radii are not constant from one environment to another. For example, the cesium ion has a radius of 181 pm only when it is surrounded by six anion neighbors. With

Table 5.7 Selected examples of exceptions to the packing predicted by the radius ratio rule

Compound	r_+/r_-	Expected packing	Actual packing
HgS	0.68	NaCl	ZnS
LiI	0.35	ZnS	NaCl
RbCl	0.99	CsCl	NaCl

eight neighbors, such as we find in the cesium chloride lattice, it has a Shannon-Prewitt radius of 188 pm. This is not a major factor in most of our calculations, but with small ions there is a very significant difference. For lithium, the four-coordinated ion has a radius of 73 pm, whereas that of the crowded six-coordinated ion is 90 pm. For consistency in this text, all ionic radii quoted are for six-coordination, except for the Period 2 elements, for which four-coordination is much more common and realistic.

Crystal Structures Involving Polyatomic Ions

Up to this point, only binary ionic compounds have been discussed, but ionic compounds containing polyatomic ions also crystallize to give specific structures. In these crystals, the polyatomic ion occupies the same site as a monatomic ion. For example, calcium carbonate forms a distorted sodium chloride structure, with the carbonate ions occupying the anion sites and the calcium ions, the cation sites.

In some cases, properties of a compound can be explained in terms of a mismatch between the anions and the cations, usually a large anion with a very small cation. One possible way of coping with this problem is for the compound to absorb moisture and form a hydrate. In the hydration process, the water molecules usually surround the tiny cation. The hydrated cation is then closer in size to the anion. Magnesium perchlorate is a good example of this arrangement. The anhydrous compound absorbs water so readily that it is used as a drying agent. In the crystal of the hydrate, the hexaaquamagnesium ion, $Mg(OH_2)_6^{2+}$, occupies the cation sites and the perchlorate ion occupies the anion sites.

Ion mismatch may cause compounds to be thermally unstable. Lithium carbonate, with a smallish, low-charge cation and a larger, high-charge anion, is an example of a thermally unstable product. The compound decomposes on heating to give lithium oxide, whereas all the other alkali metal carbonates are thermally stable:

$$Li_2CO_3(s) \xrightarrow{\Delta} Li_2O(s) + CO_2(g)$$

Some ions are so mismatched in size that they cannot form compounds under any circumstances. The effects of ion mismatch constrain large, low-charge anions to form stable compounds only with large, single-charge cations. Thus the hydrogen carbonate ion, HCO_3^-, forms solid stable compounds only with the alkali metals and the ammonium ion.

The Bond Triangle

In these last three chapters, we have discussed the three types of bonding: covalent, metallic, and ionic. Covalent bonding involves orbital overlap between specific pairs of atoms in an element or compound. If the orbital overlap is not between specific atoms but is delocalized throughout the entire crystal, then the bonding is regarded as metallic. The third alternative, ionic bonding, involves the electrostatic interaction between individual ions.

Although we categorize bonding in a substance as being of one specific type, in fact, all three types are related, and the bonding in many species is a

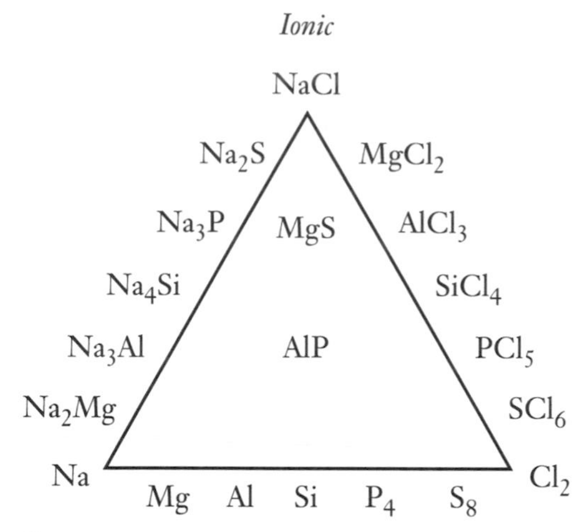

Figure 5.13 The bond triangle.

combination of two or even all three types. Figure 5.13 shows the bond triangle (or more correctly, the Van Arkel-Ketelaar triangle) for a few elements and compounds of Period 2. This figure is not to scale, but it does illustrate the continuum of bonding that seems to exist.

Looking first at the covalent-ionic side of the triangle, we start with dichlorine, a nonpolar molecule. The bonding in this molecule, involving two identical, strongly electronegative atoms, is essentially covalent. As we progress up the right side of the triangle, the electronegativity difference between the constituent atoms increases, causing the bonds to become increasingly polar. A point is reached, in this case at magnesium chloride, at which the electronegativity difference is so large that there is very little overlap of orbitals and the species can be regarded as independent ions with ionic bonds. However, as we will see in Chapter 11, there are some aspects of magnesium chemistry that can best be interpreted in terms of covalent bond contributions. Finally, sodium chloride exhibits almost "pure" ionic bonding.

The covalent-metallic axis corresponds to the change in bonding from the overlap of orbitals in particular directions (covalent) to the delocalized bonding in the metallic situation. Proceeding from right to left across the base of the triangle, the first three elements—dichlorine, octasulfur, S_8, and tetraphosphorus, P_4—are small-molecule covalent. However, at phosphorus, the trend toward less localized bonding is already apparent, for there are other allotropes in which the covalent bonding is directed toward several neighbors. As a result, the bonding in these allotropes is network covalent rather than small molecule covalent: The orbitals are no longer localized between pairs of atoms but overlap in all directions. Silicon more commonly exists as a network covalent structure, but it can be synthesized as a metallic allotrope. The transition from network covalent to metallic corresponds to the decreasing separation of the molecular orbitals from the $3s$ and $3p$ atomic orbitals to the point where they overlap, thereby allowing electrons to move freely through the entire crystal lattice. The metallic, delocalized bonding is the normal state for the low-electronegativity elements aluminum, magnesium, and sodium.

Finally, the left side of the bond triangle represents the metallic-ionic transition. From the metallic corner, we start with alloys of low-electronegativity elements. The delocalization of electrons is high; thus the alloys exhibit metallic bonding. Proceeding along the side of the triangle toward the apex, the electronegativity difference between the element pairs becomes greater and greater and the orbital overlap becomes less and less, with the electron density becoming centered on the more electronegative atom. The final situation is the ionic bond.

As we mentioned earlier, it is possible for compounds to have characteristics of all three bonding types, examples being magnesium sulfide and aluminum phosphide. Fortunately, as we discuss in later chapters, most elements and compounds appear to have properties that can be explained in terms of one bonding type or, at the most, a combination of two bonding types.

Exercises

5.1. Define the following terms: (a) polarization; (b) interstices; (c) bond triangle.

5.2. Define the following terms: (a) ion-dipole interactions; (b) radius ratio; (c) cubic arrangement.

5.3. What properties of a compound would lead you to expect that it contains ionic bonds?

5.4. Which would you expect to contain ionic bonds, $MgCl_2$ or SCl_2? Explain your reasoning.

5.5. Which one of each of the following pairs will be smaller? Explain your reasoning in each case. (a) K or K^+; (b) K^+ or Ca^{2+}; (c) Br^- or Rb^+.

5.6. Which one of each of the following pairs will be smaller? Explain your reasoning in each case. (a) Se^{2-} or Br^-; (b) O^{2-} or S^{2-}.

5.7. Which one, NaCl or NaI, would be expected to have a higher melting point? Explain your reasoning.

5.8. Which one, NaCl or KCl, would be expected to have a higher melting point? Explain your reasoning.

5.9. Compare the charge density values of the three silver ions: Ag^+, Ag^{2+}, and Ag^{3+} (Appendix 3). Which is most likely to form compounds exhibiting ionic bonding?

5.10. Compare the charge densities of the fluoride ion and the iodide ion (Appendix 3). On this basis, which would be the more polarizable?

5.11. Explain why tin(II) chloride, $SnCl_2$, has a melting point of 227°C and tin(IV) chloride, $SnCl_4$, a melting point of −33°C.

5.12. Magnesium ion and copper(II) ion have almost the same ionic radius. Which would you expect to have a lower melting point, magnesium chloride, $MgCl_2$, or copper(II) chloride, $CuCl_2$? Explain your reasoning.

5.13. Would you expect sodium chloride to dissolve in carbon tetrachloride, CCl_4? Explain your reasoning.

5.14. Suggest a reason why calcium carbonate, $CaCO_3$, is insoluble in water.

5.15. Which of sodium chloride and magnesium chloride is more likely to be hydrated in the solid phase? Explain your reasoning.

5.16. Iron(II) sulfate commonly exists as a hydrate. Predict the formula of the hydrate and explain your reasoning.

5.17. What are the key assumptions in the ionic lattice concept?

5.18. Explain the factor affecting the ionic coordination number in an ionic compound.

5.19. Why, in the study of an ionic lattice, is the anion packing considered to be the frame into which the cations fit?

5.20. Suggest the probable crystal structure of (a) barium fluoride; (b) potassium bromide; (c) magnesium sulfide. You can use comparisons or obtain ionic radii from data tables.

5.21. Use Figure 5.6 as a model to draw a partial ionic lattice diagram for the antifluorite structure of lithium oxide.

5.22. Would the hydrogen sulfate ion be more likely to form a stable solid compound with sodium ion or magnesium ion? Explain your reasoning.

5.23. Using the bond triangle concept, state the probable combinations of bonding in (a) $CoZn_3$; (b) BF_3.

5.24. Using the bond triangle concept, state the probable combinations of bonding in (a) As; (b) K_3As; (c) AsF_3.

CHAPTER

6 Inorganic Thermodynamics

$$\Delta G^\circ = \Delta H^\circ - T\Delta S^\circ$$

Descriptive inorganic chemistry is not simply a study of the chemical elements and the myriad compounds that they form. It also involves the need to explain why some compounds form and others do not. A crucial part of our understanding of inorganic chemistry relates to the energy factors involved in the formation of compounds. This topic is a branch of thermodynamics, and in this chapter we present a simplified introduction to inorganic thermodynamics.

Although much of the development of chemistry took place in Britain, France, and Germany, two Americans played important roles in the development of thermodynamics. The first of these was Benjamin Thompson, whose life would make a good movie script. Born in 1753 at Woburn, Massachusetts, he became a major in the (British) Second Colonial Regiment and subsequently spied for the British, using his chemical knowledge to send messages in invisible ink. When Boston fell to the Revolutionary forces, he fled to England and then to Bavaria, now part of Germany. For his scientific contributions to Bavaria's military forces, he was made a count and he chose the title of Count Rumford. At the time, heat was thought to be a type of fluid, and among many significant discoveries, Rumford showed conclusively that heat is a physical property of matter, not a material substance. In fact, it can be argued that he was the first thermodynamicist.

A little less than 100 years later, J. Willard Gibbs was born in New Haven, Connecticut. At Yale University, in 1863, he gained one of the first Ph.D. degrees awarded in the United States. It was Gibbs who first derived the mathematical equations that are the basis of modern thermodynamics. About that time, the concept of entropy had been proposed in Europe, and the German physicist Rudolf Clausius had summarized the laws of thermodynamics in the statement "The energy of the universe is a constant, the entropy of the universe tends to a maximum." Yet Clausius and other physicists did not appreciate the importance of the entropy concept. It was Gibbs who showed that everything from miscibility of gases to positions of chemical equilibria depend on entropy factors. In recognition of his role, the thermodynamic function free energy was assigned the symbol G and given the full name of Gibbs free energy.

Thermodynamics of the Formation of Compounds

Compounds are produced from elements by chemical reactions. For example, our table salt, sodium chloride, can be formed by the combination of the reactive metal sodium with a toxic green gas, chlorine:

$$2\ Na(s) + Cl_2(g) \rightarrow 2\ NaCl(s)$$

Because this reaction occurs without need for external "help," it is said to be a *spontaneous reaction* (although being spontaneous does not give any indication of how fast or slow the reaction may be). The reverse reaction, the decomposition of sodium chloride, is a nonspontaneous process, which is just as well, for we would not want the salt on the dining table to start releasing clouds of poisonous chlorine gas! One way to obtain sodium metal and chlorine gas back again is to pass an electric current (an external energy source) through molten sodium chloride:

$$2\ NaCl(l) \xrightarrow{\text{energy}} 2\ Na(l) + Cl_2(g)$$

The study of the causes of chemical reactions is a branch of thermodynamics. This chapter provides a simplified coverage of the topic as it relates to the formation of inorganic compounds. In the discussion, we will show that the feasibility of chemical reactions depends on two factors: enthalpy and entropy.

Enthalpy

Enthalpy is often defined as the heat content of a substance. When the products of a chemical reaction have a lower enthalpy than the reactants, the reaction releases heat to the surroundings, that is, the process is exothermic. If the products have a higher enthalpy than the reactants, then heat energy is acquired from the surroundings and the reaction is said to be endothermic. The difference between the enthalpy of the products and the enthalpy of the reactants is called the enthalpy change, ΔH.

Entropy

Entropy is often related to the degree of disorder of the substance (although the concept of entropy is really more complex). Thus the solid phase has a lower entropy than the liquid phase, whereas the gas phase, in which there is random motion, has a very high entropy. The entropy change is indicated by the symbol ΔS.

The Driving Force of a Reaction

There are two factors that cause a reaction: a decrease in enthalpy and an increase in entropy. If both factors occur in a particular chemical reaction, then it is certain to be spontaneous under all conditions. If the reaction would lead to an increase in enthalpy and a decrease in entropy, then it would be nonspontaneous under all conditions. Many reactions fit into the other two categories: a decrease in both enthalpy and entropy or an increase in both factors. For these cases, a more quantitative approach is necessary. This involves a function that combines the enthalpy and entropy factors, the Gibbs free energy, G. The relationship is $\Delta G = \Delta H - T\Delta S$, where T is the Kelvin temperature.

For a spontaneous chemical reaction, there must be a decrease in free energy (that is, ΔG must be negative). Both enthalpy and entropy values are temperature dependent, but it is the entropy factor that is directly multiplied by the Kelvin temperature. Thus a reaction with positive enthalpy and entropy changes will always become spontaneous above a certain temperature. An increase in entropy, then, is the driving force that results in the decomposition of compounds upon heating. The decomposition process results in more moles of products than of reactants, and some of the products are usually in the gas phase. For example, heating solid mercury(II) oxide gives liquid mercury and gaseous oxygen:

$$2\,HgO(s) \xrightarrow{\Delta} 2\,Hg(l) + O_2(g)$$

The process is spontaneous at high temperatures even though the process is endothermic, for the gas and liquid products will have a higher entropy than the reactants. The possible sign combinations for the thermodynamic functions are summarized in Table 6.1.

In this course, we are rarely concerned with exact numerical calculations. It is just as well, for many of the data values are only known approximately.

Table 6.1 Factors affecting the spontaneity of a reaction

ΔH	ΔS	ΔG	Result
Negative	Positive	Always negative	Spontaneous
Positive	Negative	Always positive	Nonspontaneous
Positive	Positive	Negative at high T	Spontaneous at high T
Negative	Negative	Negative at low T	Spontaneous at low T

The reason for performing calculations in this text is to try to understand why some compounds form and others do not. Thus it is often the sign and the exponent of the number that are meaningful rather than the precise numerical value itself.

To illustrate this point, we can study the formation of ammonia from its elements:

$$\tfrac{1}{2}\,N_2(g) + \tfrac{3}{2}\,H_2(g) \rightarrow NH_3(g)$$

For this reaction, the enthalpy change ΔH has a value of $-46\ kJ{\cdot}mol^{-1}$, whereas the entropy change ΔS has a value of $-0.099\ kJ{\cdot}mol^{-1}{\cdot}K^{-1}$. The negative enthalpy and entropy terms correspond to the last category in Table 6.1. Hence the reaction should be spontaneous at low temperature but not at high temperature. Inserting the values in the formula $\Delta G = \Delta H - T\Delta S$ at 298 K, the free energy change will be

$$(-46\ kJ{\cdot}mol^{-1}) - (298\ K)(-0.099\ kJ{\cdot}mol^{-1}{\cdot}K^{-1}) = -16\ kJ{\cdot}mol^{-1}$$

Thus the reaction is thermodynamically favored at room temperature. However, at 600 K, the calculation has a different result:

$$(-46\ kJ{\cdot}mol^{-1}) - (600\ K)(-0.099\ kJ{\cdot}mol^{-1}{\cdot}K^{-1}) = +13\ kJ{\cdot}mol^{-1}$$

At this temperature, the reaction will not be thermodynamically favored. In fact, the reverse reaction—the decomposition of ammonia—will occur.

Enthalpy of Formation

The most useful enthalpy data are the enthalpy of formation values. The enthalpy of formation is defined as the change in heat content when one mole of a compound is formed from its elements in their standard phases at 298 K and 100 kPa. By definition, the enthalpy of formation of an element is zero. This is an arbitrary standard, just as our geographical measure of altitude is taken as height above mean sea level rather than from the center of the Earth. The symbol for the enthalpy of formation under standard conditions is ΔH_f°. Thus we find in data tables values such as $\Delta H_f^\circ(CO_2(g)) = -394\ kJ{\cdot}mol^{-1}$. This datum indicates that 394 kJ of energy are released when one mole of carbon (graphite) reacts with one mole of oxygen gas at 298 K and a pressure of 100 kPa to give one mole of carbon dioxide:

$$C(s) + O_2(g) \rightarrow CO_2(g) \qquad \Delta H^\circ = -394\ kJ{\cdot}mol^{-1}$$

Enthalpies of formation can be combined to calculate the enthalpy change in other chemical reactions. For example, we can determine the enthalpy change when carbon monoxide burns in air to give carbon dioxide:

$$CO(g) + \tfrac{1}{2}\,O_2(g) \rightarrow CO_2(g)$$

First, we collect the necessary data from tables: $\Delta H_f^\circ(CO_2(g)) = -394\ kJ{\cdot}mol^{-1}$ and $\Delta H_f^\circ(CO(g)) = -111\ kJ{\cdot}mol^{-1}$; by definition, $\Delta H_f^\circ(O_2(g))$ is zero. The enthalpy change for the reaction can be obtained from the expression

$$\Delta H^\circ(\textit{reaction}) = \Sigma\Delta H^\circ(\textit{products}) - \Sigma\Delta H^\circ(\textit{reactants})$$

Hence

$$\Delta H^\circ(reaction) = (-394\ \text{kJ}\cdot\text{mol}^{-1}) - (-111\ \text{kJ}\cdot\text{mol}^{-1})$$
$$= -283\ \text{kJ}\cdot\text{mol}^{-1}$$

Thus the reaction is exothermic, as are almost all combustion reactions.

Bond Energies (Enthalpies)

Enthalpy of formation values are very convenient for the calculation of enthalpy changes in reactions. However, inorganic chemists are often interested in the energy within molecules, the bond enthalpy. The term is commonly called bond energy, although there are small differences between the two in terms of their definition and numerical values. We have already mentioned bond energies in the context of the strength of the covalent bond in simple diatomic molecules (see Chapter 2). *Bond energy* is defined as the energy needed to break one mole of the particular covalent bond. Energy is released when bonds are formed, and energy must be supplied when bonds are broken.

We can measure the exact bond energy for a particular pair of elements joined by a covalent bond. For example, Table 6.2 lists the bond energies in the diatomic molecules of the halogen series. If we look at elements within a group, we see that the bond energies usually decrease as one goes down the group as a result of the increase in atomic size and decrease in orbital overlap of electron density. We will see in this and later chapters that the anomalously low F–F bond energy has a major effect on fluorine chemistry.

The bond energy depends on the other atoms that are present in the molecule. For example, the value of the O–H bond energy is 492 kJ·mol^{-1} in water (HO–H) but 435 kJ·mol^{-1} in methanol, CH_3O–H. Because of this variability, data tables provide average bond energies for a particular covalent bond.

The energy of a specific bond, and hence the bond strength, increases substantially as the bond order increases. Table 6.3 shows this trend for the series of carbon-nitrogen average bond energies.

Bonds containing different elements usually have a higher energy than those containing the same element. The usual explanation is that, with two

Table 6.2 Bond energies of the diatomic molecules of the halogens

Molecule	Bond energy (kJ·mol^{-1})
F–F	158
Cl–Cl	242
Br–Br	193
I–I	151

Table 6.3 Average bond energies for various carbon-nitrogen bonds

Bond	Bond energy ($kJ{\cdot}mol^{-1}$)
C–N	305
C=N	615
C≡N	887

Table 6.4 Average bond energies for hydrogen and chlorine combinations

Bond	Bond energy ($kJ{\cdot}mol^{-1}$)
H–H	432
Cl–Cl	240
H–Cl	438

elements of different electronegativity, there will be a polarity to the bond. This polarization will result in an added electrostatic attraction between the two atoms, thereby strengthening the bond. A comparison of the hydrogen-chlorine bond energy with those of the hydrogen-hydrogen and chlorine-chlorine bonds shows that the polar bond is far stronger than either of the averages of the two nonpolar bonds (Table 6.4).

Lattice Energies (Enthalpies)

The *lattice energy* is the energy change for the formation of one mole of an ionic solid from its constituent gaseous ions (we are really considering lattice enthalpy here, but the difference is negligible). We can illustrate the process with sodium chloride. The lattice energy of sodium chloride corresponds to the energy change for

$$Na^+(g) + Cl^-(g) \rightarrow Na^+Cl^-(s)$$

The lattice energy is really a measure of the electrostatic attractions and repulsions of the ions in the crystal lattice. This series of interactions can be illustrated by the sodium chloride crystal lattice (Figure 6.1).

Surrounding the central cation, there are six anions at a distance of r, where r is the distance between centers of nearest neighbors. This is the major attractive force holding the lattice together. However, at a distance of $(2)^{1/2}r$, there are 12 cations. These will provide a repulsion factor. Adding additional layers of ions beyond the unit cell, we find that there are eight more anions at a distance of $(3)^{1/2}r$, then six more cations at $2r$. Hence the true balance of charge is represented by an infinite series of alternating attraction and repulsion terms, although the size of the contributions drops off rapidly with increasing distance.

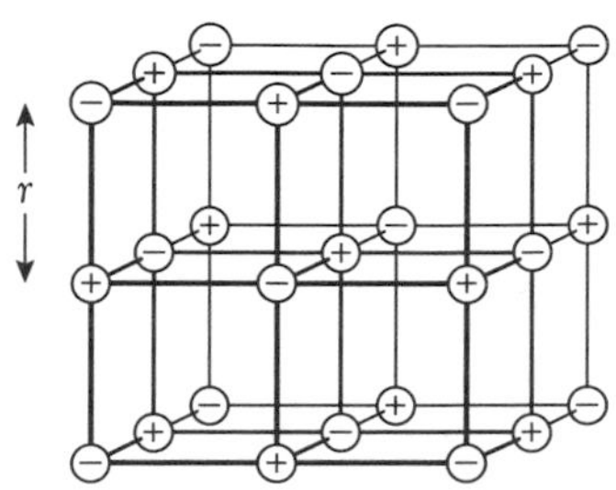

Figure 6.1 Ionic lattice diagram of the sodium chloride structure, showing the ion charge. (Adapted from A. F. Wells, *Structural Inorganic Chemistry*, 5th ed. [New York: Oxford University Press, 1984], p. 239.)

The lattice energy, then, depends on a number of factors. The most important of these is ion charge. Doubling the charge from +1 to +2 (or –1 to –2) approximately triples the lattice energy. In fact, for the series MX, MX_2, MX_3, MX_4, the lattice energies are related in the ratios of 1:3:6:10. The lattice energy is also greater if the ions are smaller, a factor that results in closer ion centers. And the crystal lattice type is also important. Each type of lattice has a unique arrangement of cations and anions that results in a unique series of attraction-repulsion terms.

Finally, it should be mentioned that all crystals have a lattice energy. For simple covalent compounds, lattice energy is attributable to intermolecular attractions; for network covalent substances, lattice energy is the energy of the covalent bonds; and for metals, it is the attractions that create the metallic bond. However, the energy term for simple covalent molecules is commonly called the *enthalpy of sublimation* rather than the lattice energy.

Enthalpies of Atomization

Finally, another useful measurement is that of *energy of atomization*. This is defined as the energy needed to produce one mole of gaseous atoms of that element from the element in its normal phase at room temperature. This energy term can be used to represent the breaking of the metallic bond in metals or the overcoming of the covalent bonds and intermolecular forces in nonmetals. For example, both of the following equations show atomization processes:

$$Cu(s) \rightarrow Cu(g)$$

$$I_2(s) \rightarrow 2\,I(g)$$

Absolute Entropy

The thermodynamic function *absolute entropy* is measured on an absolute basis. Thus, even elements have a listed value of entropy. The zero point is taken to be that of a perfect crystal of a substance at the absolute zero of temperature. We can calculate the standard entropy change for a reaction in the same way as that of the enthalpy change:

$$\Delta S°(reaction) = \Sigma S°(products) - \Sigma S°(reactants)$$

For example, we can calculate the standard entropy change for the formation of sodium chloride from sodium metal and chlorine gas:

$$Na(s) + \tfrac{1}{2}\,Cl_2(g) \rightarrow NaCl(s)$$

Hence

$$\Delta S° = [S°(NaCl(s))] - [S°(Na(s))] - \tfrac{1}{2}[S°(Cl_2(g))]$$

$$= (+72\ J{\cdot}mol^{-1}{\cdot}K^{-1}) - (+51\ J{\cdot}mol^{-1}{\cdot}K^{-1}) - \tfrac{1}{2}(+223\ J{\cdot}mol^{-1}{\cdot}K^{-1})$$

$$= -90\ J{\cdot}mol^{-1}{\cdot}K^{-1}$$

We would expect an entropy decrease for this process, because it involves the net loss of one-half mole of gas. The reaction is spontaneous at ambient temperatures, so it must be enthalpy driven. (In fact, the recorded value of the enthalpy of formation ΔH_f° of sodium chloride is -411 kJ·mol^{-1}.)

Formation of Ionic Compounds

When an ionic compound is formed from its elements, there is usually a decrease in entropy, for the ordered, solid crystalline compound has a very low entropy and the nonmetal reactant, such as oxygen or chlorine, is a high-entropy gas. For example, in the previous section we determined that the entropy change for sodium chloride was negative. Thus, for the formation of a thermodynamically stable compound from its constituent elements, a negative enthalpy change must occur; it is the driving force of the reaction.

To attempt to understand why particular compounds form and others do not, we will break the formation process of an ionic compound into a series of theoretical steps: first breaking the reactant bonds, then forming those of the products. In this way, we can identify which enthalpy factors are crucial to the spontaneity of reaction. We consider again the formation of sodium chloride:

$$Na(s) + \tfrac{1}{2}\,Cl_2(g) \rightarrow NaCl(s) \qquad \Delta H^\circ_f = -411\ kJ{\cdot}mol^{-1}$$

1. The solid sodium is converted to free (gaseous) sodium atoms. This process requires the enthalpy of atomization:

$$Na(s) \rightarrow Na(g) \qquad \Delta H^\circ = +108\ kJ{\cdot}mol^{-1}$$

2. The gaseous chlorine molecules must be dissociated into atoms. This transformation requires one-half of the bond energy of chlorine molecules:

$$\tfrac{1}{2}\,Cl_2(g) \rightarrow Cl(g) \qquad \Delta H^\circ = +121\ kJ{\cdot}mol^{-1}$$

3. The sodium atoms must then be ionized. This process requires the first ionization energy. (If we had a metal that formed a divalent cation, then we would have to add both the first and second ionization energies.)

$$Na(g) \rightarrow Na^+(g) + e^- \qquad \Delta H^\circ = +502\ kJ{\cdot}mol^{-1}$$

4. The chlorine atoms must gain electrons. This value is the electron affinity of chlorine atoms.

$$Cl(g) + e^- \rightarrow Cl^-(g) \qquad \Delta H^\circ = -354\ kJ{\cdot}mol^{-1}$$

The value for the addition of the first electron is usually exothermic, whereas addition of a second electron is usually endothermic (such as the formation of O^{2-} from O^-).

5. The free ions then associate to form a solid ionic compound. This bringing together of ions is a highly exothermic process—the lattice energy. The lattice energy can be looked on as the major driving force for the formation of an ionic compound:

$$Na^+(g) + Cl^-(g) \rightarrow NaCl(s) \qquad \Delta H^\circ = -788\ kJ{\cdot}mol^{-1}$$

All of the functions that we have mentioned are available in data tables. Some data, such as ionization energies, are based on actual measurements, but lattice energies cannot be found from actual measurements; they can be determined only from Born-Haber cycles or theoretical calculations. Furthermore, some measurements are not known to a high degree of precision. Hence we should always be cautious about drawing definitive conclusions from these calculations.

Born-Haber Cycle

It is usually easier to comprehend information if it is displayed graphically. This representation can be done for the theoretical components of the formation of an ionic compound from its elements. The "up" direction is used to indicate endothermic steps in the process, and the "down" direction corresponds to exothermic steps. The resulting diagram is called a Born-Haber cycle. Figure 6.2 shows such a cycle for the formation of sodium chloride.

These enthalpy diagrams can be used in two ways: to gain a visual image of the key enthalpy terms in the formation of the compound and to determine any one unknown enthalpy value in the thermodynamic cycle, for we know that the sum of the component terms should equal the overall enthalpy change for the formation process.

The major enthalpy input is the ionization energy, whereas the greatest release of enthalpy comes from the formation of the ionic crystal lattice. This balance, with the lattice energy exceeding the ionization energies, is common among stable ionic compounds. Magnesium fluoride, MgF_2, can be used to illustrate this point. The sum of the first and second ionization energies of the magnesium ion is +2190 $kJ \cdot mol^{-1}$, much higher than the single ionization energy of the monopositive sodium ion. However, with the small, highly charged magnesium ion, the lattice energy is also much higher at −2880 $kJ \cdot mol^{-1}$. Incorporating the other terms from the enthalpy cycle results in a relatively large, quite negative enthalpy of formation for magnesium fluoride: −1100 $kJ \cdot mol^{-1}$.

If the lattice energy increases so much with greater cation charge, why do magnesium and fluorine form MgF_2 and not MgF_3? If we estimate the

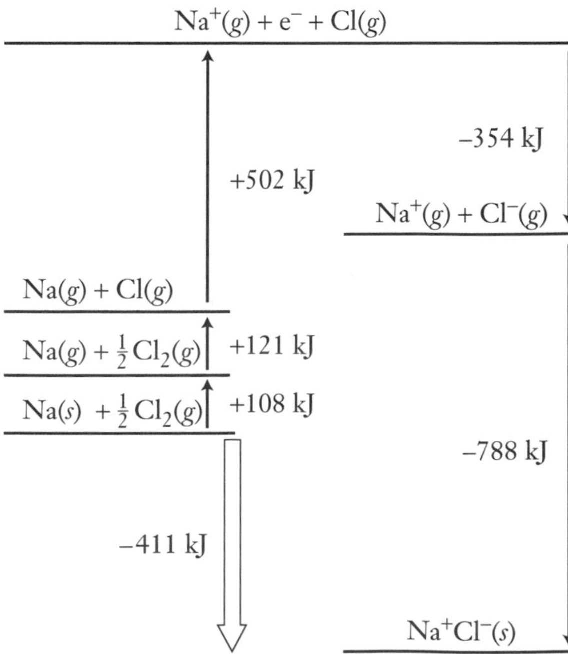

Figure 6.2 Born-Haber cycle for the formation of sodium chloride.

Table 6.5 Thermodynamic factors in the formation of three possible magnesium fluorides

Enthalpy factors (kJ·mol^{-1})	MgF	MgF_2	MgF_3
Mg atomization	+150	+150	+150
F–F bond energy	+80	+160	+240
Mg ionization (total)	+740	+2190	+9930
F electron affinity	−330	−660	−990
Lattice energy	≈−900	−2880	≈−5900
ΔH_f° (estimated)	−260	−1040	+3430

enthalpy of formation of MgF_3, the lattice energy will be much larger because of the greater electrostatic attraction. However, for magnesium, the electrons that must be ionized to give the 3+ ion are core electrons, and the third ionization energy is enormous (7740 kJ·mol^{-1})—far larger than the gain from the lattice energy. Combined with a negative entropy term, there is no possibility that the compound will exist.

Conversely, why do magnesium and fluorine form MgF_2 and not MgF? With the higher Z_{eff}, we find that the first ionization energy of magnesium is more than 200 kJ·mol^{-1} higher than that of sodium. Furthermore, the Mg^+ ion, with its one 3*s* electron remaining, is a larger cation than the Na^+ ion. Hence the lattice energy of MgF would be lower than that of NaCl. Thus the energy input needed to form MgF would be significantly higher than that needed to form sodium chloride, whereas the energy release from MgF lattice formation would be less. As a result, the overall enthalpy loss would be much smaller than that resulting from the formation of MgF_2. Table 6.5 compares the thermodynamic components of the Born-Haber cycles for the formation of MgF, MgF_2, and MgF_3. The three Born-Haber cycles can be compared graphically, where the importance of the balance between ionization energy and lattice energy becomes apparent (Figure 6.3).

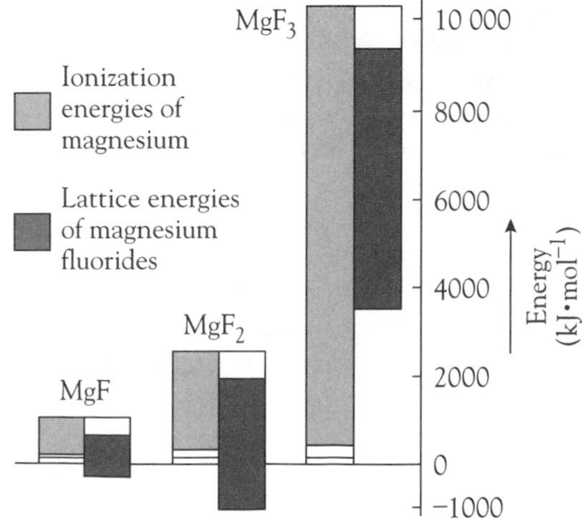

Figure 6.3 Graphical comparison of the Born-Haber cycles for three possible magnesium fluorides.

Thermodynamics of the Solution Process for Ionic Compounds

Just as the formation of a compound from its constituent elements can be considered as a series of theoretical steps, so can the solution process be broken down into several steps. For this analysis, we visualize first that the ions in the crystal lattice are dispersed into the gas phase and then, in a separate step, that water molecules surround the gaseous ions to give the hydrated ions. Thus ion-ion interactions (ionic bonds) are broken and ion-dipole interactions are formed. The degree of solubility, then, depends on the balance of these two factors, each of which has both enthalpy and entropy components.

There is one key difference between the two analyses. In the formation of a compound, we use the thermodynamic factors simply to determine whether or not a compound will form spontaneously. With the thermodynamics of the solution process, we are concerned with the degree of solubility—that is, where a compound fits on the continuum from very soluble through soluble, slightly soluble, and insoluble to very insoluble. Even for a very insoluble compound, there will be a measurable proportion of aqueous ions present in equilibrium with the solid compound.

Lattice Energy

To break the ions free from the lattice—overcoming the ionic bond—requires a large energy input. The value of the lattice energy depends on the strength of the ionic bond, and this, in turn, relates to the ion size and charge. That is, magnesium oxide, with dipositive ions, will have a much higher lattice energy than sodium fluoride with its monopositive ions (3933 $kJ{\cdot}mol^{-1}$ and 915 $kJ{\cdot}mol^{-1}$, respectively). At the same time, the entropy factor will always be highly favorable as the system changes from the highly ordered solid crystal to the disordered gas phase.

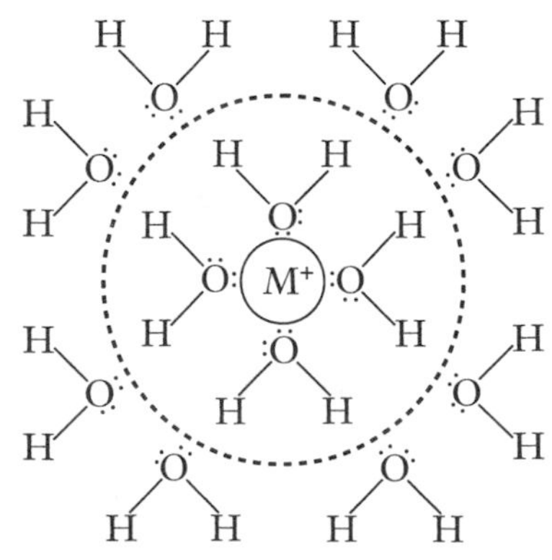

Figure 6.4 Primary and secondary hydration spheres of a metal cation. (Adapted from G. Wulfsberg, *Principles of Descriptive Inorganic Chemistry* [New York: University Science Books, 1990], p. 66.)

Energy of Hydration

In aqueous solution, the ions are surrounded by polar water molecules. A primary hydration sphere of water molecules (usually six) surrounds the cations, with the partially negative oxygen atoms oriented toward the cation. Similarly, the anion is surrounded by water molecules, with the partially positive hydrogen atoms oriented toward the anion. Beyond the first shell of water molecules, we find additional layers of oriented water molecules (Figure 6.4). The total number of water molecules that effectively surround an ion is called the *hydration number*.

The smaller and more highly charged ions will have a larger number of water molecules in hydration spheres than do larger, less highly charged ions. As a result, the effective size of a hydrated ion in solution can be very different from that in the solid phase. This size difference is illustrated in Table 6.6. It is the smaller size of the hydrated potassium ion that enables it to pass through biological membranes more readily than the larger hydrated sodium ions.

Table 6.6 Hydration effects on the size of sodium and potassium ions

Ion (pm)	Radius (pm)	Hydrated ion	Hydrated radius
Na^+	116	$Na(OH_2)_{13}{}^+$	276
K^+	152	$K(OH_2)_7{}^+$	232

Table 6.7 Enthalpy of hydration and charge density for three isoelectronic cations

Ion	Hydration enthalpy ($kJ{\cdot}mol^{-1}$)	Charge density ($C{\cdot}mm^{-3}$)
Na^+	−390	24
Mg^{2+}	−1890	120
Al^{3+}	−4610	364

The formation of the ion-dipole interactions of hydrated ions is highly exothermic. The value of the enthalpy of hydration is also dependent on ion charge and ion size, that is, the charge density. Table 6.7 shows the strong correlation between enthalpy of hydration and charge density for an isoelectronic series of cations.

The entropy of hydration is also negative, mainly because the water molecules surrounding the ions are in a more ordered state than they would be as free water molecules. With the small, more highly charged cations, such as magnesium and aluminum, the hydration spheres are larger than that of sodium, and hence there is a strong ordering of the water molecules around the two larger cations. For these cations, there is a very large decrease in entropy for the hydration process.

Energy Change of the Solution Process

We can use the solution process for sodium chloride to illustrate an enthalpy of solution cycle. First, the lattice must be vaporized:

$$NaCl(s) \rightarrow Na^+(g) + Cl^-(g) \qquad \Delta H^\circ = +788\ kJ{\cdot}mol^{-1}$$

Then the ions are hydrated:

$$Na^+(g) \rightarrow Na^+(aq) \qquad \Delta H^\circ = -406\ kJ{\cdot}mol^{-1}$$

$$Cl^-(g) \rightarrow Cl^-(aq) \qquad \Delta H^\circ = -378\ kJ{\cdot}mol^{-1}$$

Thus enthalpy change ΔH° for the solution process is

$$(+788) + (-406) + (-378) = +4\ kJ{\cdot}mol^{-1}$$

The process can be displayed as a diagram (Figure 6.5).

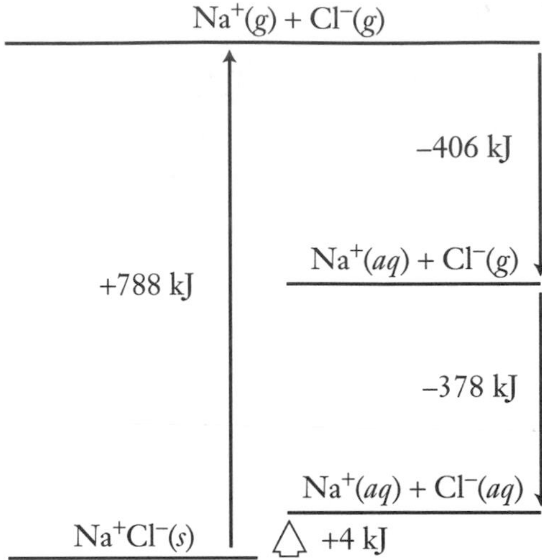

Figure 6.5 Theoretical enthalpy cycle for the solution process for sodium chloride.

The enthalpy changes are usually far larger than entropy changes at normal temperatures. However, in this case, the very large enthalpy changes essentially "cancel" each other, making the small entropy change a major factor in determining the solubility of sodium chloride. Thus we now need to do a similar calculation for the entropy factors. So that we can compare the results with the enthalpy values, we will use $T\Delta S^\circ$ data. First, the lattice must be vaporized:

$$NaCl(s) \rightarrow Na^+(g) + Cl^-(g) \qquad T\Delta S^\circ = +68\ kJ{\cdot}mol^{-1}$$

Then the ions are hydrated:

$$Na^+(g) \rightarrow Na^+(aq) \qquad T\Delta S^\circ = -27\ kJ{\cdot}mol^{-1}$$

$$Cl^-(g) \rightarrow Cl^-(aq) \qquad T\Delta S^\circ = -28\ kJ{\cdot}mol^{-1}$$

Thus entropy change (as $T\Delta S^\circ$) for the solution process is

$$(+68) + (-27) + (-28) = +13\ kJ{\cdot}mol^{-1}$$

The process can be displayed as a diagram (Figure 6.6).

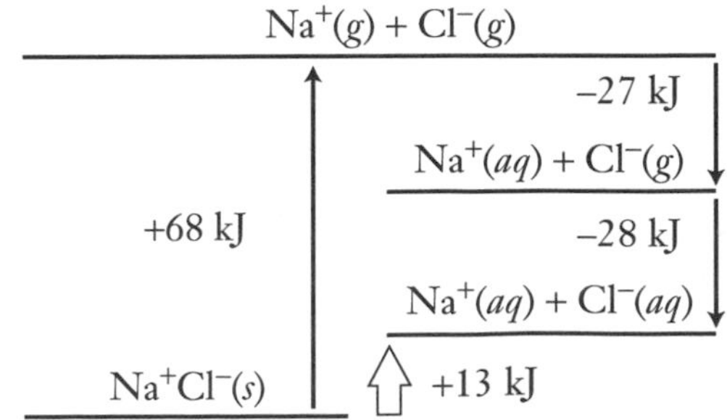

Figure 6.6 Theoretical entropy (as $T\Delta S^\circ$) cycle for the solution process for sodium chloride.

Hot and Cold Packs

Certain types of hot and cold packs employ solution reactions. The most common cold pack utilizes solid ammonium nitrate and water. When the dividing partition is broken, ammonium nitrate solution forms. This process is highly endothermic:

$$NH_4NO_3(s) \rightarrow NH_4^+(aq) + NO_3^-(aq) \qquad \Delta H^\circ = +26 \text{ kJ}\cdot\text{mol}^{-1}$$

The endothermicity must be a result of comparatively strong cation-anion attractions in the crystal lattice and comparatively weak ion-dipole attractions to the water molecules in solution. If the enthalpy factor is positive but the compound is still very soluble, the driving force must be a large increase in entropy. In fact, there is such an increase: +110 $\text{J}\cdot\text{mol}^{-1}\cdot\text{K}^{-1}$. In the solid the crystal has a low entropy, whereas in solution the ions are mobile. At the same time, the large ion size and low charge result in little ordering of the surrounding water molecules. Thus it is an increase in entropy that drives the endothermic solution process of ammonium nitrate.

One type of hot pack uses anhydrous solid calcium chloride and water. When the dividing partition is broken, calcium chloride solution forms. This process is highly exothermic:

$$CaCl_2(s) \rightarrow Ca^{2+}(aq) + 2\ Cl^-(aq) \qquad \Delta H^\circ = -82 \text{ kJ}\cdot\text{mol}^{-1}$$

With a 2+ charge cation, the lattice energy is high (about 2200 $\text{kJ}\cdot\text{mol}^{-1}$); but at the same time, the enthalpy of hydration of the calcium ion is extremely high (–1560 $\text{kJ}\cdot\text{mol}^{-1}$) and that of the chloride ion is not insignificant (–384 $\text{kJ}\cdot\text{mol}^{-1}$). The sum of these energies yields an exothermic process. By contrast, there is a slight decrease in entropy (–56 $\text{J}\cdot\text{mol}^{-1}\cdot\text{K}^{-1}$) as the small, highly charged cation is surrounded in solution by a very ordered sphere of water molecules, thereby diminishing the entropy of the water in the process. This reaction, then, is enthalpy driven.

After calculating the free energy change for the solution process, we see that it is the net entropy change that favors solution whereas the net enthalpy change does not; and it is the former that is greater than the latter. Hence, as we know, sodium chloride is quite soluble in water.

$$\begin{aligned} \Delta G^\circ &= \Delta H^\circ - T\Delta S^\circ \\ &= (+4 \text{ kJ}\cdot\text{mol}^{-1}) - (+13 \text{ kJ}\cdot\text{mol}^{-1}) \\ &= -9 \text{ kJ}\cdot\text{mol}^{-1} \end{aligned}$$

Formation of Covalent Compounds

To study the thermodynamics of covalent compound formation, it is possible to construct a cycle similar to the Born-Haber cycle that we used for ionic

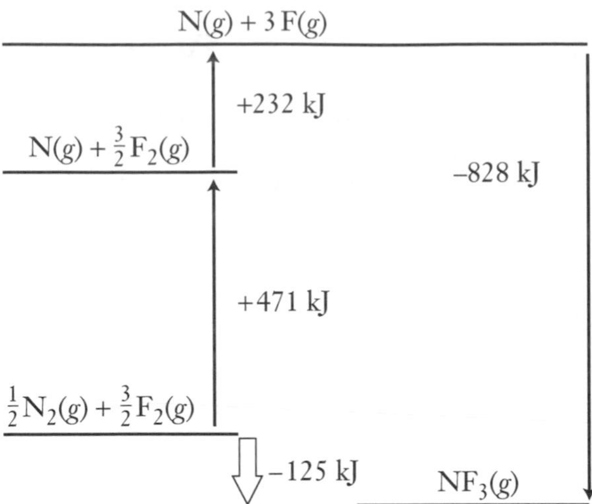

Figure 6.7 Theoretical enthalpy cycle for the formation of nitrogen trifluoride.

compounds. However, there is a major difference. The cycle does not involve ion formation; instead, we are concerned with covalent bond energies. The process can be illustrated for the formation of nitrogen trifluoride. Once again, the calculation will focus on the enthalpy terms, and the entropy factor will be considered later.

$$\tfrac{1}{2}\,N_2(g) + \tfrac{3}{2}\,F_2(g) \rightarrow NF_3(g) \qquad \Delta H_f^\circ = -125\ \text{kJ}\cdot\text{mol}^{-1}$$

1. The dinitrogen triple bond is broken. This cleavage requires one-half of the N≡N bond energy:

$$\tfrac{1}{2}\,N_2(g) \rightarrow N(g) \qquad \Delta H^\circ = +471\ \text{kJ}\cdot\text{mol}^{-1}$$

2. The difluorine single bond is broken. For the stoichiometry, three-halves of the F–F bond energy is required:

$$\tfrac{3}{2}\,F_2(g) \rightarrow 3\,F(g) \qquad \Delta H^\circ = +232\ \text{kJ}\cdot\text{mol}^{-1}$$

3. The nitrogen-fluorine bonds are formed. This process releases three times the N–F bond energy as three moles of bonds are being formed:

$$N(g) + 3\,F(g) \rightarrow NF_3(g) \qquad \Delta H^\circ = -828\ \text{kJ}\cdot\text{mol}^{-1}$$

The enthalpy diagram for the formation of nitrogen trifluoride is shown in Figure 6.7.

Turning to the entropy factor, in the formation of nitrogen trifluoride from its elements, there is a net decrease of one mole of gas. Thus a decrease in entropy would be expected. In fact, this is the case, and the overall entropy change is $-140\ \text{J}\cdot\text{mol}^{-1}\cdot\text{K}^{-1}$. The resulting free energy change is

$$\begin{aligned}\Delta G_f^\circ &= \Delta H_f^\circ - T\Delta S_f^\circ \\ &= (-125\ \text{kJ}\cdot\text{mol}^{-1}) - (298\ \text{K})(-0.140\ \text{kJ}\cdot\text{mol}^{-1}\cdot\text{K}^{-1}) \\ &= -83\ \text{kJ}\cdot\text{mol}^{-1}\end{aligned}$$

a value indicating that the compound is quite stable thermodynamically.

Thermodynamic versus Kinetic Factors

Thermodynamics is concerned with the feasibility of reaction, the position of equilibrium, and the stability of a compound. There is no information about the rate of reaction—the field of kinetics. The rate of a reaction is, to a large extent, determined by the activation energy for the reaction; that is, the energy barrier involved in the pathway for compound formation. This concept is illustrated in Figure 6.8.

A very simple example of the effect of activation energy is provided by the two more common allotropes of carbon, graphite and diamond. Diamond is thermodynamically unstable with respect to graphite:

$$C(\textit{diamond}) \rightarrow C(\textit{graphite}) \qquad \Delta G^\circ = -3 \text{ kJ·mol}^{-1}$$

Yet, of course, diamonds in diamond rings do not crumble to a black powder on a daily basis. They do not because an extremely high activation energy is required to rearrange the covalent bonds from the tetrahedral arrangement in diamond to the planar arrangement in graphite. Furthermore, all forms of carbon are thermodynamically unstable with respect to oxidation to carbon dioxide in the presence of dioxygen. Once again, it is the high activation energy that prevents diamonds in rings and the graphite ("lead") in pencils from bursting into flame:

$$C(s) + O_2(g) \rightarrow CO_2(g) \qquad \Delta G^\circ = -390 \text{ kJ·mol}^{-1}$$

We can actually make use of kinetics to alter the product of a chemical reaction. A particularly important example is the combustion of ammonia. Ammonia burns in air to form dinitrogen and water vapor:

$$4\,NH_3(g) + 3\,O_2(g) \rightarrow 2\,N_2(g) + 6\,H_2O(g)$$

This is the thermodynamically favored path, with a free energy change of -1306 kJ·mol^{-1}. When the combustion is performed in the presence of a catalyst, the activation energy of a competing reaction, that to produce nitrogen

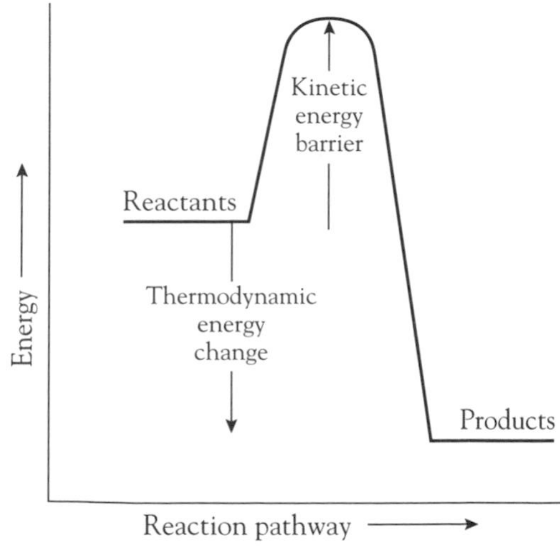

Figure 6.8 Kinetic and thermodynamic energy factors in a chemical reaction.

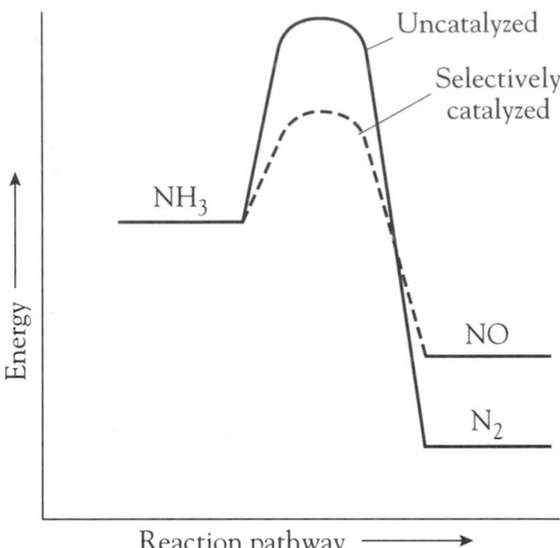

Figure 6.9 Diagram (not to scale) of the kinetic and thermodynamic energy factors in the two pathways for the combustion of ammonia.

monoxide, is, in fact, lower than that of the reaction producing dinitrogen gas:

$$4\,NH_3(g) + 5\,O_2(g) \rightarrow 4\,NO(g) + 6\,H_2O(g)$$

The latter reaction, which is a key step in the industrial preparation of nitric acid, occurs even though the free energy change for the reaction is only -958 kJ·mol^{-1}. Thus we are using kinetics to control the products of reaction and overriding the thermodynamically preferred path (Figure 6.9).

It is also possible to synthesize compounds that have a positive free energy of formation. For example, trioxygen (ozone) and all the oxides of nitrogen have positive free energies of formation. The synthesis of such substances is feasible if there is a pathway involving a net decrease in free energy and if the decomposition of the compound is kinetically slow.

An interesting example is provided by nitrogen trichloride. We saw in the previous section that nitrogen trifluoride is thermodynamically stable. In contrast, nitrogen trichloride is thermodynamically unstable; yet it exists:

$$\tfrac{1}{2}\,N_2(g) + \tfrac{3}{2}\,F_2(g) \rightarrow NF_3(g) \qquad \Delta G^\circ = -84\ \text{kJ·mol}^{-1}$$

$$\tfrac{1}{2}\,N_2(g) + \tfrac{3}{2}\,Cl_2(g) \rightarrow NCl_3(l) \qquad \Delta G^\circ \approx +250\ \text{kJ·mol}^{-1}$$

To understand this difference, we need to compare the key terms in each energy cycle. First of all, the reduction in the number of moles of gas from reactants to product means that, in both cases, the entropy term will be negative. Hence, for a spontaneous process, the enthalpy change must be negative.

In the synthesis of nitrogen trifluoride, the fluorine-fluorine bond to be broken is very weak (155 kJ·mol^{-1}), whereas the nitrogen-fluorine bond to be formed is very strong (276 kJ·mol^{-1}). As a result, the enthalpy of formation of nitrogen trifluoride is quite negative. The chlorine-chlorine bond

(242 kJ·mol^{-1}) is stronger than that of the fluorine-fluorine bond, and the nitrogen-chlorine bond (188 kJ·mol^{-1}) in nitrogen trichloride is weaker than the nitrogen-fluorine bond in nitrogen trifluoride. As a result, the enthalpy change for the formation of nitrogen trichloride is positive (Figure 6.10) and, with a negative entropy change giving a positive $-T\Delta S$ term, the free energy change will be positive.

How is it possible to prepare such a compound? The reaction between ammonia and dichlorine to give nitrogen trichloride and hydrogen chloride has a slightly negative free energy change as a result of the formation of strong hydrogen-chlorine bonds:

$$NH_3(g) + 3\ Cl_2(g) \rightarrow NCl_3(l) + 3\ HCl(g)$$

The thermodynamically unstable nitrogen trichloride decomposes violently when warmed:

$$2\ NCl_3(l) \rightarrow N_2(g) + 3\ Cl_2(g)$$

Thermodynamics, then, is a useful tool for understanding chemistry. At the same time, we should always be aware that kinetic factors can cause the product to be other than the most thermodynamically stable one (as in the case of the oxidation of ammonia). In addition, it is sometimes possible to synthesize compounds that have a positive free energy of formation, provided the synthetic route involves a net decrease in free energy and the decomposition of the compound is kinetically slow (as we saw for the existence of nitrogen trichloride).

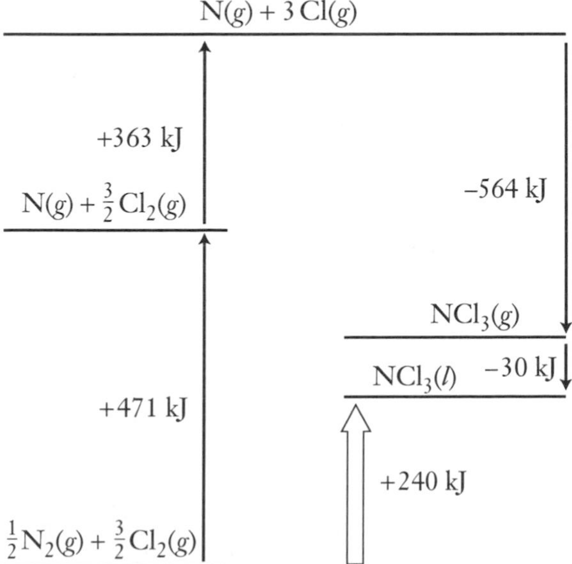

Figure 6.10 Theoretical enthalpy cycle for the formation of nitrogen trichloride.

Exercises

6.1. Define the following terms: (a) spontaneous process; (b) entropy; (c) standard enthalpy of formation.

6.2. Define the following terms: (a) enthalpy; (b) average bond energy; (c) enthalpy of hydration.

6.3. For the formation of solid calcium oxide from solid calcium and gaseous oxygen, what is the probable sign of the entropy change? What, then, must be the sign of the enthalpy change if the formation of the product occurs spontaneously? Do not consult data tables.

6.4. At very high temperatures, water will decompose to hydrogen and oxygen gas. Explain why this is to be expected in terms of the free energy equation.

6.5. Using enthalpy of formation and absolute entropy values from data tables, determine the enthalpy, entropy, and free energy of reaction for the following reaction. Use this information to identify whether the reaction is spontaneous at standard temperature and pressure.

$$H_2(g) + \tfrac{1}{2}\,O_2(g) \rightarrow H_2O(l)$$

6.6. Using enthalpy of formation and absolute entropy values from data tables, determine the enthalpy, entropy, and free energy of reaction for the following reaction. Use this information to identify whether the reaction is spontaneous at standard temperature:

$$\tfrac{1}{2}\,N_2(g) + O_2(g) \rightarrow NO_2(g)$$

6.7. Which one of the N–N or N=N bonds will be stronger? Do not look at data tables. Explain your reasoning.

6.8. The molecules of dinitrogen and carbon monoxide are isoelectronic. Yet the bond energy for the C≡O bond (1072 kJ·mol^{-1}) is stronger than that of the N≡N bond (942 kJ·mol^{-1}). Suggest an explanation.

6.9. Construct a Born-Haber cycle for the formation of aluminum fluoride. Do not perform any calculation.

6.10. Construct a Born-Haber cycle for the formation of magnesium sulfide. Do not perform any calculation.

6.11. Calculate the enthalpy of formation of calcium oxide by using a Born-Haber cycle. Obtain all necessary information from data tables. Compare the value that you obtain with the actual measured value of $\Delta H_f(CaO(s))$. Then calculate a similar cycle, assuming that calcium oxide is Ca^+O^- rather than $Ca^{2+}O^{2-}$. Take the lattice energy of Ca^+O^- to be 800 kJ·mol^{-1}. Discuss why the second scenario is less favored in enthalpy terms.

6.12. Construct Born-Haber cycles for the theoretical compounds $NaCl_2$ and $NaCl_3$. Calculate the enthalpy of formation for each of these three compounds by using information from data tables and the following estimated values:

theoretical lattice energy, $NaCl_2 = -2500$ kJ·mol^{-1}
theoretical lattice energy, $NaCl_3 = -5400$ kJ·mol^{-1}

Compare the cycles and suggest why $NaCl_2$ and $NaCl_3$ are not the preferred products.

6.13. Magnesium chloride is very soluble in water, whereas magnesium oxide is very insoluble in water. Offer an explanation for this difference in terms of the theoretical steps of the solution process. Do not use data tables.

6.14. Use lattice energy and enthalpy of hydration values from data tables to determine the enthalpy of solution of (a) lithium chloride; (b) magnesium chloride. Explain the major difference in the two values in terms of the theoretical steps.

6.15. Construct an energy diagram similar to a Born-Haber cycle for the formation of carbon tetrafluoride. Then calculate the enthalpy of formation from the steps, using numerical values from data tables. Finally, compare your value with the tabulated value of $\Delta H_f^\circ(CF_4(g))$.

6.16. Construct an energy diagram similar to a Born-Haber cycle for the formation of sulfur hexafluoride. Then calculate the enthalpy of formation from the steps, using numerical values from data tables. Finally, compare your value with the tabulated value of $\Delta H_f^\circ(SF_6(g))$.

6.17. Using enthalpy of formation and absolute entropy values from data tables, determine the free energy of formation for the following reactions:

$$S(s) + O_2(g) \rightarrow SO_2(g)$$

$$S(s) + 1\tfrac{1}{2}\,O_2(g) \rightarrow SO_3(g)$$

(a) Account for the sign of the entropy change in the formation of sulfur trioxide.

(b) Which combustion leads to the greater decrease in free energy (that is, which reaction is thermodynamically preferred)?

(c) Which of the oxides of sulfur is most commonly discussed?

(d) Suggest an explanation to account for the conflict between your answers to parts (b) and (c).

6.18. In the discussion of thermodynamic and kinetic factors, we compared the two reactions:

$$4\,NH_3(g) + 3\,O_2(g) \rightarrow 2\,N_2(g) + 6\,H_2O(g)$$

$$4\,NH_3(g) + 5\,O_2(g) \rightarrow 4\,NO(g) + 6\,H_2O(g)$$

Without consulting data tables, answer the following:

(a) Is there any major difference in entropy factors between the two reactions? Explain.

(b) Considering your answer to part (a) and the fact that the free energy for the second reaction is less negative than for the first, deduce the sign of the enthalpy of formation of nitrogen monoxide, NO.

CHAPTER

7

Hydrogen

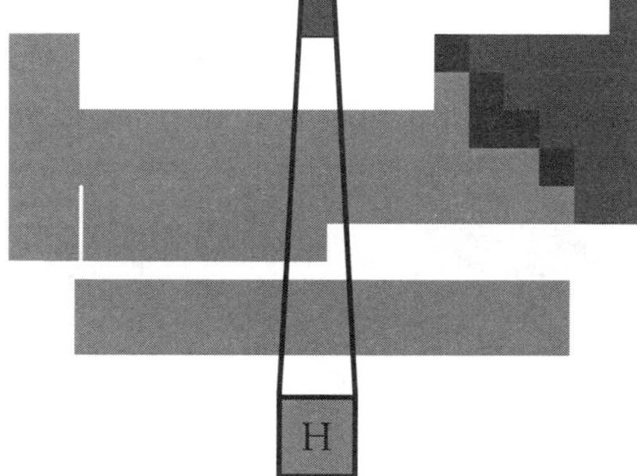

There is only one element in the periodic table that does not belong to any particular group—hydrogen. This element has a unique chemistry. Furthermore, its three isotopes differ so much in their molar masses that the physical and chemical properties of the isotopes are measurably different.

Although hydrogen was described about 200 years ago, the existence of different isotopes of hydrogen is a more recent discovery. In 1931 some very precise measurements of atomic mass indicated that there might be different isotopes of hydrogen. Harold C. Urey at Columbia University decided to try to separate them by applying the concept that the boiling point of a species depends partially on its molar mass. Urey evaporated about 5 L of liquid hydrogen, hoping that the last 2 mL would contain a larger than usual proportion of any higher molar mass isotope. The results proved him correct; the residue had a molar mass double that of normal hydrogen. This form of hydrogen was named deuterium.

Frederick Soddy, who had devised the concept of isotopes, refused to believe that deuterium was an isotope of hydrogen. The reason for his lack of acceptance was his own definition of isotopes: that isotopes are nonseparable. But Urey had separated the two forms of hydrogen, so Soddy argued that they could not be isotopes, preferring to believe that Urey was incorrect rather than question his own definition. Except for Soddy's negative opinion, Urey received considerable recognition for his discovery, culminating in the Nobel prize for chemistry in 1934. Ironically, the earlier atomic mass measurements were subsequently shown to be in error. In particular, they did not provide any evidence for the existence of hydrogen isotopes. Thus Urey's search, although successful, was based on erroneous information.

Isotopes of Hydrogen

The isotopes of hydrogen are particularly important in chemistry. Because the relative mass differences between hydrogen's isotopes are so large, there is a significant dissimilarity in physical properties and, to a lesser extent, in chemical behavior among them. Natural hydrogen contains three isotopes: protium, or "common" hydrogen, which contains zero neutrons (abundance 99.985 percent); deuterium, which contains one neutron (abundance 0.015 percent); and radioactive tritium, which contains two neutrons (abundance 10^{-15} percent). In fact, this is the only set of isotopes for which special symbols are used: H for protium, D for deuterium, and T for tritium. As the molar mass of the isotopes increases, there is a significant increase in both the boiling point and the bond energy (Table 7.1).

Deuterium and tritium bonds with other elements are also stronger than those of common hydrogen. For example, when water is electrolyzed to give hydrogen gas and oxygen gas, it is the O–H covalent bonds that are broken more readily than O–D bonds. As a result, the remaining liquid contains a higher and higher proportion of "heavy" water, deuterium oxide. When 30 L of water are electrolyzed down to a volume of 1 mL, the remaining liquid is about 99 percent pure deuterium oxide. Normal water and "heavy" water, D_2O, differ in all their physical properties; for example, deuterium oxide melts at 3.8°C and boils at 101.4°C. The density of deuterium oxide is about 10 percent higher than protium oxide at all temperatures. As a result, heavy water ice cubes will sink in "light" water at 0°C. Deuterium oxide is used

Table 7.1 Physical properties of the isotopes of hydrogen

Isotope	Molar mass ($g{\cdot}mol^{-1}$)	Boiling point (K)	Bond energy ($kJ{\cdot}mol^{-1}$)
H_2	2.02	20.6	436
D_2	4.03	23.9	443
T_2	6.03	25.2	447

widely as a solvent so that the hydrogen atoms in solute molecules can be studied without their properties being "swamped" by those in the aqueous solvent. Reaction pathways involving hydrogen atoms also can be studied by using deuterium-substituted compounds.

Tritium is a radioactive isotope with a half-life of about 12 years. With such a short half-life, we might expect that none survives naturally; in fact, tritium is constantly being formed by the impact of cosmic rays on atoms in the upper atmosphere. One pathway for its production involves the impact of a neutron on a nitrogen atom:

$$^{14}_{7}N + ^{1}_{0}n \rightarrow ^{12}_{6}C + ^{3}_{1}T$$

The isotope decays to give the rare isotope of helium, helium-3:

$$^{3}_{1}T \rightarrow ^{3}_{2}He + ^{0}_{-1}e$$

There is a significant demand for tritium. It is sought for medical purposes, where it is useful as a tracer. In its radioactive decay, the isotope emits low-energy electrons (β rays) but no harmful γ rays. The electrons can be tracked by a counter and cause minimal tissue damage. The most significant consumers of tritium are the military forces of the countries possessing hydrogen (more accurately, tritium) bombs. To extract the traces of tritium that occur in water would require the processing of massive quantities of water. An easier synthetic route entails the bombardment of lithium-6 by neutrons in a nuclear reactor:

$$^{6}_{3}Li + ^{1}_{0}n \rightarrow ^{4}_{2}He + ^{3}_{1}T$$

Tritium's short half-life creates a problem for military scientists of nuclear powers because, over time, the tritium content of nuclear warheads diminishes until it is below the critical mass needed for fusion. Hence warheads have to be periodically "topped up" if they are to remain usable.

Nuclear Magnetic Resonance

One of the most useful tools for studying molecular structure is nuclear magnetic resonance (NMR). This technique involves the study of nuclear spin. As discussed in Chapter 2, electrons, protons, and neutrons have spins of $\pm\frac{1}{2}$. In an atom, there are four possible permutations of nuclear particles: even numbers of both protons and neutrons; odd number of protons and even number of neutrons; even number of protons and odd number of neutrons; and odd numbers of both protons and neutrons. The last three categories, then, will have unpaired nucleons. This condition might be expected to occur in an enormous number of nuclei, but, as mentioned in Chapter 2, spin pairing is a major driving force for the stability of nuclei and, in fact, only 4 of the 273 stable nuclei have odd numbers of both protons and neutrons.

Unpaired nucleons can have a spin of $+\frac{1}{2}$ or $-\frac{1}{2}$; each spin state has the same energy. However, in a magnetic field, the spin can be either parallel with the field or opposed to it, and the parallel arrangement has lower energy. The splitting of (difference between) the two energy levels is very small and corresponds to the radio frequency range of the electromagnetic spectrum.

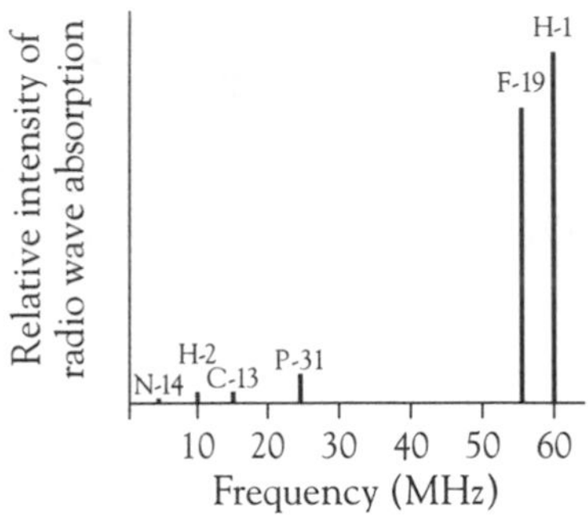

Figure 7.1 Relative intensities of the unique absorption by common isotopes in a magnetic field of 14 000 gauss.

When we focus a radio wave source on the sample with unpaired nucleons and adjust the frequency of radio waves to the energy level of the splitting, electromagnetic radiation is absorbed by the sample as unpaired nucleons reverse their spins to oppose the field, that is, move to the higher energy level. In a field of 15 000 gauss, absorption happens at 63.9 MHz (or 6.39×10^7 s^{-1}) for an isolated proton.

The relative intensity of the absorption depends very much on the identity of the nucleus. As it happens, hydrogen-1 gives the most intense absorption among the nuclei (see Figure 7.1). This is fortunate, because hydrogen is the most common element in the universe and therefore readily available for study. Even today, years after the discovery of this technique, hydrogen is the element most studied by this technique.

If this were all that NMR could do, it would not be a particularly useful technique. However, the electrons surrounding a nucleus affect the actual magnetic field experienced by the nucleus. Because the magnetic field for each environment differs from that applied by a magnet, the splitting of the energy levels and the frequency of radiation absorbed are unique for each species. Thus absorption frequency reflects the atomic environment. The difference in frequency absorbed (called the chemical shift, or simply, shift) is very small—about 10^{-6} of the signal itself. Hence we report the shifts in terms of ppm (parts per million). In addition, splitting of the transition levels can occur through interaction with neighboring odd-spin nuclei. Thus the relative locations of atoms can often be identified by NMR. This technique is a great aid to chemists, particularly organic chemists, for both identification of a compound and for the study of electron distributions within molecules. It is also used extensively in the health field under the name of magnetic resonance imaging (MRI).

Properties of Hydrogen

As stated earlier, hydrogen is a unique element, not belonging to any of the other groups in the periodic table. Some versions of the periodic table place it as a member of the alkali metals, some as a member of the halogens; others place it in both locations; and a few place it on its own. The basic reasons for and against placement of hydrogen in either Group 1 or Group 17 are summarized in Table 7.2. Throughout this book, hydrogen has a place in the periodic table all its own, thus emphasizing the uniqueness of this element. With an electronegativity higher than those of the alkali metals and lower than those of the halogens, it makes sense to place hydrogen midway between the two groups.

Dihydrogen is a colorless, odorless gas that liquefies at –253°C and solidifies at –259°C. Hydrogen gas is not very reactive, partly because of the high H–H covalent bond energy (436 $kJ \cdot mol^{-1}$). This bond is stronger than the bonds hydrogen has with most other nonmetals; for example, the H–S bond energy is only 347 $kJ \cdot mol^{-1}$. Recall that only when the bond energies of the products are similar to or greater than those of the reactants are spontaneous reactions likely. One such reaction is combustion of dihydrogen with dioxygen to produce water. If hydrogen gas and oxygen gas are mixed and sparked, the reaction is explosive:

Table 7.2 Reasons for and against placing hydrogen in Group 1 or 17

	Argument for placement	Argument against placement
Alkali metal group	Forms monopositive ion, H^+ (H_3O^+)	Is not a metal
	Has a single *s* electron	Does not react with water
Halogen group	Is a nonmetal	Rarely forms mononegative ion, H^-
	Forms a diatomic molecule	Is comparatively unreactive

$$2\ H_2(g) + O_2(g) \rightarrow 2\ H_2O(g)$$

The reaction has to be enthalpy driven because there is a decrease in entropy. If we add the bond energies, we see that the strong O–H bond (464 kJ·mol^{-1}) makes the reaction thermodynamically feasible (Figure 7.2).

Dihydrogen reacts with the halogens, the rate of reaction decreasing down the group. Its reaction with difluorine is violent; the product is hydrogen fluoride:

$$H_2(g) + F_2(g) \rightarrow 2\ HF(g)$$

The reaction of dihydrogen with dinitrogen is very slow in the absence of a catalyst. (This reaction is discussed more fully in Chapter 14.)

$$3\ H_2(g) + N_2(g) \rightleftharpoons 2\ NH_3(g)$$

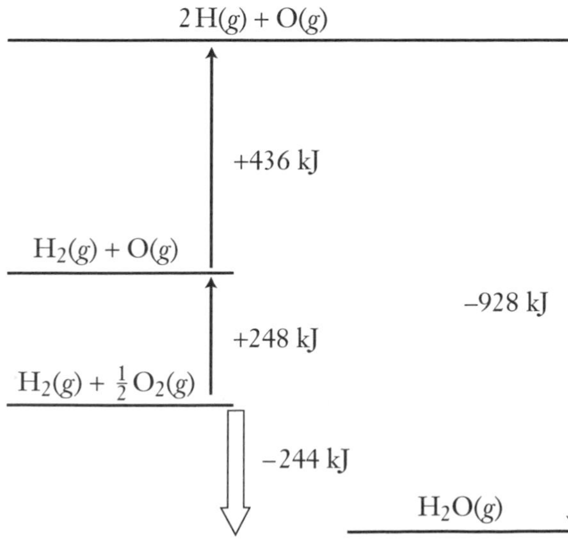

Figure 7.2 Theoretical enthalpy cycle for the formation of water.

At high temperatures, dihydrogen reduces many metal oxides to the metallic element. Thus copper(II) oxide is reduced to copper metal:

$$CuO(s) + H_2(g) \xrightarrow{\Delta} Cu(s) + H_2O(g)$$

In the presence of a catalyst (usually powdered palladium or platinum), dihydrogen will reduce carbon-carbon double and triple bonds to single bonds. For example, ethene, C_2H_4, is reduced to ethane, C_2H_6:

$$H_2C{=}CH_2(g) + H_2(g) \rightarrow H_3C{-}CH_3(g)$$

The reduction with dihydrogen is used to convert unsaturated liquid fats (edible oils), which have numerous carbon-carbon double bonds, to saturated and partially saturated solid fats (margarines), which contain fewer carbon-carbon double bonds.

Preparation of Dihydrogen

In the laboratory, hydrogen gas can be generated by the action of dilute acids on many metals. A particularly convenient reaction is that between zinc and dilute hydrochloric acid:

$$Zn(s) + 2\ HCl(aq) \rightarrow ZnCl_2(aq) + H_2(g)$$

There are several different routes of industrial synthesis, one of these being the steam reformer process. In the first step of this process, the endothermic reaction of natural gas (methane) with steam at high temperatures gives carbon monoxide and hydrogen gas. It is difficult to separate the two products because the mixture must be cooled below –205°C before the carbon monoxide will condense. To overcome this problem and to increase the yield of hydrogen gas, the mixture is cooled, additional steam is injected, and the combination passed over a different catalyst system. Under these conditions, the carbon monoxide is oxidized in an exothermic reaction to carbon dioxide, and the added water is reduced to hydrogen:

$$CH_4(g) + H_2O(g) \xrightarrow{\text{Ni, 800°C}} CO(g) + 3H_2(g)$$

$$CO(g) + H_2O(g) \xrightarrow{Fe_2O_3/Cr_2O_3,\ 400°C} CO_2(g) + H_2(g)$$

The carbon dioxide can be separated from hydrogen gas in several ways. One is to cool the products below the condensation temperature of carbon dioxide (–78°C), which is much higher than that of dihydrogen (–253°C). However, this process still requires large-scale refrigeration systems. Another route involves passage of the gas mixture through a solution of potassium carbonate. Carbon dioxide is an acid oxide, unlike carbon monoxide, which is neutral. Carbon dioxide reacts with the carbonate ion and water to give two moles of the hydrogen carbonate ion. Later, the potassium hydrogen carbonate solution can be removed and heated to regenerate the potassium carbonate:

$$K_2CO_3(aq) + CO_2(g) + H_2O(l) \rightarrow 2\ KHCO_3(aq)$$

For most purposes, the purity of the dihydrogen obtained from thermochemical processes is satisfactory. However, very pure hydrogen (at least 99.9 percent) is generated by an electrochemical route, the electrolysis of a sodium hydroxide or potassium hydroxide solution. The reaction produces dioxygen at the anode and dihydrogen at the cathode:

Cathode $2\ H_2O(l) + 2\ e^- \rightarrow 2\ OH^-(aq) + H_2(g)$

Anode $2\ OH^-(aq) \rightarrow H_2O(l) + \frac{1}{2}O_2(g) + 2\ e^-$

Hydrides

Binary compounds of hydrogen are given the generic name of hydrides. Hydrogen, which forms binary compounds with most elements, has an electronegativity only slightly above the median value of all the elements in the periodic table. As a result, it behaves as a weakly electronegative nonmetal, forming ionic compounds with very electropositive metals and covalent compounds with all the nonmetals. In addition, it forms metallic hydrides with some of the transition metals. The distribution of the three major types of hydrides is shown in Figure 7.3.

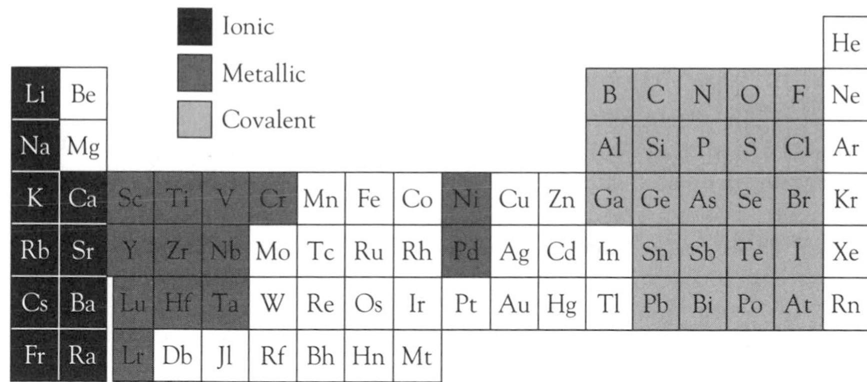

Figure 7.3 Three common types of hydrides: ionic, covalent, and metallic. Only the main and transition groups are shown. A few lanthanoids and actinoids also form metallic hydrides (not shown).

Ionic Hydrides

All the ionic hydrides are white solids and are formed only by the most electropositive metals. These ionic crystals contain the metal cation and the hydride ion (H^-). Proof of the existence of this anion comes from the electrolysis of lithium hydride in molten lithium chloride. During this process, hydrogen is evolved at the anode, the positive electrode:

$$2\ H^-(LiCl) \rightarrow H_2(g) + 2\ e^-$$

All these hydrides are very reactive; for example, dihydrogen is produced in the presence of any moisture:

$$LiH(s) + H_2O(l) \rightarrow LiOH(aq) + H_2(g)$$

The hydrides can be used as reducing agents. Thus, when heated, sodium hydride reacts with silicon tetrachloride to produce silane, SiH_4, a colorless, flammable gas:

$$SiCl_4(g) + 4\ NaH(s) \rightarrow SiH_4(g) + 4\ NaCl(s)$$

Covalent Hydrides

Hydrogen forms compounds containing covalent bonds with all the nonmetals (except the noble gases) and with very weakly electropositive metals such as gallium and tin. Almost all the simple covalent hydrides are gases at room temperature. There are three subcategories of covalent hydrides:

Those in which the hydrogen atom is nearly neutral

Those in which the hydrogen atom is substantially positive

Those in which the hydrogen atom is slightly negative, specifically the electron-deficient boron compounds

The majority of the covalent hydrides are those in which the hydrogen is nearly neutral. Because of their low polarity, the sole intermolecular force between these molecules is dispersion; as a result, these covalent hydrides are gases with low boiling points. Typical examples of these hydrides are tin(IV) hydride, SnH_4 (b.p. –52°C), and phosphane, PH_3 (b.p. –90°C).

The largest group of covalent hydrides contains carbon—the hydrocarbons—and comprises the alkanes, the alkenes, the alkynes, and the aromatic hydrocarbons. Many of the hydrocarbons are large molecules in which the intermolecular forces are strong enough to allow them to be liquids or solids at room temperature. All the hydrocarbons are thermodynamically unstable toward oxidation. For example, methane reacts spontaneously with dioxygen to give carbon dioxide and water:

$$CH_4(g) + 2\ O_2(g) \rightarrow CO_2(g) + 2\ H_2O(g) \qquad \Delta G^\circ = -800\ \text{kJ·mol}^{-1}$$

The process is very slow unless the mixture is ignited; that is, the reaction has a high activation energy barrier.

Ammonia, water, and hydrogen fluoride belong to the second category of covalent hydrides—hydrogen compounds containing positively charged hydrogen atoms (discussed in Chapter 3). These compounds differ from the other covalent hydrides in their abnormally high melting and boiling points. This property is illustrated by the boiling points of the Group 17 hydrides (Figure 7.4).

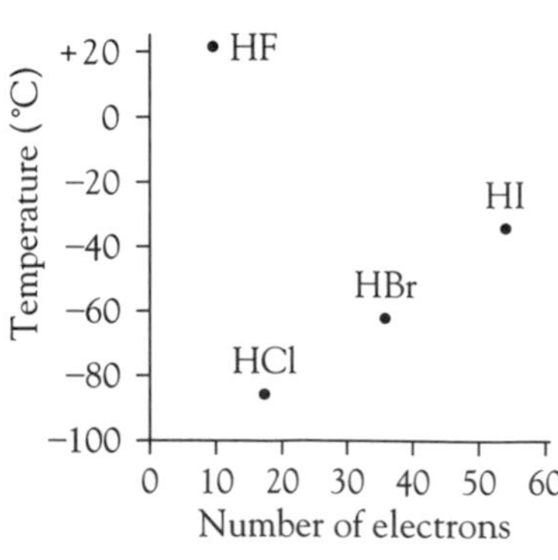

Figure 7.4 Boiling points of the Group 17 hydrides.

The positively charged hydrogen in these compounds is attracted by an electron pair on another atom to form a weak bond that is known as a hydrogen bond but is more accurately called a *protonic bridge*. Even though, as an intermolecular force, protonic bridging is very strong, it is still weak compared with a covalent bond. For example, the $H_2O\cdots HOH$ protonic bridge has a bond energy of 22 kJ·mol^{-1}, compared with 464 kJ·mol^{-1} for the O–H covalent bond. In introductory chemistry texts, protonic bridges are regarded as very strong dipole-dipole interactions occurring as a result of the very

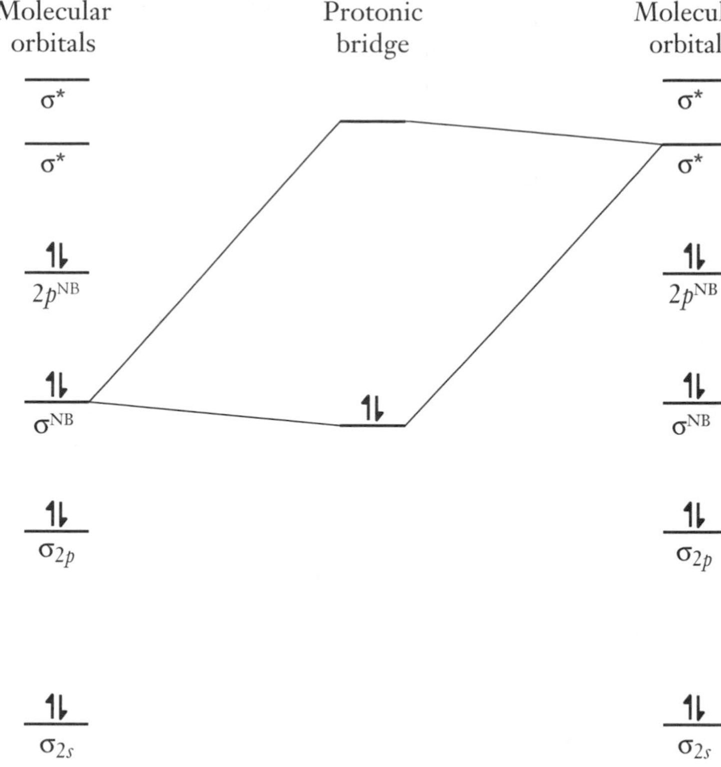

Figure 7.5 The overlap of a filled nonbinding σ^{NB} molecular orbital on one water molecule with an empty σ^* orbital on a neighboring water molecule produces a protonic bridge bond.

polar covalent bonds in the bridged molecules. However, according to this concept of electrostatic attraction, hydrogen chloride should also show this effect, and it does not (to any significant extent).

Molecular orbital theory provides a more valid representation of protonic bridging. We can visualize molecular orbitals that are formed not within a single water molecule but between neighboring molecules. The observation that hydrogen bond lengths are usually much shorter than the sum of the van der Waals radii of the two atoms supports this model. Furthermore, we find that the stronger the protonic bridge, the weaker the O–H covalent bond. Thus, the two bonds are strongly interrelated.

Figure 3.47 (see Chapter 3) is the molecular orbital diagram of the water molecule. This diagram shows a filled nonbonding orbital, σ^{NB}, and two empty antibonding orbitals, σ^*. The protonic bridge can form when the lower empty antibonding orbital of one water molecule mixes with the filled nonbonding orbital of a second water molecule. This mixing leads to the formation of a new pair of molecular orbitals in which the electron pair has lower energy (Figure 7.5). It is this lower energy bonding molecular orbital that is the protonic bridge.

Three elements form large numbers of hydrides: carbon, boron, and silicon. The hydrides of carbon and silicon contain "normal" covalent bonds, but the bonds in the boron hydrides are unusual because some hydrogen atoms link, or bridge, pairs of boron atoms. Boron has only three outer electrons, hence the expected formulation of BH_3 would not satisfy the octet rule for boron. The utilization of hydrogen atoms as bridges means, however, that one electron pair can satisfy the bonding requirements of two boron atoms. These bonds are not protonic bridges, but *hydridic bridges*, because the hydrogen possesses a partial negative charge (recall that boron is less electronegative than hydrogen). The reversed polarity of the bond results in a high chemical reactivity for these compounds. The simplest member of the series is diborane, B_2H_6. Each terminal hydrogen atom forms a normal two-electron bond with a boron atom. Each boron atom then has one electron left, and this is paired with one of the bridging hydrogen atoms (Figure 7.6).

Figure 7.6 Electron pair arrangement in diborane, B_2H_6.

The shape of the molecule can be described as approximately tetrahedral around each boron atom, with the bridging hydrogen atoms in what are sometimes called "banana bonds." The hydridic bonds behave like weak covalent bonds (Figure 7.7).

Figure 7.7 Geometry of the diborane molecule.

The bonding in a diborane molecule can be described in terms of hybridization concepts. According to these concepts, the four bonds, separated by almost equal angles, would correspond to sp^3 hybridization. Three of the four hybrid orbitals will contain single electrons from the boron atom. Two of these half-filled orbitals would then be involved in bonding with the terminal hydrogen atoms. This arrangement would leave one empty and one half-filled hybrid orbital.

To explain how we make up the eight electrons in the sp^3 orbital set, we consider that the single half-filled hybrid sp^3 orbitals of the two borons overlap with each other and with the 1*s* orbital of a bridging hydrogen atom at the same time. This arrangement will result in a single orbital that is capable of containing two electrons but is shared by three atoms (Figure 7.8). An identical arrangement forms the other B–H–B bridge. The bonding electron distribution of diborane is shown in Figure 7.9.

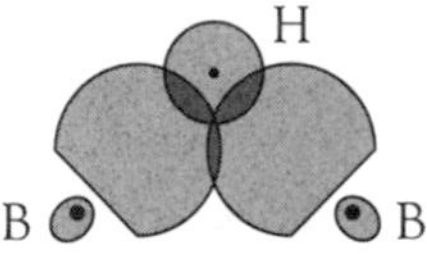

Figure 7.8 Overlap of the sp^3 hybrid orbitals of the two boron atoms with 1*s* orbital of the bridging hydrogen atom.

Alternatively, we can consider the molecular orbital explanation. The detailed molecular orbital diagram for this eight-atom molecule is complex. Although molecular orbitals relate to the molecule as a whole, it is sometimes possible to identify molecular orbitals that are involved primarily in one particular bond. In this case, we find that the mixing of the orbital wave functions of the atoms in each bridge bond results in the formation of three molecular orbitals. When we compare the energies of the atomic orbitals with those of the molecular orbitals, we find that one molecular orbital is lower in energy [σ(*bonding*)], one is higher in energy [σ(*antibonding*)], and the third has an energy level equivalent to the mean energy of the three component atomic orbitals [σ(*nonbonding*)] (see Chapter 3). The bridging hydrogen atom contributes one electron, and each boron atom contributes one-half electron. This arrangement fills the bonding orbital between the three atoms (Figure 7.10). Because there is one bonding orbital shared between two pairs of atoms, the bond order for each B–H component must be $\frac{1}{2}$. The same arguments apply to the other bridge. From bond energy measurements, we do indeed find each B–H bridging bond to be about half the strength of a terminal B–H bond, although it is still in the energy range of a true covalent bond, unlike the much weaker protonic bridges in hydrogen-bonded molecules. Of equal importance, the set of molecular orbitals shows that the structure makes maximum use of the few boron electrons. The presence of more electrons would not strengthen the bond because these electrons would enter nonbonding molecular orbitals.

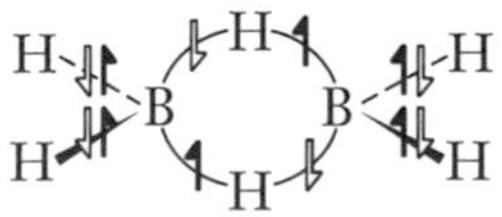

Figure 7.9 The electron pairing that is consistent with sp^3 hybridization for each boron atom and with the two-electron, three-atom B–H–B bridging bonds. The electrons contributed by the hydrogen atoms are the open half-headed arrows.

Diborane is a highly reactive, toxic, colorless gas. It catches fire in air and explodes when mixed with pure dioxygen. The extremely exothermic reaction produces diboron trioxide and steam:

$$B_2H_6(g) + 3\ O_2(g) \rightarrow B_2O_3(s) + 3\ H_2O(g)$$

The hydride also reacts with any trace of moisture to give boric acid and hydrogen gas:

$$B_2H_6(g) + 6\ H_2O(l) \rightarrow 2\ H_3BO_3(aq) + 3\ H_2(g)$$

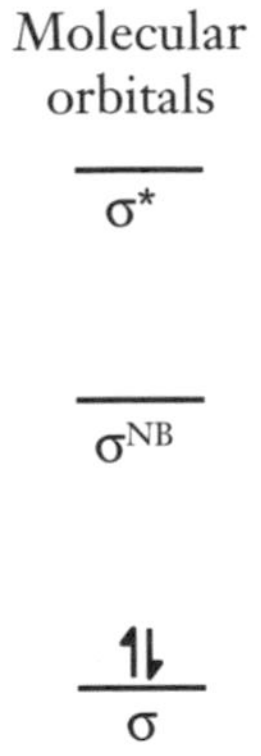

There are two main series of boranes: one with the generic formula B_nH_{n+4}, the *nido*-boranes, such as decaborane, $B_{10}H_{14}$; the other with the generic formula B_nH_{n+6}, the *arachno*-boranes, such as tetraborane, B_4H_{10} (Figure 7.11). In each of the boranes, there are bridging hydrogen atoms and, except for diborane, direct boron-boron bonds as well. All of the compounds have positive ΔG_f° values; that is, they are thermodynamically unstable.

Boranes were once considered as possible rocket fuels because they burn very exothermically. In fact, for equal mass, only dihydrogen produces more heat upon combustion. However, the cost of synthesis on a large scale was prohibitive. Today the main interest in these compounds is the study of the unique structures that boranes form. In addition to the many boron-hydrogen molecules, there is an equally large array of boron-hydrogen anions. Figure 7.12 shows the structure of the $B_2H_7^-$ ion, which, unlike diborane, has a single boron-hydrogen bridge.

An enormous number of borane compounds containing other elements have been synthesized. These include the carboranes, boranes that incorporate carbon atoms in the borane skeletons, and metallocarboranes, boron-carbon-hydrogen compounds that contain a metal, for example, the $[Fe(C_2B_9H_{11})_2]^{2-}$ ion (Figure 7.13). There may be major uses of the boranes someday, but it is curiosity about their unique chemistry that currently drives chemists to study these compounds.

Molecular orbitals

σ^*

σ^{NB}

⇅

σ

Figure 7.10 The molecular orbitals that are involved in the hydridic bridge.

Metallic (*d*-Block) Hydrides

Some transition metals form a third class of hydrides, the metallic hydrides. These compounds are often nonstoichiometric; for example, the highest hydrogen-titanium ratio is found in a compound with the formula $TiH_{1.9}$.

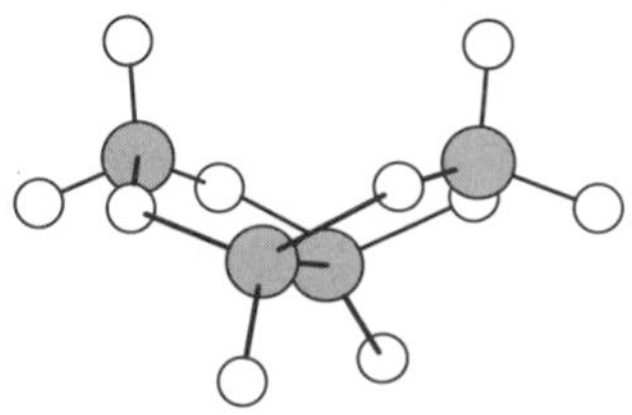

Figure 7.11 Structure of tetraborane, B_4H_{10}. The boron atoms are shaded.

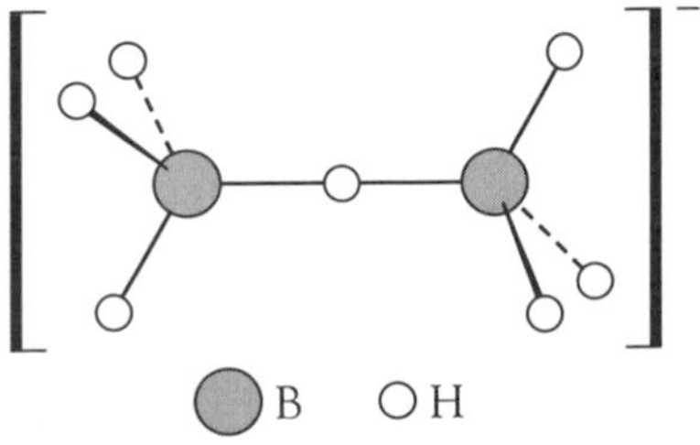

Figure 7.12 Structure of the $B_2H_7^-$ anion.

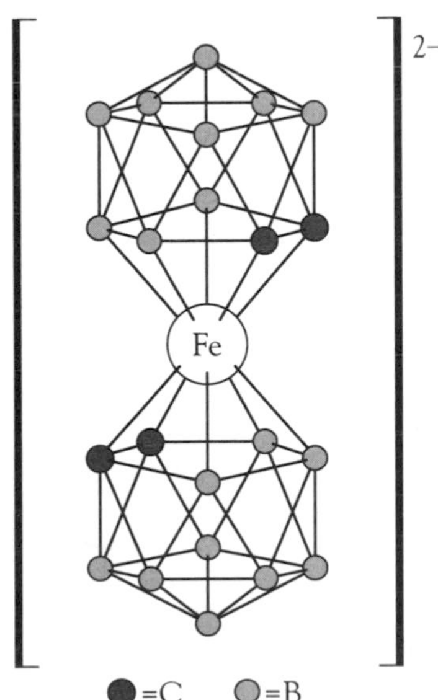

Figure 7.13 The $[Fe(C_2B_9H_{11})_2]^{2-}$ ion. For clarity, the hydrogen atoms attached to each boron atom and carbon atom are not shown.

Dihydrogen as a Fuel

The reaction of dihydrogen with dioxygen is exothermic, producing 286 kJ·mol^{-1}, and the sole product is water. As a result, there is much interest in hydrogen's potential as a fuel. Up to now, the most important application has been in space flight. The three main engines on the space shuttle utilize this reaction, and the external fuel tank contains over 5×10^5 L of liquid oxygen and about 1.4×10^6 L of liquid hydrogen. Because hydrogen is always the lowest density gas at any pressure, an enormous volume of hydrogen would be required to operate more conventional vehicles such as automobiles. Conveying it as a compressed gas would require large thick-walled containers, and this requirement would substantially increase the mass of the vehicle, thereby further increasing the fuel requirements. Liquid hydrogen would be more efficient, but it has to be stored at a temperature below −253°C. Evaporation would be a real problem. The reversible absorption of dihydrogen by some interstitial hydrides has actually been used experimentally for buses in the United States and in Europe. The dihydrogen is pumped into a tank filled with a hydrogen-absorbing alloy. Passage of an electric current through the metal releases the hydrogen gas, which is then conveyed to a traditional internal combustion engine.

Dihydrogen seems to have greater potential for city-scale storage and supply of power. The peak production times of energy-generating methods such as wind, water, and wave power do not coincide with our times of maximum need. We require ways of storing electrical energy produced in the middle of the night and of regenerating it in the periods of highest demand. This can be accomplished by electrolyzing a solution of sodium hydroxide or potassium hydroxide:

Cathode $2\,H_2O(l) + 2\,e^- \rightarrow 2\,OH^-(aq) + H_2(g)$

Anode $2\,OH^-(aq) \rightarrow H_2O(l) + \frac{1}{2}O_2(g) + 2\,e^-$

giving an overall reaction of

$$H_2O(l) \rightarrow H_2(g) + \tfrac{1}{2}O_2(g)$$

Provided water is added to the cell, gas production can be performed continuously because the alkali metal hydroxide is not consumed in the process. The dihydrogen and dioxygen are stored, and then, at times of high electrical demand, the reverse process can be performed in a fuel cell or a thermoelectric plant, where the hydrogen gas and oxygen gas are recombined to form water, with the release of electrical energy. Pilot plants are already using this technology.

The nature of these compounds is complex. Thus the titanium hydride mentioned above is now believed to consist of $(Ti^{4+})(H^-)_{1.9}(e^-)_{2.1}$. It is the free electrons that account for the metallic luster and high electrical conductivity of these compounds. The density of the metal hydride is often less than that of the pure metal because of structural changes in the metallic crystal lattice, and the compounds are usually brittle. The electrical conductivity of the metallic hydrides is generally lower than that of the parent metal as well.

Most metallic hydrides can be prepared by warming the metal with hydrogen under high pressure. At high temperatures, the hydrogen is released as dihydrogen gas again. Many alloys (for example, Ni_5La) can absorb and release copious quantities of hydrogen. Their proton densities exceed that of liquid hydrogen, a property that makes them of great interest because of their potential use in the storage of hydrogen.

Water and Hydrogen Bonding

Hydrogen bonding between water molecules is of particular importance for life on this planet. Without hydrogen bonding, water would melt at about –100°C and boil at about –90°C. Hydrogen bonding results in another very rare property of water—the liquid phase is denser than the solid phase. For most substances, the molecules are packed closer in the solid phase than in the liquid, so the solid has a higher density than the liquid has. Thus a solid usually settles to the bottom as it starts to crystallize from the liquid phase. Were this to happen with water, those lakes, rivers, and seas in parts of the world where temperatures drop below freezing would freeze from the bottom up. Fishes and other marine organisms would be unlikely to survive in such environments.

Fortunately, ice is less dense than liquid water, so in subzero temperatures, a layer of insulating ice forms over the surface of lakes, rivers, and oceans, keeping the water beneath in the liquid phase. The cause of this abnormal behavior is the open structure of ice, which is due to the network of hydrogen bonds (Figure 7.14).

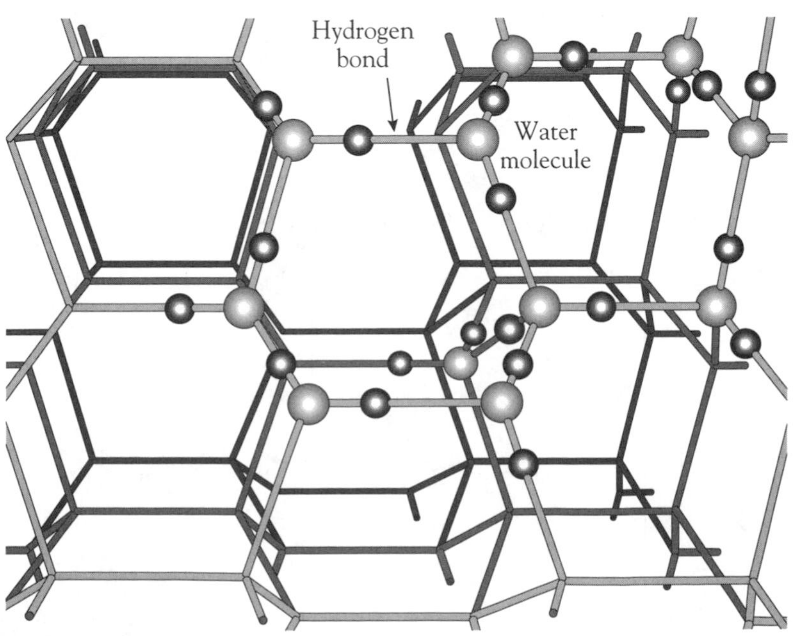

Figure 7.14 A representation of part of the ice structure, showing the open framework. The larger circles represent the oxygen atoms. (From G. Rayner-Canham et al., *Chemistry: A Second Course* [Don Mills, ON: Addison-Wesley, 1989], p. 165.)

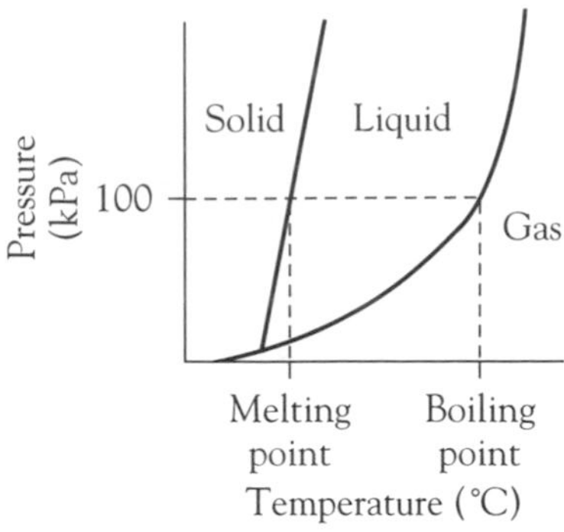

Figure 7.15 An idealized phase diagram.

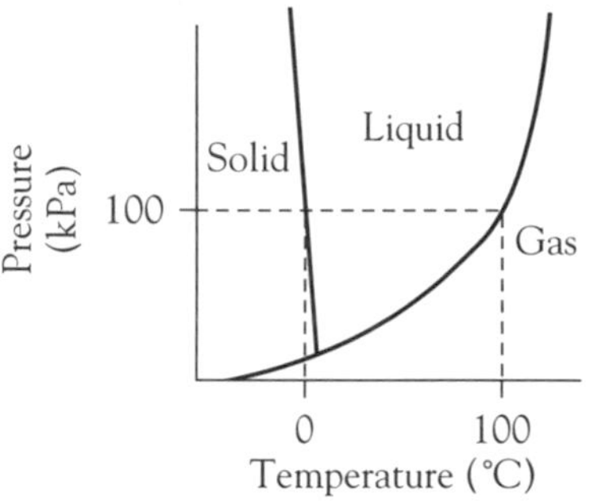

Figure 7.16 Phase diagram for water (not to scale).

Upon melting, some of these hydrogen bonds are broken and the open structure partially collapses. This change increases the liquid's density. The density reaches a maximum at 4°C, at which point the increase in density due to the collapsing of the hydrogen-bonded clusters of water molecules is overtaken by the decrease in density due to the increasing molecular motion resulting from the rise in temperature.

To appreciate another unusual property of water, we must look at a phase diagram. A phase diagram displays the thermodynamically stable phase of an element or compound with respect to pressure and temperature. Figure 7.15 shows an idealized phase diagram. The regions between the solid lines represent the phase that is thermodynamically stable under those particular combinations of pressure and temperature. The standard melting and boiling points can be determined by considering the phase changes at standard pressure, 100 kPa. When a horizontal dashed line is projected from the point of standard pressure, the temperature at which that line crosses the solid-liquid boundary is the melting point, and the temperature at which the line crosses the liquid-gas line is the boiling point. For almost all substances, the solid-liquid line has a positive slope. This trend means that application of a sufficiently high pressure to the liquid phase will cause the substance to solidify.

Water, however, has an abnormal phase diagram because the density of ice is lower than that of liquid water. The Le Châtelier principle indicates that the denser phase is favored by increasing pressure. So, for water, application of pressure to the less dense solid phase causes it to melt to the denser liquid phase. It is this anomalous behavior that enables us to ice-skate, because the pressure of the blade is enough to melt the ice surface, providing a low-friction liquid layer (Figure 7.16) down to about –30°C.

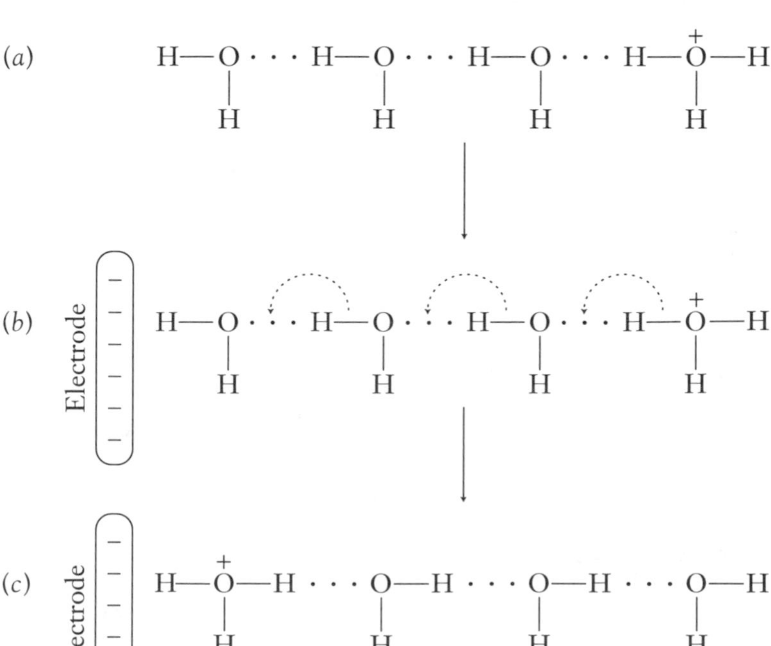

Figure 7.17 The high electrical conductivity of the hydronium ion is explained by the switching of the alternating covalent bonds and hydrogen bonds as the positive charge is attracted toward the negative electrode.

Furthermore, as a result of hydrogen bonding, solutions of hydronium ion, H_3O^+, or of hydroxide ion, OH^-, have much higher electrical conductivities than do solutions of the same concentration of any other ion. Ionic conductivity is a measure of the rate at which ions travel through the solution. The abnormally high values for these two ions can be explained in terms of hydrogen bonding. If one pictures part of the hydronium ion environment, the ion is hydrogen bonded to neighboring water molecules (Figure 7.17a). When a negative charge is placed in the solution, the hydronium ion does not move; instead, the covalent bonds and hydrogen bonds switch places (Figure 7.17b). Thus the hydronium ion that conveys the charge to the electrode is not the one that possessed the charge initially (Figure 7.17c).

Biological Aspects of Hydrogen Bonding

Hydrogen is a key element in living organisms. In fact, the existence of life depends on two particular properties of hydrogen: the closeness of the electronegativities of carbon and hydrogen and the ability of hydrogen to form hydrogen bonds when covalently bonded to nitrogen or oxygen. The low polarity of the carbon-hydrogen bond contributes to the stability of organic compounds in our chemically reactive world. Biological processes also rely on both polar and nonpolar surfaces, the best example of the latter being the lipids. It is important to realize that nonpolar sections of biological molecules, usually containing just carbon and hydrogen atoms, are just as significant as their polar regions.

Hydrogen bonding is a vital part of all biomolecules. Proteins are held in shape by hydrogen bonds that form cross-links between chains. The strands of DNA and RNA, the genetic material, are held together by hydrogen bonds as well. But more than that, the hydrogen bonds in the double helices are not random; they form between specific pairs of organic bases. These pairs are preferentially hydrogen bonded because the two components fit together to give particularly close approaches of the hydrogen atoms involved in the hydrogen bonding. This bonding is illustrated in Figure 7.18 for the interaction between two particular base units, thymine and adenine.

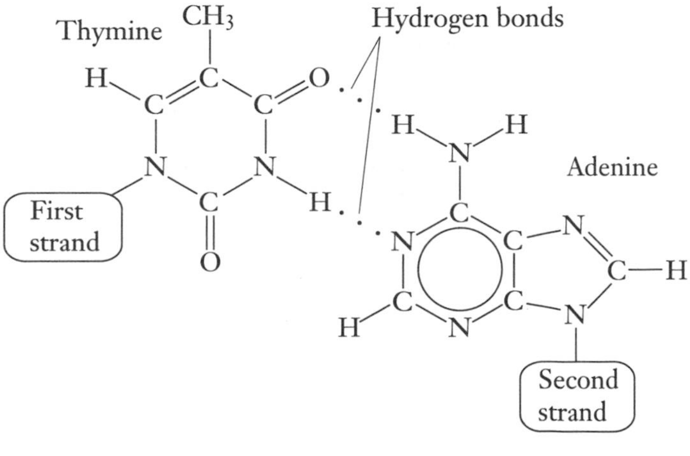

Figure 7.18 Hydrogen bond interaction between thymine and adenine fragments of the two strands in a DNA molecule.

It is the specific matching that results in the precise ordering of the components in the DNA and RNA chains, a system that allows those molecules to reproduce themselves almost completely error-free.

All proteins depend on hydrogen bonding for their function as well. Proteins consist mainly of one or more strands of linked amino acids. But to function, most proteins must form a compact shape. To do this, the protein strand loops and intertwines with itself, being held in place by hydrogen bonds cross-linking one part of the strand to another.

Exercises

7.1. Define the following terms: (a) protonic bridge; (b) hydridic bridge.

7.2. Define the following terms: (a) borane; (b) phase diagram.

7.3. An ice cube at 0°C is placed in some liquid water at 0°C. The ice cube sinks. Suggest an explanation.

7.4. Which of the following isotopes can be studied by nuclear magnetic resonance: carbon-12, oxygen-16, oxygen-17?

7.5. When we study the NMR spectrum of a compound, why are the absorption frequencies expressed as ppm?

7.6. Explain why hydrogen is not placed with the alkali metals in the periodic table.

7.7. Explain why hydrogen is not placed with the halogens in the periodic table.

7.8. Explain why hydrogen gas is comparatively unreactive.

7.9. Is the reaction of dihydrogen with dinitrogen to produce ammonia entropy driven or enthalpy driven? Do not consult data tables. Explain your reasoning.

7.10. Write chemical equations for the reaction between

(a) tungsten(VI) oxide, WO_3, and dihydrogen, with heating
(b) hydrogen gas and chlorine gas
(c) aluminum metal and dilute hydrochloric acid
(d) tetraborane, B_4H_{10}, and dioxygen

7.11. Write chemical equations for the following reactions:

(a) the heating of potassium hydrogen carbonate
(b) ethyne, $HC{\equiv}CH$, and dihydrogen
(c) the heating of lead(IV) oxide with hydrogen gas
(d) calcium hydride with water

7.12. Show that the combustion of methane

$$CH_4(g) + 2\,O_2(g) \rightarrow CO_2(g) + 2\,H_2O(l)$$

is indeed spontaneous by calculating the standard molar enthalpy, entropy, and free energy of combustion from enthalpy of formation and absolute entropy values. Use a data table.

7.13. Construct a theoretical enthalpy cycle (similar to Figure 7.2) for the formation of ammonia from its elements. Obtain bond energy information and the standard enthalpy of formation of ammonia from data tables. Compare your diagram with that in Figure 7.2 and comment on the differences.

7.14. What is the major difference between ionic and covalent hydrides in terms of physical properties?

7.15. Discuss the three types of covalent hydrides.

7.16. Calculate an approximate value for the bond energy for the bridging B–H bond in diborane if $\Delta H_f^\circ(B_2H_6(g)) = +36\ \text{kJ·mol}^{-1}$, the H–H bond energy is 432 kJ·mol^{-1}, the normal B–H bond energy is about 390 kJ·mol^{-1}, and the enthalpy of atomization of boron is 590 kJ·mol^{-1}. In comparison with the normal B–H bond energy, what does this suggest about the bond order? Is this result compatible with the bond order (per bond) deduced from the molecular orbital diagram (Figure 7.10)?

7.17. Calculate the standard enthalpy of formation of diboron trioxide, given

$$\Delta H_f^\circ(B_2H_6(g)) = +36\ \text{kJ·mol}^{-1}$$

$$\Delta H_f^\circ(H_2O(g)) = -242\ \text{kJ·mol}^{-1}$$

$$\Delta H^\circ_{\text{combustion}}(B_2H_6(g)) = -2165\ \text{kJ·mol}^{-1}$$

7.18. Construct a diagram similar to that of Figure 7.17 to explain why the hydroxide ion has a very high electrical conductivity.

7.19. What are the two properties of hydrogen that are crucial to the existence of life?

CHAPTER

8

Acids and Bases

$$pK_a + pK_b = pK_w$$

Brønsted-Lowry Theory

Acid-Base Equilibrium Constants

Brønsted-Lowry Acids

Brønsted-Lowry Bases

Acid-Base Reactions of Oxides

Lewis Theory

Pearson Hard-Soft Acid-Base Concepts

Superacids

The study of inorganic acids and bases overlaps the fields of inorganic, analytical, and physical chemistry. In this chapter, we focus on the structural and theoretical aspects of acid-base behavior. Most of the chapter is devoted to the Brønsted-Lowry interpretation of acid-base properties and a smaller proportion to the Lewis concepts.

When we study theories, we often forget that new theories are devised when the former theories no longer explain all the known facts. The theories of acids and bases are an excellent example of this progression of knowledge. A simple theory of acids and bases was first devised by Svante Arrhenius in 1884. According to the Arrhenius theory, acids contain hydrogen ions and bases contain hydroxide ions. However, there were two major flaws in this theory: the solvent problem and the salt problem.

The Arrhenius theory assumes that the solvent has no influence on acid-base properties. If hydrogen chloride is dissolved in water (to give hydrochloric acid), the solution conducts electricity, but if it is dissolved in a nonpolar solvent, such as benzene, C_6H_6, the solution does not conduct an electric current. The difference in properties of hydrogen chloride in the two solvents means that the solvent does affect the behavior of a solute.

The second problem of the Arrhenius theory concerns the behavior of salts. Salts should be neutral compounds, yet there are many salts that contradict this rule. For example, solutions of phosphate ion and carbonate ion are basic, whereas those of ammonium ion are slightly acidic and those of aluminum ion are very acidic. To add to the confusion, a solution of sodium dihydrogen phosphate is acidic but that of disodium hydrogen phosphate is basic.

To provide a more realistic model of acid-base behavior, Thomas Lowry in England and Johannes Brønsted in Denmark independently devised a theory that involved the solvent in the acid-base phenomenon. Even though there have been newer and more sophisticated theories of acid-base behavior, Brønsted-Lowry theory still provides the most convenient framework for understanding acids and bases.

Brønsted-Lowry Theory

According to Brønsted-Lowry theory, acids are proton donors and bases are proton acceptors. The language is somewhat misleading, for it is more accurately a competition for the proton between two chemical substances (with the base winning). The acid does not "donate" the proton any more willingly than you would "donate" your wallet or purse to a mugger. However, the central feature of the theory is the importance of the solvent, which self-ionizes by its own acid-base reaction. Thus water undergoes a slight *self-ionization* (also called *autoionization*) to give the hydronium ion and the hydroxide ion:

$$H_2O(l) + H_2O(l) \rightleftharpoons H_3O^+(aq) + OH^-(aq)$$

Like all equilibria, the self-ionization of water is temperature dependent. Thus the 1.0×10^{-14} value of the ion product, $[H_3O^+][OH^-]$, used in many acid-base calculations is only true at 25°C. In fact, the constant has a value of 1.2×10^{-15} at 0°C and 4.8×10^{-13} at 100°C. Furthermore, the value of the ion product is different when there are other solutes in the water; thus, at 25°C, it is about 1.5×10^{-14} for blood.

During autoionization, the water molecule that donates the hydrogen ion is an acid and the water molecule that accepts the hydrogen ion is a base. When we consider the reverse process, we see that the hydronium ion acts as a hydrogen ion donor (an acid) and the hydroxide ion is a hydrogen ion acceptor (a base). Two species that differ in formula by a hydrogen ion are called a conjugate acid-base pair. In this case, water is the conjugate base of the hydronium ion and water is the conjugate acid of the hydroxide ion. The ability of a substance to act either as an acid or as a base is called *amphiprotic behavior*.

The Brønsted-Lowry theory obviously depends on the actual existence of the hydronium ion. Brønsted and Lowry proposed their theory in 1923, and the first evidence for the existence of the hydronium ion came a year later when it was shown that crystals of perchloric acid monohydrate, $HClO_4{\cdot}H_2O$, had the same appearance as crystals of ammonium perchlorate, $NH_4^+{\cdot}ClO_4^-$. This similarity in appearance suggested a similarity in formula. The obvious conclusion was that the solid perchloric acid contained a hydronium ion that was analogous to the ammonium ion and that the actual structure of solid perchloric acid was $OH_3^+{\cdot}ClO_4^-$ (or more conventionally, $H_3O^+{\cdot}ClO_4^-$). Decades later, when crystal structures of acid hydrates were obtained, the proposal was shown to be correct. The shapes of the ammonium ion and hydronium ion are shown in Figure 8.1. Later still, evidence was obtained that the hydronium ion is present in solution. In fact, the hydronium ion is now known to be hydrogen bonded to three neighboring water molecules; thus it is more correctly written as $H_9O_4^+$. However, for simplicity, we usually ignore the three molecules of hydration.

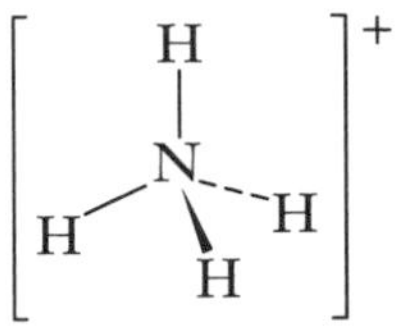

Acid or base behavior, then, usually depends on a chemical reaction with the solvent—in this case, water. This behavior can be illustrated for hydrofluoric acid:

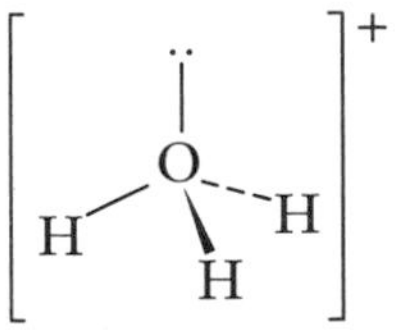

$$HF(aq) + H_2O(l) \rightleftharpoons H_3O^+(aq) + F^-(aq)$$

In this reaction, water functions as a base, and the fluoride ion is the conjugate base of hydrofluoric acid. Similarly, ammonia can be used as an example of a typical base, reacting with water (which, in this case, acts like an acid) to give the conjugate acid, the ammonium ion:

$$NH_3(aq) + H_2O(l) \rightleftharpoons NH_4^+(aq) + OH^-(aq)$$

(a)

(b)

Figure 8.1 (*a*) The ammonium ion and (*b*) the hydronium ion.

The reaction between a strong acid and a strong base can be represented simply as the reaction between the hydronium ion of the acid and the hydroxide of the base. This reaction is the reverse of the self-ionization of water. Because water self-ionizes to such a small extent, the reaction between hydronium ion and hydroxide ion goes essentially to completion:

$$H_3O^+(aq) + OH^-(aq) \rightarrow 2\ H_2O(l)$$

In aqueous solution, the strongest possible acid is the hydronium ion and the strongest possible base is the hydroxide ion. Thus, if a stronger base, such as the oxide ion, O^{2-}, is placed in water, it immediately reacts to give hydroxide ion:

$$O^{2-}(aq) + H_2O(l) \rightarrow 2\ OH^-(aq)$$

A stronger acid, such as perchloric acid, $HClO_4$, will similarly decompose to give hydronium ion:

$$HClO_4(aq) + H_2O(l) \rightarrow H_3O^+(aq) + ClO_4^-(aq)$$

Brønsted-Lowry acid-base chemistry need not be performed in water—any solvent system containing ionizable hydrogen will suffice. For example, liquid ammonia undergoes a self-ionization to produce ammonium ion (the acid) and amide ion, NH_2^- (the base):

$$NH_3(l) + NH_3(l) \rightleftharpoons NH_4^+(NH_3) + NH_2^-(NH_3)$$

Acid-base reactions can be performed in this solvent, although with a boiling point of –33°C, it is only used for specialized purposes. An example of an acid-base reaction in ammonia is that between ammonium chloride and sodium amide, $NaNH_2$, to give sodium chloride and ammonia.

$$NH_4Cl(NH_3) + NaNH_2(NH_3) \rightarrow NaCl(s) + 2\ NH_3(l)$$

This reaction parallels the acid-base reaction in water between hydrochloric acid and sodium hydroxide. The similarity can be seen more clearly if we write hydrochloric acid as "hydronium chloride," $H_3O^+{\cdot}Cl^-$:

$$H_3OCl(aq) + NaOH(aq) \rightarrow NaCl(aq) + 2\ H_2O(l)$$

Acid-Base Equilibrium Constants

The strength of an acid is a measure of how easily a hydrogen ion can be removed from the substance. The usual index of this is the position of equilibrium with water, in other words, the value of the equilibrium constant. In the case of an acid, the constant is identified as the acid ionization constant, K_a. For a general acid, HA, the equilibrium can be written as

$$HA(aq) + H_2O(l) \rightleftharpoons H_3O^+(aq) + A^-(aq)$$

The corresponding acid ionization expression would be

$$K_a = \frac{[H_3O^+][A^-]}{[HA]}$$

Because the values of acid ionization constants can involve very large or very small exponents, the most useful quantitative measure of acid strength is the pK_a, where

$$pK_a = -(\log_{10} K_a)$$

The stronger the acid, the more negative the pK_a. Typical values are shown in Table 8.1. Throughout this text, the equilibrium constants used are those measured at the standard temperature of 25°C.

The relevant constant for bases is identified as the base ionization constant, K_b. For a general base, A^-, the equilibrium can be written as

$$A^-(aq) + H_2O(l) \rightleftharpoons HA(aq) + OH^-(aq)$$

The corresponding base ionization expression would be

Table 8.1 Acid ionization constants of various inorganic acids

Acid	HA	A^-	K_a (at 25°C)	pK_a
Perchloric acid	$HClO_4$	ClO_4^-	10^{10}	–10
Hydrochloric acid	HCl	Cl^-	10^2	–2
Hydrofluoric acid	HF	F^-	3.5×10^{-4}	3.45
Ammonium ion	NH_4^+	NH_3	5.5×10^{-10}	9.26

Table 8.2 Base ionization constants of various inorganic bases

Base	A^-	HA	K_b (at 25°C)	pK_b
Phosphate ion	PO_4^{3-}	HPO_4^{2-}	4.7×10^{-2}	1.33
Ammonia	NH_3	NH_4^+	1.8×10^{-5}	4.74
Hydrazine	N_2H_4	$N_2H_5^+$	8.5×10^{-7}	6.07

$$K_b = \frac{[HA][OH^-]}{[A^-]}$$

Similarly, pK_b is defined as

$$pK_b = -(\log_{10} K_b)$$

Typical values are shown in Table 8.2.

There is a mathematical relationship between the acid ionization constant K_a of an acid and the base ionization constant K_b of its conjugate base. The product of the two terms equals the ion product constant for water, K_w:

$$K_w = [K_a] \times [K_b]$$

This can be expressed more conveniently in logarithmic form as

$$pK_w = pK_a + pK_b \qquad \text{where } pK_w = 14.00 \text{ at } 25°C$$

Thus the stronger the base, the weaker the conjugate acid. Conversely, a strong acid will have a weak conjugate base.

Brønsted-Lowry Acids

For an inorganic chemist, it is trends in the strengths of acids that are interesting. The common acids with positive K_a exponents (negative pK_a values), such as hydrochloric acid, nitric acid, sulfuric acid, and perchloric acid, are all regarded as strong acids. Those with negative K_a exponents (positive pK_a values), such as nitrous acid, hydrofluoric acid, and most of the other inorganic acids, are weak acids; that is, there are appreciable proportions of the molecular acid present in solution.

In water, all the strong acids seem equally strong, undergoing close to 100 percent ionization; that is, water acts as a leveling solvent (as we saw earlier in the chapter, the hydronium ion is the strongest possible acid in aqueous solution). To qualitatively identify the stronger acid, we dissolve the acids in a base weaker than water. A weaker base—often a pure weak acid—will function as a differentiating solvent for acids. This test can be illustrated by the equilibrium for perchloric acid in hydrofluoric acid:

$$HClO_4(HF) + HF(l) \rightleftharpoons H_2F^+(HF) + ClO_4^-(HF)$$

The weaker acid, hydrofluoric acid, functions in this case as a proton acceptor (base) for the stronger perchloric acid. However, because hydrofluoric acid is a weaker base than water, the equilibrium does not lie completely to the right, as does that for the reaction of perchloric acid with water.

The experiment can be repeated with the other strong acids; and the strongest acid is the acid that causes the equilibrium to lie furthest to the right. Of the common acids, this is perchloric acid.

Binary Acids

The most common binary acids are the hydrohalic acids, whose pK_a values are shown in Table 8.3. With a positive pK_a, hydrofluoric acid is clearly a much weaker acid than the other three. The others are all strong acids, and they ionize almost completely. Hydroiodic acid is the strongest; hence it has the most negative pK_a.

If we write the equation for the ionization process, using HX to represent any hydrohalic acid, we see that the main changes are the breaking of the H–X bond and the formation of an additional O–H bond as the water molecule becomes the hydronium ion:

$$HX(aq) + H_2O(l) \rightleftharpoons H_3O^+(aq) + X^-(aq)$$

The values of the various H–X bond energies are also given in Table 8.3, and the O–H bond energy is 459 $kJ{\cdot}mol^{-1}$. Because the tendency of any reaction is toward the formation of the stronger bond, it is apparent that ionization is not energetically favored for hydrofluoric acid but is favored for the other hydrohalic acids. In fact, the bond energy differences correlate remarkably well with the trends in acid strength. However, it should be kept in mind that this is a simplistic explanation; the comparative enthalpies and entropies of hydration of the ions and molecules must be included in any quantitative calculations of acid behavior.

Oxyacids

Oxyacids are ternary acids containing oxygen. For all the common inorganic acids, the ionizable hydrogen atoms are covalently bonded to oxygen atoms. For example, nitric acid, HNO_3, is more appropriately written as $HONO_2$.

Table 8.3 Correlation between the acid strengths of the hydrohalic acids and the energies of the hydrogen-halogen bonds

Acid	pK_a	Bond energy ($kJ{\cdot}mol^{-1}$)
$HF(aq)$	+3	565
$HCl(aq)$	−7	428
$HBr(aq)$	−9	362
$HI(aq)$	−10	295

In a series of oxyacids of one element, there is a correlation between acid strength and the number of oxygen atoms. Thus nitric acid is a strong acid ($pK_a = -1.4$), whereas nitrous acid, HONO, is a weak acid ($pK_a = +3.3$). Electronegativity arguments can be used to provide an explanation. Oxyacids are like the hydrohalic acids in that their acid strength depends on the weakness of the covalent bond between the ionizable hydrogen atom and its neighbor. For the oxyacids, the greater the number of highly electronegative oxygen atoms in the molecule, the more the electron density is pulled away from the hydrogen atom and the weaker the hydrogen-oxygen bond. As a result, an acid with numerous oxygen atoms is more easily ionized and hence stronger. This tendency is illustrated in Figure 8.2.

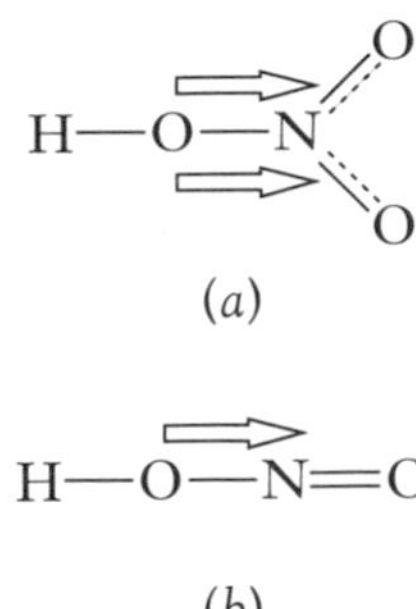

Figure 8.2 Nitric acid (*a*) is a stronger acid than nitrous acid (*b*) because the electron flow away from the H–O bond is greater in nitric acid.

The considerable dependence of acid strength on the number of oxygen atoms can actually be used in a semiquantitative fashion. If the formula of an oxyacid is written as $(HO)_nXO_m$, then when $m = 0$, the value of pK_a for the first ionization is about 8; for $m = 1$, it is about 2; for $m = 2$, it is about –1; and for $m = 3$, it is about –8.

Polyprotic Acids

There are several acids, including sulfuric acid and phosphoric acid, that have more than one ionizable hydrogen atom. The successive ionizations always proceed to a lesser and lesser extent. This trend can be illustrated by the two ionization steps for sulfuric acid:

$$H_2SO_4(aq) + H_2O(l) \rightarrow H_3O^+(aq) + HSO_4^-(aq) \qquad pK_a = -2$$

$$HSO_4^-(aq) + H_2O(l) \rightleftharpoons H_3O^+(aq) + SO_4^{2-}(aq) \qquad pK_a = +1.9$$

The first step proceeds essentially to completion; hence sulfuric acid is identified as a strong acid. The equilibrium for the second step lies slightly to the left at common acid concentrations. Thus, in an aqueous solution of sulfuric acid, the hydrogen sulfate ion, HSO_4^-, is one of the major species.

Yet when we add a dipositive (or tripositive) metal ion, it is the metal sulfate, not the hydrogen sulfate, that crystallizes. The reason for this lies in the comparative lattice energies. As we discussed in Chapter 5, the lattice energy depends to a significant extent on the ionic charges, that is, the electrostatic attraction between the ions. Hence the lattice energy of a crystal containing a 2+ cation and a 2– anion is greater than that of a crystal containing a 2+ cation and two 1– anions. For example, the lattice energy of magnesium fluoride, MgF_2, is 2.9 $MJ{\cdot}mol^{-1}$, whereas that of magnesium oxide, MgO, is 3.9 $MJ{\cdot}mol^{-1}$. Thus the formation of the solid metal sulfate will be about 1 $MJ{\cdot}mol^{-1}$ more exothermic than that of the solid metal hydrogen sulfate. Reaction is favored by a decrease in enthalpy; thus it is the sulfate rather than the hydrogen sulfate that is formed by dipositive and tripositive metal ions. As the sulfate ion is removed by precipitation, more is generated from the supply of hydrogen sulfate ions, in accord with the Le Châtelier principle:

$$HSO_4^-(aq) + H_2O(l) \rightarrow H_3O^+(aq) + SO_4^{2-}(aq)$$

It is only the monopositive ions, such as the alkali metal ions, that form stable crystalline compounds with the hydrogen sulfate ion. In these cases, the large, low-charge cations can form a stable lattice with the low-charge polyatomic anion.

Generally, the salts of acid anions behave like acids themselves. In fact, sodium hydrogen sulfate is used as a household cleaner because its solution is acidic and the reagent itself is an easily handled powder. Dissolving the compound in water yields the free hydrogen sulfate ion and, hence, hydronium ion:

$$HSO_4^-(aq) + H_2O(l) \rightleftharpoons H_3O^+(aq) + SO_4^{2-}(aq)$$

Acidic Metal Ions

Solutions of some metal ions are very acidic, the two most common examples being the aluminum and iron(III) ions. These are both small, highly charged cations, and they exist in solution as the hexahydrates: $[Al(OH_2)_6]^{3+}$ and $[Fe(OH_2)_6]^{3+}$. Their behavior in solution is illustrated for the iron(III) ion:

$$[Fe(OH_2)_6]^{3+}(aq) + H_2O(l) \rightleftharpoons H_3O^+(aq) + [Fe(OH_2)_5(OH)]^{2+}(aq) \quad pK_a = 3.3$$

Brønsted-Lowry Bases

After the hydroxide ion itself, ammonia is the next most important Brønsted-Lowry base. This compound reacts with water to produce the hydroxide ion, and it is the production of the hydroxide ion that makes ammonia solutions a useful glass cleaner (the hydroxide ion reacts with fat molecules to form water-soluble salts):

$$NH_3(aq) + H_2O(l) \rightleftharpoons NH_4^+(aq) + OH^-(aq)$$

However, there are many other common bases—the conjugate bases of weak acids. It is these anions that are present in many metal salts and yield basic solutions when the salts are dissolved in water. These conjugate bases include the phosphate ion and the sulfide ion, both strong bases; and the fluoride ion, a weak base:

$$PO_4^{3-}(aq) + H_2O(l) \rightleftharpoons HPO_4^{2-}(aq) + OH^-(aq) \quad pK_b = 1.35$$

$$S^{2-}(aq) + H_2O(l) \rightleftharpoons HS^-(aq) + OH^-(aq) \quad pK_b = 2.04$$

$$F^-(aq) + H_2O(l) \rightleftharpoons HF(aq) + OH^-(aq) \quad pK_b = 10.55$$

A second equilibrium step occurs to a significant extent for the sulfide ion, and this reaction is the cause of the hydrogen sulfide smell that can always be detected above sulfide ion solutions:

$$HS^-(aq) + H_2O(l) \rightleftharpoons H_2S(aq) + OH^-(aq) \quad pK_b = 6.96$$

The equilibrium for the phosphate ion also lies far to the right, so here, too, the second equilibrium step occurs to an appreciable extent:

$$HPO_4^{2-}(aq) + H_2O(l) \rightleftharpoons H_2PO_4^-(aq) + OH^-(aq) \qquad pK_b = 6.79$$

Thus the hydrogen phosphate ion acts as a base, not as an acid as might be expected. The third equilibrium reaction for the phosphate ion does not proceed to the right to any meaningful degree:

$$H_2PO_4^-(aq) + H_2O(l) \rightleftharpoons H_3PO_4(aq) + OH^-(aq) \qquad pK_b = 11.88$$

Sodium phosphate is effective as a household cleaner partly because of the very high pH of the solution from the first and second steps of the reaction with water and the resulting reaction of hydroxide ion with fat molecules.

It is important to note that the conjugate bases of strong acids do not react to any significant extent with water; that is, they are very weak bases (or in other words, the base ionization equilibrium lies far to the left). Therefore solutions of the nitrate and halide ions (except fluoride) are essentially pH neutral, and those of the sulfate ion are very close to neutral.

Acid-Base Reactions of Oxides

As we will see in Chapter 15, oxides can be classified as basic (most metal oxides), acidic (generally nonmetal oxides), amphoteric (the "weak" metal oxides with both acidic and basic properties), and neutral (a few nonmetal oxides). In this section we look specifically at the acid-base reactions involving oxides.

The most typical of the oxide reactions is the reaction of an acidic oxide with a basic oxide to form a salt. For example, sulfur dioxide (an acidic oxide) is a major waste product from metal smelters and other industrial processes. Traditionally, it was released into the atmosphere, but now a number of acid-base reactions have been devised to remove the acidic gas from the waste emissions. The simplest of these reacts with basic calcium oxide to give solid calcium sulfite:

$$CaO(s) + SO_2(g) \rightarrow CaSO_3(s)$$

The calcium sulfite produced in the neutralization reaction then oxidizes in the presence of air to calcium sulfate:

$$2\ CaSO_3(s) + O_2(g) \rightarrow 2\ CaSO_4(s)$$

In the industrial process, cheap powdered limestone (calcium carbonate) is mixed with the very hot sulfur dioxide–containing gas. Thus the calcium oxide is formed just prior to reaction with the sulfur dioxide:

$$CaCO_3(s) \xrightarrow{\Delta} CaO(s) + CO_2(g)$$

Acidic oxides often react with bases; for example, carbon dioxide reacts with sodium hydroxide solution to produce sodium carbonate:

$$CO_2(g) + 2\ NaOH(aq) \rightarrow Na_2CO_3(aq) + H_2O(l)$$

Conversely, many basic oxides react with acids; for example, zinc oxide reacts with nitric acid to form zinc nitrate:

$$ZnO(s) + 2\ HNO_3(aq) \rightarrow Zn(NO_3)_2(aq) + H_2O(l)$$

To determine an order of acidity among acidic oxides, the free energy of reaction of the various acidic oxides with the same base can be compared. The larger the free energy change, the stronger the acidic oxide. Let us use calcium oxide as the common basic oxide:

$$CaO(s) + CO_2(g) \rightarrow CaCO_3(s) \qquad \Delta G^\circ = -134\ kJ{\cdot}mol^{-1}$$

$$CaO(s) + N_2O_5(g) \rightarrow Ca(NO_3)_2(s) \qquad \Delta G^\circ = -272\ kJ{\cdot}mol^{-1}$$

$$CaO(s) + SO_3(g) \rightarrow CaSO_4(s) \qquad \Delta G^\circ = -347\ kJ{\cdot}mol^{-1}$$

Thus sulfur trioxide is the most acidic of the three acidic oxides. We can do an analogous test to find the most basic of several basic oxides, by comparing the free energy of reaction of various basic oxides with an acid (in this case, water):

$$Na_2O(s) + H_2O(l) \rightarrow 2\ NaOH(s) \qquad \Delta G^\circ = -142\ kJ{\cdot}mol^{-1}$$

$$CaO(s) + H_2O(l) \rightarrow Ca(OH)_2(s) \qquad \Delta G^\circ = -59\ kJ{\cdot}mol^{-1}$$

$$Al_2O_3(s) + 3\ H_2O(l) \rightarrow 2\ Al(OH)_3(s) \qquad \Delta G^\circ = -2\ kJ{\cdot}mol^{-1}$$

These calculations suggest that sodium oxide is the most basic of the three basic oxides and aluminum oxide, the least. Again it is important to recall that the thermodynamic values predict the energetic feasibility rather than the rate of reaction (the kinetic feasibility).

Traditionally, geochemists classified silicate rocks on an acid-base scale. Such rocks contain metal ions, silicon, and oxygen, and we can think of them as a combination of metal oxides (basic) and silicon dioxide (acidic). We consider a rock such as granite, with more than 66 percent silicon dioxide, to be acidic; those with 52–66 percent SiO_2, intermediate; those with 45–52 percent SiO_2, such as basalt, basic; and those with less than 45 percent, ultrabasic. For example, the mineral olivine, a common component of ultrabasic rocks, has a chemical composition $MgFeSiO_4$, which can be thought of as a combination of oxides, $MgO{\cdot}FeO{\cdot}SiO_2$. The percentage by mass of silicon dioxide is 35 percent; hence this mineral is classified as ultrabasic. In general, the acidic silicate rocks tend to be light in color (granite is pale gray), whereas the basic rocks are dark (basalt is black).

Lewis Theory

The Brønsted-Lowry theory worked well for the study of reactions in hydrogen-containing solvents, and it is still used in such circumstances. However, it soon became apparent that acid-base concepts could be applied to reactions where hydrogen transfer did not occur. For these reactions, the Lewis definition is much more appropriate. G. N. Lewis defined an acid as an electron pair acceptor and a base as an electron pair donor. The classic example of a Lewis acid-base reaction is that between boron trifluoride and ammonia. Using electron-dot diagrams, we can see that boron trifluoride is the Lewis acid and ammonia is the Lewis base (Figure 8.3).

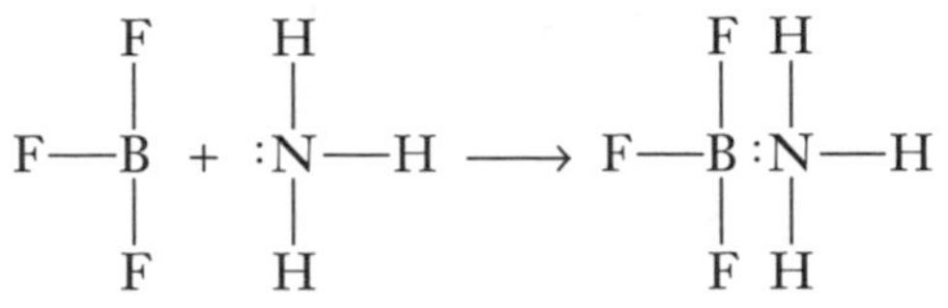

```
     F      H                 F H
     |      |                 | |
F—B  +  :N—H  ⟶  F—B:N—H
     |      |                 | |
     F      H                 F H
```

Figure 8.3 The reaction of boron trifluoride with ammonia.

The combination of a Lewis acid and a Lewis base can be interpreted in terms of molecular orbital theory. We assume that the energy level of the lowest energy unoccupied molecular orbital (LUMO) of the Lewis acid is similar to that of the highest energy occupied molecular orbital (HOMO) of the Lewis base. A combination molecular orbital in which the electron pair from the Lewis base is reduced in energy can then form (Figure 8.4).

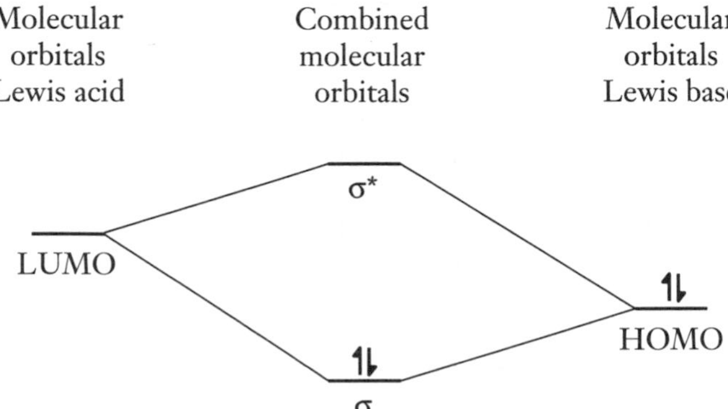

Figure 8.4 Overlap of an empty molecular orbital (LUMO) of the Lewis acid with a full molecular orbital (HOMO) of the Lewis base.

There are other Lewis acid-base reactions that more closely parallel Brønsted-Lowry reactions, and some of these apply to nonaqueous solvents. Any pure liquid that has a measurable electrical conductivity must contain ions, one such liquid being bromine trifluoride, BrF_3. It has been shown that the conductivity of this compound results from an autoionization to give the BrF_2^+ cation and the BrF_4^- anion:

$$2\ BrF_3(l) \rightleftharpoons BrF_2^+(\mathit{BrF_3}) + BrF_4^-(\mathit{BrF_3})$$

Compounds containing either the cation or the anion are known, for example, $(BrF_2)(SbF_6)$ and $Ag(BrF_4)$, the former being an acid and the latter a base in the BrF_3 solvent system. With bromine trifluoride as the solvent, they can be reacted together in a Lewis neutralization reaction:

$$(BrF_2)(SbF_6)(\mathit{BrF_3}) + Ag(BrF_4)(\mathit{BrF_3}) \rightarrow Ag(SbF_6)(\mathit{BrF_3}) + 2\ BrF_3(l)$$

Pearson Hard-Soft Acid-Base Concepts

R. G. Pearson proposed that Lewis acids and bases could be categorized as either "hard" or "soft." That is, a soft acid or base is one in which the valence

Table 8.4 Classification of Lewis acids and bases

Class	Acids	Bases
Hard	H^+, Li^+, Na^+, K^+ Mg^{2+}, Ca^{2+}	H_2O, NH_3 F^-, Cl^-, OH^-, NO_3^-, ClO_4^- CO_3^{2-}, O^{2-}, SO_4^{2-} PO_4^{3-}
Borderline	Fe^{2+}, Co^{2+}, Ni^{2+} Cu^{2+}, Zn^{2+}, Pb^{2+}	NO_2^-, Br^- SO_3^{2-}
Soft	Cu^+, Ag^+ Hg_2^{2+}, Hg^{2+}, Cd^{2+}	CO CN^-, I^- S^{2-}

electrons are easily polarized or removed, whereas a hard acid or base is one that does not have valence electrons or whose valence electrons have a low polarizability. Table 8.4 shows a classification of various common ions. Generally, the metals on the left side of the periodic table are hard. This category corresponds to the low-electronegativity (high-electropositivity) metals. The borderline cases are mostly found among the transition metals; and the main group metals on the right side of the periodic table provide the soft acids. The two categories of acids also display a strong charge dependence: For a particular metal, the low-charge ions are "soft" and the high-charge ions are "hard."

The key use of the Pearson classification is for predicting the reactions of various elements: Hard acids prefer to react with hard bases and soft acids prefer to react with soft bases. The hard-soft categorization also can be used to predict bonding preferences and to fashion a synthesis of abnormal oxidation states in metals. For example, although chromium commonly forms the Cr^{3+} ion, it is possible to use the oxide ion (a hard base) to form the CrO_4^{2-} ion in which chromium has an oxidation state of +6 (a hard acid). Conversely, the very low chromium oxidation state of –2 (a soft acid) can be produced by using carbon monoxide (a soft base) to synthesize the $Cr(CO)_5^{2-}$ ion.

A second example of the application of the hard-soft concept involves a polyatomic anion, the thiocyanate ion, $(N{=}C{=}S)^-$. The nitrogen end of the ion is "hard," whereas the sulfur end is "soft." The ion forms covalent compounds with metals in the borderline and soft categories. With the soft metals, it is the sulfur end that is bonded to the metal ion; whereas with the borderline metals, it is generally the nitrogen that is bonded to the metal ion.

The hard-soft concept can also be applied in geochemistry. In particular, it can be used to explain the presence of the various minerals found on Earth. For example, the common ore of aluminum is aluminum oxide, Al_2O_3 (hard acid, hard base), whereas the common ore of mercury is mercury(II) sulfide, HgS (soft acid, soft base).

It is sometimes possible to predict whether or not a reaction will occur by applying the premise of like combinations being favored over unlike

Superacids

A superacid can be defined as an acid that is stronger than 100 percent sulfuric acid. In fact, chemists have synthesized superacids that are from 10^7 to 10^{19} times stronger than sulfuric acid. There are four categories of superacids: Brønsted, Lewis, conjugate Brønsted-Lewis, and solid superacids. A common Brønsted superacid is perchloric acid. When perchloric acid is mixed with pure sulfuric acid, the sulfuric acid actually acts like a base:

$$\mathrm{HClO_4}(H_2SO_4) + \mathrm{H_2SO_4}(l) \rightleftharpoons \mathrm{H_3SO_4^+}(H_2SO_4) + \mathrm{ClO_4^-}(H_2SO_4)$$

Fluorosulfuric acid, HSO_3F, is the strongest Brønsted superacid; it is more than 1000 times more acidic than sulfuric acid. This superacid, which is an ideal solvent because it is liquid from −89°C to +164°C, has the structure shown below.

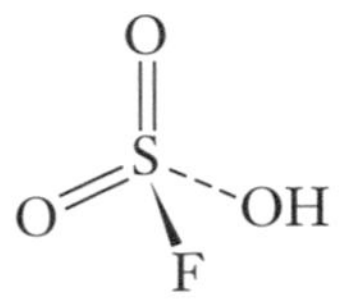

A Brønsted-Lewis superacid is a mixture of a powerful Lewis acid and a strong Brønsted-Lowry acid. The most potent combination is a 10 percent solution of antimony pentafluoride, SbF_5, in fluorosulfuric acid. The addition of antimony pentafluoride increases the acidity of the fluorosulfuric acid several thousand times. The reaction between the two acids is very complex, but the super-hydrogen-ion donor present in the mixture is the $H_2SO_3F^+$ ion. This acid mixture will react with many substances, such as hydrocarbons, that do not react with normal acids. For example, propene, C_3H_6, reacts with this ion to give the propyl cation:

$$\mathrm{C_3H_6}(HSO_3F) + \mathrm{H_2SO_3F^+}(HSO_3F) \rightarrow \mathrm{C_3H_7^+}(HSO_3F) + \mathrm{HSO_3F}(l)$$

The solution of antimony pentafluoride in fluorosulfuric acid is commonly called "Magic Acid." The name originated in the Case Western Reserve University laboratory of George Olah, a pioneer in the field of superacids. A researcher working with Olah put a small piece of Christmas candle left over from a lab party into the acid and found that it dissolved rapidly. He studied the resulting solution and found that the long-chain hydrocarbon molecules of the paraffin wax had added hydrogen ions, and the resulting cations had rearranged themselves to form branched-chain molecules. This unexpected finding suggested the name "Magic Acid," and it is now a registered trade name for the compound. This family of superacids is used in the petroleum industry for the conversion of the less important straight-chain hydrocarbons to the more valuable branched-chain molecules, which are needed to produce high-octane gasoline.

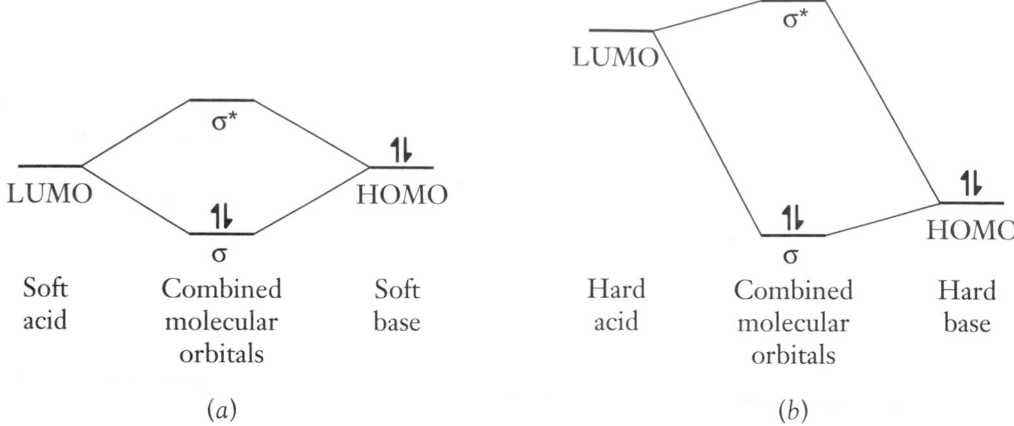

Figure 8.5 Representation of comparative orbital energies for (*a*) the soft acid–soft base combination and (*b*) the hard acid–hard base combination.

combinations. This approach is illustrated in the following high-temperature gas-phase reaction:

$$\underset{\text{soft-hard}}{HgF_2(g)} + \underset{\text{hard-soft}}{BeI_2(g)} \rightarrow \underset{\text{hard-hard}}{BeF_2(g)} + \underset{\text{soft-soft}}{HgI_2(g)}$$

The Pearson approach can be related to the earlier discussions of ionic and covalent bonding (Chapters 3 and 5). The hard acid–hard base combination is really the pairing of a low-electronegativity cation with a high-electronegativity anion, properties that result in ionic behavior. Conversely, the soft acids are the metals that lie close to the nonmetal border and have very high electronegativities. These metallic ions will form covalent bonds with the soft base ions such as sulfide.

We can express this theory in molecular orbital terms. The combination of soft acid with soft base is analogous to the situation in which the LUMO of the Lewis acid has about the same energy as the HOMO of the Lewis base. The combination of orbitals results in a greatly reduced energy for the electron pair coming from the base (Figure 8.5a), thus yielding a strong covalent bond. For the hard acid and hard base, there is a great mismatch of LUMO and HOMO energies, so there is little energy advantage to be gained from orbital interaction (Figure 8.5b). Hence covalent bonding will not occur to any significant extent, and an ionic interaction results instead.

Exercises

8.1. Define the following terms: (a) conjugate acid-base pairs; (b) self-ionization; (c) amphiprotic.

8.2. Define the following terms: (a) acid ionization constant; (b) leveling solvent; (c) polyprotic.

8.3. Write equilibrium equations to represent the reaction of chloramine, $ClNH_2$, a base, with water.

8.4. Write equilibrium equations to represent the reaction of fluorosulfuric acid, HSO_3F, with water.

8.5. Pure sulfuric acid can be used as a solvent. Write an equilibrium to represent the self-ionization reaction.

8.6. When liquid ammonia is used as a solvent, what is (a) the strongest acid? (b) the strongest base?

8.7. Hydrogen fluoride is a strong acid when dissolved in liquid ammonia. Write a chemical equation to represent this process.

8.8. Hydrogen fluoride behaves like a base when dissolved in pure sulfuric acid. Write a chemical equation to represent the acid-base equilibrium and identify the conjugate acid-base pairs.

8.9. Identify the conjugate acid-base pairs in the following equilibrium:

$$HSeO_4^-(aq) + H_2O(l) \rightleftharpoons H_3O^+(aq) + SeO_4^{2-}(aq)$$

8.10. Identify the conjugate acid-base pairs in the following equilibrium:

$$HSeO_4^-(aq) + H_2O(l) \rightleftharpoons OH^-(aq) + H_2SeO_4(aq)$$

8.11. Which will be the stronger acid, sulfurous acid, $H_2SO_3(aq)$, or sulfuric acid, $H_2SO_4(aq)$? Use electronegativity arguments to explain your reasoning.

8.12. Hydrogen selenide, H_2Se, is a stronger acid than hydrogen sulfide. Use bond strength arguments to explain why.

8.13. Addition of copper(II) ion to a hydrogen phosphate, HPO_4^{2-}, solution results in precipitation of copper(II) phosphate. Use chemical equations to suggest an explanation.

8.14. The hydrated zinc ion, $Zn(OH_2)_6^{2+}$, forms an acidic solution. Use a chemical equation to suggest an explanation.

8.15. A solution of the cyanide ion, CN^-, is a strong base. Write a chemical equation to explain this. What can you deduce about the properties of hydrocyanic acid, HCN?

8.16. The weak base, hydrazine, H_2NNH_2, can form a diprotic acid, $^+H_3NNH_3^+$. Write chemical equations to depict the two equilibrium steps. When hydrazine is dissolved in water, which of the three hydrazine species will be present in the lowest concentration?

8.17. When dissolved in water, which of the following salts will give neutral, acidic, or basic solutions: (a) potassium fluoride; (b) ammonium chloride? Explain your reasoning.

8.18. When dissolved in water, which of the following salts will give neutral, acidic, or basic solutions: (a) aluminum nitrate; (b) sodium iodide? Explain your reasoning.

8.19. When two sodium salts, NaX and NaY, are dissolved in water to give solutions of equal concentration, the pH values obtained are 7.3 and 10.9, respectively. Which is the stronger acid, HX or HY? Explain your reasoning.

8.20. The pK_b values of the bases A^- and B^- are 3.5 and 6.2, respectively. Which is the stronger acid, HA or HB? Explain your reasoning.

8.21. From free energy of formation values, determine the free energy of reaction of magnesium oxide with water to give magnesium hydroxide and hence deduce whether magnesium oxide is more or less basic than calcium oxide.

8.22. From free energy of formation values, determine the free energy of reaction of silicon dioxide with calcium oxide to give calcium silicate, Ca_2SiO_4, and hence deduce whether magnesium oxide is more or less acidic than carbon dioxide.

8.23. Nitrosyl chloride, NOCl, can be used as a nonaqueous solvent. It undergoes the following self-ionization:

$$NOCl(l) \rightleftharpoons NO^+(NOCl) + Cl^-(NOCl)$$

Identify which of the ions is a Lewis acid and which is a Lewis base. Also, write a balanced equation for the reaction between $(NO)(AlCl_4)$ and $[(CH_3)_4N]Cl$.

8.24. Will either of the following high-temperature gas-phase reactions be feasible? Give your reasoning in each case.

(a) $CuBr_2(g) + 2\ NaF(g) \rightarrow CuF_2(g) + 2\ NaI(g)$

(b) $TiI_4(g) + 2\ TiF_2(g) \rightarrow TiF_4(g) + 2\ TiI_2(g)$

CHAPTER

9 Oxidation and Reduction

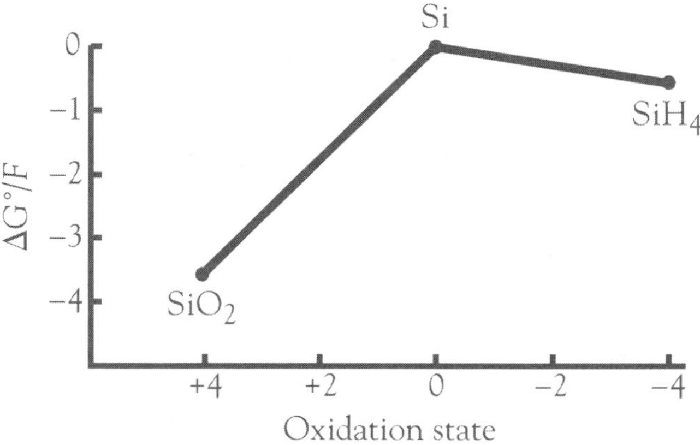

A large proportion of chemical reactions involves changes in oxidation state. In this chapter, we show how oxidation numbers can be determined and then we examine redox reactions. The redox properties of a particular oxidation state of an element can be displayed graphically, giving us information about the thermodynamic stability of that compound or ion.

In the history of chemistry, one of the most vehement disputes concerned the nature of oxidation. The story really begins in 1718 when Georg Stahl, a German chemist, proposed that metals form from oxides that are heated with charcoal (carbon) because the metals absorb a substance that he named "phlogiston." Conversely, he proposed that the heating of a metal in air to form its oxide caused the release of phlogiston to the atmosphere.

Fifty-four years later, the French chemist Louis-Bernard Guyton de Morveau performed careful experiments showing that, during combustion, metals *increase* in weight. However, the existence of phlogiston was so well established among chemists that he interpreted the results as meaning that phlogiston had a negative weight. It was his colleague, Antoine Lavoisier, who was willing to throw out the phlogiston concept and propose that combustion was due to the addition of oxygen to the metal (oxidation) and that the formation of a metal from an oxide corresponded to the loss of oxygen (reduction).

The editors of the French science journal were phlogistonists, so they would not publish this proposal. Thus Lavoisier, with his new convert, de Morveau, and others, had to establish their own journal to publish the "new chemistry." It was the overthrow of the phlogiston theory that caused chemists to realize that elements were the fundamental substances in chemistry—and modern chemistry was born.

Redox Terminology

Many inorganic reactions are redox reactions; and, like so many fields of chemistry, the study of oxidation and reduction has its own vocabulary and definitions. Traditionally, oxidation and reduction were each defined in three different ways, as shown in Table 9.1. In modern chemistry, we use more general definitions of oxidation and reduction:

Oxidation: Increase in oxidation number

Reduction: Decrease in oxidation number

Table 9.1 Traditional definitions of oxidation and reduction

Oxidation	Reduction
Gain of oxygen atoms	Loss of oxygen atoms
Loss of hydrogen atoms	Gain of hydrogen atoms
Loss of electrons	Gain of electrons

Oxidation Number Rules

Of course, now we have to define an oxidation number (also called oxidation state). Oxidation numbers are simply theoretical values used to simplify electron bookkeeping. We assign these values to the common elements on the basis of a simple set of rules:

1. The oxidation number, N_{ox}, of an atom as an element is zero.

2. The oxidation number of a monatomic ion is the same as its ion charge.

3. The algebraic sum of the oxidation numbers in a neutral polyatomic compound is zero; in a polyatomic ion, it is equal to the ion's charge.

4. In combinations of elements, the more electronegative element has its characteristic negative oxidation number (for example, −3 for nitrogen, −2 for oxygen, −1 for chlorine), and the more electropositive element has a positive oxidation number.

5. Hydrogen usually has an oxidation number of +1 (except with more electropositive elements, when it is −1).

For example, to find the oxidation number of sulfur in sulfuric acid, H_2SO_4, we can use rule 3 to write

$$2[N_{ox}(H)] + [N_{ox}(S)] + 4[N_{ox}(O)] = 0$$

Because oxygen usually has an oxidation number of −2 (rule 4) and hydrogen, +1 (rule 5), we write

$$2(+1) + [N_{ox}(S)] + 4(-2) = 0$$

Hence $[N_{ox}(S)] = +6$.

Now let us deduce the oxidation number of iodine in the ion ICl_4^-. For this, we can use rule 3 to write

$$[N_{ox}(I)] + 4[N_{ox}(Cl)] = -1$$

Chlorine is more electronegative than iodine, so chlorine will have the conventional negative oxidation number of −1 (rule 4). Thus

$$[N_{ox}(I)] + 4(-1) = -1$$

Hence $[N_{ox}(I)] = +3$.

Determination of Oxidation Numbers from Electronegativities

Memorizing rules does not necessarily enable us to understand the concept of oxidation number. Furthermore, there are numerous polyatomic ions and molecules for which there is no obvious way of applying the "rules." We can use several ways to derive oxidation numbers, but the following method is particularly useful for cases in which there are two atoms of the same element that appear in a molecule or ion but have different chemical environments. This method assigns each atom its own oxidation number, not just a mean value.

To assign oxidation numbers to covalently bonded atoms, we draw the electron-dot formula of the molecule and refer to the electronegativity values of the elements involved (Figure 9.1). Although the electrons in a polar covalent bond are unequally shared, for the purpose of assigning oxidation numbers, we assume that they are completely "owned" by the more electronegative atom. Then we compare how many outer (valence) electrons an atom "possesses" in its molecule or ion with the number it has as a free monatomic element. The difference—number of valence electrons possessed by free atom minus number of valence electrons "possessed" by molecular or ionic atom—is the oxidation number.

The hydrogen chloride molecule will serve as an example. Figure 9.1 shows that chlorine has a higher electronegativity than hydrogen, so we assign the bonding electrons to chlorine. A chlorine atom in hydrogen chloride will "have" one more electron in its outer set of electrons than a neutral chlorine atom has. Hence we assign it an oxidation number of 7 minus 8, or

H 2.2					He –
B 2.0	C 2.5	N 3.0	O 3.4	F 4.0	Ne –
	Si 1.9	P 2.2	S 2.6	Cl 3.2	Ar –
	Ge 2.0	As 2.2	Se 2.6	Br 3.0	Kr 3.0
			Te 2.1	I 2.7	Xe 2.6
				At 2.2	Rn –

Figure 9.1 Pauling electronegativity values of various nonmetals and semimetals.

–1. The hydrogen atom has "lost" its one electron; thus it has an oxidation number of 1 minus 0, or +1. This assignment is illustrated as

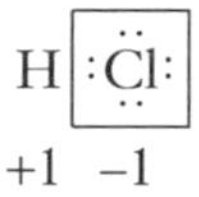

H :Cl:

+1 –1

When we construct a similar electron-dot diagram for water, we see that each hydrogen atom has an oxidation number of 1 minus 0, or +1, and the oxygen atom, 6 minus 8, or –2, as rules 4 and 5 state.

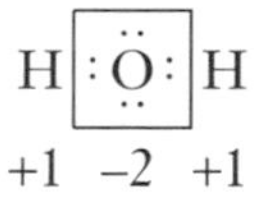

H :O: H

+1 –2 +1

In hydrogen peroxide, oxygen has an "abnormal" oxidation state. This is easy to comprehend if we realize that when pairs of atoms of the same electronegativity are bonded together, we must assume that the bonding electron pair is split between them. In this case, each oxygen atom has an oxidation number of 6 minus 7, or –1. The hydrogen atoms are still each +1.

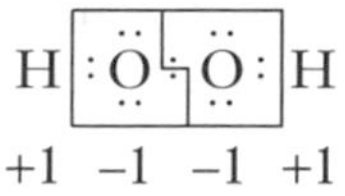

H :O:O: H

+1 –1 –1 +1

This method can be applied to molecules containing three (or more) different elements. Hydrogen cyanide, HCN, illustrates the process. Nitrogen is more electronegative than carbon, so it "possesses" the electrons participating in the C–N bond. And carbon is more electronegative than hydrogen, so it "possesses" the electrons of the H–C bond.

H | :C | :::N:

+1 +2 −3

Polyatomic ions can be treated in the same way as neutral molecules are. We can use the simple electron-dot structure of the sulfate ion to illustrate this. (However, in Chapter 15 we will see that the bonding in the sulfate ion is actually more complex.) Following the same rules of assigning bonding electrons to the more electronegative atom, we assign an oxidation number of −2 for each oxygen. But sulfur has six outer electrons in the neutral atom and none in this structure. Hence, according to the rules, the sulfur atom is assigned an oxidation number of 6 minus 0, or +6. Note that this same oxidation number was assigned earlier to the sulfur atom in sulfuric acid.

2−

:Ö:

:Ö: S :Ö:

:Ö:

As already mentioned, two atoms of the same element can actually have different oxidation numbers within the same molecule. The classic example of this situation is the thiosulfate ion, $S_2O_3^{2-}$, which has sulfur atoms that are in different environments. Each oxygen atom has an oxidation number of −2. But, according to the rule stated earlier, the two equally electronegative sulfur atoms divide the two electrons participating in the S–S bond. Hence, the central sulfur atom has an oxidation number of +5 (6 minus 1), and the other sulfur atom has an oxidation number of −1 (6 minus 7). These assignments correlate with this ion's chemical reactions, in which the two sulfur atoms behave differently.

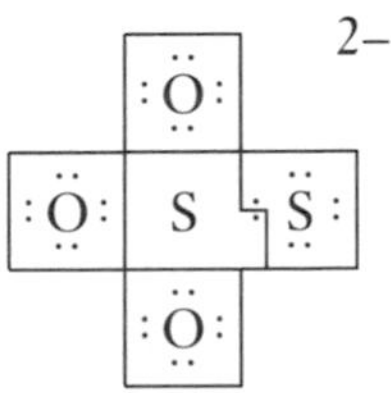

Rule 1 states that the oxidation number of any element is 0. How do we arrive at this value? In a molecule consisting of two identical atoms, such as difluorine, the electrons in the covalent bond will always be shared equally. Thus we divide the shared electrons between the fluorine atoms. Each fluorine atom had seven valence electrons as a free atom and now it still has seven; hence 7 minus 7 is 0.

:F̤ F̤:

0 0

The Difference Between Oxidation Number and Formal Charge

In Chapter 3 we mentioned the concept of formal charge as a means of identifying feasible electron-dot structures for covalent molecules. To calculate formal charge, we divided the bonding electrons equally between the constituent atoms. The favored structures were generally those with the lowest formal charges. For example, the electron-dot diagram for carbon monoxide is shown below with the electrons allocated according to the rules for formal charge.

:C : : :O:

Formal charge ⊖ ⊕

However, to determine oxidation numbers, which can quite often have large numerical values, we assign the bonding electrons to the atom with higher electronegativity. According to this method, the atoms in carbon monoxide are assigned electrons as shown below.

:C | : : :O:

Oxidation number +2 −2

Periodic Variations of Oxidation Numbers

There are patterns to the oxidation numbers of the main group elements; in fact, they are one of the most systematic periodic trends. This pattern can be seen in Figure 9.2, which shows the oxidation numbers of the most common compounds of the first 25 main group elements (*d*-block elements have been omitted). The most obvious trend is the stepwise increase in the positive oxidation number as we progress from left to right across the periods. An atom's

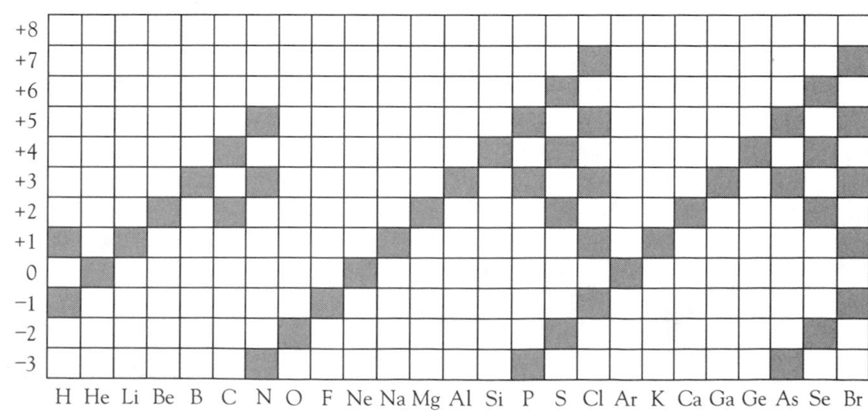

Figure 9.2 Common oxidation numbers in the compounds of the first 25 main group elements.

Table 9.2 Oxidation number of chlorine in common oxyanions

Ion	Oxidation number
ClO^-	+1
ClO_2^-	+3
ClO_3^-	+5
ClO_4^-	+7

maximum positive oxidation number is equal to the number of electrons in its outer orbital set. For example, aluminum, with an electron configuration of $[Ne]3s^23p^1$ has an oxidation number of +3. Electrons in inner orbital sets do not enter into the calculation for main group elements. Hence the maximum oxidation number for bromine, which has an electron configuration of $[Ar]4s^23d^{10}4p^5$, is +7, a value corresponding to the sum of the electrons in the 4*s* and 4*p* orbitals.

Many of the nonmetals and semimetals exhibit more than one oxidation number. For example, in its different compounds, nitrogen assumes every oxidation number between −3 and +5. The common oxidation states of nonmetals, however, tend to decrease in units of two. This pattern can be seen in the oxidation numbers chlorine has in the various oxyanions it forms (Table 9.2).

Redox Equations

In a redox reaction, one substance is oxidized and another is reduced. This process is sometimes easy to see; for example, when a rod of copper metal is placed in a silver nitrate solution, shiny crystals of silver metal are formed on the copper surface and the solution turns blue. In this case, the oxidation number of copper has increased from 0 to +2 and that of silver has decreased from +1 to 0:

$$Cu(s) + 2\ Ag^+(aq) \rightarrow Cu^{2+}(aq) + Ag(s)$$

We can think of the process as two separate half-reactions, the loss of electrons by the copper metal and the gain of electrons by the silver ions:

$$Cu(s) \rightarrow Cu^{2+}(aq) + 2\ e^-$$

$$2\ Ag^+(aq) + 2\ e^- \rightarrow 2\ Ag(s)$$

A more complex example is the reaction between hydrogen sulfide gas and a solution of iron(III) ions, which yields the products solid sulfur, iron(II) ions, and hydrogen (or hydronium) ions:

$$H_2S(g) + 2\ Fe^{3+}(aq) \rightarrow S(s) + 2\ Fe^{2+}(aq) + 2\ H^+(aq)$$

In this case, it is necessary to comb through the formulas and calculate the oxidation number of each element. It is easy to see that the iron has been reduced from +3 to +2, but it requires a quick calculation to see that the oxidation state of sulfur has changed from –2 to 0. Thus we can write the half-reactions as

$$H_2S(g) \rightarrow S(s) + 2\ H^+(aq) + 2\ e^-$$

$$2\ Fe^{3+}(aq) + 2\ e^- \rightarrow 2\ Fe^{2+}(aq)$$

We have just seen that a redox equation can be divided into oxidation and reduction half-reactions. It is equally useful to perform the reverse process, that is, to construct balanced redox reactions from the oxidation and reduction components. For example, a solution of purple permanganate ion oxidizes an iron(II) ion solution to iron(III) ion in acid solution, itself being reduced to colorless manganese(II) ion. We can write the following skeletal (unbalanced) equation:

$$MnO_4^-(aq) + Fe^{2+}(aq) \rightarrow Mn^{2+}(aq) + Fe^{3+}(aq)$$

The first step is to identify the two half-reactions:

$$Fe^{2+}(aq) \rightarrow Fe^{3+}(aq)$$

$$MnO_4^-(aq) \rightarrow Mn^{2+}(aq)$$

We can balance the iron half-reaction first because it is very simple, requiring just one electron:

$$Fe^{2+}(aq) \rightarrow Fe^{3+}(aq) + e^-$$

But in the reduction equation, we have oxygen atoms on the left, but none on the right. We can remedy this by adding the appropriate number of water molecules to the side lacking oxygen atoms:

$$MnO_4^-(aq) \rightarrow Mn^{2+}(aq) + 4\ H_2O(l)$$

This addition has balanced the oxygen atoms, but it has introduced hydrogen atoms in the process. To balance these, we add hydrogen ions to the left-hand side:

$$MnO_4^-(aq) + 8\ H^+(aq) \rightarrow Mn^{2+}(aq) + 4\ H_2O(l)$$

Finally, we balance the charges by adding electrons as needed:

$$MnO_4^-(aq) + 8\ H^+(aq) + 5\ e^- \rightarrow Mn^{2+}(aq) + 4\ H_2O(l)$$

Before adding the two half-reactions, the number of electrons required for the reduction must match the number of electrons produced during the oxidation. In this case, we achieve this balance by multiplying the iron oxidation half-reaction by 5:

$$5\ Fe^{2+}(aq) \rightarrow 5\ Fe^{3+}(aq) + 5\ e^-$$

The final balanced reaction will be

$$5\ Fe^{2+}(aq) + MnO_4^-(aq) + 8\ H^+(aq) \rightarrow 5\ Fe^{3+}(aq) + Mn^{2+}(aq) + 4\ H_2O(l)$$

Now consider the disproportionation reaction of dichlorine to chloride ion and chlorate ion in basic solution:

$$Cl_2(aq) \rightarrow Cl^-(aq) + ClO_3^-(aq)$$

A disproportionation reaction occurs when some ions (or molecules) are oxidized while others of the same species are reduced. In the case of dichlorine, some of the chlorine atoms are oxidized, changing their oxidation number from 0 to +5; the remainder are reduced, changing from 0 to –1. As before, we can construct the two half-reactions:

$$Cl_2(aq) \rightarrow Cl^-(aq)$$

$$Cl_2(aq) \rightarrow ClO_3^-(aq)$$

Now consider the reduction half-reaction first. The initial step is to balance the number of atoms of chlorine:

$$Cl_2(aq) \rightarrow 2\ Cl^-(aq)$$

Then we can balance for charge:

$$Cl_2(aq) + 2\ e^- \rightarrow 2\ Cl^-(aq)$$

For the oxidation half-reaction, the number of chlorine atoms also has to be balanced:

$$Cl_2(aq) \rightarrow 2\ ClO_3^-(aq)$$

Then, because the reaction takes place in basic solution, we add twice the number of moles of hydroxide ion on one side as there are oxygen atoms on the other:

$$Cl_2(aq) + 12\ OH^-(l) \rightarrow 2\ ClO_3^-(aq)$$

We balance the added hydrogen with water molecules:

$$Cl_2(aq) + 12\ OH^-(l) \rightarrow 2\ ClO_3^-(aq) + 6\ H_2O(l)$$

Next, we balance the equation with respect to charge by adding electrons:

$$Cl_2(aq) + 12\ OH^-(l) \rightarrow 2\ ClO_3^-(aq) + 6\ H_2O(l) + 10\ e^-$$

The reduction half-reaction must be multiplied by 5 to balance the electrons:

$$5\ Cl_2(aq) + 10\ e^- \rightarrow 10\ Cl^-(aq)$$

The sum of the two half-reactions will be

$$6\ Cl_2(aq) + 12\ OH^-(aq) \rightarrow 10\ Cl^-(aq) + 2\ ClO_3^-(aq) + 6\ H_2O(l)$$

Finally, the coefficients of the equation can be divided through by 2:

$$3\ Cl_2(aq) + 6\ OH^-(aq) \rightarrow 5\ Cl^-(aq) + ClO_3^-(aq) + 3\ H_2O(l)$$

Ammonium Nitrate

Ammonium nitrate contains nitrogen in two different oxidation states, –3 in the ammonium cation and +5 in the nitrate anion. The substance is mostly used as a nitrogen-rich fertilizer, but it is also a very potent explosive. Heating it leads to the exothermic formation of dinitrogen oxide, N_2O, in which the nitrogen has an oxidation number of +1:

$$NH_4NO_3(s) \xrightarrow{\Delta} 2\ H_2O(g) + N_2O(g)$$

Ammonium nitrate is hygroscopic. That is, it absorbs moisture in humid conditions and forms a sticky mass that hardens when the humidity drops. This water uptake happened on a large scale in 1921 at a fertilizer factory in Germany. Lacking sufficient chemical knowledge, the workers decided to use dynamite to break up 4500 tonnes of solidified ammonium nitrate–sulfate mixture. The resulting explosion of the ammonium nitrate destroyed the whole factory and killed 561 people.

To prevent the hygroscopic problem, the Tennessee Valley Authority devised a way of coating the granules with wax. This practice solved the stickiness, but it had one disadvantage. Any organic substance can be oxidized to carbon dioxide and water vapor. The ammonium nitrate–wax combination proved to be an even better explosive than pure ammonium nitrate, because in the combustion of these reactants nitrogen is reduced to dinitrogen, oxidation number 0. The "leftover" oxygen then forms more water and carbon dioxide by "combining" with the wax, a hydrocarbon:

$$NH_4NO_3(s) \xrightarrow{\Delta} 2\ H_2O(g) + N_2(g) + \text{"O"}$$

$$C_nH_{2n+2}(s) + (3n + 1)\ \text{"O"} \rightarrow n\ CO_2(g) + (n + 1)\ H_2O(g)$$

The accidental fire on a ship carrying these wax-coated pellets killed at least 500 people in Texas City, Texas, in 1947. Clay is now used to safely coat the ammonium nitrate pellets, and bulk quantities of the compound are stored and shipped in tightly sealed containers.

About 1955, the North American blasting explosives industry recognized the potential of the ammonium nitrate–hydrocarbon mixture. As a result, a mixture of ammonium nitrate with fuel oil has become very popular with the industry. It is actually quite safe, because the ammonium nitrate and fuel oil can be stored separately until use, and a detonator is then employed to initiate the explosion. It is this mixture that was probably used in the Oklahoma City, Oklahoma, bombing in 1995.

Quantitative Aspects of Half-Reactions

The relative oxidizing or reducing power of a half-reaction can be determined from the half-cell potential, which is the potential of the half-reaction relative to the potential of a half-reaction in which hydrogen ion (1 mol·L^{-1})

is reduced to hydrogen gas (100 kPa pressure on a black platinum surface). This reference half-reaction is assigned a standard potential, $E°$, of zero:

$$2\ H^+(aq) + 2\ e^- \rightarrow H_2(g) \qquad E° = 0.00\ V$$

For a redox reaction to be spontaneous, the sum of its half-cell reduction potentials must be positive. For example, consider the reaction of copper metal with silver ion, which we discussed earlier. The values of the half-cell reduction potentials are

$$Cu^{2+}(aq) + 2\ e^- \rightarrow Cu(s) \qquad E° = +0.34\ V$$

$$Ag^+(aq) + e^- \rightarrow Ag(s) \qquad E° = +0.80\ V$$

Adding the silver ion reduction potential to the copper metal oxidation potential

$$2\ Ag^+(aq) + 2\ e^- \rightarrow 2\ Ag(s) \qquad E° = +0.80\ V$$

$$Cu(s) \rightarrow Cu^{2+}(aq) + 2\ e^- \qquad E° = -0.34\ V$$

gives a positive cell potential:

$$2\ Ag^+(aq) + Cu(s) \rightarrow 2\ Ag(s) + Cu^{2+}(aq) \qquad E° = +0.46\ V$$

The more positive the half-cell reduction potential, the stronger the oxidizing power of the species. For example, difluorine is an extremely strong oxidizing agent (or electron acceptor):

$$\tfrac{1}{2}\ F_2(g) + e^- \rightarrow F^-(aq) \qquad E° = +2.80\ V$$

Conversely, the lithium ion has a very negative reduction potential:

$$Li^+(aq) + e^- \rightarrow Li(s) \qquad E° = -3.04\ V$$

For lithium, the reverse half-reaction is spontaneous; hence lithium metal is a very strong reducing agent (or electron provider):

$$Li(s) \rightarrow Li^+(aq) + e^- \qquad E° = +3.04\ V$$

However, it must always be kept in mind that half-cell potentials are concentration dependent. Thus it is possible for a reaction to be spontaneous under certain conditions but not under others. The variation of potential with concentration is given by the Nernst equation:

$$E = E° - \frac{RT}{nF} \ln \frac{[\text{products}]}{[\text{reactants}]}$$

where R is the ideal gas constant (8.31 $V{\cdot}C{\cdot}mol^{-1}{\cdot}K^{-1}$), T is the temperature in kelvins, n is the moles of transferred electrons according to the redox equation, F is the Faraday constant ($9.65 \times 10^4\ C{\cdot}mol^{-1}$), and $E°$ is the potential under standard conditions of 1 $mol{\cdot}L^{-1}$ for species in solution and 100 kPa pressure for gases.

To see the effects of nonstandard conditions, consider the permanganate ion to manganese(II) ion half-cell. This half-cell is represented by the half-reaction we balanced earlier:

$$MnO_4^-(aq) + 8\ H^+(aq) + 5\ e^- \rightarrow Mn^{2+}(aq) + 4\ H_2O(l) \qquad E^\circ = +1.70\ V$$

The corresponding Nernst equation will be

$$E = +1.70\ V - \frac{RT}{5F} \ln \frac{[Mn^{2+}]}{[MnO_4^-][H^+]^8}$$

Suppose the pH is increased to 4.00 (that is, $[H^+]$ is reduced to 1.0×10^{-4} $mol{\cdot}L^{-1}$), but the concentrations of permanganate ion and manganese(II) ion are kept at 1.0 $mol{\cdot}L^{-1}$. Under the new conditions (first, solving for $RT/5F$), the half-cell potential becomes

$$E = +1.70\ V - 5.13 \times 10^{-3}\ V \ln \frac{(1.00)}{(1.00)(1.0 \times 10^{-4})^8}$$

$$E = +1.70\ V - 5.13 \times 10^{-3}\ V \ln(1.0 \times 10^{32})$$

$$E = +1.70\ V - 0.38\ V = +1.32\ V$$

Thus permanganate ion is a significantly weaker oxidizing agent in less acid solutions. Notice, however, that the effect is substantial only because in the Nernst equation the hydrogen ion concentration is raised to the eighth power; as a result, the potential is exceptionally sensitive to pH. Fortunately, for the qualitative comparisons of half-cell potentials used in this text, standard values will suffice.

Electrode Potentials as Thermodynamic Functions

As we have just seen in the equation for the silver ion–copper metal reaction, electrode potentials are not altered when coefficients of equations are changed. The potential is the force driving the reaction, and it is localized either at the surface of an electrode or at a point where two chemical species come in contact. Hence the potential does not depend on stoichiometry. Potentials are simply a measure of the free energy of the process. The relationship between free energy and potential is

$$\Delta G^\circ = -nFE^\circ$$

where ΔG° is the standard free energy change, n is the moles of electrons, F is the Faraday constant, and E° is the standard electrode potential. The Faraday constant is usually expressed as $9.65 \times 10^4\ C{\cdot}mol^{-1}$, but for use in this particular formula, it is best written in units of joules: $9.65 \times 10^4\ J{\cdot}V^{-1}{\cdot}mol^{-1}$. For the calculations in this section, however, it is even more convenient to express the free energy change as the product of moles of electrons and half-cell potentials. When we do that we do not need to evaluate the Faraday constant.

To illustrate this point, let us repeat the previous calculation of the copper-silver reaction, using free energies instead of just standard potentials:

$$2\ Ag^+(aq) + 2\ e^- \rightarrow 2\ Ag(s) \qquad \Delta G^\circ = -2(F)(+0.80) = -1.60F$$

$$Cu(s) \rightarrow Cu^{2+}(aq) + 2\ e^- \qquad \Delta G^\circ = -2(F)(-0.34) = +0.68F$$

The free energy change for the process, then, is $(-1.60F + 0.68F)$, or $-0.92F$. Converting this value back to a standard potential gives

$$E^\circ = -\Delta G^\circ/nF = -(-0.92F)/2F = +0.46\text{ V}$$

or the same value obtained by simply adding the standard potentials.

But suppose we want to combine two half-cell potentials to derive the value for an unlisted half-cell potential; then the shortcut of using standard electrode potentials does not work. Note that we are adding half-reactions to get another half-reaction, not a balanced redox reaction. The number of electrons in the two reduction half-reactions will not balance. Consequently, we must work with free energies. As an example, we can determine the half-cell potential for the reduction of iron(III) ion to iron metal,

$$Fe^{3+}(aq) + 3\ e^- \rightarrow Fe(s)$$

given the values for the reduction of iron(III) ion to iron(II) ion and from iron(II) ion to iron metal:

$$Fe^{3+}(aq) + e^- \rightarrow Fe^{2+}(aq) \qquad E^\circ = +0.77\text{ V}$$

$$Fe^{2+}(aq) + 2\ e^- \rightarrow Fe(s) \qquad E^\circ = -0.44\text{ V}$$

First, we calculate the free energy change for each half-reaction:

$$Fe^{3+}(aq) + e^- \rightarrow Fe^{2+}(aq) \qquad \Delta G^\circ = -1(F)(+0.77) = -0.77F$$

$$Fe^{2+}(aq) + 2\ e^- \rightarrow Fe(s) \qquad \Delta G^\circ = -2(F)(-0.44) = +0.88F$$

Adding the two equations results in the "cancellation" of the Fe^{2+} species. Hence the free energy change for

$$Fe^{3+}(aq) + 3\ e^- \rightarrow Fe(s)$$

will be $(-0.77F + 0.88F)$, or $+0.11F$. Converting this ΔG° value back to potential for the reduction of ion (III) to iron metal gives

$$E^\circ = -\Delta G^\circ/nF = -(+0.11F)/3F = -0.04\text{ V}$$

Latimer (Reduction Potential) Diagrams

It is easier to interpret data when they are displayed in the form of a diagram. The standard reduction potentials for a related set of species can be displayed in a reduction potential diagram, or what is sometimes called a Latimer diagram. The various iron oxidation states in acid solution are shown here in such a diagram:

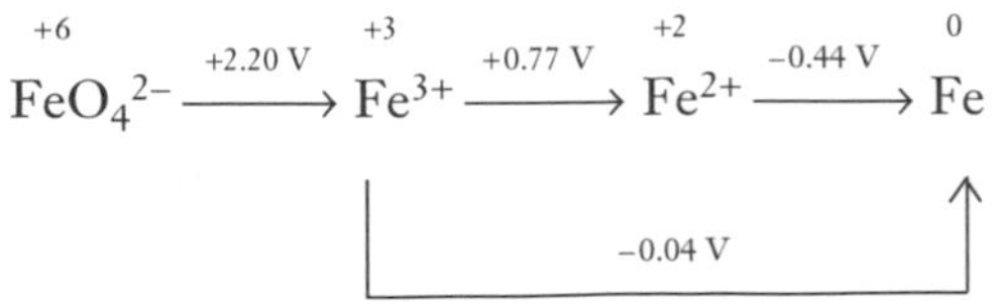

$$\overset{+6}{FeO_4^{2-}} \xrightarrow{+2.20\text{ V}} \overset{+3}{Fe^{3+}} \xrightarrow{+0.77\text{ V}} \overset{+2}{Fe^{2+}} \xrightarrow{-0.44\text{ V}} \overset{0}{Fe}$$

$$Fe^{3+} \xrightarrow{-0.04\text{ V}} Fe$$

The diagram includes the three common oxidation states of iron (+3, +2, 0) and the uncommon oxidation state of +6. The number between each pair of species is the standard reduction potential for the reduction half-reaction involving those species. Notice that although the species are indicated, to use the information we have to write the corresponding complete half-reaction. For the simple ions, writing the half-reaction is very easy. For example, for the reduction of iron(III) ion to iron(II) ion, we simply write

$$Fe^{3+}(aq) + e^- \rightarrow Fe^{2+}(aq) \qquad E° = +0.77\ V$$

However, for the reduction of ferrate ion, FeO_4^{2-}, we have to balance the oxygen with water, then the hydrogen in the added water with hydrogen ion, and finally the charge with electrons:

$$FeO_4^{2-}(aq) + 8\ H^+(aq) + 3\ e^- \rightarrow Fe^{3+}(aq) + 4\ H_2O(l) \qquad E° = +2.20\ V$$

Latimer diagrams display the redox information about a series of oxidation states in a very compact form. More than that, they enable us to predict the redox behavior of the species. For example, the high positive value between the ferrate ion and the iron(III) ion indicates that the ferrate ion is a strong oxidizing agent (that is, it is very easily reduced). A negative number indicates that the species to the right is a reducing agent. In fact, iron metal can be used as a reducing agent, itself being oxidized to the iron(II) ion.

Let us look at another example of a reduction potential diagram, that for oxygen in acid solution:

$$\overset{0}{O_2} \xrightarrow{+0.68\ V} \overset{-1}{H_2O_2} \xrightarrow{+1.78\ V} \overset{-2}{H_2O}$$

$$O_2 \xrightarrow{+1.23\ V} H_2O$$

With a reduction potential of +1.78 V, hydrogen peroxide is a strong oxidizing agent with respect to water. For example, hydrogen peroxide will oxidize iron(II) ion to iron(III) ion:

$$H_2O_2(aq) + 2\ H^+(aq) + 2\ e^- \rightarrow 2\ H_2O(l) \qquad E° = +1.78\ V$$

$$Fe^{2+}(aq) \rightarrow Fe^{3+}(aq) + e^- \qquad E° = -0.77\ V$$

The diagram tells us something else about hydrogen peroxide. The sum of the potentials for the reduction and oxidation of hydrogen peroxide is positive (+1.78 V – 0.68 V). This value indicates that hydrogen peroxide will disproportionate:

$$H_2O_2(aq) + 2\ H^+(aq) + 2\ e^- \rightarrow 2\ H_2O(l) \qquad E° = +1.78\ V$$

$$H_2O_2(aq) \rightarrow O_2(g) + 2\ H^+(aq) + 2\ e^- \qquad E° = -0.68\ V$$

Summing the two half-equations gives the overall equation:

$$2\ H_2O_2(aq) \rightarrow 2\ H_2O(l) + O_2(g) \qquad E° = +1.10\ V$$

Even though the disproportionation is thermodynamically spontaneous, it is kinetically very slow. The decomposition happens rapidly, however, in the presence of a catalyst such as iodide ion or many transition metal ions.

In all the examples used so far in this section, the reactions occur in acid solution. The values are sometimes quite different in basic solution, because of the presence of different chemical species at high pH. For example, as the diagram at the beginning of this section shows, iron metal is oxidized in acid solution to the soluble iron(II) cation:

$$Fe(s) \rightarrow Fe^{2+}(aq) + 2\ e^-$$

However, in basic solution, the iron(II) ion immediately reacts with the hydroxide ion present in high concentration to give insoluble iron(II) hydroxide:

$$Fe(s) + 2\ OH^-(aq) \rightarrow Fe(OH)_2(s) + 2\ e^-$$

Thus the Latimer diagram for iron in basic solution (shown below) contains several different species from the diagram under acid conditions and, as a result, the potentials are different:

$$\overset{+6}{FeO_4^{2-}} \xrightarrow{+0.9\ V} \overset{+3}{Fe(OH)_3} \xrightarrow{-0.56\ V} \overset{+2}{Fe(OH)_2} \xrightarrow{-0.89\ V} \overset{0}{Fe}$$

We see that in basic solution iron(II) hydroxide is easily oxidized to iron(III) hydroxide (+0.56 V), and the ferrate ion is now a very weak oxidizing agent (+0.9 V in basic solution, 2.20 V in acid solution).

Although Latimer diagrams are useful for identifying reduction potentials for specific redox steps, they can become very complex. For example, a diagram for the five species of manganese has 10 potentials relating the various pairs of the five species. It is tedious to sort out the information that is stored in such a complex diagram. For this reason, it is more useful to display the oxidation states and their comparative energies as a two-dimensional graph. This is the topic of the next section.

Frost (Oxidation State) Diagrams

It is preferable to display the information about the numerous oxidation states of an element as an oxidation state diagram, or a Frost diagram, as it is sometimes called. Such a diagram enables us to extract information about the properties of different oxidation states visually, without the need for calculations. A Frost diagram shows the relative free energy (rather than potential) on the vertical axis and the oxidation state on the horizontal axis. Note that we denote energy as $-nE°$; thus energy values are usually plotted in units of volts times moles of electrons for that redox step (V·mol e⁻). We obtain the same value by dividing the free energy by the Faraday constant, $\Delta G°/F$. For consistency, the element in oxidation state 0 is considered to have zero free energy. Lines connect species of adjacent oxidation states.

From the Latimer diagram for oxygen shown here, we can construct a Frost diagram for oxygen species in acid solution (Figure 9.3):

$$\overset{0}{O_2} \xrightarrow{+0.68\ V} \overset{-1}{H_2O_2} \xrightarrow{+1.78\ V} \overset{-2}{H_2O}$$

$$O_2 \xrightarrow{+1.23\ V} H_2O$$

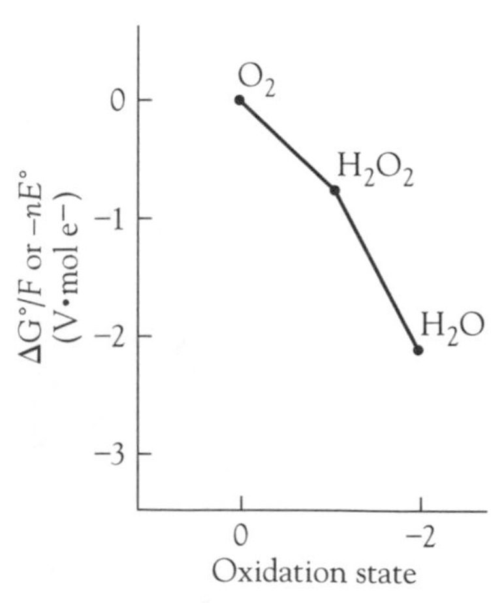

Figure 9.3 Frost diagram for oxygen in acid solution.

The first point will simply be 0, 0 for dioxygen because its free energy is taken to be 0 when its oxidation state is 0. The point for hydrogen peroxide will then be –1, –0.68 because the oxidation state for oxygen in hydrogen peroxide is –1 and its free energy is –1 times the product of the moles of electrons (1) and the half-cell reduction potential (+0.68 V). Finally, the point for water will be at –2, –2.46 because the oxygen has an oxidation state of –2 and the free energy of the oxygen in water will be $-(1 \times 1.78)$ units below the hydrogen peroxide point. This diagram enables us to obtain a visual image of the redox chemistry of oxygen in acid solution. Water, at the lowest point on the plot, must be the most thermodynamically stable. Hydrogen peroxide, on a convex curve, will disproportionate.

All the features of a Frost diagram can be appreciated by studying the redox chemistry of manganese (Figure 9.4). From this diagram, we can draw the following conclusions:

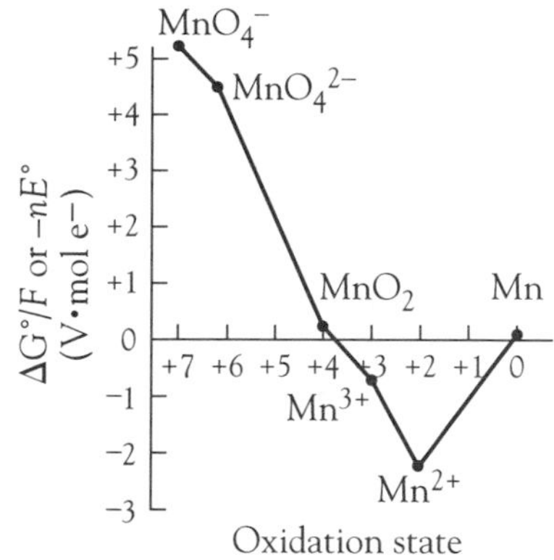

Figure 9.4 Frost diagram for manganese in acid solution.

1. More thermodynamically stable states will be found lower in the diagram. Thus manganese(II) is the most stable (from a redox perspective) of all the manganese species.

2. A species on a convex curve (such as the manganate ion, MnO_4^{2-}, and the manganese(III) ion) will tend to disproportionate.

3. A species on a concave curve (such as manganese(IV) oxide, MnO_2) will not disproportionate.

4. A species that is high and on the left of the plot (such as the permanganate ion, MnO_4^-) will be strongly oxidizing.

5. A species that is high and on the right of the plot will be strongly reducing. Thus manganese metal is moderately reducing.

However, interpretation of Frost diagrams has caveats. First, the diagram represents the comparative free energy for standard conditions, that is, a solution of concentration 1 $mol \cdot L^{-1}$ at pH 0 (a hydrogen ion concentration of 1 $mol \cdot L^{-1}$). If the conditions are changed, then the energy will be different, and the relative stabilities might be different.

As pH changes, the potential of any half-reaction that includes the hydrogen ion also changes. But, even more important, often the actual species involved will change. For example, the aqueous manganese(II) ion does not exist at high pH values. Under these conditions, insoluble manganese(II) hydroxide, $Mn(OH)_2$, is formed. It is this compound, not Mn^{2+}, that appears on the Frost diagram for manganese(II) in basic solution.

Finally, we must emphasize that the Frost diagrams are thermodynamic functions and do not contain information about the rate of decomposition of a thermodynamically unstable species. Potassium permanganate, $KMnO_4$, is a good example. Even though the reduction of permanganate ion to a more stable lower oxidation state of manganese(II) ion is favored, it is kinetically slow (except in the presence of a catalyst). Thus we can still work with permanganate ion solutions.

Ellingham Diagrams and the Extraction of Metals

Frost diagrams are very useful for the study of reactions in aqueous solution. However, one of the most important types of redox reactions is usually

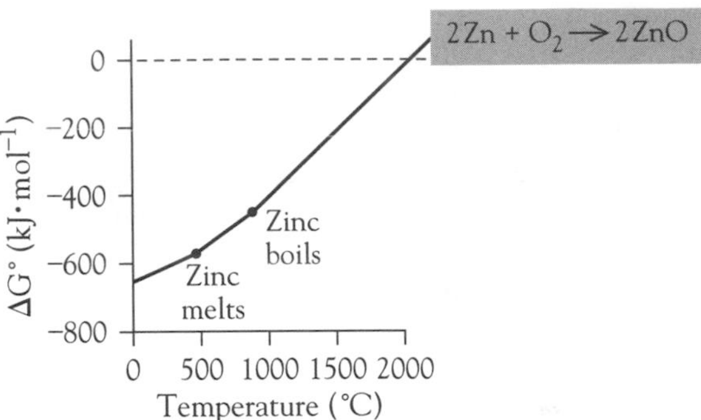

Figure 9.5 Free energy of oxidation of zinc as a function of temperature.

performed in the solid, liquid, and gaseous phases—the reduction of metal compounds to the pure metal.

For most metallic elements, the oxides are more thermodynamically stable than the metals themselves over the range of normal working temperatures. For example, zinc metal will spontaneously (although slowly) oxidize to zinc oxide at room temperature:

$$2\ Zn(s) + O_2(g) \rightarrow 2\ ZnO(s) \qquad \Delta G^\circ(298\ K) = -636\ kJ \cdot mol^{-1}$$

However, recalling the formula given in Chapter 6,

$$\Delta G^\circ = \Delta H^\circ - T\Delta S^\circ$$

we can identify what factors actually lead to the spontaneity of the reaction. Because the number of moles of gas changes from 1 on the left to 0 on the right, the entropy change of this reaction must be negative. Hence the driving force for this reaction has to be the enthalpy factor. And, in fact, when we check the thermodynamic tables, we see that the enthalpy of formation of zinc oxide is very negative. That is, the oxidation is highly exothermic.

The entropy term, $T\Delta S^\circ$, involves the Kelvin temperature, so increasing the temperature will cause ΔG° to become less and less negative (the actual values of both ΔH° and ΔS° change slightly with temperature as well, but we will ignore this complication here). Finally, at a sufficiently high temperature, ΔG° will become 0; and above that temperature, it will have a positive value. In other words, the converse process, the reduction of zinc oxide to zinc metal, will become spontaneous. Figure 9.5, a plot of free energy per mole of dioxygen against Celsius temperature, displays this information.

The slope of the curve steepens slightly above the melting point of zinc (lower dot on the line) and even more above the boiling point of zinc (upper dot on the line). Above the melting point of zinc, two moles of solid zinc oxide are being produced from one mole of gas and two moles of liquid; above the boiling point of zinc, two moles of solid zinc oxide form from three moles of gas (one of dioxygen and two of zinc). Hence the entropy decrease at these temperatures and above will be greater. Even then, the temperature will have to be extremely high (about 2000°C) before the $T\Delta S^\circ$ term exceeds the ΔH° term. In fact, the requisite temperature is so high that this reaction does not represent a realistic means of obtaining zinc metal from the oxide. Furthermore, the dioxygen and gaseous zinc would have to be separated before cooling, or the reverse reaction would occur, and we would end up with zinc oxide again.

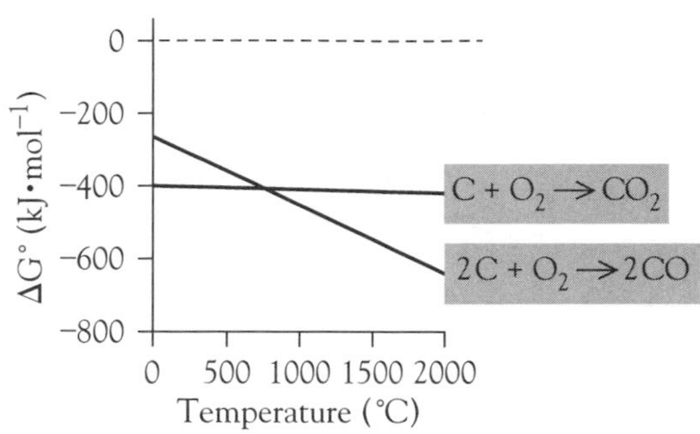

Figure 9.6 Free energy of oxidation of carbon as a function of temperature.

There is a way around the problem. It is possible to couple the reduction reaction, which has a positive free energy, with an oxidation reaction that has a greater negative free energy to give a net negative free energy value for the combined reaction. The most useful oxidation reaction with a negative free energy is that of carbon. The coupled reaction has the right thermodynamic characteristics, as we will see, and carbon is a very inexpensive industrial reagent. The temperature dependence of this reaction is illustrated in Figure 9.6.

Up to 710°C, the oxidation of carbon to carbon dioxide is thermodynamically preferred:

$$C(s) + O_2(g) \rightarrow CO_2(g)$$

This slope of the line for the free energy change of the reaction is very close to zero because there is one mole of gas on each side of the equation. The line representing the free energy change during the oxidation to produce carbon monoxide, however, has a steep negative slope because the reaction produces two moles of gas for every mole consumed:

$$2\ C(s) + O_2(g) \rightarrow 2\ CO(g)$$

Thus the production of carbon monoxide becomes thermodynamically preferred above 710°C, and, because both reactions are kinetically fast, the latter is the actual reaction observed above this temperature.

Figure 9.7 shows the two plots of Figures 9.5 and 9.6 superimposed. We can see that the lines cross at about 900°C. At this temperature, the oxidation

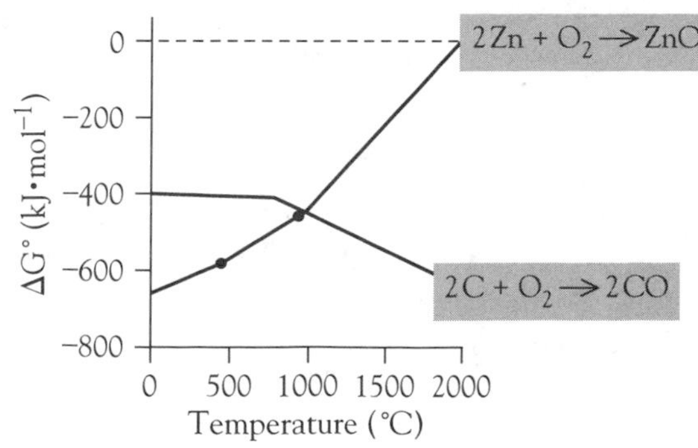

Figure 9.7 Superimposed plots of free energy of oxidation of carbon and zinc as a function of temperature.

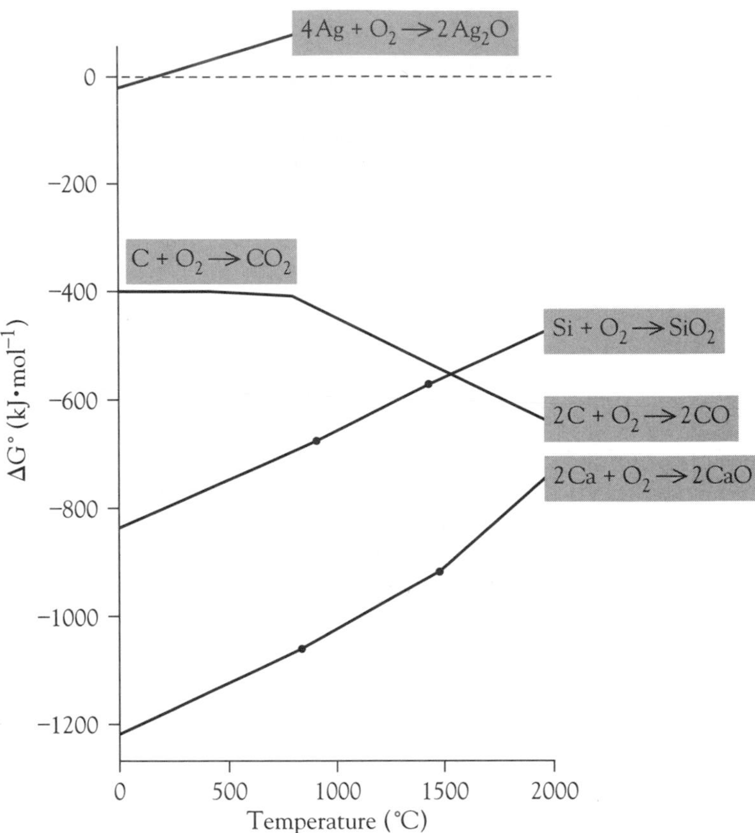

Figure 9.8 Ellingham diagram for silver, carbon, silicon, and calcium. Points indicate melting and boiling points of elements.

of carbon becomes more negative than the reduction of zinc oxide is positive. Thus the oxidation of carbon can cause the reduction of zinc oxide above this temperature:

$$ZnO(s) + C(s) \rightarrow Zn(g) + CO(g) \qquad T > 1000°C$$

Note that all the thermodynamic calculations relate to standard conditions of pressure. In an industrial smelter, conditions are far from this; consequently, the temperature that we have calculated is only an approximate guide to the actual minimum temperature for the reduction process.

It was the chemist H. G. T. Ellingham who first recognized how useful plots of free energy as a function of temperature were for investigating the conditions under which redox reactions were feasible. As a result, these plots are usually referred to as Ellingham diagrams. Figure 9.8 shows an Ellingham diagram for oxides of calcium, carbon, silicon, and silver.

We can see that

1. After the silver plot crosses the $\Delta G° = 0$ line, the formation of silver(I) oxide is no longer spontaneous (about 300°C). Above this temperature, the reverse reaction, the decomposition of silver(I) oxide to silver metal is spontaneous ($\Delta G°$ negative).

2. The silicon plot crosses that of carbon at about 1500°. Above this temperature, the free energy of formation of silicon dioxide is less than that of carbon monoxide. Hence the sum of the free energy of decomposition of silicon dioxide coupled with the free energy of formation of carbon monoxide will result in a net negative value. In other words, silicon dioxide can be reduced to silicon by using carbon as a reductant above this temperature (below this temperature, the reverse reaction pair is spontaneous).

3. The calcium plot does not cross the carbon plot at temperatures that are feasible in conventional smelters. Hence thermochemical methods are not practical for extracting calcium metal. In fact, an electrolytic process is used to produce most calcium.

There are Ellingham diagrams for the reduction of most oxides, sulfides, and chlorides. As a result, the feasibility of smelting processes can be identified simply by looking at the appropriate Ellingham plot rather than by laboratory testing; this alternative represents a significant saving in time and money.

Pourbaix Diagrams

Earlier in this chapter, we saw how a Frost diagram could be used to compare the thermodynamic stabilities of different oxidation states of an element. Frost diagrams can be constructed for both acid (pH = 0) and basic (pH = 14) conditions. It would be useful to be able to identify the thermodynamically stable species at any particular permutation of half-cell potential, *E*, and pH. A French chemist, M. Pourbaix, devised such a plot; hence they are usually named after him. Figure 9.9 shows a Pourbaix diagram for various sulfur systems.

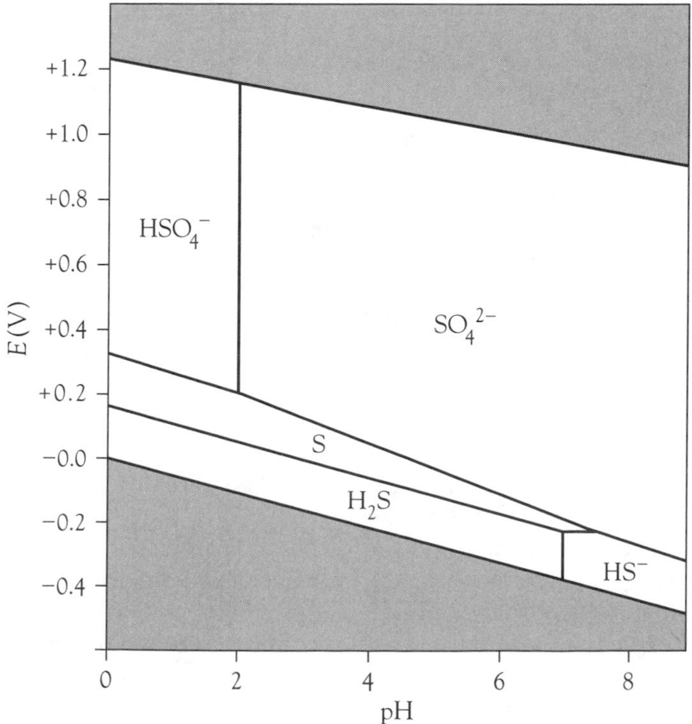

Figure 9.9 Pourbaix diagram showing the thermodynamically stable sulfur species in aqueous solution as a function of potential, *E*, and pH. (Adapted from Gunter Faure, *Principles and Applications of Inorganic Geochemistry* [New York: Macmillan, 1991], p. 334.)

Each Pourbaix diagram has upper and lower limits. Because in this plot we are considering aqueous solutions, water itself will oxidize to oxygen gas above the HSO_4^-/SO_4^{2-} boundary:

$$3\ H_2O(l) \rightarrow \tfrac{1}{2}\ O_2(g) + 2\ H_3O^+(aq) + 2\ e^-$$

Conversely, below the H_2S/HS^- boundary, water will be reduced to hydrogen gas:

$$2\ H_2O(l) + 2\ e^- \rightarrow H_2(g) + 2\ OH^-(aq)$$

Within those boundaries, the sulfate ion is the predominant species over most of the range of pH and *E*. Because the hydrogen sulfate ion is the conjugate base of a fairly strong acid, only below about pH 2 is the HSO_4^- ion preferred:

$$SO_4^{2-}(aq) + H^+(aq) \rightleftharpoons HSO_4^-(aq)$$

Over most of the pH range, a more reducing environment results in the conversion of sulfate and hydrogen sulfate to elemental sulfur:

$$SO_4^{2-}(aq) + 8\ H^+(aq) + 6\ e^- \rightarrow S(s) + 4\ H_2O(l)$$

In stronger reducing potentials, the sulfur is, in turn, reduced to hydrogen sulfide:

$$S(s) + 2\ H^+(aq) + 2\ e^- \rightarrow H_2S(aq)$$

Notice that aqueous hydrogen sulfide is the predominant reduced species. The reason relates to the weakness of this acid. Only in basic conditions will the hydrogen sulfide ion predominate:

$$H_2S(aq) + OH^-(aq) \rightarrow HS^-(aq) + H_2O(l)$$

The sulfide ion, S^{2-}, would exist as the preferred species only beyond the pH limits of this particular diagram.

It is easier to identify the major aqueous species under different pH and *E* conditions from a Pourbaix diagram, but the study of the relative stabilities of different oxidation states is best accomplished from a Frost diagram. It is important to realize that Pourbaix diagrams display the thermodynamically preferred species. As a result of kinetic factors, other species can often exist.

Biological Aspects

Many biological processes, for example, photosynthesis and respiration, involve oxidation and reduction. Many plants rely on bacteria-filled nodules to convert the dinitrogen in the air to the ammonium ion that the plants require. This complex process, known as nitrogen fixation, involves the reduction of nitrogen from an oxidation state of 0 to a –3 oxidation state.

In all biological systems, we have to consider both the potential, *E*, and acidity, pH, simultaneously when trying to decide what species of an element might be present (and we have to consider kinetic factors as well). Thus Pourbaix diagrams have a particular importance for bioinorganic chemistry and inorganic geochemistry. Figure 9.10 shows the limits of pH and *E* that we find in natural waters. The upper dashed line, representing water in contact with the atmosphere, corresponds to a partial pressure of dioxygen of 20 kPa, the oxygen gas pressure at sea level. Rain tends to be slightly acidic as a result of the absorption of carbon dioxide from the atmosphere:

$$CO_2(g) + H_2O(l) \rightleftharpoons H_3O^+(aq) + HCO_3^-(aq)$$

Depending on the geology of an area, the water in streams tends to be closer to neutral; seawater is slightly basic. Open water is rarely more basic than pH 9 because of the carbonate–hydrogen carbonate buffer system that is present:

$$CO_3^{2-}(aq) + H_2O(l) \rightleftharpoons HCO_3^-(aq) + OH^-(aq)$$

All surface waters, however, are oxidizing as a result of the high partial pressure of dissolved oxygen.

In a lake or river in which there is a high level of plant or algal growth, the level of oxygen is less. As a result, these waters have a lower potential. The lowest positive potentials occur in environments with a high biological activity and no atmospheric contact, typically bogs and stagnant lakes. Under these conditions, anaerobic bacteria flourish, the level of dissolved dioxygen will be close to 0, and the environment will be highly reducing. Bogs are also

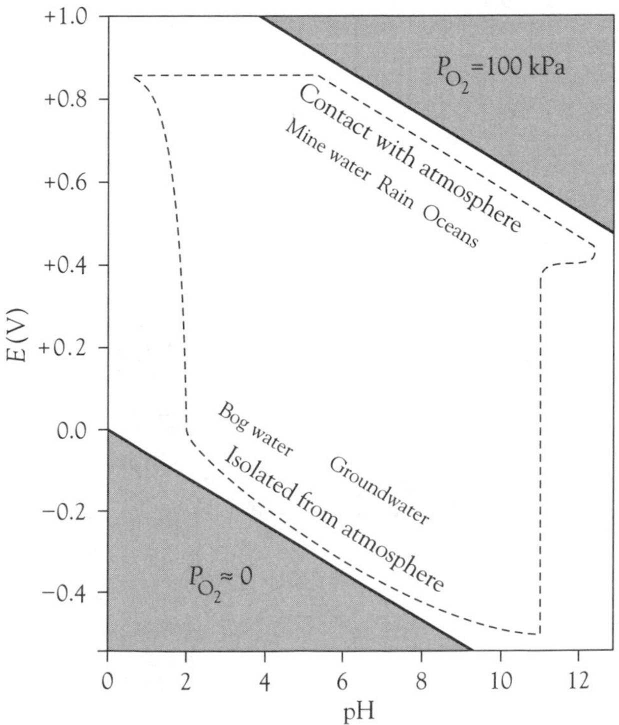

Figure 9.10 Pourbaix diagram showing the limits of *E* and pH conditions in natural waters (dashed line). (Adapted from Gunter Faure, *Principles and Applications of Inorganic Geochemistry* [New York: Macmillan, 1991], p. 324.)

often highly acidic because of the decaying vegetation they contain. Referring back to Figure 9.9, we see that the predominant sulfur compound under reducing conditions is hydrogen sulfide. It is this gas, produced by reduction of sulfate and other sulfur species, that is responsible for much of the foul odor from bogs and marshes.

Exercises

9.1. Define the following terms: (a) oxidizing agent; (b) Ellingham diagrams.

9.2. Define the following terms: (a) Frost diagrams; (b) Pourbaix diagrams.

9.3. Using the oxidation state rules, determine the oxidation number of phosphorus in (a) P_4O_6; (b) H_3PO_4; (c) Na_3P; (d) PH_4^+; (e) $POCl_3$.

9.4. Using the oxidation state rules, determine the oxidation number of chlorine in (a) ICl_2^-; (b) ClO_2; (c) Cl_2O; (d) Cl_2O_7; (e) HCl.

9.5. Using electron-dot diagrams, determine the oxidation number of sulfur in each of the following compounds: (a) H_2S; (b) SCl_2; (c) H_2S_2; (d) SF_6; (e) COS (structure O=C=S).

9.6. Using electron-dot diagrams, determine the formal charges and the oxidation numbers for each element in $SOCl_2$.

9.7. What are the likely oxidation states of iodine in its compounds?

9.8. What would you predict to be the highest oxidation state of xenon in its compounds? What other oxidation states are likely?

9.9. Identify the changes in oxidation states in the following equations:

(a) $Mg(s) + FeSO_4(aq) \rightarrow Fe(s) + MgSO_4(aq)$

(b) $2\ HNO_3(aq) + 3\ H_2S(aq) \rightarrow 2\ NO(g) + 3\ S(s) + 4\ H_2O(l)$

9.10. Identify the changes in oxidation states in the following equations:

(a) $NiO(s) + C(s) \rightarrow Ni(s) + CO(g)$

(b) $2\ MnO_4^-(aq) + 5\ H_2SO_3(aq) + H^+(aq) \rightarrow 2\ Mn^{2+}(aq) + 5\ HSO_4^-(aq) + 3\ H_2O(l)$

9.11. Write a half-reaction for the following reduction in acidic solution:

$H_2MoO_4(aq) \rightarrow Mo^{3+}(aq)$

9.12. Write a half-reaction for the following oxidation in acidic solution:

$NH_4^+(aq) \rightarrow NO_3^-(aq)$

9.13. Write a half-reaction for the following oxidation in basic solution:

$S^{2-}(aq) \rightarrow SO_4^{2-}(aq)$

9.14. Write a half-reaction for the following oxidation in basic solution:

$N_2H_4(aq) \rightarrow N_2(g)$

9.15. Balance the following redox reactions in acidic solution:

(a) $Fe^{3+}(aq) + I^-(aq) \rightarrow Fe^{2+}(aq) + I_2(s)$

(b) $Ag(s) + Cr_2O_7^{2-}(aq) \rightarrow Ag^+(aq) + Cr^{3+}(aq)$

9.16. Balance the following redox reactions in acidic solution:

(a) $HBr(aq) + HBrO_3(aq) \rightarrow Br_2(aq)$

(b) $HNO_3(aq) + Cu \rightarrow NO_2(g) + Cu^{2+}(aq)$

9.17. Balance the following redox reactions in basic solution:

(a) $Ce^{4+}(aq) + I^-(aq) \rightarrow Ce^{3+}(aq) + IO_3^-(aq)$

(b) $Al(s) + MnO_4^-(aq) \rightarrow MnO_2(s) + Al(OH)_4^-(aq)$

9.18. Balance the following redox reactions in basic solution:

(a) $V(s) + ClO_3^-(aq) \rightarrow HV_2O_7^{3-}(aq) + Cl^-(aq)$

(b) $S_2O_4^{2-}(aq) + O_2(g) \rightarrow SO_4^{2-}(aq)$

9.19. Use standard reduction potentials to determine which of the following reactions will be spontaneous under standard conditions:

(a) $SO_2(aq) + MnO_2(s) \rightarrow Mn^{2+}(aq) + SO_4^{2-}(aq)$

(b) $2\ H^+(aq) + 2\ Br^-(aq) \rightarrow H_2(g) + Br_2(aq)$

(c) $Ce^{4+}(aq) + Fe^{2+}(aq) \rightarrow Ce^{3+}(aq) + Fe^{3+}(aq)$

9.20. Use standard reduction potentials to determine which of the following disproportionation reactions will be spontaneous under standard conditions:

(a) $2\ Cu^+(aq) \rightarrow Cu^{2+}(aq) + Cu(s)$

(b) $3\ Fe^{2+}(aq) \rightarrow 2\ Fe^{3+}(aq) + Fe(s)$

9.21. For the two half-reactions

$Al^{3+}(aq) + 3\ e^- \rightarrow Al(s) \qquad E° = -1.67\ V$

$Au^{3+}(aq) + 3\ e^- \rightarrow Au(s) \qquad E° = +1.46\ V$

(a) Identify the half-reaction that would provide the strongest oxidizing agent.

(b) Identify the half-reaction that would provide the strongest reducing agent.

9.22. Calculate the half-reaction potential for the reaction

$Au^{3+}(aq) + 2\ e^- \rightarrow Au^+(aq)$

given

$Au^{3+}(aq) + 3\ e^- \rightarrow Au(s) \qquad E° = +1.46\ V$

$Au^+(aq) + e^- \rightarrow Au(s) \qquad E° = +1.69\ V$

9.23. Calculate the half-cell potential for the reduction of lead(II) ion to lead metal in a saturated solution of lead(II) sulfate, concentration $1.5 \times 10^{-5}\ mol{\cdot}L^{-1}$.

9.24. Calculate the half-cell potential for the reduction of permanganate ion to manganese(IV) at a pH of 9.00 (all other ions being at standard concentration).

9.25. The following Latimer potential diagram shows bromine species in acidic conditions.

$$\overset{+7}{BrO_4^-} \xrightarrow{+1.82\ V} \overset{+5}{BrO_3^-} \xrightarrow{+1.49\ V} \overset{+1}{HBrO} \xrightarrow{+1.59\ V} \overset{0}{Br_2} \xrightarrow{+1.07\ V} \overset{-1}{Br^-}$$

(a) Identify which species are unstable with respect to disproportionation.

(b) Determine the half-potential for the reduction of the bromate ion, BrO_3^-, to bromine.

9.26. The following Latimer potential diagram shows bromine species in basic conditions.

$$\overset{+7}{BrO_4^-} \xrightarrow{+0.99\ V} \overset{+5}{BrO_3^-} \xrightarrow{+0.54\ V} \overset{+1}{BrO^-} \xrightarrow{+0.45\ V} \overset{0}{Br_2} \xrightarrow{+1.07\ V} \overset{-1}{Br^-}$$

(a) Identify which species are unstable with respect to disproportionation.

(b) Determine the half-potential for the reduction of the bromate ion, BrO_3^-, to bromine.

(c) Explain why the bromine to bromide half-potential has the same value in both acidic and basic solutions.

9.27. The Frost diagram below shows lead species (connected by a solid line) and silicon species (connected by a dashed line).

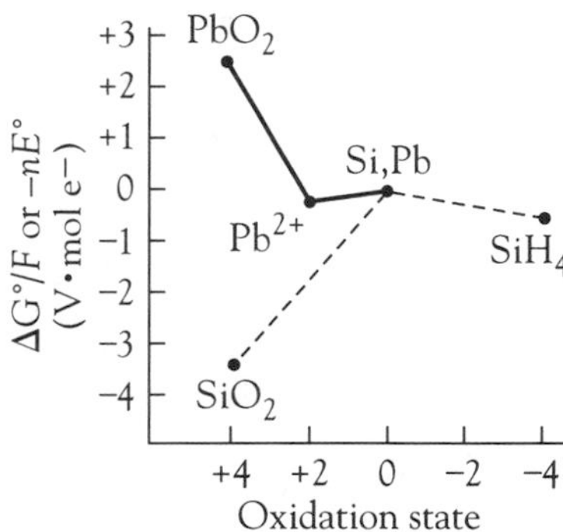

(a) Identify a strong oxidizing agent.

(b) Which is the most thermodynamically stable lead species?

(c) Which is the most thermodynamically stable silicon species?

(d) Which species could potentially disproportionate?

9.28. Construct a Frost diagram for cerium and discuss the relative stability of the oxidation states given

$Ce^{3+}(aq) + 3\ e^- \rightarrow Ce(s) \qquad E° = -2.33\ V$

$Ce^{4+}(aq) + e^- \rightarrow Ce^{3+}(aq) \qquad E° = +1.70\ V$

9.29. The Ellingham diagram on page 160 shows the free energy changes for oxides of mercury, carbon, hydrogen, magnesium, and aluminum.

(a) Which metal oxide can be reduced to its metallic state simply by heating?

(b) At a very high temperature, it is feasible to reduce one metal oxide with carbon. Which is it?

(c) Which metal oxide cannot be reduced to its metal by means of carbon reduction?

(d) Why is it that the slope of the magnesium–magnesium oxide line changes so dramatically beyond the second phase change?

(e) Why is hydrogen gas rarely used for metal reduction?

9.30. Using data table values of $\Delta H_f°$ and $S°$, calculate approximate $\Delta G_f°$ values for the formation of uranium(IV) oxide at 0°C, 500°C, 1000°C, 1500°C, and 2000°C (both uranium and its oxide are solid throughout the temperature range). Plotting the values on the figure in Exercise 29, suggest how uranium metal might be produced from uranium(IV) oxide.

9.31. From the Pourbaix diagram in Figure 9.9, identify the thermodynamically preferred sulfur species at a pH of 7.0 and an E of 0.0 V.

9.32. From the Pourbaix diagram in Figure 9.9, write a half-equation for the oxidation that will occur at a pH of 1.0 as the E is increased from +0.2 V to +0.4 V.

9.33. Before the industrial revolution, a major source of iron was peat bogs. Suggest why "bog iron" should exist.

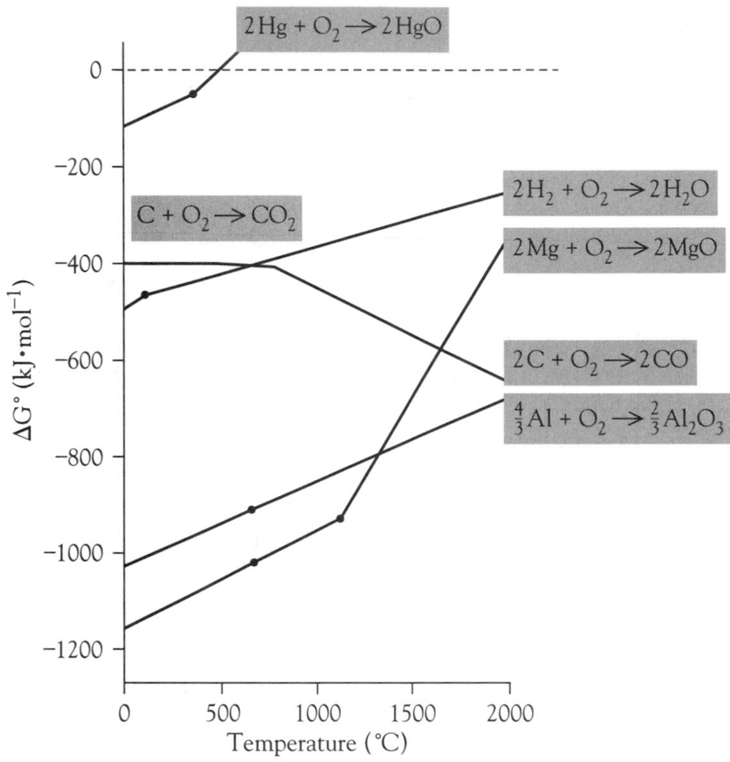

Ellingham diagram for problem 9.29.

CHAPTER

10

The Group 1 Elements: The Alkali Metals

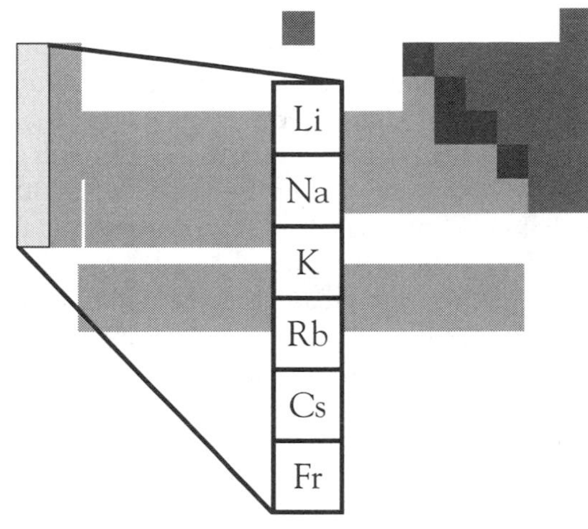

Group Trends

Common Features of Alkali Metal Compounds

Solubility of Alkali Metal Salts

Flame Colors

Lithium

Sodium

Potassium

Oxides

Hydroxides

Sodium Chloride

Potassium Chloride

Sodium Carbonate

Sodium Hydrogen Carbonate

Ammonia Reaction

Ammonium as a Pseudo–Alkali Metal Ion

Similarities Between Lithium and the Alkaline Earth Metals

Biological Aspects

Alkali Metal Anions

Salt Substitutes

Lithium and Mental Health

Metals are usually thought of as being dense and nonreactive. The alkali metals, however, are actually the opposite of this characterization, being of low density and very high chemical reactivity.

Compounds of the alkali metals have been known since ancient times. However, the alkali metal cations are extremely difficult to reduce, and it was not until after electric power was harnessed that the metals themselves could be extracted. A British scientist, Humphry Davy, electrolyzed molten potassium hydroxide in 1807 to extract the first of the alkali metals. Davy obtained such acclaim for his extraction of these metals from their salts that a rhyme was written about him:

> Sir Humphry Davy
> Abominated gravy
> Lived in the odium
> Of having discovered sodium.

Nationalism has often become interwoven with chemistry. When Napoleon heard of Davy's discovery, he was extremely angry that the French chemists had not been first. But, by coincidence, it was a French scientist, Marguerite Perey, who in 1939 isolated the one alkali metal that exists only as radioactive isotopes. She named the element francium after her native country—Napoleon would have been delighted!

Group Trends

All of the alkali metals are shiny, silver-colored metals. Like the other metals, they have high electrical and thermal conductivities. But in other respects, they are very atypical. For example, the alkali metals are very soft, and they become softer as one progresses down the group. Thus lithium can be cut with a knife, whereas potassium can be "squashed" like soft butter. Most metals have high melting points, but those of the alkali metals are very low and become lower as the elements in Group 1 become heavier, with cesium melting just above room temperature. In fact, the combination of high thermal conductivity and low melting point makes sodium useful as a heat transfer material in some nuclear reactors. The softness and low melting points of the alkali metals can be attributed to the very weak metallic bonding in these elements. For a "typical" metal, the enthalpy of atomization is in the range of 400 to 600 $kJ{\cdot}mol^{-1}$; but as can be seen from Table 10.1, those of the alkali metals are much lower. In fact, there is a correlation between both softness and low melting point and a small enthalpy of atomization.

Even more atypical are the densities of the alkali metals. Most metals have densities between 5 and 15 $g{\cdot}cm^{-3}$, but those of the alkali metals are far less (Table 10.2). In fact, lithium has a density one-half that of water!

With such a low density, lithium would be ideal for making unsinkable (although soft!) ships, except for one other property of the alkali metals—their high chemical reactivity. The metals are usually stored under oil, because when they are exposed to air, a thick coating of oxidation products covers the lustrous surface of the metal very rapidly. For example, lithium is oxidized to lithium oxide, which in turn reacts with carbon dioxide to give lithium carbonate:

$$4\,Li(s) + O_2(g) \rightarrow 2\,Li_2O(s)$$

$$Li_2O(s) + CO_2(g) \rightarrow Li_2CO_3(s)$$

Table 10.1 Melting points and enthalpies of atomization of the alkali metals

Element	Melting point (°C)	$\Delta H_{atomization}$ ($kJ{\cdot}mol^{-1}$)
Li	180	162
Na	98	108
K	64	90
Rb	39	82
Cs	29	78

Table 10.2 Densities of the alkali metals

Element	Density ($g{\cdot}cm^{-3}$)
Li	0.53
Na	0.97
K	0.86
Rb	1.53
Cs	1.87

Reaction with water is more dramatic. Lithium bubbles quietly to produce the hydroxide and hydrogen gas. Sodium melts, skating around on the water surface as a silvery globule, and the hydrogen that is produced usually burns. For the heavier members of the group, the reaction is extremely violent; explosions often occur when rubidium and cesium encounter water. The explosions are the result of the ignition of the dihydrogen-dioxygen gas mixture by the hot metal surface. The equation for the reaction of water with potassium is

$$2\,K(s) + 2\,H_2O(l) \rightarrow 2\,KOH(aq) + H_2(g)$$

Because they are so much more reactive than the "average" metal, the alkali metals are sometimes referred to as the *supermetals*.

Common Features of Alkali Metal Compounds

All the Group 1 elements are metals. As a result, all the members of the group have common features.

Ionic Character

The alkali metal ions always have an oxidation number of +1, and most of their compounds are stable, ionic solids. The compounds are colorless unless they contain a colored anion such as chromate or permanganate. Even for these highly electropositive elements, the bonds in their compounds with nonmetals have a small covalent component.

Stabilization of Large Low-Charge Anions

Because the cations of the alkali metals (except for that of lithium) have among the lowest charge densities, they are able to stabilize large low-charge anions. For example, the ions of sodium through cesium are the only cations that form solid hydrogen carbonate salts.

Ion Hydration

The higher the charge density of the ion, the more strongly it is hydrated. As we mentioned earlier, the alkali metals have very low charge densities compared with those of other metals, but the values are high enough for lithium

Alkali Metal Anions

It is so easy to become locked into preconceptions. Everyone "knows" that the alkali metals "want" to lose an electron and form cations. In fact, this is not true. In a very simplistic explanation, we can say that the element with which the alkali metal reacts has the greater electron affinity and hence "drags" the electron off the alkali metal. Left to itself, the alkali metal would prefer to complete its *s* orbital set by gaining an electron.

It was a Michigan State chemist, James Dye, who realized that the alkali metals have such positive electron affinities that it might just be possible to stabilize the alkali metal anion. After a number of attempts, he found a complex organic compound of formula $C_{20}H_{36}O_6$ that could just contain a sodium cation within its structure. He was hoping that, by adding this compound to a sample of sodium metal, some of the sodium atoms would pass their *s* electrons to neighboring sodium atoms, to produce sodium anions. This happened, as predicted:

$$2\ Na(s) + C_{20}H_{36}O_6 \rightarrow [Na(C_{20}H_{36}O_6)]^{+} \cdot Na^{-}$$

The metallic-looking crystals were shown to contain the sodium anion, but the compound was found to be very reactive with almost everything. So, to the present day, this compound is no more than a laboratory curiosity. But its existence does remind us to question even the most commonly held beliefs.

and sodium ions to favor the formation of a few hydrated salts. An extreme example is lithium hydroxide, which forms an octahydrate, $LiOH \cdot 8H_2O$. With the lowest charge densities of all, few potassium salts are hydrated, and hydration is also very rare for rubidium and cesium salts.

The low charge densities are reflected in the trend in hydration enthalpy among the alkali metals (Table 10.3). The values are very low (for example, that of the Mg^{2+} ion is 1920 $kJ \cdot mol^{-1}$), and the values decrease as radius increases down the group.

Table 10.3 Hydration enthalpies of the alkali metal ions

Ion	Hydration enthalpy ($kJ \cdot mol^{-1}$)
Li^+	519
Na^+	406
K^+	322
Rb^+	301
Cs^+	276

Solubility

Almost all the compounds of the alkali metals are soluble in water, although they are soluble to different extents. For example, a saturated solution of lithium chloride has a concentration of 14 mol·L^{-1}, whereas a saturated solution of lithium carbonate has a concentration of only 0.18 mol·L^{-1}. We explore the reason for the differences in the next section.

Solubility of Alkali Metal Salts

It is the solubility of all the common alkali metal salts that makes them so useful as reagents in the chemistry laboratory. Whether it is a nitrate, a phosphate, or a fluoride anion that we need, we can always count on the alkali metal salt to enable us to make a solution of the required anion. Yet the solubilities cover a wide range of values. This variability is illustrated by the solubilities of the sodium halides (Table 10.4).

Table 10.4 Solubilities of the sodium halides at 25°C

Compound	Solubility (mol·L^{-1})
NaF	0.099
NaCl	0.62
NaBr	0.92
NaI	1.23

To explain this solubility trend, we need to look at the energy cycle involved in the formation of a solution from the solid. As we discussed in Chapter 6, the solubility of a compound is dependent on the enthalpy terms, the lattice energy, and the enthalpy of hydration of the cation and anion, together with the corresponding entropy changes (shown in Figure 10.1). For

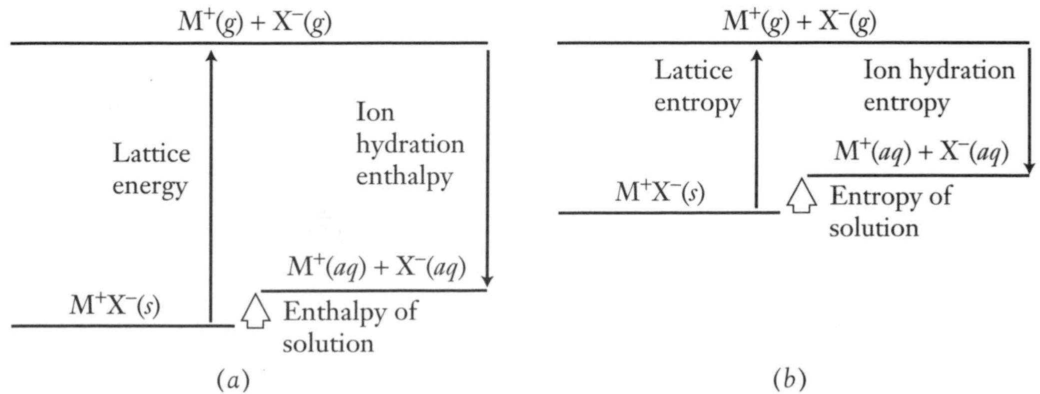

Figure 10.1 Enthalpy cycle (*a*) and entropy cycle (*b*) for the solution of an ionic compound. M^+ is the alkali metal ion and X^- is the anion.

Table 10.5 Enthalpy factors in the solution process for the sodium halides

Compound	Lattice energy ($kJ\cdot mol^{-1}$)	Hydration enthalpy ($kJ\cdot mol^{-1}$)	Net enthalpy change ($kJ\cdot mol^{-1}$)
NaF	+930	–929	+1
NaCl	+788	–784	+4
NaBr	+752	–753	–1
NaI	+704	–713	–9

the salt to be appreciably soluble, the free energy, $\Delta G°$, should be negative, where

$$\Delta G° = \Delta H° - T\Delta S°$$

If we look at the enthalpy terms (Table 10.5), we see that for each sodium halide, the lattice energy is almost exactly balanced by the sum of the cation and anion hydration enthalpies. In fact, the error in these experimental values is larger than the calculated differences. As a result, we can only say that the lattice energy and hydration enthalpy terms are essentially equal.

When we calculate the entropy changes (Table 10.6), we find that for all the salts except sodium fluoride, the entropy gained by the ions as they are freed from the crystal lattice is numerically larger than the entropy lost when these gaseous ions are hydrated in solution. When we combine these two very small net effects to obtain the free energy change for the solution process, amazingly, we do obtain a trend parallel to that of the measured solubilities (Table 10.7). Furthermore, if we plot the solubilities of the salts one anion forms with different alkali metal cations as a function of the ionic radius of the alkali metal ions, in most cases we get a smooth curve. This curve may have a positive or negative slope (or in some cases, reach a minimum in the middle of the series). To illustrate such trends, the solubilities of alkali metal fluorides and iodides are shown in Figure 10.2.

We can understand the different curves in Figure 10.2 by focusing on the lattice energies. Although there is a strong dependence of lattice energy on

Table 10.6 Entropy factors in the solution process for the sodium halides, expressed as $T\Delta S$ values

Compound	Lattice entropy ($kJ\cdot mol^{-1}$)	Hydration entropy ($kJ\cdot mol^{-1}$)	Net entropy change ($kJ\cdot mol^{-1}$)
NaF	+72	–74	–2
NaCl	+68	–55	+13
NaBr	+68	–50	+18
NaI	+68	–45	+23

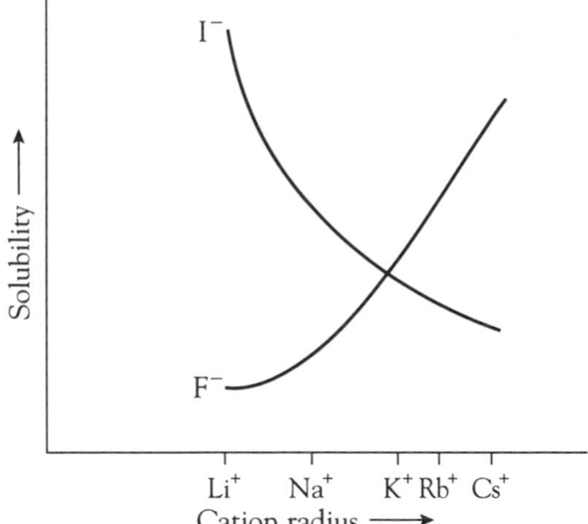

Figure 10.2 Solubility of alkali metal fluorides and iodides as a function of alkali metal ion radius.

Table 10.7 Calculated free energy change for the solution process for the sodium halides

Compound	Enthalpy change ($kJ \cdot mol^{-1}$)	Entropy change ($kJ \cdot mol^{-1}$)	Free energy change ($kJ \cdot mol^{-1}$)
NaF	+1	−2	+3
NaCl	+4	+13	−11
NaBr	−1	+18	−19
NaI	−9	+23	−32

ionic charge, there is a secondary relationship to the cation to anion ratio; that is, a significant mismatch in ionic sizes will lead to a lower than expected lattice energy. Table 10.8 shows the ionic radii of the cations lithium and cesium and the anions fluoride and iodide. Thus lithium iodide, the ions of which have very different sizes, is much more soluble than lithium fluoride, the ions of which have similar sizes. Conversely, cesium iodide, the ions of which have similar sizes, is much less soluble than cesium fluoride, in which there is a large mismatch in ionic size.

Table 10.8 Selected ionic radii

Cation	Radius (pm)	Anion	Radius (pm)
Li^+	90	F^-	119
Cs^+	181	I^-	206

Table 10.9 Alkali metals and their flame colors

Metal	Color
Lithium	Crimson
Sodium	Yellow
Potassium	Lilac
Rubidium	Red-violet
Cesium	Blue

Flame Colors

Almost all the alkali metal compounds are water-soluble, so precipitation tests cannot be used to detect the presence of an alkali metal. Fortunately, each of the alkali metals produces a characteristic flame color when a sample of an alkali metal salt is placed in a flame (see Table 10.9). In the process, energy from the combustion reactions of the fuel is transferred to the metal salt that is placed in the flame. This transfer causes electrons in the alkali metal atoms to be raised to excited states. The energy is released in the form of visible radiation as the electron returns to the ground state. Each alkali metal undergoes its own unique electron transitions. For example, the yellow color of sodium is a result of the energy (photon) emitted when an electron drops from the $3p^1$ orbital to the $3s^1$ orbital of a neutral sodium atom, the ion having acquired its valence electron from the combustion reactions in the flame (Figure 10.3).

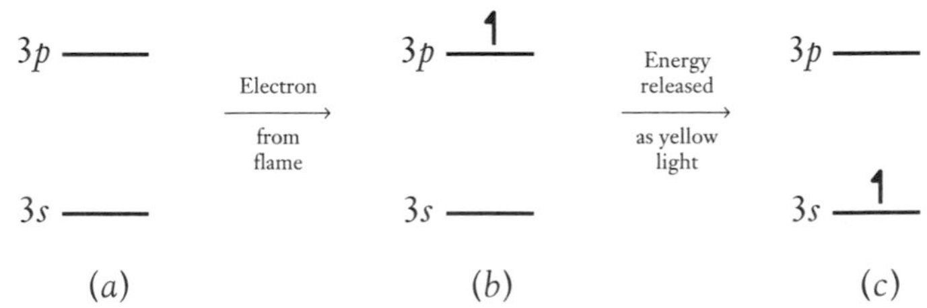

Figure 10.3 In a flame, the sodium ion (*a*) acquires an electron in the $3p$ orbital (*b*). As the electron drops from the excited $3p$ state to the ground $3s$ state (*c*), the energy is released as yellow light.

Lithium

With a density of about half that of water, lithium is the least dense of all the elements that are solids at room temperature and pressure. The metal has a bright silvery appearance; but when a surface is exposed to moist air, it very rapidly turns black. Like the other alkali metals, lithium reacts with the dioxygen in air. It is the only alkali metal, and one of a very few elements in the entire periodic table, to react with dinitrogen. Breaking the triple bond in the dinitrogen molecule requires an energy input of 945 $kJ{\cdot}mol^{-1}$. To balance this energy uptake, the lattice energy of the product must be very high. Of the alkali metals, only the lithium ion, which has the greatest charge density of the group, forms a nitride with a sufficiently high lattice energy:

$$6\ Li(s) + N_2(g) \rightarrow 2\ Li_3N(s)$$

The nitride is reactive, however, forming ammonia when added to water:

$$Li_3N(s) + 3\ H_2O(l) \rightarrow 3\ LiOH(aq) + NH_3(g)$$

Liquid lithium is the most corrosive material known. For example, if a sample of lithium is melted in a glass container, it reacts spontaneously with the glass to produce a hole in the container; the reaction is accompanied by the emission of an intense, greenish white light. In addition, lithium has the most negative standard reduction potential of any element:

$$Li^+(aq) + e^- \rightarrow Li(s) \qquad E° = -3.05\ V$$

That is, the metal itself releases more energy than any other element when it is oxidized to its ion (+3.05 V). Yet, of the alkali metals, it has the least spectacular reaction with water. As discussed in Chapter 6, we must not confuse spontaneity, which depends on the free energy change, with rate of reaction, which is controlled by the height of the activation energy barrier. In this particular case, we must assume that the activation energy for the reaction with water is greatest for lithium; hence lithium has the slowest reaction rate of all the alkali metals.

The comparatively high charge density of lithium is responsible for several other important ways in which lithium's chemistry differs from that of the rest of the alkali metals. In particular, there is an extensive organometallic chemistry of lithium in which the bonding is definitely covalent. Even for common salts, such as lithium chloride, their high solubilities in many solvents of low polarity, particularly ethanol and acetone, indicate a high degree of covalency in the bonding.

Lithium used to be a laboratory curiosity, but no more. The metal itself, because of its very low density, is used in aerospace alloys. For example, alloy LA 141, which consists of 14 percent lithium, 1 percent aluminum, and 85 percent magnesium, has a density of only 1.35 $g{\cdot}cm^{-3}$, almost exactly half that of aluminum, the most commonly used low-density metal. Lithium is also favored for battery technology. With its high reduction potential, it is currently used in compact high-voltage cells. Moreover, it is strongly favored to replace the lead-acid battery for electric vehicle propulsion. Having a density one-twentieth that of lead, substantial mass savings are possible once the very challenging task of devising a reversible (rechargeable) lithium cycle is perfected.

There are hundreds of different redox cycles under investigation as possible battery combinations, but one of the most interesting employs the lithium half-cell and an aqueous electrolyte, lithium nitrate solution. To prevent the lithium metal from reacting with the water, the lithium metal atoms are formed within the holes of a metal oxide lattice. This insertion of a "guest" atom into a "host" solid, a process accompanied by only small, reversible changes in structure, is known as intercalation; and the resulting product is called an intercalation compound. In this environment, the lithium reduction potential is altered dramatically from that under "normal" conditions:

$$Li^{+}(aq) + e^{-} \rightarrow Li(s) \qquad E° = -3.0\ V$$

The reduction potential of the preceding "normal" half-reaction is dependent on the identity of the metal oxide. For example, lithium atoms formed within a vanadium(IV) oxide lattice have a positive reduction potential, whereas those formed within a manganese(IV) oxide lattice have a slightly negative reduction potential:

$$Li^{+}(aq) + e^{-} \rightarrow Li(Mn_2O_4) \qquad E° = +1.0\ V$$

$$Li^{+}(aq) + e^{-} \rightarrow Li(VO_2) \qquad E° = -0.5\ V$$

It is this potential difference of 1.5 V between the two lithium environments that drives the cell. During cell discharge, the following reactions occur:

$$Li(VO_2) \rightarrow Li^{+}(aq) + e^{-} \qquad E° = +0.5\ V$$

$$Li^{+}(aq) + e^{-} \rightarrow Li(Mn_2O_4) \qquad E° = +1.0\ V$$

Recharging the battery causes the reverse reactions to occur.

The largest industrial use of lithium is in lithium greases—in fact, more than 60 percent of all automotive greases contain lithium. The compound used is lithium stearate, $C_{17}H_{35}COOLi$, which is mixed with oil to give a water-resistant, greaselike material that does not harden in cold temperatures yet is stable at high temperatures.

We mentioned earlier that lithium forms an extensive range of covalent compounds with carbon. One in particular, butyllithium, LiC_4H_9, is a useful reagent in organic chemistry. It can be prepared by treating lithium metal with chlorobutane, C_4H_9Cl, in a hydrocarbon solvent such as hexane, C_6H_{12}:

$$2\ Li(s) + C_4H_9Cl(C_6H_{12}) \rightarrow LiC_4H_9(C_6H_{12}) + LiCl(s)$$

After removing the lithium chloride by filtration, the solvent can be removed by distillation; liquid butyllithium remains in the distillation vessel. This compound has to be handled carefully, because it spontaneously burns when exposed to the dioxygen in air.

Sodium

Sodium is the alkali metal for which there is the highest industrial demand. Like all the alkali metals, the pure element does not exist naturally because of its very high reactivity. The silvery metal is manufactured by the Downs process, in which sodium chloride (m.p. 801°C) is electrolyzed in the molten state. The electrolysis is done in a cylindrical cell with a central graphite anode and a surrounding steel cathode (Figure 10.4). A mixture of calcium chloride and sodium chloride is used to reduce the melting point and hence lower the temperature at which the cell needs to be operated. Although calcium chloride itself has a melting point of 772°C, a mixture of 33 percent sodium chloride and 67 percent calcium chloride gives a melting point of

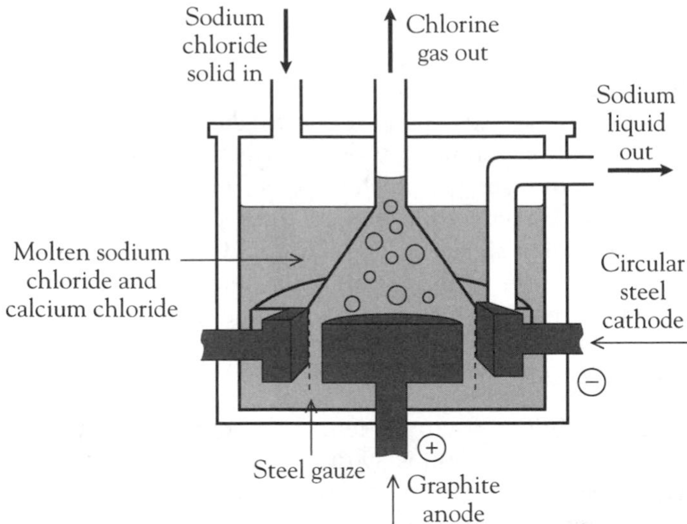

Figure 10.4 Downs cell.

about 580°C. It is the lower melting point of the mixture that makes the process commercially feasible.

The two electrodes are separated by a cylindrical steel gauze diaphragm, so that the molten sodium, which floats to the top of the cathode compartment, will be kept away from the gaseous chlorine formed at the anode:

$$Na^{+}(NaCl) + e^{-} \rightarrow Na(l)$$

$$2\ Cl^{-}(NaCl) \rightarrow Cl_2(g) + 2\ e^{-}$$

The sodium metal produced contains about 0.2 percent calcium metal. Cooling the metal mixture to 110°C allows the calcium impurity (m.p. 842°C) to solidify and sink into the melt. The pure sodium (m.p. 98°C) remains liquid and can be pumped into cooled molds, where it solidifies.

Sodium metal is required for the synthesis of a large number of sodium compounds, but it has two major uses, the first of which is the extraction of other metals. The easiest way to obtain many of the rarer metals such as thorium, zirconium, tantalum, and titanium is by the reduction of their compounds with sodium. For example, titanium can be obtained by reducing titanium(IV) chloride with sodium metal:

$$TiCl_4(l) + 4\ Na(s) \rightarrow Ti(s) + 4\ NaCl(s)$$

The sodium chloride can then be washed away from the pure titanium metal.

The second major use of sodium metal is in the production of the gasoline additive tetraethyllead (TEL). Although TEL is now banned from gasolines in North America because of its toxicity and the lead pollution resulting from its use, it is still employed throughout much of the world to boost the octane rating of cheap gasolines. The synthesis of TEL uses the reaction between a lead-sodium alloy and ethyl chloride:

$$4\ NaPb(s) + 4\ C_2H_5Cl(g) \rightarrow (C_2H_5)_4Pb(l) + 3\ Pb(s) + 4\ NaCl(s)$$

Potassium

The potassium found in the natural environment is slightly radioactive because it contains about 0.012 percent of the radioactive isotope potassium-40. In fact, a significant proportion of the radiation generated within our bodies comes from this isotope, which has a half-life of 1.3×10^9 years.

The industrial extraction of potassium metal is accomplished by chemical means. Extraction in an electrolytic cell would be too hazardous because of the extreme reactivity of the metal. The chemical process involves the reaction of sodium metal with molten potassium chloride at 850°C:

$$Na(l) + KCl(l) \rightleftharpoons K(g) + NaCl(l)$$

Although the equilibrium lies to the left, at this temperature potassium is a gas (b.p. 766°C; b.p. for sodium is 890°C)! Thus the Le Châtelier principle can be used to drive the reaction to the right by pumping the green potassium gas from the mixture as it is formed.

We have already mentioned that alkali metal salts exhibit a wide range of solubilities. In particular, the least soluble are those with the greatest similarity in ion size. Thus a very large anion would form the least soluble salts with the larger cations of Group 1. This concept holds for the very large hexanitritocobaltate(III) anion, $[Co(NO_2)_6]^{3-}$. Its salts with lithium and sodium are soluble, whereas those with potassium, rubidium, and cesium are insoluble. Thus if a solution is believed to contain either sodium or potassium ion, addition of the hexanitritocobaltate(III) ion can be used as a test. A bright yellow precipitate indicates the presence of potassium ion:

$$3\ K^+(aq) + [Co(NO_2)_6]^{3-}(aq) \rightarrow K_3[Co(NO_2)_6](s)$$

Another very large anion that can be used in a precipitation test with the larger alkali metals is the tetraphenylborate ion, $[B(C_6H_5)_4]^-$:

$$K^+(aq) + [B(C_6H_5)_4]^-(aq) \rightarrow K[B(C_6H_5)_4](s)$$

Oxides

Most metals in the periodic table react with dioxygen gas to form oxides containing the oxide ion, O^{2-}. However, of the alkali metals, only lithium forms a normal oxide when it reacts with oxygen:

$$4\ Li(s) + O_2(g) \rightarrow 2\ Li_2O(s)$$

Sodium reacts with dioxygen to give sodium dioxide(2–), Na_2O_2 (commonly called sodium peroxide),

$$2\ Na(s) + O_2(g) \rightarrow Na_2O_2(s)$$

which contains the dioxide(2–) ion, O_2^{2-} (often called the peroxide ion). The notation "2–" simply indicates the charge on the ion, and it avoids the need for learning the many prefixes that used to be employed for that purpose. This is the first time that we have used parenthetical Arabic numbers in naming, a method recommended by the American Chemical Society for use

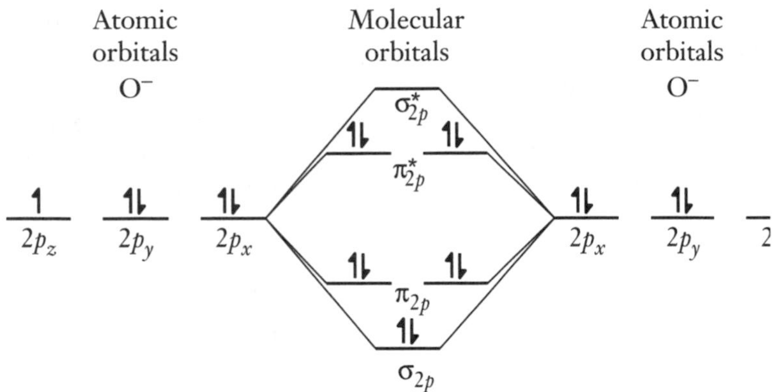

Figure 10.5 Filling of the molecular orbitals derived from the $2p$ orbitals for the dioxide(2–) ion.

whenever there is more than one possible ionic charge (as we shall see shortly).

Sodium dioxide(2–) is diamagnetic, and the oxygen-oxygen bond length is about 149 pm, much longer than the 121 pm in the dioxygen molecule. We can explain the diamagnetism and the weaker bond by constructing the part of the molecular orbital diagram derived from the $2p$ atomic orbitals (Figure 10.5). This diagram shows that three bonding orbitals and two antibonding orbitals are occupied. All the electrons are paired and the net bond order is 1 rather than 2, the bond order in the dioxygen molecule (see Chapter 3).

The other three alkali metals react with an excess of dioxygen to form dioxides(1–) (traditionally named superoxides) containing the paramagnetic dioxide(1–) ion, O_2^-:

$$K(s) + O_2(g) \rightarrow KO_2(s)$$

The oxygen-oxygen bond length in these ions (133 pm) is less than that in the dioxide(2–) but slightly greater than that in dioxygen itself. We can also explain these different bond lengths in terms of the molecular orbital filling (Figure 10.6). The dioxide(1–) ion possesses three bonding pairs and $1\frac{1}{2}$ antibonding pairs. The net bond order in a dioxygen(1–) ion is $1\frac{1}{2}$, between the bond order of 1 in the dioxide(2–) ion and the bond order of 2 in the

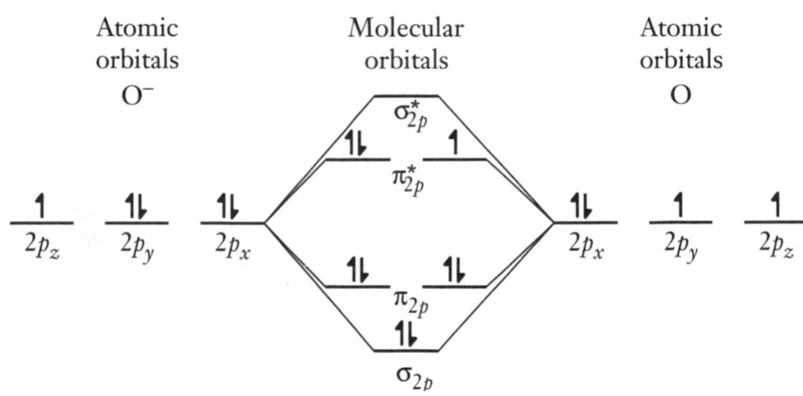

Figure 10.6 Filling of the molecular orbitals derived from the $2p$ orbitals for the dioxide(1–) ion.

dioxygen molecule. We can explain the ready formation of both the dioxide(1–) and the dioxide(2–) ions by postulating that the least polarizing cations (those with low charge density) stabilize these large polarizable anions.

All the Group 1 oxides react vigorously with water to give the metal hydroxide solution. In addition, sodium dioxide(2–) generates hydrogen peroxide, and the dioxides(1–) produce hydrogen peroxide and oxygen gas:

$$Li_2O(s) + H_2O(l) \rightarrow 2\ LiOH(aq)$$

$$Na_2O_2(s) + 2\ H_2O(l) \rightarrow 2\ NaOH(aq) + H_2O_2(aq)$$

$$2\ KO_2(s) + 2\ H_2O(l) \rightarrow 2\ KOH(aq) + H_2O_2(aq) + O_2(g)$$

Potassium dioxide(1–) is used in space capsules, submarines, and some types of self-contained breathing equipment because it absorbs exhaled carbon dioxide (and moisture) and releases dioxygen gas:

$$4\ KO_2(s) + 2\ CO_2(g) \rightarrow 2\ K_2CO_3(s) + 3\ O_2(g)$$

$$K_2CO_3(s) + CO_2(g) + H_2O(g) \rightarrow 2\ KHCO_3(s)$$

Hydroxides

The solid hydroxides are white, translucent solids that absorb moisture from the air until they dissolve in the excess water—a process known as *deliquescence*. The one exception is lithium hydroxide, which forms the stable octahydrate, $LiOH{\cdot}8H_2O$. Alkali metal hydroxides are all extremely hazardous because the hydroxide ion reacts with skin protein to destroy the skin surface. Sodium hydroxide and potassium hydroxide are supplied as pellets, and these are produced by filling molds with the molten compound. As solids or in solution, they also absorb carbon dioxide from the atmosphere:

$$2\ NaOH(aq) + CO_2(g) \rightarrow Na_2CO_3(aq) + H_2O(l)$$

The alkali metal hydroxides are convenient sources of the hydroxide ion because they are very water-soluble. When hydroxide ion is needed as a reagent, its source is chosen on the basis of either cost or solubility. In inorganic chemistry, sodium hydroxide (caustic soda) is most commonly used as the source of hydroxide ion because it is the least expensive metal hydroxide. Potassium hydroxide (caustic potash) is preferred in organic chemistry because it has a higher solubility in organic solvents than sodium hydroxide has.

Preparation of Sodium Hydroxide

Sodium hydroxide is the sixth most important inorganic chemical in terms of quantity produced. It is prepared by the electrolysis of brine (aqueous sodium chloride). The three common types of electrolytic cells are the diaphragm cell, the membrane cell, and the mercury cathode cell. These cells use prodigious quantities of electricity, the first two types running at between 30 000 and 150 000 A; the mercury cell requires up to 400 000 A.

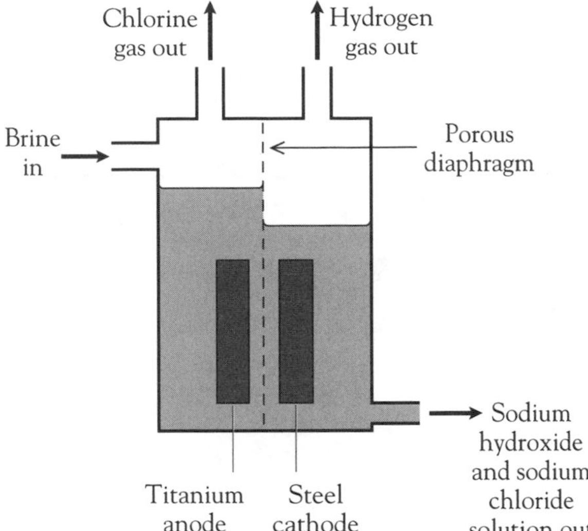

Figure 10.7 Diaphragm cell.

In the diaphragm cell, water is reduced to hydrogen gas and hydroxide ion at the cathode, and chloride ion is oxidized to chlorine gas at the anode (although some water is oxidized to oxygen gas as well):

$$2\ H_2O(l) + 2\ e^- \rightarrow H_2(g) + 2\ OH^-(aq)$$

$$2\ Cl^-(aq) \rightarrow Cl_2(g) + 2\ e^-$$

The essential design feature (Figure 10.7) is the diaphragm or separator, which prevents the hydroxide ion produced at the cathode from coming into contact with the chlorine gas produced at the anode. This separator, which has pores that are large enough to allow the brine to pass through, used to be made of asbestos, but it is now made of a Teflon® mesh. During the electrolysis, the cathode solution, which consists of a mixture of 11 percent sodium hydroxide and 16 percent sodium chloride, is removed continuously. The harvested solution is evaporated, a process that causes the less soluble sodium chloride to crystallize. The final product is a solution of 50 percent sodium hydroxide and about 1 percent sodium chloride. This composition is quite acceptable for most industrial purposes.

The membrane cell functions like the diaphragm cell except that the anode and cathode solutions are separated by a micropore polymer membrane that is permeable only to cations—in this case, the sodium ion. Thus the chloride ions from the brine cannot enter the cathode compartment, nor can the hydroxide ions produced in the cathode compartment escape in the opposite direction. As a result, the sodium hydroxide solution produced contains no more than 50 ppm chloride ion contaminant. Unfortunately, the membrane is very expensive, and it can become clogged by trace impurities of calcium ion, which form insoluble calcium hydroxide on the membrane surface.

The mercury cathode cell, as its name implies, uses liquid mercury as the cathode (Figure 10.8). At the anode, chlorine gas is produced; but at the cathode, sodium ion is reduced to sodium metal:

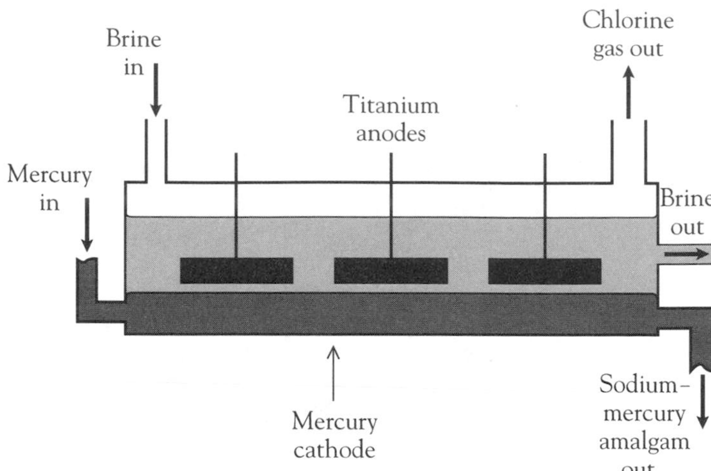

Figure 10.8 Mercury cathode cell.

$$Na^+(aq) + e^- \rightarrow Na(Hg)$$

Sodium reduction occurs because the mercury electrode surface inhibits any half-reaction that produces a gas, raising the electrode potential of such a reaction above its standard value. Thus in this environment the "expected" reduction of hydrogen ion to hydrogen gas actually requires a higher potential than the reduction of sodium ion. The surface phenomenon responsible for this result is called overvoltage.

The sodium-mercury amalgam is pumped to a separate chamber, where water is reacted with the amalgam on a graphite surface:

$$2\ Na(Hg) + 2\ H_2O(l) \rightarrow 2\ NaOH(aq) + H_2(g)$$

The sodium hydroxide solution produced by this route is pure and concentrated. Hence the mercury cathode cell is the preferred source of pure, solid sodium hydroxide. Unfortunately, there is a loss of mercury into the environment during the operation of this process. Traditionally, loss of mercury was regarded as just another business expense, and there was little or no concern about the ultimate fate of the mercury. Much of it ended up in rivers, where it caused severe pollution. Now, of course, we recognize the folly of this approach, and a considerable research effort has been undertaken to completely eliminate mercury losses from the cells. The comparative advantages and disadvantages of the cell types are listed in Table 10.10.

Uses of Sodium Hydroxide

About 30 percent of the sodium hydroxide produced is used as a reagent in organic chemical plants, and about 20 percent is used for the synthesis of other inorganic chemicals. Another 20 percent is consumed in the pulp and paper industry, and the remaining 30 percent is used in hundreds of other ways.

Sodium hydroxide is the most important base in the chemistry laboratory. It also has a number of household uses, where it is commonly referred

Table 10.10 Advantages and disadvantages of the industrial processes for the production of sodium hydroxide

Process	Advantages	Disadvantages
Diaphragm process	Utilizes less pure brine	Dilute, chloride-contaminated product (11% NaOH)
	Lower electrical consumption, because process is more efficient	Chlorine often oxygen-contaminated
		Asbestos concerns
Membrane process	Produces uncontaminated product	Maximum concentration, 35% NaOH
	Lower electrical consumption	Chlorine often oxygen-contaminated
	No mercury or asbestos problems	Very high purity brine required
		High cost and short lifetime of membranes
Mercury process	Produces pure, high concentration sodium hydroxide (50%)	Higher electrical consumption
	Produces pure chlorine gas	Needs purer brine than diaphragm process
		Mercury containment problems

to as lye. The most direct application takes advantage of its reaction with greases, particularly those in ovens (such as Oven-Off®) or those clogging drains (such as Drano®). In some commercial drain-treatment products, aluminum metal is mixed with the sodium hydroxide. Upon addition to water, the following chemical reaction occurs, producing the aluminate ion and hydrogen gas. The hydrogen gas bubbles cause the liquid to churn vigorously, enhancing the contact of grease with fresh sodium hydroxide solution, an action that dissolves the plug more quickly:

$$2\ Al(s) + 2\ OH^-(aq) + 6\ H_2O(l) \rightarrow 2\ [Al(OH)_4]^-(aq) + 3\ H_2(g)$$

Sodium hydroxide is also used in the food industry, mainly to provide hydroxide ion for breaking down proteins. For example, potatoes are sprayed with sodium hydroxide solution to soften and remove the skins before processing. (Of course, they are washed thoroughly before the next processing step!) Olives have to be soaked in sodium hydroxide solution to soften the flesh enough to make them edible. Grits, too, are processed with sodium hydroxide solution. The most unusual application is in the manufacture of pretzels. The dough is coated with a thin layer of sodium hydroxide solution before salt crystals are applied. The sodium hydroxide appears to function as a cement, attaching the salt crystals firmly to the dough surface. In the baking process, carbon dioxide is released, thereby converting the sodium hydroxide to harmless sodium carbonate:

$$2\ NaOH(s) + CO_2(g) \rightarrow Na_2CO_3(s) + H_2O(g)$$

Sodium Chloride

Common salt has played a major role in the history of civilization. Salt was one of the earliest commodities to be traded, and Roman soldiers were partially paid in salt (*sal*)—hence our term for wages, salary. In central Europe during the Middle Ages, the Catholic Church controlled the salt mines, a source of wealth and power. Centuries later, the salt taxes in France were part of the cause of the French Revolution.

Seawater is a 3 percent solution of sodium chloride, together with many other minerals. It has been calculated that the sea contains 19 million cubic meters of salt—about one and a half times the volume of all North America above sea level. The salt produced by using the sun's energy to evaporate seawater used to be a major source of income for some Third World countries, such as the Turks and Caicos Islands. Unfortunately, production of salt by this method is no longer economically competitive, and the ensuing loss of income and employment has caused serious economic problems for these countries.

Even today, salt is a vital commodity. More sodium chloride is used for chemical manufacture than any other mineral, with world consumption exceeding 150 million tonnes per year. Today almost all commercially produced sodium chloride is extracted from vast underground deposits, often hundreds of meters thick. These beds were produced when large lakes evaporated to dryness hundreds of millions of years ago. About 40 percent of the rock salt is mined like coal, and the remainder is extracted by pumping water into the deposits and pumping out the saturated brine solution.

Potassium Chloride

Like sodium chloride, potassium chloride (commonly called potash) is recovered from ancient dried lake deposits, many of which are now deep underground. About half of the world's reserves of potassium chloride lie under the Canadian provinces of Saskatchewan, Manitoba, and New Brunswick. As the ancient lakes dried, all their soluble salts crystallized. Hence the deposits are not of pure potassium chloride but also contain crystals of sodium chloride,

Salt Substitutes

We need about 3 g of sodium chloride per day, but in Western countries our daily diet usually contains between 8 and 10 g. Provided we have sufficient liquid intake, this level of consumption presents no problem. However, for those with high blood pressure, a decrease in sodium ion intake has been shown to cause a reduction in blood pressure. To minimize sodium ion intake, there are a number of salt substitutes on the market that taste salty but do not contain the sodium ion. Most of these contain potassium chloride and other compounds that mask the bitter, metallic aftertaste of the potassium ion. One enterprising producer of pure household salt claims its product contains "33 percent less sodium." This claim is technically true; and it is accomplished by producing hollow salt crystals. These have a bulk density 33 percent less than the normal cubic crystals. Hence a spoonful of these salt crystals will contain 33 percent less of both sodium and chloride ions! Provided you sprinkle your food with the same volume of salt, it will obviously have the desired effect; but for the same degree of saltiness, you will need 50 percent more by volume of the product than regular salt.

potassium magnesium chloride hexahydrate, $KMgCl_3{\cdot}6H_2O$, magnesium sulfate monohydrate, $MgSO_4{\cdot}H_2O$, and many other salts.

To separate the components, several different routes are used. One employs the differences in solubility: The mixture is dissolved in water and then the salts crystallize out in sequence as the water evaporates. However, this process requires considerable amounts of energy to vaporize the water. A second route involves adding the mixture of crystals to saturated brine. When air is blown through the slurry, the potassium chloride crystals adhere to the bubbles. The potassium chloride froth is then skimmed off the surface. The sodium chloride crystals sink to the bottom and can be dredged out.

The third route is most unusual, because it is an electrostatic process. The solid is ground to a powder, and an electric charge is imparted to the crystals by a friction process. The potassium chloride crystals acquire a charge that is the opposite of that of the other minerals. The powder is then poured down a tower containing two highly charged drums. The potassium chloride adheres to one drum, from which it is continuously removed, and the other salts adhere to the oppositely charged drum. Unfortunately, the reject minerals from potash processing have little use, and their disposal is a significant problem.

There is just one use for all this potassium chloride—as fertilizer. Potassium ion is one of the three essential elements for plant growth (nitrogen and phosphorus being the other two), and about 4.5×10^7 tonnes of potassium chloride are used worldwide for this purpose every year, so it is a major chemical product.

Sodium Carbonate

The alkali metals (and ammonium ion) form the only soluble carbonates. Sodium carbonate, the most important of the alkali metal carbonates, exists in the anhydrous state (soda ash), as a monohydrate, $Na_2CO_3{\cdot}H_2O$, and most commonly the decahydrate, $Na_2CO_3{\cdot}10H_2O$ (washing soda). The large transparent crystals of the decahydrate effloresce (lose water of crystallization) in dry air to form a powdery deposit of the monohydrate:

$$Na_2CO_3{\cdot}10H_2O(s) \rightarrow Na_2CO_3{\cdot}H_2O(s) + 9\ H_2O(g)$$

Preparation of Sodium Carbonate

Sodium carbonate is the ninth most important inorganic compound in terms of quantity used. For over half a century, sodium bicarbonate (and from it, the carbonate) were made by the Solvay, or ammonia-soda, process. However, the environmental problems associated with the reactions used in this process have made it preferable to mine underground deposits. By far the largest quantity in the world, 4.5×10^{10} tonnes, is found in Wyoming. The mineral, known as trona, contains about 90 percent of a mixed carbonate-hydrogen carbonate, $Na_2CO_3{\cdot}NaHCO_3{\cdot}2H_2O$, which is commonly called sodium sesquicarbonate. *Sesqui* means "one and one-half," and it is the number of sodium ions per carbonate unit in the mineral. Sodium sesquicarbonate is not a mixture of the two compounds but a single compound in which the crystal lattice contains alternating carbonate and hydrogen carbonate ions interspersed with sodium ions and water molecules in a 1:1:3:2 ratio, that is, $Na_3(HCO_3)(CO_3){\cdot}2H_2O$.

In the monohydrate process of extraction, trona is mined like coal about 400 m underground, crushed, and then heated (calcined) in rotary kilns. This treatment converts the sesquicarbonate to carbonate:

$$2\ [Na_2CO_3{\cdot}NaHCO_3{\cdot}2H_2O(s)] \xrightarrow{\Delta} 3\ Na_2CO_3(s) + 5\ H_2O(g) + CO_2(g)$$

The resulting sodium carbonate is dissolved in water and the insoluble impurities filtered off. The sodium carbonate solution is then evaporated to dryness, thereby producing sodium carbonate monohydrate. Heating this product in a rotary kiln gives the anhydrous sodium carbonate:

$$Na_2CO_3{\cdot}H_2O(s) \xrightarrow{\Delta} Na_2CO_3(s) + H_2O(g)$$

The Californian salt lake deposits contain about 6×10^8 tonnes of sodium carbonate, and this also is mined. However, even though the last Solvay plant in the United States closed in 1986, many other countries have to rely on this chemical process for their sodium carbonate supplies. In fact, about 70 percent of the world's supply of this reagent still comes from the Solvay process. For this reason, it is worth discussing the process. Besides, there is some interesting chemistry involved in the procedure. The overall reaction uses two low-cost reagents, and it appears to be quite simple:

$$2\ NaCl(aq) + CaCO_3(s) \rightleftharpoons Na_2CO_3(aq) + CaCl_2(aq)$$

However, the equilibrium position for this reaction lies far to the left. That is, the converse reaction will occur: Calcium chloride will react with sodium carbonate to give a precipitate of calcium carbonate and a solution of sodium chloride. It is fortunate that the equilibrium does favor the left side. Otherwise the White Cliffs of Dover would have long ago dissolved in the salt water of the English Channel!

To force the reverse reaction and obtain the desired product of sodium carbonate, an indirect, multistep procedure has to be used. In the first step, carbon dioxide is forced into a solution saturated with sodium chloride and ammonia. The carbon dioxide reacts with the ammonia to give ammonium ions and hydrogen carbonate ions:

$$CO_2(g) + NH_3(aq) + H_2O(l) \rightarrow NH_4^+(aq) + HCO_3^-(aq)$$

The solution now contains sodium and ammonium cations and chloride and hydrogen carbonate anions. Upon cooling, the relatively low solubility of sodium hydrogen carbonate in cold water causes it to crystallize:

$$HCO_3^-(aq) + Na^+(aq) \rightarrow NaHCO_3(s)$$

The solid sodium hydrogen carbonate is filtered off and then gently heated to give the carbonate:

$$2\ NaHCO_3(s) \rightarrow Na_2CO_3(s) + H_2O(g) + CO_2(g)$$

The commercial success of the process depends on the recovery of ammonia by the reaction:

$$2\ NH_4^+(aq) + 2\ Cl^-(aq) + Ca(OH)_2(s) \rightarrow 2\ NH_3(g) + CaCl_2(aq) + 2\ H_2O(l)$$

The calcium hydroxide and carbon dioxide used in the process are both obtained by heating limestone strongly:

$$CaCO_3(s) \xrightarrow{\Delta} CaO(s) + CO_2(g)$$

$$CaO(s) + H_2O(l) \rightarrow Ca(OH)_2(s)$$

Summing these six equations gives the equation for the overall process:

$$2\ NaCl(aq) + CaCO_3(s) \rightarrow Na_2CO_3(aq) + CaCl_2(aq)$$

The problem with the Solvay process is the amount of the by-product calcium chloride that is produced. The demand for calcium chloride is much less than the supply from this reaction. Furthermore, the process is quite energy intensive, making it more expensive than the simple method of extraction from trona.

Uses of Sodium Carbonate

About 50 percent of the U.S. production of sodium carbonate is used in glass manufacture. In the process, the sodium carbonate is reacted with silicon dioxide (sand) and other components at about 1500°C. The actual formula of the product depends on the stoichiometric ratio of reactants (the process is discussed in more detail in Chapter 13). The key reaction is the formation of a sodium silicate and carbon dioxide:

$$Na_2CO_3(l) + x\ SiO_2(s) \rightarrow Na_2O{\cdot}xSiO_2(l) + CO_2(g)$$

Sodium carbonate also is used to remove alkaline earth metal ions from water supplies by converting them to their insoluble carbonates, a process called water "softening." The most common ion that needs to be removed is calcium. Very high concentrations of this ion are found in water supplies that have come from limestone or chalk geological formations:

$$CO_3^{2-}(aq) + Ca^{2+}(aq) \rightarrow CaCO_3(s)$$

Sodium Hydrogen Carbonate

The alkali metals, except for lithium, form the only solid hydrogen carbonates (commonly called bicarbonates). Once again, the notion that low charge density cations stabilize large low-charge anions can be used to explain the existence of these hydrogen carbonates.

Sodium hydrogen carbonate is less water-soluble than sodium carbonate. Thus it can be prepared by bubbling carbon dioxide through a saturated solution of the carbonate:

$$Na_2CO_3(aq) + CO_2(g) + H_2O(l) \rightarrow 2\ NaHCO_3(s)$$

Heating sodium hydrogen carbonate causes it to decompose back to sodium carbonate:

$$2\ NaHCO_3(s) \xrightarrow{\Delta} Na_2CO_3(s) + CO_2(g) + H_2O(g)$$

This reaction provides one application of sodium hydrogen carbonate, the major component in dry powder fire extinguishers. The sodium hydrogen carbonate powder itself smothers the fire, but, in addition, the solid decomposes to give carbon dioxide and water vapor, themselves fire-extinguishing gases.

The main use of sodium hydrogen carbonate is in the food industry, to cause bakery products to rise. It is commonly used as a mixture (baking powder) of sodium hydrogen carbonate and calcium dihydrogen phosphate, $Ca(H_2PO_4)_2$, with some starch added as a filler. The calcium dihydrogen phosphate is acidic and, when moistened, reacts with the sodium hydrogen carbonate to generate carbon dioxide:

$$2\ NaHCO_3(s) + Ca(H_2PO_4)_2(s) \rightarrow Na_2HPO_4(s) + CaHPO_4(s) + 2\ CO_2(g) + 2\ H_2O(g)$$

Ammonia Reaction

The alkali metals themselves have the interesting property of dissolving in liquid ammonia to yield solutions that are deep blue when dilute. These solutions conduct current electrolytically, and the main current carrier in the solution is thought to be the solvated electron that is a product of the ionization of the sodium atoms:

$$Na(s) \rightleftharpoons Na^+(NH_3) + e^-(NH_3)$$

When concentrated by evaporation, the solutions have a bronze color and behave like a liquid metal. On long standing, or more rapidly in the presence of a transition metal catalyst, the solutions decompose to yield the amide salt, $NaNH_2$, and hydrogen gas:

$$2\ Na^+(NH_3) + 2\ NH_3(l) + 2\ e^- \rightarrow 2\ NaNH_2(NH_3) + H_2(g)$$

Ammonium as a Pseudo–Alkali Metal Ion

Even though the ammonium ion is a polyatomic cation containing two nonmetals, it behaves in many respects like an alkali metal ion. In particular, all of its salts are soluble, as are those of the alkali metals. These ions all behave similarly because the ammonium ion is a large low-charge cation just like those of the alkali metals. In fact, the radius of the ammonium ion (151 pm) is very close to that of the potassium ion (152 pm).

However, the similarity does not extend to all chemical reactions that these ions undergo. For example, heating sodium nitrate gives sodium nitrite and oxygen gas, but heating ammonium nitrate results in decomposition of the cation and anion to give dinitrogen oxide and water:

$$2\ NaNO_3(s) \xrightarrow{\Delta} 2\ NaNO_2(s) + O_2(g)$$

$$NH_4NO_3(s) \xrightarrow{\Delta} N_2O(g) + 2\ H_2O(g)$$

Similarities Between Lithium and the Alkaline Earth Metals

As was noted earlier, the chemistry of lithium is strikingly different from that of the other alkali metals, more closely resembling that of the alkaline earth elements (see Chapter 11). Some of these similarities with the neighboring group are listed below.

1. The hardness of lithium metal is greater than that of the alkali metals but similar to those of the alkaline earth metals.

2. Like the alkaline earth metals but unlike the alkali metals, lithium forms a normal oxide, not a dioxide(2–) or a dioxide(1–).

3. Lithium is the only alkali metal to form a nitride, whereas the alkaline earth metals all form nitrides.

4. Similarly, lithium is the only alkali metal to react with carbon to form a dicarbide(2–), Li_2C_2 (commonly called lithium acetylide), whereas the alkaline earth metals all form dicarbides(2–).

5. Three lithium salts—the carbonate, the phosphate, and the fluoride—have very low solubilities. These anions also form insoluble salts with the alkaline earth metals.

Table 10.11 Radii and charge densities for the alkali metals and the alkaline earth metal ions

Ion	Charge density ($C \cdot mm^{-3}$)	Ion	Charge density ($C \cdot mm^{-3}$)
Li^+	98	—	—
Na^+	24	Mg^{2+}	120
K^+	11	Ca^{2+}	52
Rb^+	8	Sr^{2+}	33
Cs^+	6	Ba^{2+}	23

6. Lithium forms organometallic compounds similar to those of magnesium.

7. Many lithium salts exhibit a high degree of covalency in their bonding. This bonding is similar to that of magnesium.

8. The lithium and magnesium carbonates decompose to give the appropriate metal oxide, carbon dioxide, and dioxygen. The carbonates of the other alkali metals do not decompose when heated.

The relationship between lithium and the alkaline earth metals is sometimes referred to as a *diagonal relationship*—that is, the resemblance of the properties of a Period 2 element to those of the Period 3 element to its lower right. For lithium, this element is magnesium.

How can we explain this? Because the lithium ion is smaller than the other alkali metals, its positive ion charge is concentrated in a smaller volume; hence the ion has a greater polarizing power. An examination of the charge densities of the elements in Groups 1 and 2 (Table 10.11) reveals that the charge density of lithium is much closer to that of magnesium than to those of the other alkali metals. Hence similarity in charge density may explain the similarity in the chemical behaviors of lithium and magnesium.

Table 10.11 also reveals that sodium and barium have very similar charge densities. There are indeed similarities in their behaviors. For example, upon reaction with dioxygen, both elements form a dioxide(2–), Na_2O_2 and BaO_2.

Biological Aspects

We tend to forget that both sodium and potassium ions are essential to life. For example, we need at least a gram of sodium ion per day in our diet. However, because of our addiction to salt on foods, the intake of many people is as much as five times that value. Excessive intake of potassium ion is rarely a problem. In fact, potassium deficiency is much more common; thus it is important to ensure that we include in our diets potassium-rich foods such as bananas and coffee.

The main use of the alkali metal ions is to balance the negative charge associated with many of the protein units in the body. They also help to maintain the osmotic pressure within cells, preventing them from collapse. Of particular importance, the concentrations of potassium and sodium ions

Lithium and Mental Health

The saga of the use of lithium ion to treat a mental disorder is an example of how a discovery is made through a combination of accident (serendipity) and observation. In 1938 an Australian psychiatrist, J. Cade, was studying the effects of a large organic anion on animals. To increase the dosage, he needed a more soluble salt. For large anions, the solubilities of the alkali metal ions increase as their radius decreases; hence he chose the lithium salt. However, when he administered this compound, the animals started to show behavioral changes. He realized that the lithium ion itself must have had an effect on the workings of the brain. Further studies showed that the lithium ion had a profound effect on manic depressive patients. To this day, lithium ion is the safest and most effective treatment for manic depression, although careful dosage and monitoring are crucial because too much lithium ion can cause cardiac arrest (a blood level of 1×10^{-3} mol·L^{-1} is optimal). Its mechanism of action is still not well understood, but it may work by affecting the sodium-potassium or magnesium-calcium balance in the body.

Ironically, the discovery of the health effects of lithium could have been made much earlier, because it had been well known in folk medicine that water from certain lithium-rich British springs helped alleviate the disorder. More recently, a study in Texas showed that locations having lower levels of hospital admissions with manic depression correlated with higher levels of lithium ion in the local drinking water.

are strikingly different inside and outside of cells (Table 10.12). It is this difference in total alkali metal ion concentrations inside and outside cells that produces an electrical potential across the cell membrane. This potential is essential for the functioning of nerves and muscles. The cell walls contain biological ion pumps that selectively move potassium ions into the cells and remove excessive concentrations of sodium ions. When we "go into shock" as a result of an accident, it is a massive leakage of the alkali metal ions through the cell walls that causes the phenomenon.

A number of antibiotics seem to be effective because they have the ability to transfer specific ions across cell membranes. These organic molecules have holes in the middle that are just the right size to accommodate an ion with a particular ionic radius. For example, valinomycin has an aperture that is just right for holding a potassium ion but too large for a sodium ion. Thus the drug functions by selectively transporting potassium ions across biological membranes.

Table 10.12 Concentrations of ions (mmol·L^{-1})

Ion	[Na^+]	[K^+]
Red blood cells	11	92
Blood plasma	160	10

Exercises

10.1. Write balanced chemical equations for each of the following reactions:
(a) sodium metal with water
(b) rubidium metal with dioxygen
(c) solid potassium hydroxide with carbon dioxide
(d) heating solid sodium nitrate

10.2. Write balanced chemical equations for each of the following reactions:
(a) lithium metal with dinitrogen
(b) solid cesium dioxide(1–) with water
(c) heating solid sodium hydrogen carbonate
(d) heating solid ammonium nitrate

10.3. In what ways do the alkali metals resemble "typical" metals? In what ways are they very different?

10.4. Which is the least reactive alkali metal? Why is this unexpected on the basis of standard oxidation potentials? What explanation can be provided?

10.5. Describe three of the common features of the chemistry of the alkali metals.

10.6. An alkali metal, designated as M, forms a hydrated sulfate, $M_2SO_4{\cdot}10H_2O$. Is the metal more likely to be sodium or potassium? Explain your reasoning.

10.7. Suggest a possible reason why sodium hydroxide is much more water-soluble than sodium chloride.

10.8. In the box on lithium and mental health, we mention that the researcher used the lithium salt in his experiments with a large organic anion because the compound was more water-soluble. Why would you expect this?

10.9. The Downs cell is used for the preparation of sodium metal.
(a) Why can't the electrolysis be performed in aqueous solution?
(b) Why is calcium chloride added?

10.10. Why is it important to use a temperature of about 850°C in the extraction of potassium metal?

10.11. Describe both the advantages and disadvantages of the diaphragm cell for the production of sodium hydroxide.

10.12. Several of the alkali metal compounds have common names. Give the systematic name corresponding to (a) caustic soda; (b) soda ash; (c) washing soda.

10.13. Several of the alkali metal compounds have common names. Give the systematic name corresponding to (a) caustic potash; (b) trona; (c) lye.

10.14. Explain what is meant by (a) efflorescence; (b) diagonal relationship.

10.15. Explain what is meant by (a) supermetals; (b) deliquescent.

10.16. Write the chemical equations for the reactions involved in the Solvay synthesis of sodium carbonate. What are the two major problems with this process?

10.17. Explain briefly why only the alkali metals form solid, stable hydrogen carbonate salts.

10.18. Explain briefly why the ammonium ion is often referred to as a pseudo-alkali metal.

10.19. List five ways in which lithium resembles the alkaline earth elements.

10.20. Suggest two reasons why potassium dioxide(1–), not cesium dioxide(1–), is used in the air recirculation systems of spacecraft.

10.21. Where are the sodium ions and potassium ions located with respect to living cells?

10.22. In this chapter, we have ignored the radioactive member of Group 1, francium. On the basis of group trends, suggest the key properties of francium and its compounds.

CHAPTER

11

The Group 2 Elements: The Alkaline Earth Metals

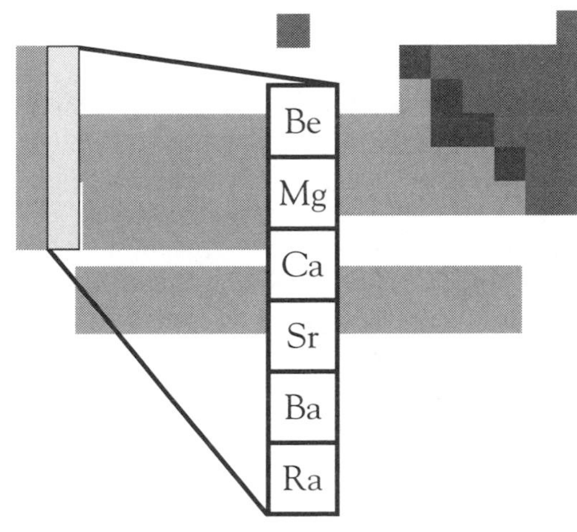

Although harder, denser, and less reactive than the alkali metals, the alkaline earth metals are more reactive and of lower density than a "typical" metal.

The last of the alkaline earth metals to be extracted from its compounds was radium. Marie Curie and André Debierne accomplished this task in 1910, delighting in the bright glow from this element, not realizing that it was the result of the element's intense and dangerous radiation. During the 1930s, cabaret shows sometimes featured dancers painted with radium salts so that they would literally glow in the dark. Some of the dancers may have died of radiation-related diseases, never being aware of the cause. Even quite recently, it was possible to purchase watches with hands and digits painted with radium-containing paint, its glow enabling the owner to read the time in the dark. (Safer substitutes are now available.)

Table 11.1 Densities of the common alkaline earth metals

Element	Density ($g{\cdot}cm^{-3}$)
Mg	1.74
Ca	1.55
Sr	2.63
Ba	3.62

Group Trends

In this section, we consider the properties of magnesium, calcium, strontium, and barium. Beryllium is discussed separately, because it behaves chemically more like a semimetal. The properties of radium, the radioactive member of the group, are not known in as much detail.

The alkaline earth metals are silvery and of fairly low density. As with the alkali metals, density generally increases with increasing atomic number (Table 11.1). The alkaline earth metals have stronger metallic bonding than do the alkali metals, a characteristic that is evident from the significantly greater enthalpies of atomization (Table 11.2). The metallic bonding of the alkaline earth metals is also reflected in both their higher melting points and their greater hardness. Although the density increases down the group (in parallel with that of the alkali metals), the melting points and enthalpies of atomization change very little. The ionic radii increase down the group and are smaller than those of the alkali metals (Figure 11.1).

The alkaline earth metals are less chemically reactive than are the alkali metals, but they are still more reactive than the majority of the other metallic elements. For example, calcium, strontium, and barium all react with cold water, barium reacting the most vigorously of all:

$$Ba(s) + 2\ H_2O(l) \rightarrow Ba(OH)_2(aq) + H_2(g)$$

As with the alkali metals, reactivity increases as mass increases within the group. Thus magnesium does not react with cold water, but it will react slowly with hot water to produce magnesium hydroxide and hydrogen gas.

Table 11.2 Melting points of the common alkaline earth metals

Element	Melting point (°C)	$\Delta H_{atomization}$ ($kJ{\cdot}mol^{-1}$)
Mg	649	149
Ca	839	177
Sr	768	164
Ba	727	175

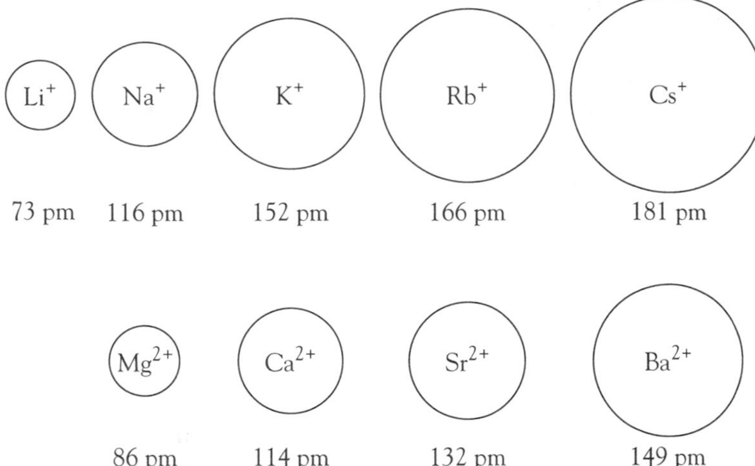

Figure 11.1 Comparison of the ionic radii of the alkali and alkaline earth metals.

Common Features of Alkaline Earth Metal Compounds

We again exclude beryllium from the discussion because its properties are very different from those of the other members of Group 2.

Ionic Character

The alkaline earth metal ions always have an oxidation number of +2, and their compounds are mainly stable, colorless, ionic solids—unless a colored anion is present. The bonds in alkaline earth metal compounds are mostly ionic in character, but covalent behavior is particularly evident in compounds of magnesium. (Covalency dominates the chemistry of beryllium.)

Ion Hydration

The salts of the alkaline earth metals are almost always hydrated. Table 11.3 shows the maximum number of molecules of water of crystallization in some common alkaline earth metal compounds. As the charge density of the metal becomes smaller, so does the number of molecules of water.

Table 11.3 Maximum waters of hydration for common alkaline earth metal salts

Element	MCl_2	$M(NO_3)_2$	MSO_4
Mg	12	9	12
Ca	6	4	2
Sr	6	4	0
Ba	2	0	0

Solubility of Alkaline Earth Metal Salts

Whereas all the common Group 1 salts are water-soluble, many of those of Group 2 are insoluble. Generally it is the compounds with mononegative anions, such as chloride and nitrate, that are soluble, whereas those with more than one negative charge, such as carbonate and phosphate, are insoluble. There are also a few anions that show striking trends in solubility. In particular, the sulfates change from soluble to insoluble down the group, whereas the hydroxides change from insoluble to soluble.

In Chapter 10 we discussed the solubility of alkali metal halides in terms of thermodynamic functions. For the alkaline earth metals, the values of each function differ dramatically from those of the alkali metals, yet the net changes in entropy and enthalpy for the solution process are little different.

First let us consider the enthalpy factors involved. The initial step of our enthalpy cycle is vaporization of the crystal lattice. For a salt of a dipositive cation, about three times the energy will be needed to vaporize the lattice as is needed for a monopositive cation, because there are much greater electrostatic attractions in the dipositive cation salts (2+ charge with 1–, versus 1+ with 1–). Furthermore, per mole, three ions must be separated rather than two. However, the enthalpy of hydration of the dipositive ions will also be much greater than those of the monopositive alkali metal ions. As a result of the higher charge densities of the Group 2 ions, the water molecules are more strongly attracted to the "naked" cation, so there is a much greater release of energy when they form a solvation sphere around it. For example, the enthalpy of hydration of the magnesium ion is –1921 $kJ{\cdot}mol^{-1}$, whereas that of the sodium ion is –435 $kJ{\cdot}mol^{-1}$. Enthalpy data for magnesium chloride and sodium chloride are compared in Table 11.4. As these figures indicate, when (anhydrous) magnesium chloride is dissolved in water, the solution process is noticeably exothermic.

Now let us consider the entropy factors (Table 11.5). The lattice entropy of magnesium chloride is almost exactly one and a half times that of sodium chloride, reflecting the fact that three gaseous ions rather than two are being produced. However, because the magnesium ion has a much higher charge density, the entropy of hydration for the magnesium ion is significantly more negative than that for the sodium ion. There is a much more ordered environment around the magnesium ion, which is surrounded by the strongly held layers of water molecules. Thus, overall, the entropy factors do not favor the solution process for magnesium chloride. Recall that for sodium chloride, it was the entropy factor that favored solution.

Table 11.4 Enthalpy factors in the solution process for magnesium chloride and sodium chloride

Compound	Lattice energy ($kJ{\cdot}mol^{-1}$)	Hydration enthalpy ($kJ{\cdot}mol^{-1}$)	Net enthalpy change ($kJ{\cdot}mol^{-1}$)
$MgCl_2$	+2526	–2659	–133
NaCl	+788	–784	+4

Table 11.5 Entropy factors in the solution process for magnesium chloride and sodium chloride, expressed as $T\Delta S$ values

Compound	Lattice entropy ($kJ \cdot mol^{-1}$)	Hydration entropy ($kJ \cdot mol^{-1}$)	Net entropy change ($kJ \cdot mol^{-1}$)
$MgCl_2$	+109	−143	−34
NaCl	+68	−55	+13

Table 11.6 Calculated free energy changes for the solution process for magnesium chloride and sodium chloride

Compound	Enthalpy change ($kJ \cdot mol^{-1}$)	Entropy change ($kJ \cdot mol^{-1}$)	Free energy change ($kJ \cdot mol^{-1}$)
$MgCl_2$	−133	−34	−99
NaCl	+4	+13	−11

When we combine the enthalpy and entropy terms—keeping in mind that all of the data values have associated errors—we see that the solubility process results primarily from very small differences in very large energy terms (Table 11.6). Furthermore, for magnesium chloride, enthalpy factors favor solution and entropy factors oppose them, a situation that is the converse of that for sodium chloride.

It is the much higher lattice energy that partially accounts for the insolubility of the salts containing di- and trinegative ions. As the charge increases, so does the electrostatic attraction that must be overcome in the lattice vaporization step. At the same time, there are fewer ions (two for the metal sulfates compared with three for the metal halides); hence the total ion hydration enthalpy will be less than that for the salts with mononegative ions. The combination of these two factors, then, is responsible for the low solubility.

Beryllium

Beryllium is a unique member of Group 2. The element is steel gray and hard; it has a high melting temperature and a low density. It also has a high electrical conductivity, so it is definitely a metal. Because of beryllium's resistance to corrosion, its low density, high strength, and nonmagnetic behavior, beryllium alloys are often used in precision instruments such as gyroscopes. A minor but crucial use is in the windows of X-ray tubes. Absorption of X-rays increases with the square of the atomic number, and beryllium has the lowest atomic number of all the air-stable metals. Hence it is one of the most transparent materials for the X-ray spectrum.

The source of beryllium is the gemstone beryl, $Be_3Al_2Si_6O_{18}$, which occurs in various colors because of trace amounts of impurities. When it is a light blue-green, beryl is called aquamarine; when it is deep green, it is called emerald. The green color is due to the presence of about 2 percent chromium(III) ion in the crystal structure. Of course, emeralds are not used for the production of metallic beryllium; the very imperfect crystals of colorless or brown beryl are used instead.

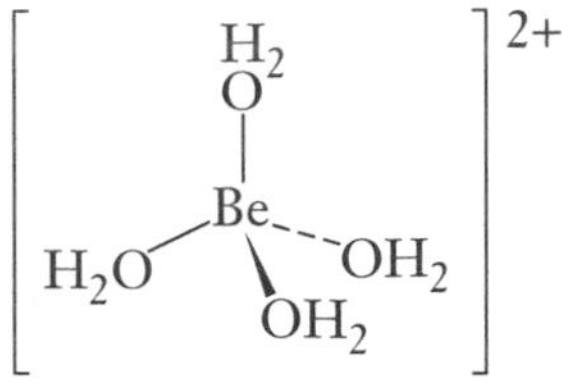

Beryllium compounds have a sweet taste and are extremely poisonous. When new compounds were prepared in the nineteenth century, it was quite common to report taste as well as melting point and solubility! Inhalation of the dust of beryllium compounds results in a chronic condition known as berylliosis.

The chemistry of beryllium is significantly different from that of the other Group 2 elements because covalent bonding predominates in its compounds. The very small beryllium cation has such a high charge density ($1100\ C{\cdot}mm^{-3}$) that it polarizes any approaching anion, and overlaps of electron density occur. Hence there are no crystalline compounds or solutions containing a free Be^{2+} ion. The only ionic species of beryllium are those in which the ion charge can be delocalized over several atoms. The most common example is the tetraaquaberyllium ion, $Be(OH_2)_4{}^{2+}$, in which the four oxygen atoms of the water molecules are covalently bonded to the beryllium ion. Four coordination is the norm for beryllium, because only *s* and *p* orbitals are available to this Period 2 element for covalent bonding (Figure 11.2).

Figure 11.2 Tetrahedral shape of the $[Be(OH_2)_4]^{2+}$ ion.

Evidence that beryllium is a "borderline" metal comes from the reactions of beryllium oxide. Metal oxides generally react with acids to give cations but not with bases to form oxyanions. However, beryllium does both; that is, it is *amphoteric*. Thus beryllium oxide reacts with hydronium ion to form the tetraaquaberyllium ion, $[Be(OH_2)_4]^{2+}$, and with hydroxide ion to form the tetrahydroxoberyllate ion, $[Be(OH)_4]^{2-}$:

$$H_2O(l) + BeO(s) + 2\ H_3O^+(aq) \rightarrow [Be(OH_2)_4]^{2+}(aq)$$

$$H_2O(l) + BeO(s) + 2\ OH^-(aq) \rightarrow [Be(OH)_4]^{2-}(aq)$$

Magnesium

Magnesium is found in nature as one component in a number of mixed-metal salts such as carnallite, $MgCl_2{\cdot}KCl{\cdot}6H_2O$, and dolomite, $MgCO_3{\cdot}CaCO_3$. These compounds are not simply mixtures of salts but are pure ionic crystals in which the alternating sizes of the cations confer on the crystal lattice a greater stability than that conferred by either cation alone. Thus carnallite contains arrays of chloride anions with interspersed potassium and magnesium cations and water molecules in a ratio of 3:1:1:6, that is, $KMgCl_3{\cdot}6H_2O$.

Magnesium is the third most common ion in seawater (after sodium and chloride), and seawater is a major industrial source of this metal. In fact, 1 km^3 of seawater contains about 1 million tonnes of magnesium ion. With $10^8\ km^3$ of seawater on this planet, there is more than enough magnesium for our needs. The Dow Chemical extraction process is based on the fact that magnesium hydroxide has a lower solubility than calcium hydroxide does. Thus a suspension of finely powdered calcium hydroxide is added to the seawater, causing magnesium hydroxide to form:

$$Ca(OH)_2(s) + Mg^{2+}(aq) \rightarrow Ca^{2+}(aq) + Mg(OH)_2(s)$$

The hydroxide is then filtered off and mixed with hydrochloric acid. The resulting neutralization reaction gives a solution of magnesium chloride:

$$Mg(OH)_2(s) + 2\ HCl(aq) \rightarrow MgCl_2(aq) + 2\ H_2O(l)$$

The solution is evaporated to dryness, and the residue is placed in an electrolytic cell similar to the Downs cell used for the production of sodium. The magnesium collects on the surface of the cathode compartment and is siphoned off. The chlorine gas produced at the anode is reduced back to hydrogen chloride, which is then used to react with more magnesium hydroxide:

$$Mg^{2+}(MgCl_2) + 2\ e^- \rightarrow Mg(l)$$

$$2\ Cl^-(MgCl_2) \rightarrow Cl_2(g) + 2\ e^-$$

Magnesium metal oxidizes slowly in air at room temperature but very vigorously when heated. The burning magnesium gives an intense white light. The combustion of magnesium powder was used in early photography (and to a lesser extent today) as a source of illumination:

$$2\ Mg(s) + O_2(g) \rightarrow 2\ MgO(s)$$

The combustion reaction is so vigorous that it cannot be extinguished by using a conventional fire extinguisher material such as carbon dioxide. Burning magnesium even reacts with carbon dioxide to give magnesium oxide and carbon:

$$2\ Mg(s) + CO_2(g) \rightarrow 2\ MgO(s) + C(s)$$

To extinguish reactive metal fires, such as those of magnesium, a class D fire extinguisher must be used (classes A, B, and C are used to fight conventional fires). Class D fire extinguishers contain either graphite or sodium chloride. Graphite produces a solid coating of metal carbide over the combusting surface and effectively smothers the reaction. Sodium chloride melts at the temperature of the burning magnesium and forms an inert liquid layer over the metal surface; so it too prevents oxygen from reaching the metal.

Over half of the approximately 4×10^5 tonnes of magnesium metal produced worldwide is used in aluminum-magnesium alloys. The attraction of these alloys is primarily their low density. Magnesium, with a density less than twice that of water ($1.74\ g{\cdot}cm^{-3}$), is the lowest density construction metal. Such alloys are particularly important wherever the low density provides significant energy savings: in aircraft, railroad passenger cars, rapid transit vehicles, and bus bodies. For a period of time in the 1970s, these alloys were used in the superstructure of warships because the lower mass of the ship allowed higher speeds. However, during the Falkland Islands War, the Royal Navy discovered a major disadvantage of this alloy—its flammability when subjected to missile attack. The U.S. Navy had already experienced accidents with the same alloy. An appreciation of the high reactivity of the alkaline earth metals might have prevented these mishaps.

Even though magnesium is a chemically reactive metal, its reactivity is less than would be expected on the basis of its standard reduction potential of

–2.37 V, because a thin coating of magnesium oxide rapidly forms over any metal surface exposed to the atmosphere. This coating protects the rest of the metal from attack. As we mentioned earlier, ignition with a source of heat or an electric current produces an intense white flame as the metal oxidizes to magnesium oxide.

In its chemistry, magnesium differs from the remaining Group 2 metals. For example, heating calcium, strontium, or barium chlorides causes release of the bound water molecules as steam, leaving the anhydrous metal chloride behind. For example

$$CaCl_2{\cdot}2H_2O(s) \rightarrow CaCl_2(s) + 2\ H_2O(l)$$

However, magnesium chloride monohydrate decomposes when heated to give magnesium chloride hydroxide and hydrogen chloride gas:

$$MgCl_2{\cdot}H_2O(s) \rightarrow Mg(OH)Cl(s) + HCl(g)$$

Magnesium readily forms compounds containing covalent bonds. This behavior can be explained in terms of its comparatively high charge density (120 $C{\cdot}mm^{-3}$; calcium's charge density is 52 $C{\cdot}mm^{-3}$). For example, magnesium metal reacts with organic compounds called halocarbons (or alkyl halides) such as bromoethane, C_2H_5Br, in a solvent such as ethoxyethane, $(C_2H_5)_2O$, commonly called ether. The magnesium atom inserts itself between the carbon and halogen atoms, forming covalent bonds to its neighbors:

$$C_2H_5Br(ether) + Mg(s) \rightarrow C_2H_5MgBr(ether)$$

This class of compounds is referred to as *Grignard reagents*, and they are used extensively as intermediates in synthetic organic chemistry.

Calcium and Barium

Both of these elements are grayish metals that react slowly with the oxygen in air at room temperature but burn vigorously when heated. Calcium burns to give only the oxide:

$$2\ Ca(s) + O_2(g) \rightarrow 2\ CaO(s)$$

whereas barium forms some dioxide(2–) in excess oxygen:

$$2\ Ba(s) + O_2(g) \rightarrow 2\ BaO(s)$$

$$Ba(s) + O_2(g) \rightarrow BaO_2(g)$$

The formation of barium peroxide can be explained in terms of the charge density of barium ion (23 $C{\cdot}mm^{-3}$), which is as low as that of sodium (24 $C{\cdot}mm^{-3}$). Cations with such a low charge density are able to stabilize polarizable ions like the dioxide(2–) ion.

Whereas beryllium is transparent to X-rays, barium and calcium, both with high atomic numbers, are strong absorbers of this part of the electromagnetic spectrum. It is the calcium ion in bones that causes the skeleton to show up dark on X-ray film. The elements in the soft tissues do not absorb

X-rays, a property that presents a problem when one wants to visualize the stomach and intestine. Because barium ion is such a good X-ray absorber, swallowing a solution containing barium should be an obvious way of imaging these organs. There is one disadvantage—barium ion is extremely poisonous. Fortunately, barium forms an extremely insoluble salt, barium sulfate. This compound is so insoluble (2.4×10^{-3} g·L^{-1}) that a slurry in water can be safely swallowed, the organs X-rayed, and the compound later safely excreted.

Oxides

As mentioned earlier, the Group 2 metals burn in air to yield the normal oxides, except for the member of the group with the lowest charge density—barium—which also forms some barium peroxide. Magnesium oxide is insoluble in water, whereas the other alkaline earth metal oxides react with water to form the respective hydroxide. For example, strontium oxide forms strontium hydroxide:

$$SrO(s) + H_2O(l) \rightarrow Sr(OH)_2(s)$$

Magnesium oxide has a very high melting point, 2825°C, so bricks of this compound are useful as industrial furnace linings. Such high-melting materials are known as *refractory compounds*. Crystalline magnesium oxide is an unusual compound, because it is a good conductor of heat but a very poor conductor of electricity, even at high temperatures. It is this combination of properties that results in its crucial role in electric kitchen range elements. It conducts the heat rapidly from a very hot coil of resistance wire to the metal exterior of the element without allowing any of the electric current to traverse the same route.

Calcium oxide, commonly called quicklime, is produced in enormous quantities, particularly for use in steel production (see Chapter 20). It is formed by heating calcium carbonate very strongly (over 1170°C):

$$CaCO_3(s) \xrightarrow{\Delta} CaO(s) + CO_2(g)$$

Calcium oxide, a solid with a very high melting point, is unusual in a different way: When a flame is directed against blocks of calcium oxide, the blocks glow with a bright white light. This phenomenon is called *thermoluminescence*. Before the introduction of electric light, theaters were lighted by these glowing chunks of calcium oxide, hence the origin of the phrase "being in the limelight" for someone who attains a prominent position. Thorium(IV) oxide, ThO_2, exhibits a similar property, hence its use in the mantles of gas-fueled camping lights.

Calcium oxide reacts with water to form calcium hydroxide, a product that is referred to as hydrated lime or slaked lime:

$$CaO(s) + H_2O(l) \rightarrow Ca(OH)_2(s)$$

Hydrated lime is sometimes used in gardening to neutralize acid soils; however, it is not a wise way of accomplishing this because an excess of calcium hydroxide will make the soil too basic:

$$Ca(OH)_2(s) + 2\ H^+(aq) \rightarrow Ca^{2+}(aq) + 2\ H_2O(l)$$

Powdered limestone can be used more safely as a soil neutralizing agent:

$$CaCO_3(s) + 2\ H^+(aq) \rightarrow Ca^{2+}(aq) + CO_2(g) + H_2O(l)$$

Hydroxides

Whereas magnesium hydroxide is almost completely insoluble in water, calcium and strontium hydroxides are slightly soluble, and barium hydroxide is very soluble (see Table 11.7).

The insolubility of magnesium hydroxide gives it an important household use—as a stomach antacid. To neutralize excess stomach acid, one could consume hydroxide ion. However, a solution of hydroxide ion is extremely corrosive and would cause severe and painful burns if ingested. Instead, pure, finely ground solid magnesium hydroxide can be mixed with water to form a slurry called "milk of magnesia." The low solubility of the magnesium hydroxide means that there is a negligible concentration of free hydroxide ion in the suspension. When the compound reaches the stomach, the magnesium hydroxide neutralizes the excess hydrogen ion:

$$Mg(OH)_2(s) + 2\ H^+(aq) \rightarrow Mg^{2+}(aq) + 2\ H_2O(l)$$

A saturated solution of calcium hydroxide is referred to as limewater. The solution is one of the simplest confirmatory tests for carbon dioxide. Bubbling the gas through a calcium hydroxide solution first gives a white precipitate of calcium carbonate. Continued passage of carbon dioxide through the solution causes the precipitate to disappear as a solution of calcium hydrogen carbonate forms:

$$Ca(OH)_2(aq) + CO_2(g) \rightarrow CaCO_3(s) + H_2O(l)$$

$$CaCO_3(s) + H_2O(l) + CO_2(g) \rightarrow Ca(HCO_3)_2(aq)$$

It is this second step that has led to the slow deterioration of marble sculptures in the streets and parks of Europe, a process accelerated by the acid oxides in the industrially polluted air.

Table 11.7 Solubilities of the alkaline earth metal hydroxides

Hydroxide	Solubility ($g{\cdot}L^{-1}$)
Mg	0.0001
Ca	1.2
Sr	10
Ba	47

Calcium Carbonate

Calcium is the fifth most abundant element on Earth. It is found largely as calcium carbonate in the massive deposits of chalk, limestone, and marble that occur worldwide. Chalk was formed in the seas, mainly during the Cretaceous period, about 135 million years ago, from the calcium carbonate skeletons of countless marine organisms. Limestone was formed in the same seas, but as a simple precipitate, because the solubility of calcium carbonate was exceeded in those waters:

$$Ca^{2+}(aq) + CO_3^{2-}(aq) \rightleftharpoons CaCO_3(s)$$

Marble was formed when deposits of limestone became buried deep in the Earth's crust, where the combination of heat and pressure caused the limestone to melt. The molten calcium carbonate cooled again as it was pushed back up to the surface, eventually solidifying into the dense solid form that we call marble. Very pure calcium carbonate occurs in two different crystal forms, calcite and the much rarer Iceland spar. The latter crystal form is unusual in that it transmits two images of any object placed under it. The two images appear because the crystal has two different indices of refraction.

Caves like Carlsbad Caverns and Mammoth Cave occur in beds of limestone. These structures are formed when rainwater seeps into cracks in the limestone. During the descent of rain through the atmosphere, carbon dioxide dissolves in it. The reaction of this dissolved acid oxide with the calcium carbonate produces a solution of calcium hydrogen carbonate:

$$CaCO_3(s) + CO_2(aq) + H_2O(l) \rightarrow Ca(HCO_3)_2(aq)$$

The solution is later washed away, leaving a hole in the rock.

As was mentioned in Chapter 10, only the alkali metals have a charge density low enough to stabilize the large polarizable hydrogen carbonate ion. Hence, when the water evaporates from the solution of calcium hydrogen carbonate, the compound immediately decomposes back to solid calcium carbonate:

$$Ca(HCO_3)_2(aq) \rightarrow CaCO_3(s) + CO_2(g) + H_2O(l)$$

It is deposited calcium carbonate that forms the stalagmites growing up from a cave floor and the stalactites descending from the roof of the cave.

Calcium carbonate is another popular antacid. Even though it has the marginal advantage of increasing an individual's intake of an essential element, it has one disadvantage: The reaction with stomach acid produces carbon dioxide gas:

$$CaCO_3(s) + 2\,H^+(aq) \rightarrow Ca^{2+}(aq) + CO_2(g) + H_2O(l)$$

Curiously, calcium and magnesium ions have opposite effects on the human system; in other words, calcium ion is a constipating agent, whereas magnesium ion is a laxative. Certain antacids contain a mixture of the two ions to cancel out the effects of each. When traveling, it is advisable to drink low-

How Was Dolomite Formed?

One of the great mysteries of geochemistry is how the mineral dolomite was formed. Dolomite is found in vast deposits, including the whole of the Dolomite mountain range in Europe. The chemical structure is $CaMg(CO_3)_2$; that is, it consists of carbonate ions interspersed with alternating calcium and magnesium ions. Of particular interest, many of the world's hydrocarbon (oil) deposits are found in dolomite deposits. Yet this composition does not form readily. If you mix solutions of calcium ions, magnesium ions, and carbonate ions in the laboratory, you merely obtain a mixture of calcium carbonate crystals and magnesium carbonate crystals. For 200 years, geochemists have struggled with the problem of how such enormous deposits were formed. To form dolomite, temperatures of over 150°C are required—not typical conditions on the surface of the Earth! Furthermore, magnesium ion concentrations in seawater are far lower than those of calcium ion. The most popular idea is that beds of limestone were formed first and then buried deep in the Earth. Water rich in magnesium ion is then postulated to have circulated through pores in the rock, selectively replacing some of the calcium ions with magnesium ions. For this to happen uniformly throughout thousands of cubic kilometers of rock seems unlikely, but at this time, it is the best explanation that we have.

mineral-content bottled water rather than local tap water, because the tap water might well be higher in one or the other of the alkaline earth metal ions than your system has become used to at home, thereby causing undesirable effects. (Of course, in certain parts of the world, there is also the danger of more serious health problems from tap water supplies, such as bacterial and viral infections.)

Calcium carbonate is a common dietary supplement prescribed to help maintain bone density. In the form of powdered limestone (commonly called agricultural lime), it is added to farmland to increase the pH by reacting with acids in the soil. Attempts are being made to reduce the effects of acid rain on lake waters by adding large quantities of powdered limestone.

Cement

About 1500 B.C., it was first realized that a paste of calcium hydroxide and sand (mortar) could be used to bind bricks or stones together in the construction of buildings. The material slowly picked up carbon dioxide from the atmosphere, thereby converting the calcium hydroxide back to the hard calcium carbonate from which it had been made:

$$Ca(OH)_2(s) + CO_2(g) \rightarrow CaCO_3(s) + H_2O(g)$$

Between 100 B.C. and A.D. 400, the Romans perfected the use of lime mortar to construct buildings and aqueducts, many of which are still standing. They also made the next important discovery: that mixing volcanic ash with the lime mortar gave a far superior product. This material was the precursor of our modern cements.

The production of cement is one of the largest modern chemical industries. Worldwide production is about 700 million tonnes, with the United States producing about 10 percent of that figure. Cement is made by grinding together limestone and shales (a mixture of aluminosilicates) and heating the mixture to about 1500°C. The chemical reaction releases carbon dioxide and partially melts the components to form solid lumps called clinker. The clinker is ground to a powder, and a small quantity of calcium sulfate is mixed in. This mixture is known as *Portland cement*. Chemically, its main components are 26 percent dicalcium silicate, Ca_2SiO_4; 51 percent tricalcium silicate, Ca_3SiO_5; and 11 percent tricalcium aluminate, $Ca_3Al_2O_6$. When water is added, a number of complex hydration reactions take place. A typical idealized reaction can be represented as

$$2\,Ca_2SiO_4(s) + 4\,H_2O(l) \rightarrow Ca_3Si_2O_7{\cdot}3H_2O(s) + Ca(OH)_2(s)$$

The hydrated silicate, called tobermorite gel, forms strong crystals that adhere by means of strong silicon-oxygen bonds to the sand and aggregate (small rocks) that are mixed with the cement. Because the other product in this reaction is calcium hydroxide, the mixture should be treated as a corrosive material while it is hardening.

Calcium Chloride

Anhydrous calcium chloride is a white solid that absorbs moisture very readily (an example of deliquescence). As a result, it is sometimes used as a drying agent in the chemistry laboratory. The reaction to form the hexahydrate, $CaCl_2{\cdot}6H_2O$, is very exothermic, and this property is exploited commercially. One type of instant hot packs consists of two inner pouches, one containing water and the other, anhydrous calcium chloride. Squeezing the pack breaks the inner partition between the pouches and allows the exothermic hydration reaction to occur.

Anhydrous calcium chloride, instead of sodium chloride, is also used for melting ice. Calcium chloride works in two ways. First, its reaction with water is highly exothermic; and second, calcium chloride forms a freezing mixture that substantially reduces the melting point. Calcium chloride is very water-soluble: A mixture of 30 percent calcium chloride and 70 percent water by mass (the eutectic, or minimum, freezing mixture) will remain liquid down to −55°C, a temperature much lower than the −18°C produced by the best sodium chloride and water mixture. Another advantage of using the calcium salt is that the calcium ion causes less damage to plants than does the sodium ion.

The concentrated calcium chloride solution has a very "sticky" feel, and this property leads to another of its applications: It is sprayed on unpaved road surfaces to minimize dust problems. It is much less environmentally hazardous than oil, the other substance commonly used. The concentrated solution is also very dense, and for this reason, it is sometimes used to fill tires of earth-moving equipment to give them a higher mass and hence better traction.

Magnesium and Calcium Sulfates

Magnesium sulfate is found as the heptahydrate, $MgSO_4{\cdot}7H_2O$. It was named Epsom salts after the town in England where it was first discovered. Like all magnesium salts, it has a laxative effect, and this was once its main use. During the nineteenth century, one British hospital was using 2.5 tonnes per year on its patients! A few bottled mineral waters, such as Vichy, have high magnesium ion content. These should obviously be consumed in fairly small volumes.

Calcium sulfate is found as the dihydrate, $CaSO_4{\cdot}2H_2O$, known as gypsum. Mineral deposits of pure, high-density gypsum, called alabaster, have been used for delicate sculptures. It is also used in some brands of blackboard chalk. When heated to about 100°C, the hemihydrate, plaster of Paris, is formed:

$$CaSO_4{\cdot}2H_2O(s) \rightarrow CaSO_4{\cdot}\tfrac{1}{2}H_2O(s) + 1\tfrac{1}{2}\,H_2O(l)$$

This white powdery solid slowly reacts with water to form long interlocking needles of calcium sulfate dihydrate. It is the strong, meshing crystals of gypsum that give plaster casts their strength. A more correct common name would be "gypsum casts."

One of the major uses of gypsum is in the fire-resistant wallboard used for interior walls in houses and offices. Its nonflammability and low cost are two reasons for choosing this material. But why gypsum and not chalk? The answer lies with the gypsum dehydration reaction that yields the hemihydrate. In a fire, this reaction occurs. Because it is an endothermic process—to the extent of 446 $kJ{\cdot}mol^{-1}$—it absorbs energy from the fire. Furthermore, each mole of liquid water produced absorbs the enthalpy of vaporization of water—another 44 $kJ{\cdot}mol^{-1}$—as it becomes gaseous water. Finally, the gaseous water acts as an inert gas, decreasing the supply of dioxygen to the fire. Chalk, being anhydrous, offers no equivalent reactions.

Calcium Carbide

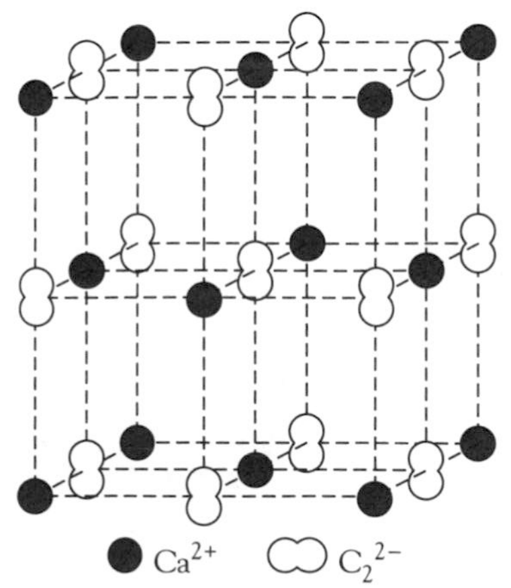

Figure 11.3 Crystal structure of calcium dicarbide(2–), which closely resembles the sodium chloride crystal structure.

Calcium forms an industrially important compound with carbon. Even though it is commonly called calcium carbide, the compound does not contain the carbide ion, C^{4-}. Instead, it contains the dicarbide(2–) ion, C_2^{2-}, which is commonly called the acetylide ion. The compound adopts the sodium chloride crystal structure, with the whole dicarbide ion occupying the anion site (Figure 11.3).

Calcium dicarbide is prepared by heating carbon (coke) and calcium oxide at about 2000°C in an electric furnace:

$$CaO(s) + 3\,C(s) \xrightarrow{\Delta} CaC_2(s) + CO(g)$$

Worldwide production has dropped from about 10 million tonnes in the 1960s to about 5 million tonnes in the 1990s—China is now the main producer—as the chemical industry has shifted to the use of oil and natural gas as the starting point for synthesizing organic compounds. The major use of the carbide process is to produce ethyne (acetylene) for oxyacetylene welding:

$$CaC_2(s) + 2\ H_2O(l) \rightarrow Ca(OH)_2(s) + C_2H_2(g)$$

Historically, miners' lamps and early automobile lamps used the combustion of the ethyne from the water-carbide reaction to provide light. Some cave explorers still use carbide lamps because they are so reliable and give such an intense light. The very exothermic reaction with dioxygen gives carbon dioxide and water vapor:

$$2\ C_2H_2(g) + 5\ O_2(g) \rightarrow 4\ CO_2(g) + 2\ H_2O(g)$$

Another important reaction of calcium carbide is that with atmospheric nitrogen, one of the few simple chemical methods of breaking the strong nitrogen-nitrogen triple bond. In the process, calcium carbide is heated in an electric furnace with nitrogen gas at about 1100°C:

$$CaC_2(s) + N_2(g) \xrightarrow{\Delta} CaCN_2(s) + C(s)$$

The cyanamide ion $[N{=}C{=}N]^{2-}$ is isoelectronic with carbon dioxide, and it also has the same linear structure. Calcium cyanamide is a starting material for the manufacture of several organic compounds, including melamine plastics. It is also used as a slow-release nitrogen-containing fertilizer:

$$CaCN_2(s) + 3\ H_2O(l) \rightarrow CaCO_3(s) + 2\ NH_3(aq)$$

Similarities Between Beryllium and Aluminum

Once again there are striking similarities between the first member of one group and the subsequent members of the next group. Let us compare beryllium and aluminum, which resemble each other in three ways:

1. In air, both metals form tenacious oxide coatings that protect the interior of the metal sample from attack.
2. Both elements are amphoteric, forming anions (beryllates and aluminates) in a reaction with concentrated hydroxide ion.
3. Both form carbides (Be_2C and Al_4C_3) that react with water to form methane, whereas the dicarbides(2–) of the other Group 2 elements react with water to form ethyne.

However, there are some major differences between the chemical properties of beryllium and aluminum. One of the most apparent differences is in the formula of the hydrated ions each forms. Beryllium forms the $[Be(OH_2)_4]^{2+}$ ion, whereas aluminum forms the $[Al(OH_2)_6]^{3+}$ ion. The lower coordination number of the beryllium may be accounted for in two ways: The beryllium atom has no *d* orbitals available for bonding, and the ion is physically too small to accommodate six surrounding water molecules at a bonding distance.

We can explain the big difference in behavior between beryllium and the other alkaline earth metals and beryllium's similarity to aluminum by looking at the charge density values (Table 11.8), as we did for lithium's relationship to the alkaline earth elements. There is clearly a greater similarity in the

Table 11.8 Charge densities for the alkaline earth metal ions and the aluminum ion

Period	Ion	Charge density ($C{\cdot}mm^{-3}$)	Ion	Charge density ($C{\cdot}mm^{-3}$)
2	Be^{2+}	1100	—	—
3	Mg^{2+}	120	Al^{3+}	364
4	Ca^{2+}	52		
5	Sr^{2+}	33		
6	Ba^{2+}	23		

charge densities of beryllium and aluminum than there is in the charge densities of beryllium with the other members of Group 2.

As discussed in Chapter 5, the ionic radii of the Period 2 elements are those of the tetrahedrally coordinate ions, whereas those for the larger, later-period ions are for the octahedrally coordinate ions. However, if we calculate the charge density of the aluminum ion by using the tetrahedrally coordinated radius for aluminum of 53 pm (which is not unreasonable for this very small ion), rather than the octahedrally coordinated value of 68 pm, a charge density of 770 $C{\cdot}mm^{-3}$ is obtained—a value even more comparable to that of beryllium.

Biological Aspects

The most important aspect of the biochemistry of magnesium is its role in photosynthesis. Magnesium-containing chlorophyll, using energy from the sun, converts carbon dioxide and water into sugars and oxygen:

$$6\,CO_2(g) + 6\,H_2O(l) \rightarrow C_6H_{12}O_6(aq) + 6\,O_2(g)$$

Without the oxygen from the chlorophyll reaction, this planet would still be blanketed in a dense layer of carbon dioxide; and without the sugar energy source, it would have been difficult for life to progress from plants to herbivorous animals. Interestingly, the magnesium ion seems to be used for its particular ion size and for its low reactivity. It sits in the middle of the chlorophyll molecule, holding the molecule in a specific configuration. Magnesium has only one possible oxidation number, +2. Thus the electron transfer reactions involved in photosynthesis can proceed without interference from the metal ion.

Both magnesium and calcium ions are present in body fluids. Mirroring the alkali metals, magnesium ion is concentrated within cells, whereas calcium ion is concentrated in the intracellular fluids. Calcium ions are important in blood clotting, and they are required to trigger the contraction of muscles, such as those that control the beating of the heart. In fact, certain types of muscle cramps can be prevented by increasing the intake of calcium ion.

Calcium ion is present in the external skeleton of creatures such as shellfish and corals. The material used is calcium carbonate. And creatures with interior skeletons, such as mammals and reptiles, use calcium hydroxide phosphate (apatite), $Ca_5(OH)(PO_4)_3$. Both these compounds are highly insoluble. A major health concern today is the low calcium intake among teenagers. Low levels of calcium lead to larger pores in the bone structure, and these weaker structures mean easier bone fracturing and a higher chance of osteoporosis in later life.

Exercises

11.1. Write balanced chemical equations for the following processes:
(a) heating calcium in dioxygen
(b) heating calcium carbonate
(c) evaporating a solution of calcium hydrogen carbonate
(d) heating calcium oxide with carbon

11.2. Write balanced chemical equations for the following processes:
(a) adding strontium to water
(b) passing sulfur dioxide over barium oxide
(c) heating calcium sulfate dihydrate
(d) adding strontium dicarbide to water

11.3. For the alkaline earth metals (except beryllium), which will (a) have the most insoluble sulfate? (b) have the softest metal?

11.4. For the alkaline earth metals (except beryllium), which will (a) have the most insoluble hydroxide? (b) have the greatest density?

11.5. Explain why entropy factors favor the solution of sodium chloride but not that of magnesium chloride.

11.6. Explain why the salts of alkaline earth metals with mononegative ions tend to be soluble whereas those with dinegative ions tend to be insoluble.

11.7. What are the two most important common features of the Group 2 elements?

11.8. Explain why the solid salts of magnesium tend to be highly hydrated.

11.9. Why does the hydrated beryllium ion have the formula $Be(OH_2)_4^{2+}$, whereas that of magnesium is $Mg(OH_2)_6^{2+}$?

11.10. How does the chemistry of magnesium differ from that of the lower members of the Group 2 metals? Suggest an explanation.

11.11. Explain briefly how caves are formed in limestone deposits.

11.12. What are the main raw materials for the manufacture of cement?

11.13. Summarize the industrial process for the extraction of magnesium from seawater.

11.14. How is calcium cyanamide obtained from calcium oxide?

11.15. Several of the alkaline earth metal compounds have common names. Give the systematic name for (a) lime; (b) milk of magnesia; (c) Epsom salts.

11.16. Several of the alkaline earth metal compounds have common names. Give the systematic name for (a) dolomite; (b) marble; (c) gypsum.

11.17. Why is lead commonly used as a shielding material for X-rays?

11.18. The dissolving of anhydrous calcium chloride in water is a very exothermic process. However, dissolving calcium chloride hexahydrate causes a very much smaller heat change. Explain this observation.

11.19. Discuss briefly the similarities between beryllium and aluminum.

11.20. In this chapter, we have ignored the radioactive member of the group, francium. On the basis of group trends, suggest the key properties of francium and its compounds.

11.21. Describe briefly the importance of magnesium ion to life on Earth.

11.22. What is the calcium-containing structural material in animals?

11.23. From tabulated data, calculate the enthalpy and entropy change when plaster of Paris is formed from gypsum. Calculate the temperature at which the process of dehydration becomes significant, that is, when $\Delta G^\circ = 0$.

CHAPTER

12

The Group 13 Elements

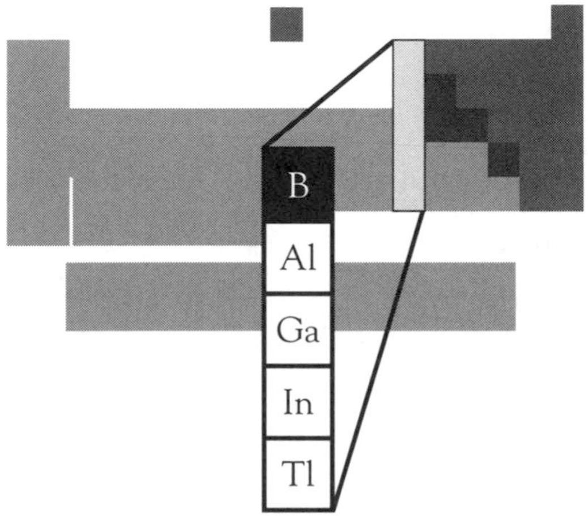

Boron and aluminum are the only members of this group that are of major importance. Aluminum is one of the most widely used metals, and its chemistry will be the focus of much of this chapter.

The German chemist Friedrich Wöhler was among the first to prepare pure aluminum metal. He did so by heating potassium metal with aluminum chloride; aluminum was produced in a single replacement reaction:

$$3\ K(l) + AlCl_3(s) \rightarrow Al(s) + 3\ KCl(s)$$

Before he could do this, he had to obtain stocks of the very reactive potassium metal. Because he did not have a battery that was powerful enough to generate the potassium metal electrochemically, he devised a chemical route that used intense heat and a mixture of potassium hydroxide and charcoal. He and his sister, Emilie Wöhler, shared the exhausting work of pumping the bellows to keep the mixture hot enough to produce the potassium. So expensive was aluminum in the mid-nineteenth century that Emperor Napoleon III used aluminum tableware for special state occasions.

Group Trends

Boron exhibits mostly nonmetallic behavior and is classified as a semimetal, whereas the other members of Group 13 are metals. But even the metals have no simple pattern in melting points, although their boiling points do show a decreasing trend as the mass of the elements increases (Table 12.1). The reason for this lack of order is that each element in the group is organized a different way in the solid phase. For example, in one of its four allotropes, boron forms clusters of 12 atoms. Each cluster has a geometric arrangement called an icosahedron (see Figure 12.1). Aluminum adopts a face-centered cubic structure, but gallium forms a unique structure containing pairs of atoms. Indium and thallium each form other, different structures. It is only when the elements are melted and the crystal arrangements destroyed that we see, from the decreasing boiling points as the group is descended, that the metallic bond becomes weaker.

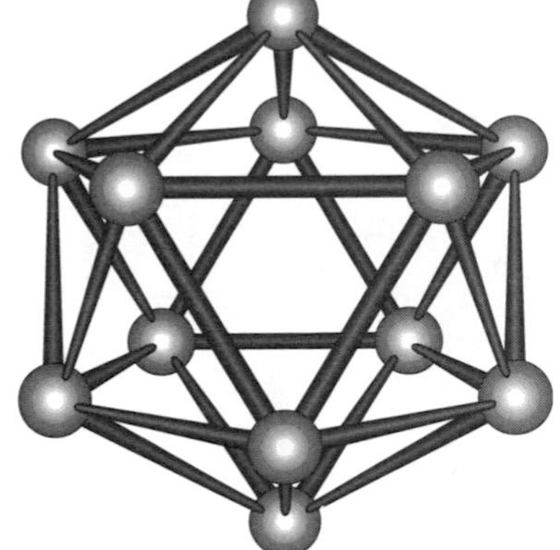

Figure 12.1 Icosahedral arrangement of boron.

As we would expect, boron, classified as a semimetal, favors covalent bond formation. However, covalency is common among the metallic members of the group as well. The reason for the covalent behavior can be attributed to the high charge and small radius of each metal ion. The resulting high charge density of Group 13 ions is sufficient to polarize almost any approaching anion enough to produce a covalent bond (Table 12.2).

The only way to stabilize the ionic state of Group 13 elements is to hydrate the metal ion. For aluminum, the enormous hydration enthalpy of the tripositive ion, $-4665\ kJ{\cdot}mol^{-1}$, is almost enough on its own to balance the sum of the three ionization energies, $+5137\ kJ{\cdot}mol^{-1}$. Thus the aluminum compounds that we regard as ionic do not contain the aluminum ion, Al^{3+}, as such, but the hexaaquaaluminum ion, $[Al(OH_2)_6]^{3+}$.

It is in Group 13 that we first encounter elements possessing more than one oxidation state. Aluminum has the +3 oxidation state, whether the bonding is ionic or covalent. However, gallium, indium, and thallium have a second oxidation state of +1. For gallium and indium, the +3 state predominates,

Table 12.1 Melting and boiling points of the Group 13 elements

Element	Melting point (°C)	Boiling point (°C)
B	2180	3650
Al	660	2467
Ga	30	2403
In	157	2080
Tl	303	1457

Table 12.2 Charge densities of Period 3 metal ions

Group	Ion	Charge density ($C{\cdot}mm^{-3}$)
1	Na^+	24
2	Mg^{2+}	120
13	Al^{3+}	364

whereas the +1 state is most common for thallium. At this point, it is appropriate to note that formulas can sometimes be deceiving. Gallium forms a chloride, $GaCl_2$, a compound implying that a +2 oxidation state exists. However, the actual structure of this compound is now established as $[Ga]^+[GaCl_4]^-$; thus the compound actually contains gallium in both +1 and +3 oxidation states.

Boron

Boron is the only element in Group 13 that is not classified as a metal. In Chapter 2 we classified it as a semimetal. However, on the basis of its extensive oxyanion and hydride chemistry, it is equally valid to consider it a nonmetal.

Boron is a rare element in the Earth's crust, but fortunately there are several large deposits of its salts. These deposits, which are found in locations that once had intense volcanic activity, consist of the salts borax and kernite, which are conventionally written as $Na_2B_4O_7{\cdot}10H_2O$ and $Na_2B_4O_7{\cdot}4H_2O$, respectively. Total annual worldwide production of boron compounds amounts to over 3 million tonnes. The world's largest deposit is found at Boron, California; it covers about 10 km^2, with beds of kernite up to 50 m thick. The actual structure of borate ions is much more complex than the simple formulas would indicate. For example, borax actually contains the $[B_4O_5(OH)_4]^{2-}$ ion, shown in Figure 12.2.

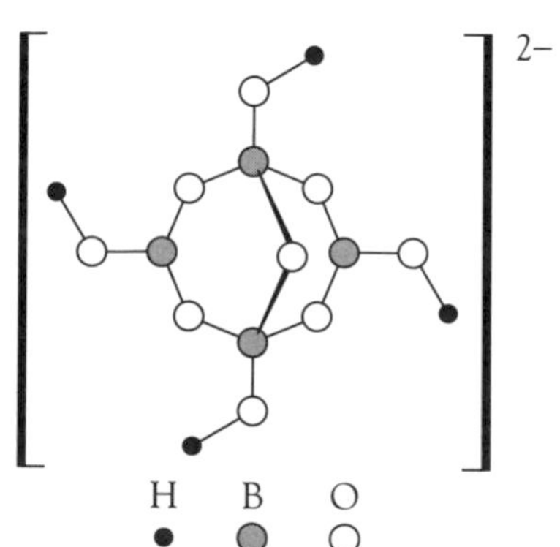

Figure 12.2 Actual structure of the borate ion in borax.

About 35 percent of boron production is used in the manufacture of borosilicate glass. Conventional soda glass suffers from thermal shock; that is, when a piece of glass is heated strongly, the outside becomes hot and tries to expand while the inside is still cold because glass is such a poor conductor of heat. As a result of stress between the outside and the inside, the glass cracks. When the sodium ions in the glass structure are replaced by boron atoms, the glass expansion (thermal expansivity) is less than half that of conventional glass. As a result, containers made of borosilicate glass (sold under trademarks such as Pyrex®) are capable of being heated without great danger of cracking.

In the early part of the twentieth century, the major use for boron compounds was as a cleaning agent called borax. This use has now dropped behind that for glassmaking, consuming only 20 percent of production. In detergent formulations, it is no longer borax but sodium peroxoborate, $NaBO_3$, that is used. Once again, the simple formula does not show the true

Inorganic Fibers

In our everyday lives, the fibers we encounter are usually organic, for example, nylon and polyester. These materials are fine for clothing and similar purposes, but most organic fibers have the disadvantages of low melting points, flammability, and low strengths. For materials that are strong and unaffected by high temperatures, inorganic materials fit the specifications best. Some inorganic fibers are well known, for example, asbestos and fiberglass. However, it is the elements boron, carbon, and silicon that currently provide some of the toughest materials for our high-technology world. Carbon fiber is the most widely used—not just for tennis rackets and fishing rods but for aircraft parts as well. The Boeing 767 was the first commercial plane to make significant use of carbon fiber; in fact, about 1 tonne is incorporated into the structure of each aircraft. Aircraft constructed with newer technology, such as the Airbus 320, contain a much higher proportion of carbon fibers.

Fibers of boron and silicon carbide, SiC, are becoming increasingly important in the search for tougher, less fatigue-prone materials. The boron fibers are prepared by reducing boron trichloride with hydrogen gas at about 1200°C:

$$2\ BCl_3(g) + 3\ H_2(g) \rightarrow 2\ B(g) + 6\ HCl(g)$$

The gaseous boron can then be condensed onto carbon or tungsten microfibers. For example, boron is deposited onto tungsten fibers of 15 μm until the diameters of the coated fibers are about 100 μm. The typical inorganic fiber prices are several hundred dollars per kilogram; so even though production of each type is mostly in the hundreds of tonnes range, inorganic fiber production is already a billion-dollar business.

structure of the ion, which is $[B_2(O_2)_2(OH)_4]^{2-}$ (Figure 12.3). This ion acts as an oxidizing agent as a result of the two peroxo groups (–O–O–) linking the boron atoms. About 5×10^5 tonnes of sodium peroxoborate are produced every year for European detergent manufacturing companies. It is a particularly effective oxidizing (bleaching) agent at the water temperatures used in European washing machines (90°C), but it is ineffective at the water temperatures usually used in North American washing machines (70°C). In North America, hypochlorites (see Chapter 16) are used instead.

The peroxoborate ion is prepared by the reaction of hydrogen peroxide with borax in base:

$$[B_4O_5(OH)_4]^{2-}(aq) + 4\ H_2O_2(aq) + 2\ OH^-(aq) \rightarrow 2\ [B_2(O_2)_2(OH)_4]^{2-}(aq) + 3\ H_2O(l)$$

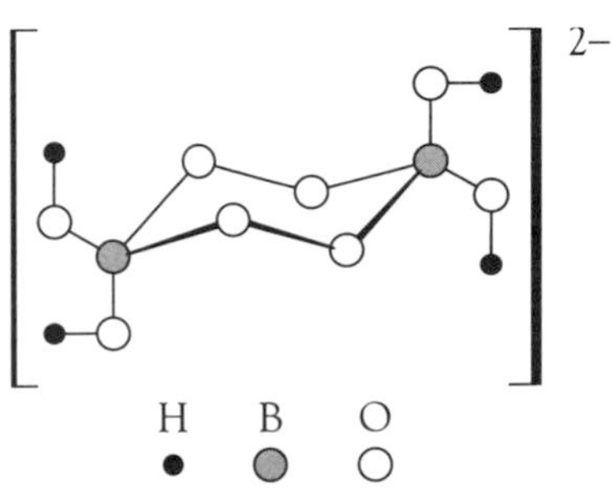

Figure 12.3 Structure of the peroxoborate ion.

Boron is a vital component of nuclear power plants because it is a strong absorber of neutrons. Boron-containing control rods are lowered into reactors to maintain the nuclear reaction at a steady rate. Borates are used as wood preservatives and as a fire retardant in fabrics. Borates are also used as a flux in soldering. In this latter application, the borates melt on the hot pipe surface and react with metal oxide coatings, such as copper(II) oxide on copper pipes. The metal borates (such as copper(II) borate) can be easily removed to give a clean metal surface for the soldering.

Boron Trifluoride

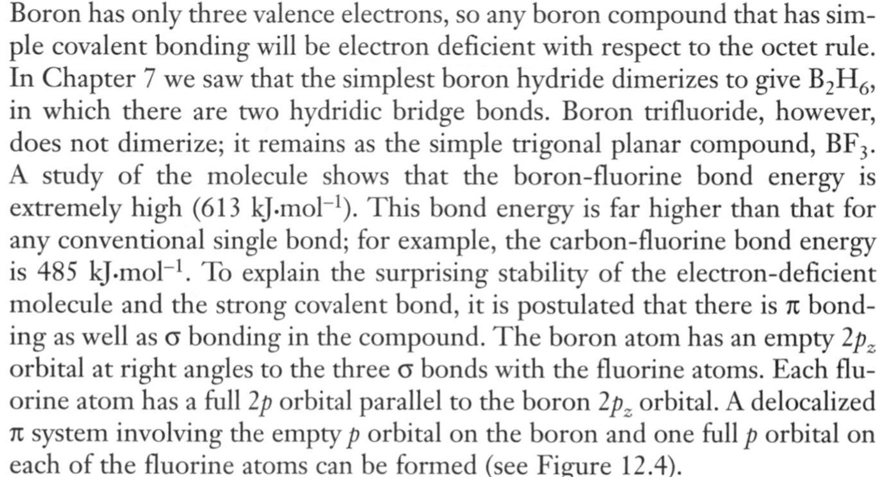

Boron has only three valence electrons, so any boron compound that has simple covalent bonding will be electron deficient with respect to the octet rule. In Chapter 7 we saw that the simplest boron hydride dimerizes to give B_2H_6, in which there are two hydridic bridge bonds. Boron trifluoride, however, does not dimerize; it remains as the simple trigonal planar compound, BF_3. A study of the molecule shows that the boron-fluorine bond energy is extremely high (613 $kJ{\cdot}mol^{-1}$). This bond energy is far higher than that for any conventional single bond; for example, the carbon-fluorine bond energy is 485 $kJ{\cdot}mol^{-1}$. To explain the surprising stability of the electron-deficient molecule and the strong covalent bond, it is postulated that there is π bonding as well as σ bonding in the compound. The boron atom has an empty $2p_z$ orbital at right angles to the three σ bonds with the fluorine atoms. Each fluorine atom has a full $2p$ orbital parallel to the boron $2p_z$ orbital. A delocalized π system involving the empty p orbital on the boron and one full p orbital on each of the fluorine atoms can be formed (see Figure 12.4).

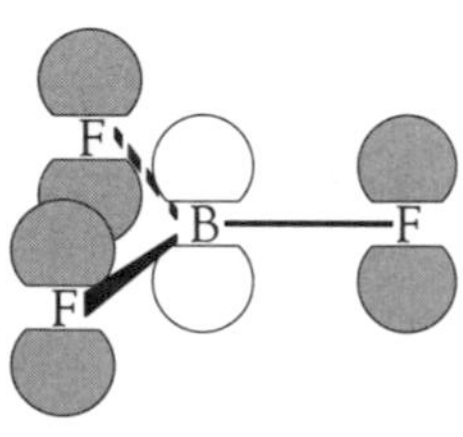

Figure 12.4 Proposed π bonding in boron trifluoride, involving the full p orbitals (shaded) on the fluorine atoms and the empty p_z orbital on the boron atom.

There is experimental evidence to support this explanation: When boron trifluoride reacts with fluoride ion to form the tetrahedral tetrafluoroborate ion, BF_4^-, the B–F bond length increases from 130 pm in boron trifluoride to 145 pm in the tetrafluoroborate ion. This lengthening would be expected because the $2s$ and three $2p$ orbitals of the boron in the tetrafluoroborate ion are used to form four σ bonds. Hence there are no orbitals available for π bonding in the tetrafluoroborate ion, and so the B–F bond in this ion would be a "pure" single bond.

By using the vacant $2p_z$ orbital, boron trifluoride can behave as a powerful Lewis acid. The classic illustration of this behavior is the reaction between boron trifluoride and ammonia, where the nitrogen lone pair acts as the electron pair donor:

$$BF_3(g) + {:}NH_3(g) \rightarrow F_3B{:}NH_3(s)$$

About 4000 tonnes of boron trifluoride are used industrially in the United States every year as a Lewis acid and as a catalyst in organic reactions.

Boron Trichloride

The chloride of boron is the first covalent chloride that we encounter as we cross the periodic table. As such, it is quite typical. Metal (ionic) chlorides are solids that dissolve in water to form hydrated cations and anions. However, the typical covalent chloride is a gas or liquid at room temperature and reacts

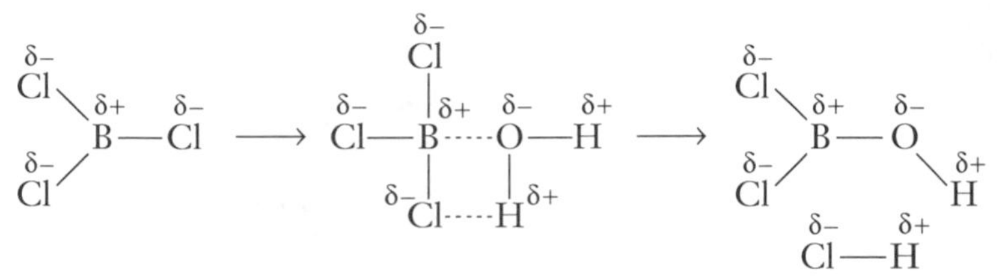

Figure 12.5 First step of the postulated mechanism for hydrolysis of boron trichloride.

violently with water. For example, bubbling boron trichloride (a gas above 12°C) into water produces boric acid and hydrochloric acid:

$$BCl_3(g) + 3\ H_2O(l) \rightarrow H_3BO_3(aq) + 3\ HCl(aq)$$

We can predict the products of these reactions in terms of the relative electronegativities of the two atoms. In this case, the electronegativity of chlorine is much greater than that of boron. Hence, as a water molecule approaches the boron trichloride molecule, we can picture the partially positive hydrogen being attracted to the partially negative chlorine atom while the partially negative oxygen atom is attracted to the partially positive boron atom (Figure 12.5). A bond shift occurs, and one chlorine atom is replaced by a hydroxyl group. When this process happens two more times, the result is boric acid.

Sodium Tetrahydridoborate

The only other species of boron used on a large scale is the tetrahydridoborate ion, BH_4^-. Most hydrides, except for those of carbon, are flammable, unstable compounds. This anion, however, can even be recrystallized from cold water as the sodium salt. Its crystal structure is interesting because the salts this anion forms adopt the sodium chloride structure, with the whole BH_4^- ion occupying the same sites as the chloride ion does. Sodium tetrahydridoborate is of major importance as a mild reducing agent, particularly in organic chemistry, where it is used to reduce aldehydes to primary alcohols and ketones to secondary alcohols without reducing other functional groups such as carboxylic acids.

Boron-Nitrogen Analogs of Carbon Compounds

Boron has one less valence electron than carbon, and nitrogen has one more. Thus for many years chemists have tried to make analogs of carbon compounds that contain alternating boron and nitrogen atoms. The most interesting of these are analogs of the pure forms of carbon. The two common allotropes of carbon are graphite, the lubricant, and diamond, the hardest naturally occurring substance known. Unfortunately, both carbon allotropes burn when heated to give carbon dioxide gas, thus precluding the use of either of these substances in high-temperature applications. Boron nitride,

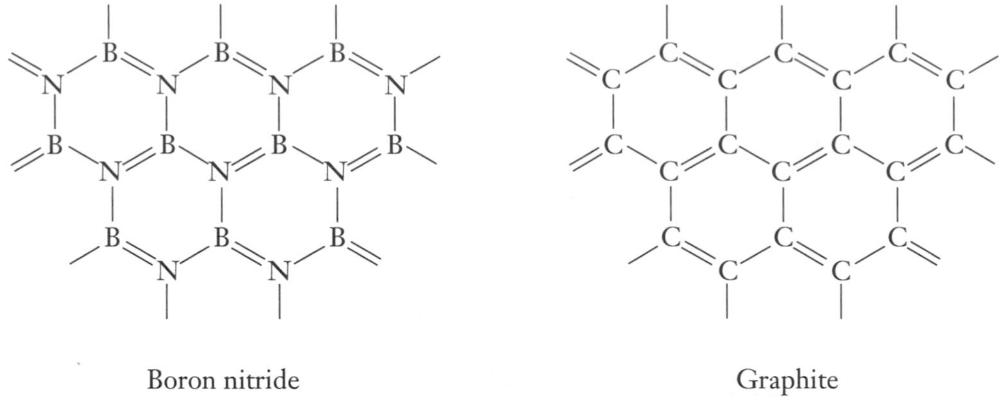

Boron nitride

Graphite

Figure 12.6 Layer structures of boron nitride and graphite.

BN, however, is the ideal substitute. The simplest method of synthesis involves heating diboron trioxide with ammonia at about 1000°C:

$$B_2O_3(s) + 2\ NH_3(g) \xrightarrow{\Delta} 2\ BN(s) + 3\ H_2O(l)$$

The product has a graphitelike structure (Figure 12.6) and is an excellent high-temperature, chemically resistant lubricant.

Unlike graphite, boron nitride is a white solid that does not conduct electricity. This difference is possibly due to differences in the way the layers in the two crystals are stacked. The layers in the graphitelike form of boron nitride are almost exactly the same distance apart as those in graphite, but the boron nitride layers are organized so that the nitrogen atoms in one layer are located directly over boron atoms in the layers above and below, and vice versa. This arrangement is logical, because the partially positive boron atoms and partially negative nitrogen atoms are likely to be electrostatically attracted to each other (Figure 12.7). By contrast, the carbon atoms in one layer of graphite are directly over the center of the carbon rings in the layers above and below.

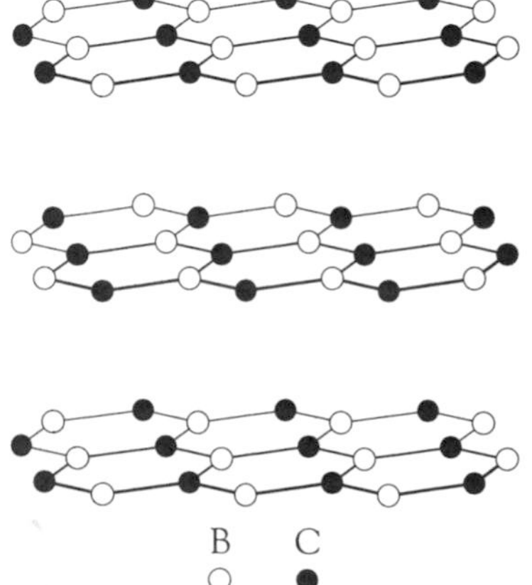

Figure 12.7 Repeating layer structure of boron nitride.

In a further analogy to carbon, application of high pressures and high temperatures converts the graphitelike allotrope of boron nitride to a diamondlike form called borazon. This form of boron nitride is similar to diamond in terms of hardness and is far superior in terms of chemical inertness at high temperatures. Hence borazon it is often used in preference to diamond as a grinding agent.

A number of compounds adopt the diamond structure. These are all 1:1 mole ratio compounds that have an identical combined number of valence electrons. For example, the total number of valence electrons in boron and nitrogen is eight (3 + 5), just as the number of valence electrons in carbon with carbon totals eight (4 + 4). Other compounds that can adopt diamondlike structures include gallium arsenide, GaAs (3 + 5); zinc selenide, ZnSe (2 + 6); and copper(I) bromide, CuBr (1 + 7). This electron relationship, known as the *Zintl principle*, is very useful in understanding and predicting structural similarity among very different compounds.

There is another similarity between boron-nitrogen and carbon compounds. The reaction between diborane and ammonia gives borazine, $B_3N_3H_6$, a cyclic molecule analogous to benzene, C_6H_6:

$$3\ B_2H_6(g) + 6\ NH_3(g) \rightarrow 2\ B_3N_3H_6(l) + 12\ H_2(g)$$

In fact, borazine is sometimes called "inorganic benzene" (Figure 12.8). This compound is a useful reagent for synthesizing other boron-nitrogen analogs of carbon compounds, but at this time it has no commercial applications.

In spite of similarities in boiling points, densities, and surface tensions, the polarity of the boron-nitrogen bond means that borazine is much more prone to chemical attack than is the homogeneous ring of carbon atoms in benzene. For example, hydrogen chloride reacts with borazine to give $B_3N_3H_9Cl_3$, in which the chlorine atoms bond to the more electropositive boron atoms:

$$B_3N_3H_6(l) + 3\ HCl(g) \rightarrow B_3N_3H_9Cl_3(s)$$

This compound can be reduced by sodium tetrahydroborate to give $B_3N_3H_{12}$, an analog of cyclohexane, C_6H_{12}. In fact, like cyclohexane, $B_3N_3H_{12}$ adopts the chair conformation.

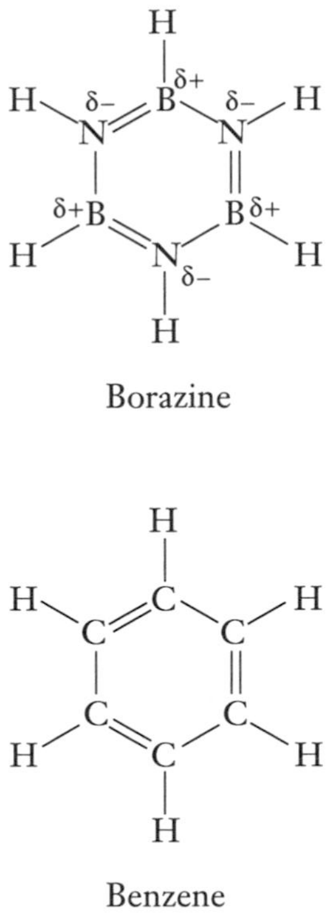

Borazine

Benzene

Figure 12.8 Comparison of the structures of borazine and benzene.

Aluminum

Because aluminum is a metal with a high negative standard reduction potential, it might be expected to be very reactive. This is indeed the case. Why, then, can aluminum be used as an everyday metal rather than be consigned to the chemistry laboratory like sodium? The answer is found in its reaction with oxygen gas. Any exposed surface of aluminum metal rapidly reacts with oxygen to form aluminum oxide, Al_2O_3. An impermeable oxide layer, between 10^{-4} and 10^{-6} mm thick, then protects the layers of aluminum atoms underneath. This can happen because the oxygen ion has an ionic radius (124 pm) similar to the metallic radius of the aluminum atom (143 pm). As a result, the surface packing is almost unchanged because the small aluminum ions (68 pm) fit into interstices in the oxide surface structure. The process is shown in Figure 12.9.

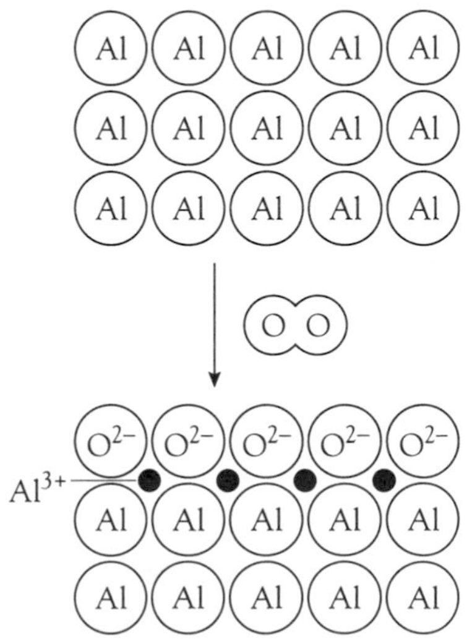

Figure 12.9 Formation of a single oxide layer on the surface of aluminum metal. The small aluminum 3+ ions are indicated by the solid circles.

To increase their corrosion resistance, aluminum products are "anodized." In other words, the aluminum product is used as the anode in an electrochemical cell, and additional aluminum oxide is deposited as an electrolytic product over the naturally formed layers. This anodized aluminum possesses an oxide layer about 0.01 mm thick, and this very thick oxide coating has the useful property of absorbing dyes and pigments so that a colored surface can be produced.

The particular attraction of aluminum as a construction metal is its low density (2.7 $g{\cdot}cm^{-3}$), second only to that of magnesium (1.7 $g{\cdot}cm^{-3}$)—disregarding the very reactive alkali metals. For instance, compare the density of aluminum with that of iron (7.9 $g{\cdot}cm^{-3}$) or gold (19.3 $g{\cdot}cm^{-3}$). Aluminum is a good conductor of heat, a property accounting for its role in cookware. It is not as good as copper, however. To spread heat more evenly from the electrical element (or gas flame), higher priced pans have a copper-coated bottom. Aluminum also is exceptional as a conductor of electricity, hence its major role in electric power lines and home wiring. The major problem with using aluminum wiring occurs at the connections. If aluminum is joined to an electrochemically dissimilar metal, such as copper, an electrochemical cell will be established in damp conditions. This development causes oxidation (corrosion) of the aluminum. For this reason, use of aluminum in home wiring is now discouraged.

Chemical Properties of Aluminum

Like other powdered metals, aluminum powder will burn in a flame to give a dust cloud of aluminum oxide:

$$4\ \text{Al}(s) + 3\ \text{O}_2(g) \rightarrow 2\ \text{Al}_2\text{O}_3(s)$$

Otherwise, aluminum is only weakly metallic in its chemical properties. For example, it is an amphoteric metal, reacting with both acid and base:

$$2\ \text{Al}(s) + 6\ \text{H}^+(aq) \rightarrow 2\ \text{Al}^{3+}(aq) + 3\ \text{H}_2(g)$$

$$2\ \text{Al}(s) + 2\ \text{OH}^-(aq) + 6\ \text{H}_2\text{O}(l) \rightarrow 2\ [\text{Al(OH)}_4]^-(aq) + 3\ \text{H}_2(g)$$

In aqueous solution, the aluminum ion is present as the hexaaquaaluminum ion, $[\text{Al(OH}_2)_6]^{3+}$, but it undergoes a hydrolysis reaction to give a solution of the hydroxopentaaquaaluminum ion, $[\text{Al(OH}_2)_5(\text{OH})]^{2+}$, and the hydronium ion, and then to the dihydroxotetraaquaaluminum ion:

$$[\text{Al(OH}_2)_6]^{3+}(aq) + \text{H}_2\text{O}(l) \rightleftharpoons [\text{Al(OH}_2)_5(\text{OH})]^{2+}(aq) + \text{H}_3\text{O}^+(aq)$$

$$[\text{Al(OH}_2)_5(\text{OH})]^{2+}(aq) + \text{H}_2\text{O}(l) \rightleftharpoons [\text{Al(OH}_2)_4(\text{OH})_2]^{+}(aq) + \text{H}_3\text{O}^+(aq)$$

Thus solutions of aluminum salts are acidic, with almost the same acid ionization constant as ethanoic (acetic) acid. The mixture in antiperspirants commonly called aluminum hydrate is, in fact, a mixture of the chloride salts of these two hydroxy ions. It is the aluminum ion in these compounds that acts to constrict pores on the surface of the skin.

Addition of hydroxide ion to aluminum ion first gives a gelatinous precipitate of aluminum hydroxide, but this product redissolves in excess hydroxide ion to give the aluminate ion:

$$[\text{Al(OH}_2)_6]^{3+}(aq) \xrightarrow{\text{OH}^-} \text{Al(OH)}_3(s) \xrightarrow{\text{OH}^-} [\text{Al(OH)}_4]^-(aq)$$

As a result, aluminum is soluble at low and high pHs but insoluble under neutral conditions (see Figure 12.10). Aluminum hydroxide is used in a number of antacid formulations. Like other antacids, the compound is an insoluble base that will neutralize excess stomach acid:

$$\text{Al(OH)}_3(s) + 3\ \text{H}^+(aq) \rightarrow \text{Al}^{3+}(aq) + 3\ \text{H}_2\text{O}(l)$$

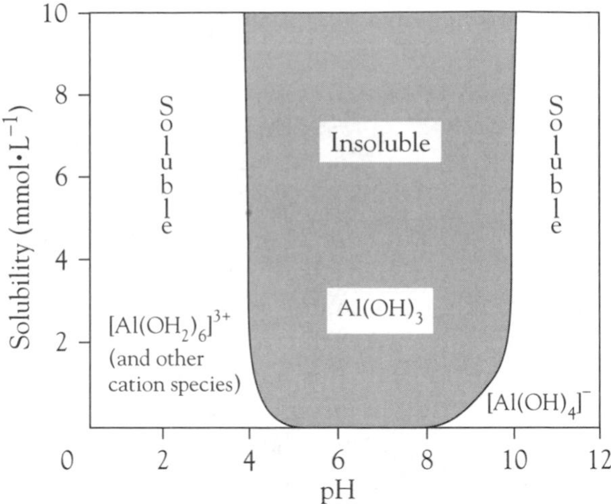

Figure 12.10 Aluminum solubility as a function of pH.

Production of Aluminum

The discovery of an electrolytic method by the French chemist Henri Sainte-Claire Deville and the decreasing cost of electricity caused the price of metallic aluminum to drop dramatically in the late nineteenth century. However, the production of the metal on the large scale required a method that would use an inexpensive, readily available ore. This route was found independently in 1886 by three young chemists; one in France, Paul Héroult, and two in the United States, Charles and Julia Hall. Hence the process is known as the Hall-Héroult process. In fact, it was Charles's sister, Julia, who had the stronger background in chemistry and who maintained detailed notes of the experiments. Yet in the accounts of the discovery, Julia's role has usually been overlooked.

Aluminum is the most abundant metal in the Earth's crust, mostly in the form of clays. To this day, there is no economical route for the extraction of aluminum from clay. However, in hot, humid environments, the more soluble ions are leached from the clay structure to leave the ore bauxite (impure hydrated aluminum oxide). Thus the countries producing bauxite are mainly those near the equator, Australia being the largest source, followed by Guinea, Brazil, Jamaica, and Suriname.

The first step in the extraction process is the purification of bauxite. This step is accomplished by digesting (heating and dissolving) the crushed ore with hot sodium hydroxide solution to give the soluble aluminate ion:

$$Al_2O_3(s) + 2\ OH^-(aq) + 3\ H_2O(l) \rightarrow 2\ [Al(OH)_4]^-(aq)$$

The insoluble materials, particularly iron(III) oxide, are filtered off as "red mud." Upon cooling, the equilibrium in the solution shifts to the left, and white aluminum oxide trihydrate precipitates:

$$2\ [Al(OH)_4]^-(aq) \rightarrow Al_2O_3{\cdot}3H_2O(s) + 2\ OH^-(aq)$$

The hydrate is heated strongly in a rotary kiln (similar to that used in cement production) to give anhydrous aluminum oxide:

$$Al_2O_3{\cdot}3H_2O(s) \rightarrow Al_2O_3(s) + 3\ H_2O(g)$$

With its high ion charges, aluminum oxide has a very large lattice energy and hence a high melting point (2040°C). But, to electrolyze the aluminum oxide, it was necessary to find an aluminum compound with a much lower melting point. Hall and Héroult simultaneously announced the discovery of this lower-melting aluminum compound, the mineral cryolite, whose chemical name is sodium hexafluoroaluminate, Na_3AlF_6. There are few naturally occurring deposits of this mineral; Greenland has the largest deposit. As a result of its rarity, almost all cryolite is manufactured. This in itself is an interesting process, because the starting point is usually a waste material, silicon tetrafluoride, SiF_4, which is produced in the synthesis of hydrogen fluoride. Silicon tetrafluoride gas reacts with water to give insoluble silicon dioxide and a solution of hexafluorosilicic acid, H_2SiF_6, a relatively safe fluorine-containing compound:

$$3\ SiF_4(g) + 2\ H_2O(l) \rightarrow 2\ H_2SiF_6(aq) + SiO_2(s)$$

The acid is then treated with ammonia to give ammonium fluoride:

$$H_2SiF_6(aq) + 6\ NH_3(aq) + 2\ H_2O(l) \rightarrow 6\ NH_4F(aq) + SiO_2(s)$$

Finally, the ammonium fluoride solution is mixed with a solution of sodium aluminate to give the cryolite and ammonia, which can be recycled:

$$6\ NH_4F(aq) + Na[Al(OH)_4](aq) + 2\ NaOH(aq) \rightarrow Na_3AlF_6(s) + 6\ NH_3(aq) + 6\ H_2O(l)$$

The detailed chemistry that occurs in the electrolytic cell is still poorly understood, but the cryolite acts as the electrolyte (Figure 12.11). The aluminum oxide is dissolved in molten cryolite at about 950°C. Molten aluminum is produced at the cathode, and the oxygen that is produced at the anode oxidizes the carbon to carbon monoxide (and some carbon dioxide):

$$Al^{3+}(Na_3AlF_6) + 3\ e^- \rightarrow Al(l)$$

$$O^{2-}(Na_3AlF_6) + C(s) \rightarrow CO(g) + 2\ e^-$$

The process is very energy intensive, requiring currents of about 3.5×10^4 A at 6 V. In fact, about 25 percent of the cost of aluminum metal is derived from its energy consumption. The production of 1 kg of aluminum consumes about 2 kg of aluminum oxide, 0.6 kg of anodic carbon, 0.1 kg of cryolite, and 16 kWh of electricity.

Aluminum production yields four by-products that create major pollution problems:

1. Red mud, which is produced from the bauxite purification and is highly basic

2. Hydrogen fluoride gas, which is produced when cryolite reacts with traces of moisture in the aluminum oxide

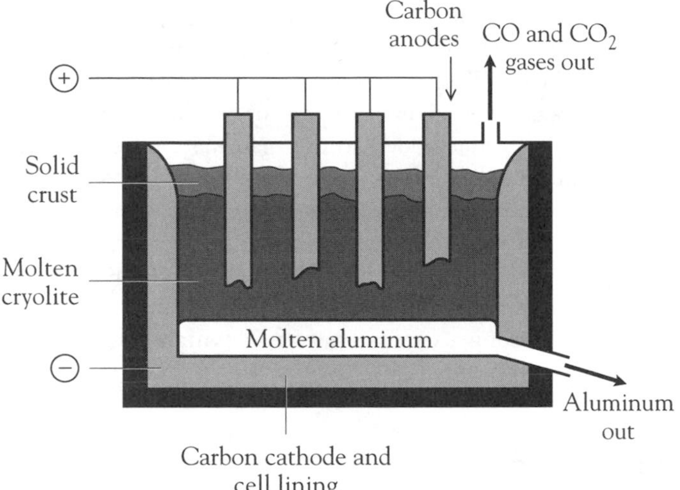

Figure 12.11 Electrolytic cell for aluminum production.

3. Oxides of carbon, which are produced at the anode
4. Fluorocarbons, which are produced by reaction of fluorine with the carbon anode

To reduce the red mud disposal problem, the slurry is poured into settling tanks, from which the liquid component, mainly sodium hydroxide solution, is removed and recycled or neutralized. The solid, mostly iron(III) oxide, can then be used as landfill or shipped to iron smelters for extraction of the iron.

The problem of what to do with the emissions of hydrogen fluoride gas has been solved to a large extent by absorbing the hydrogen fluoride in a filter bed of aluminum oxide. The product of this process is aluminum fluoride:

$$Al_2O_3(s) + 6\ HF(g) \rightarrow 2\ AlF_3(s) + 3\ H_2O(g)$$

This fluoride can be added periodically to the melt, thereby recycling the hydrogen fluoride.

A partial solution to the problem of disposing of the large volumes of the oxides of carbon that are produced is to burn the poisonous carbon monoxide, a process giving the dioxide and providing some of the heat required to operate the aluminum plant. However, the electrolytic method inevitably produces these two gases; and until an alternative, economical process is devised, aluminum production will continue to contribute carbon dioxide to the atmosphere.

For every tonne of aluminum, about 1 kg of tetrafluoromethane, CF_4, and about 0.1 kg of hexafluoroethane, C_2F_6, are produced. These compounds are second only to the chlorofluorocarbons (CFCs) as contributors to the greenhouse effect. The fluorocarbon problem has not yet been solved, and it is the focus of a major research effort by the aluminum companies.

The major producers of aluminum metal and the suppliers of bauxite are different. The largest producer of aluminum metal is the United States. The large energy requirement of the production process favors those countries with inexpensive energy sources. Thus Canada and Norway, neither of which is a bauxite producer nor large aluminum consumer, make the top five of aluminum metal producers. Both countries have low-cost hydroelectric power and deep-water ports favoring easy import of ore and export of aluminum metal. The bulk of the value added to the material comes through the processing steps. Even though the developed world relies heavily on third world countries for the raw material, the third world receives comparatively little in the way of income from the mining phase.

About 25 percent of the output of aluminum metal is used in the construction industry, and lesser proportions are used to manufacture aircraft, buses, and railroad passenger cars (18 percent); containers and packaging (17 percent); and electric power lines (14 percent).

Aluminum Halides

The aluminum halides constitute an interesting series of compounds: Aluminum fluoride melts at 1290°C; aluminum chloride sublimes at 180°C; and aluminum bromide and iodide melt at 97.5°C and 190°C, respectively. Thus the fluoride has the characteristic high melting point of an ionic compound, whereas the melting points of the bromide and iodide are typical of covalent compounds. The aluminum ion has a charge density of 364 C·mm^{-3}, so we expect all anions, except the small fluoride ion, to be polarized to the point of covalent bond formation with aluminum. In fact, aluminum fluoride does have a typically ionic crystal structure with arrays of alternating cations and anions. But the bromide and iodide both exist as dimers, Al_2Br_6 and Al_2I_6, analogous to diborane, with two bridging halogen atoms (Figure 12.12). The chloride forms an ionic-type lattice structure in the solid, which collapses in the liquid phase to give molecular Al_2Cl_6 dimers. Thus the ionic and covalent forms must be almost equal in energy. These dimers are also formed when solid aluminum chloride is dissolved in low-polarity solvents.

Figure 12.12 Structure of aluminum iodide.

Even though the anhydrous aluminum chloride appears to adopt an ionic structure in the solid phase, its reactions are more typical of a covalent chloride. This covalent behavior is particularly apparent in the solution processes of the anhydrous aluminum chloride and the hexahydrate. As mentioned above, the hexahydrate actually contains the hexaaquaaluminum ion, $[Al(OH_2)_6]^{3+}$. It dissolves quietly in water, although the solution is acidic as a result of hydrolysis. Conversely, the anhydrous aluminum chloride reacts very exothermically with water in the typical manner of a covalent chloride, producing a hydrochloric acid mist:

$$AlCl_3(s) + 3\ H_2O(l) \rightarrow Al(OH)_3(s) + 3\ HCl(g)$$

Anhydrous aluminum chloride is an important reagent in organic chemistry. In particular, it is used as a catalyst for the substitution of aromatic rings in the Friedel-Crafts reaction. The overall reaction can be written as the reaction between an aromatic compound, R–H, and an organic chloro compound, R_1–Cl. The aluminum chloride reacts as a strong Lewis acid with the chloro compound to give the carbon cation, which can then react with the aromatic compound:

$$R_1\text{–}Cl + AlCl_3 \rightarrow R_1^+ + AlCl_4^-$$

$$R\text{–}H + R_1^+ \rightarrow R\text{–}R_1 + H^+$$

$$H^+ + [AlCl_4]^- \rightarrow HCl + AlCl_3$$

Aluminum Potassium Sulfate

The traditional name for aluminum potassium sulfate is alum, and it is the only common water-soluble mineral of aluminum. As such, it has played an important role in the dyeing industry. To absorb a dye permanently onto cloth, the cloth is first soaked in a solution of alum. A layer of aluminum hydroxide is deposited on the cloth's surface, to which dye molecules readily bond. Because of its usefulness, alum has been a valuable import item from Asia since the time of the Romans. The following anecdote indicates the value placed on alum at one time. In the fifteenth century an alum mine was found on papal land in Italy. After production of alum from this mine commenced, Pope Paul II excommunicated from the Catholic Church anyone who purchased alum from "Turkish infidels" rather than from the papal mine.

Alum crystallizes from an "equimolar" mixture of potassium sulfate and aluminum sulfate; it has the formula $KAl(SO_4)_2{\cdot}12H_2O$. Alum crystals have a very high lattice stability because the sulfate anions are packed between alternating potassium and hexaaquaaluminum ions. In fact, there is a family of compounds having this type of formula. They all involve the combination of sulfate ion with a mixture of a monopositive cation and a hexahydrated tripositive ion; their general formula is $M^+[M(OH_2)_6]^{3+}(SO_4^{2-})_2{\cdot}6H_2O$. Among this large family, it is most common for the monopositive ion to be any of the alkali metals or ammonium, whereas the tripositive ion can be aluminum, iron(III), or chromium(III).

Alum is sometimes used to stop bleeding because it causes coagulation of proteins on the surface of cells without killing the cells themselves.

Spinels

Spinel itself is magnesium aluminum oxide, $MgAl_2O_4$; but of more importance are the enormous number of compounds that adopt the same crystal structure and are also called spinel. Many of these compounds have unique properties that will make them important in the chemistry of the twenty-first century. The general formula of a spinel is AB_2X_4, where A is usually a dipositive metal ion; B, usually a tripositive metal ion; and X, a dinegative anion, usually oxygen.

The framework of the unit cell of a spinel consists of 32 oxygen atoms in an almost perfect cubic close-packed arrangement. Thus the unit cell composition is actually $A_8B_{16}O_{32}$. Figure 12.13 shows part of the unit cell. The oxygen ions form a face-centered cubic array, and there are octahedral sites at the center of the cube and in the middle of each cube edge, and tetrahedral sites in the middle of each "cubelet." In the normal spinel structure, the 8 A cations occupy one-eighth of the tetrahedral sites and the 16 B cations occupy

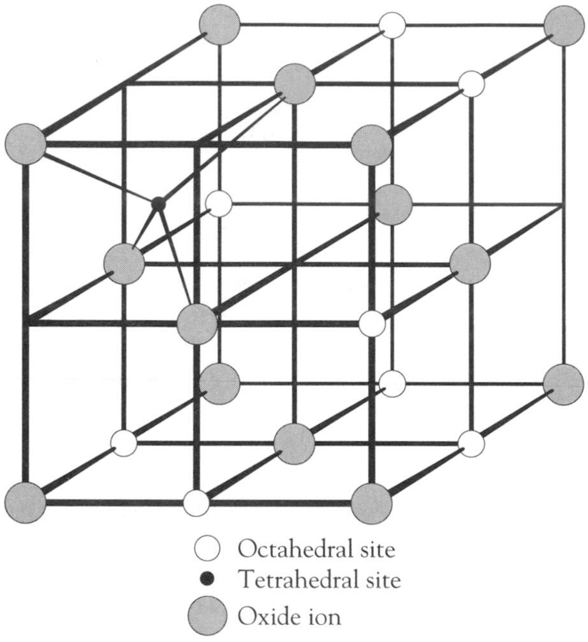

Figure 12.13 Part of the unit cell of the spinel structure showing the occupied lattice sites. Of the eight "cubelets" shown, the upper left front cubelet shows an occupied tetrahedral cation site (zinc sulfide–type), whereas the other seven cubelets have some occupied octahedral cation sites (sodium chloride–type).

one-half of the octahedral sites. Thus the unit cell can be considered to consist of "cubelets" of zinc sulfide–type tetrahedral units interspersed among "cubelets" of sodium chloride–type octahedral units.

To indicate site occupancy, we can use subscripts t and o to represent tetrahedral and octahedral cation sites; thus spinel itself can be written as $(Mg^{2+})_t(2\ Al^{3+})_o(O^{2-})_4$. There are some spinels in which the dipositive ions are located in the octahedral sites. Because there are twice as many octahedral holes as tetrahedral holes in the cubic close-packed arrangement, only half of the tripositive ions can be placed in tetrahedral sites; the remainder must occupy octahedral sites. Such compounds are called *inverse spinels*. The most common example is magnetite, Fe_3O_4, or more accurately, $Fe^{2+}(Fe^{3+})_2(O^{2-})_4$. The arrangement here is $(Fe^{3+})_t(Fe^{2+},Fe^{3+})_o(O^{2-})_4$.

We might expect that all spinels would adopt the inverse structure, for the tetrahedral holes are smaller than the octahedral holes and the tripositive cations are smaller than the dipositive cations. However, in addition to size factors, we have to consider energy factors. Because lattice energy depends on the size of the ionic charge, it is the location of the 3+ ion that is responsible for the majority of the energy. Lattice energy will be higher when the 3+ ion is an octahedral site surrounded by six anions than when it occupies a tetrahedral site and is surrounded by only four anions. Nevertheless, the inverse spinel structure is preferred by many transition metal ions because the *d*-orbital occupancy affects the energy preferences, as we will see in Chapter 18.

The interest in spinels derives from their unusual electrical and magnetic properties, particularly those in which the tripositive ion is Fe^{3+}. These compounds are known as *ferrites*. For example, it is possible to synthesize a series of compounds MFe_2O_4, where M is any combination of zinc ions and manganese ions, provided the formula $Zn_xMn_{1-x}FeO_4$ is obeyed. By choosing the appropriate ratio, very specific magnetic properties can be obtained for these zinc ferrites. We discuss ferrites in more detail in Chapter 19.

Even more peculiar is sodium-β-alumina, $NaAl_{11}O_{17}$. Though its formula does not look like that of a spinel, most of the ions fit the spinel lattice sites. The sodium ions, however, are free to roam throughout the structure. It is this property that makes the compound so interesting, because its electrical conductivity is very high, and it can act as a solid-phase electrolyte. This type of structure offers great potential for low-mass storage batteries.

Thallium and the Inert-Pair Effect

Whereas aluminum is one of our most important metals, thallium must rate as one of the least significant, with annual world production being only about 5 tonnes. Its uses are very specialized. For example, thallium(I) bromide and thallium(I) iodide are among the few substances highly transparent to long-wavelength infrared radiation; hence sheets of the compounds are used in infrared detector units. Thallium also has intriguing chemical properties. In particular, this is the first element that we have discussed so far that has two common oxidation states, +1 and +3. To account for the lower oxidation state, we use the concept of the inert-pair effect.

The Group 13 elements have an outer electron configuration of s^2p^1. As we have seen, these elements can form compounds by sharing these three valence electrons to form three covalent bonds or, rarely, lose all three electrons to ionize to the 3+ ion. The heavier members of all groups (Period 4 and beyond) possessing both *s* and *p* electrons are known to form ionic compounds in which only the *p* electrons are removed. This ionization behavior is called the inert-pair effect in recognition of the two *s* electrons that are not removed. But this term is only a name, not an explanation.

To find a reasonable explanation for the formation of these low-charge ions, we have to consider relativistic effects. The velocities of electrons in outer orbitals, particularly in the 6*s* orbital, become close to that of light. As a result, the mass of these 6*s* electrons increases and, following from this, their mean distance from the nucleus decreases. In other words, the orbital shrinks. This effect is apparent from the successive ionization energies. In Chapter 2 we saw that ionization energies usually decrease down groups, but a comparison of the first three ionization energies of aluminum and thallium shows that the ionization energy of the outer *p* electron is slightly greater, and the ionization energies of each *s* electron significantly greater, for thallium than for aluminum (Table 12.3). Recalling the Born-Haber cycles of Chapter

Table 12.3 Ionization energies of aluminum and thallium

	Ionization energy ($MJ\cdot mol^{-1}$)		
Element	First	Second	Third
Aluminum	0.58	1.82	2.74
Thallium	0.59	1.97	2.88

Table 12.4 A comparison of the properties of thallium(I) ion with those of potassium and silver ions

Properties of potassium	Properties of silver	Properties of thallium(I)
Forms dioxide(1–), not normal oxide	Forms normal oxide	Forms normal oxide
Soluble, very basic hydroxide	Insoluble hydroxide	Soluble, very basic hydroxide
Hydroxide reacts with carbon dioxide to form carbonate	Hydroxide stable	Hydroxide reacts with carbon dioxide to form carbonate
All halides soluble	Fluoride soluble; other halides insoluble	Fluoride soluble; other halides insoluble

6, the large energy input needed to form the cation must be balanced by a high lattice energy (energy output). But the thallium(III) ion is much larger than the aluminum(III) ion; hence the lattice energy of a thallium(III) ionic compound will be less than that of the aluminum analog. The combination of these two factors, particularly the higher ionization energy, leads to a decreased stability of the thallium(III) ionic state and, hence, the stabilizing of the thallium(I) ionic oxidation state.

With a very low charge density (9 $C \cdot mm^{-3}$), thallium(I) ion resembles the lower alkali metals in some respects but the silver ion in others. Table 12.4 shows the chemical similarities and differences between thallium and potassium and silver.

The thallium(I) ion is extremely toxic, because it is a water-soluble, large, low-charge cation like potassium. Thus it can infiltrate cells as a potassium mimic; and, once there, it interferes with enzyme processes.

Thallium(III) halides are known, but as would be expected from thallium(III)'s high charge density (105 $C \cdot mm^{-3}$), their behavior is typical of covalent halides. For example, thallium(III) fluoride reacts with water to give thallium hydroxide and hydrogen fluoride gas:

$$TlF_3(s) + 3\ H_2O(l) \rightarrow Tl(OH)_3(s) + 3\ HF(g)$$

Things are not always what they seem in inorganic chemistry. Thallium forms a compound of formula TlI_3, which one would assume, like the other thallium(III) halides, contains thallium in the +3 oxidation state. However, the compound is found to contain Tl^+ and I_3^- ions. To account for this, we have to look at the relevant redox potentials:

$$Tl^{3+}(aq) + 2\ e^- \rightarrow Tl^+(aq) \qquad E^\circ = +1.25\ V$$

$$I_3^-(aq) + 2\ e^- \rightarrow 3\ I^-(aq) \qquad E^\circ = +0.55\ V$$

Thus iodide will reduce thallium(III) to thallium(I), itself being oxidized to the triiodide ion, I_3^-.

Similarities Between Boron and Silicon

A comparison of boron and silicon is our third and final example of the "diagonal relationship," but this case is very different from the two other examples that we discussed in Chapters 10 and 11. In this comparison, the chemistry of both elements involves covalent bonding; thus there can be no justification in terms of charge density or any similar parameter. In fact, this relationship is not easy to understand except that both elements are on the borderline of the metal-nonmetal divide and have similar electronegativities. Some of the similarities are listed here:

1. Boron forms a solid acidic oxide, B_2O_3, like that of silicon, SiO_2, but unlike that of either aluminum, whose oxide is amphoteric, or carbon, whose oxide, CO_2, is acidic but gaseous.

2. Boric acid, H_3BO_3, is a very weak acid that is similar to silicic acid, H_4SiO_4, in some respects. It bears no resemblance to the amphoteric aluminum hydroxide, $Al(OH)_3$.

3. There are numerous polymeric borates and silicates that are constructed in similar ways, using shared oxygen atoms.

4. Boron forms a range of flammable, gaseous hydrides, just as silicon does. There is only one aluminum hydride—a solid.

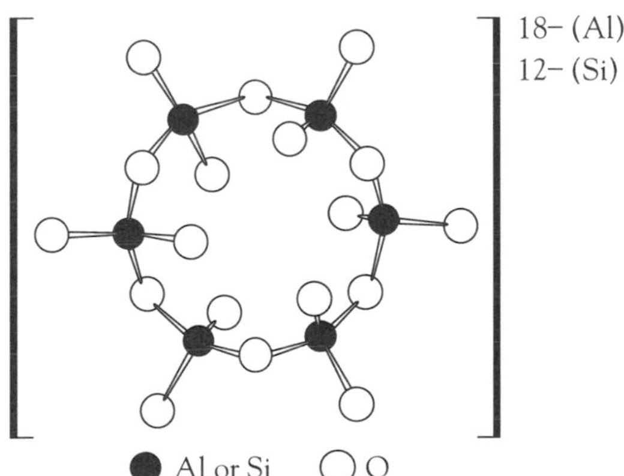

Figure 12.14 Structure of the $Al_6O_{18}^{18-}$ and $Si_6O_{18}^{12-}$ ions.

However, there are also some interesting parallels between aluminum and silicon, particularly between the complex oxyanions each forms. The most interesting of these are cyclic anions that have identical ring structures (but, of course, different charges): $[Al_6O_{18}]^{18-}$ and $[Si_6O_{18}]^{12-}$ (Figure 12.14). Calcium aluminate, $Ca_9[Al_6O_{18}]$, is a major component of Portland cement, whereas beryllium aluminum silicate, $Be_3Al_2[Si_6O_{18}]$, is the mineral beryl, which was mentioned in Chapter 11.

Biological Aspects

The Group 13 elements provide the greatest challenge to bioinorganic chemists. Boron is known to be vital to plant growth; turnips, particularly, need boron for healthy growth. Yet the biological role of boron is not known.

Many people have heard of the claimed link between aluminum and the onset of Alzheimer's disease. As this book goes to press (1995), the connection has not been established, and many scientists dispute any relationship. Nevertheless, aluminum is certainly a toxic element as far as animal life is concerned. Research has shown that the damage to fish stocks in acidified lakes is not due to the lower pH but to the higher concentrations of aluminum ion in the water that result from the lower pH (see Figure 12.10). In fact, an aluminum ion concentration of 5×10^{-6} mol·L^{-1} is sufficient to kill fish.

Human tolerance is greater, but we should still be particularly cautious of aluminum intake. Part of our dietary intake comes from aluminum-containing antacids. Tea is high in aluminum ion, but the aluminum ions form inert compounds when milk or lemon is added. It is advisable not to inhale the spray from aluminum-containing antiperspirants, because the metal ion is believed to be absorbed easily from the nasal passages directly into the bloodstream. It now appears that another essential element, silicon, reduces the hazard from aluminum. A number of researchers have found evidence that a role of silicon, in the form of soluble silicates, is to combine with aluminum ion to form insoluble, and hence harmless, aluminum silicates. Thus one element serves to protect us from the dangers of another element.

Exercises

12.1. Write balanced chemical equations for the following chemical reactions:
(a) liquid potassium metal with solid aluminum chloride
(b) solid diboron trioxide with ammonia gas at high temperature
(c) aluminum metal with hydroxide ion

12.2. Write balanced chemical equations for the following chemical reactions:
(a) liquid boron tribromide with water
(b) aluminum metal with hydrogen ion
(c) thallium(I) hydroxide solution with carbon dioxide gas

12.3. With a very high charge density, aluminum would not be expected to exist widely as a free 3+ ion, yet it does exist in the form of a hydrated 3+ ion. Explain why.

12.4. Construct an electron-dot formula for the peroxyborate ion. Deduce the oxidation number of the bridging oxygen atoms.

12.5. From bond energy data, calculate the enthalpy of formation of boron trifluoride. What two factors result in its particularly high value?

12.6. From bond energy data, calculate the enthalpy of formation of boron trichloride (gaseous). Why is the value so different from that of boron trifluoride?

12.7. Which of the following are likely to form diamondlike structures? Give your reasoning in each case: (a) aluminum phosphide; (b) silver iodide; (c) lead(II) oxide.

12.8. Explain briefly why sheets of aluminum do not oxidize completely to aluminum oxide even though aluminum is a highly reactive metal.

12.9. Explain briefly why solutions of aluminum chloride are strongly acidic.

12.10. Magnesium metal only reacts with acids, whereas aluminum reacts with both acids and bases. What does this behavior tell you about aluminum?

12.11. Describe briefly the steps in the industrial extraction of aluminum from bauxite.

12.12. Explain the potential environmental hazards from aluminum smelting.
12.13. Why are aluminum smelters sometimes located in countries other than those that produce the ore or consume much of the metal?
12.14. Contrast the bonding in the aluminum halides.
12.15. What is an alum?
12.16. Explain the difference between a spinel and an inverse spinel.
12.17. Why is thallium(I) an ionic species and thallium(III) more covalent in its behavior?
12.18. Compare and contrast the chemistry of boron and silicon.
12.19. Why is aluminum a particular environmental problem in the context of acid rain?

CHAPTER

13

The Group 14 Elements

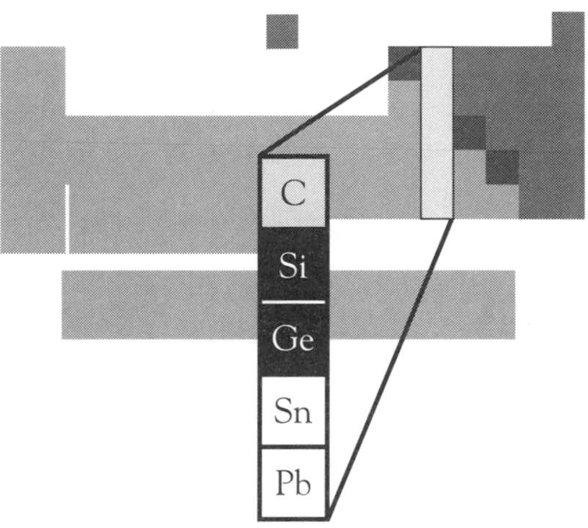

This group contains a nonmetal (carbon), two semimetals (silicon and germanium), and two weakly electropositive metals (tin and lead). Carbon has the most important chemistry of the group, but the chemistry of silicon is receiving increasing interest. The weakly metallic properties of tin and lead contrast sharply with the properties of the alkali supermetals.

It is likely that no inorganic compound has played a greater role in changing history than lead(II) ethanoate, $Pb(CH_3CO_2)_2$, also called lead(II) acetate. Lead was one of the most important elements during the Roman Empire, 2000 years ago, when about 60 000 tonnes of lead were being smelted annually to provide the elaborate piped water and plumbing system of the Romans. This level of sophistication in lifestyle was not regained until the late nineteenth century.

We know from the significant levels of lead in human bones from the Roman Empire that the inhabitants were exposed to high concentrations of this element. But it was not the plumbing that presented the major hazard to the Romans. Because they used natural yeasts, the wine that they prepared was quite acidic. To remedy this, winemakers added a sweetener, *sapa*, produced by boiling grape juice in lead pots. The sweet flavor was the result of the formation of "sugar of lead," what we now call lead(II) ethanoate. This sweetener was also used in food preparation, and about 20 percent of the recipes from the period required addition of *sapa*. The mental instability of the Roman emperors (many of whom were excessive wine drinkers) was a major contributor to the decline and fall of the Roman Empire. In fact, idiosyncrasies of many of the emperors match the known symptoms of lead poisoning. Unfortunately, the governing class of Romans never correlated their use of *sapa* with the sterility and mental disorders that plagued the rulers. Thus the course of history was probably changed by this sweet-tasting but deadly compound.

Group Trends

The first three elements of Group 14 have very high melting points, a characteristic of network covalent bonding for nonmetals and semimetals, whereas the two metals in the group have low melting points and, typical of metals, long liquid ranges (Table 13.1). All the Group 14 elements form compounds in which they *catenate* (form chains of atoms with themselves). The ability to catenate decreases down the group. Thus carbon forms carbon-carbon chains of unlimited length; silicon forms chains up to 16 atoms; germanium, 6; and tin, 2.

Now that we have reached the middle of the main groups, the nonmetallic properties are starting to predominate. In particular, multiple oxidation states become common. All members of Group 14 form compounds in which they have an oxidation number of +4. This oxidation state involves covalent bonding, even for the two metals of the group. In addition, an oxidation state of −4 exists for the three nonmetals-semimetals when they are bonded to more electropositive elements. Tin and lead also have an oxidation state of +2, which is the only oxidation state in which they form ionic compounds.

Consider the oxidation state diagram shown in Figure 13.1. For carbon and germanium, the +4 oxidation state is more thermodynamically stable

Table 13.1 Melting and boiling points of the Group 14 elements

Element	Melting point (°C)	Boiling point (°C)
C	Sublimes at 4100	
Si	1420	3280
Ge	945	2850
Sn	232	2623
Pb	327	1751

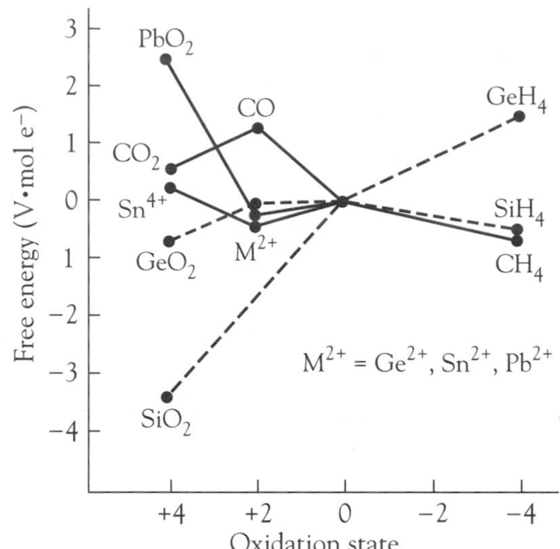

Figure 13.1 Frost diagram in acid solution for the Group 14 elements.

than is the +2 state, whereas for tin and lead, the +2 state is more stable than the +4 state. For silicon, there is no common compound in which silicon exists in a +2 oxidation state; in contrast, the +2 oxidation state of lead is the most stable and in the +4 state lead is strongly oxidizing. One of the few common examples of carbon in the +2 oxidation state is the reducing compound, carbon monoxide. Among the hydrides, the –4 oxidation state becomes less stable and more strongly reducing as one goes down the group.

Carbon

Two common allotropes of carbon have been known throughout much of recorded history, but recently, a whole new family of allotropes has been identified.

Diamond

In the diamond form of carbon, there is a network of single, tetrahedrally arranged covalent bonds (Figure 13.2). Diamond is an electrical insulator but an excellent thermal conductor, being about five times better than copper. We can understand the thermal conductivity in terms of the diamond structure. Because the giant molecule is held together by a continuous network of covalent bonds, little movement of individual carbon atoms can occur. Hence any added heat energy will be transferred as molecular motion directly across the whole diamond. Diamond also has a very high melting point of 4100°C, because an enormous amount of energy is needed to break these strong covalent bonds.

In "normal" diamond, the arrangement of the tetrahedrons is the same as that in the cubic ZnS-sphalerite ionic structure (Chapter 5). There is also an extremely rare form, lonsdaleite (named after the famous crystallographer Kathleen Lonsdale), in which the tetrahedra are arranged in the hexagonal

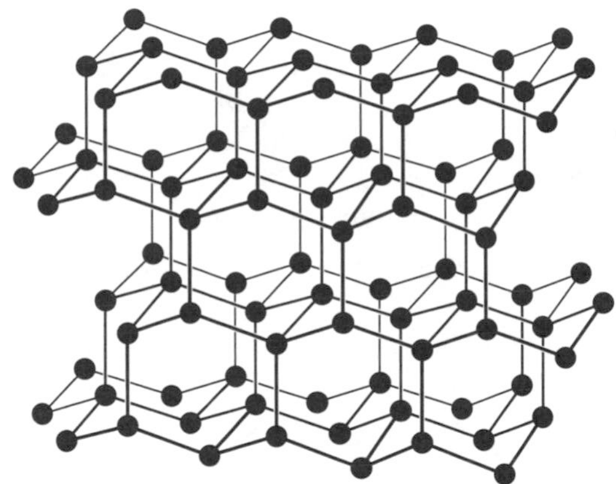

Figure 13.2 Structure of diamond.

ZnS-wurtzite structure. A crystal of lonsdaleite was first found in the Canyon Diablo meteorite in Arizona, and since then it has been synthesized by a route in which graphite is subjected to high pressure and temperature.

Natural (sphalerite-type) diamonds are found predominantly in Africa. Zaire is the largest producer (29 percent), but South Africa (17 percent of the production) still produces the most gem-quality stones. Russia is in second place with 22 percent of world production. In North America, diamonds are found in Crater of Diamonds State Park, Arkansas, but no large-scale mining operations occur there.

The density of diamond (3.5 $g{\cdot}cm^{-3}$) is much greater than that of graphite (2.2 $g{\cdot}cm^{-3}$), so a simple application of the Le Châtelier principle indicates that diamond formation from graphite is favored under conditions of high pressure. Furthermore, to overcome the considerable activation energy barrier accompanying the rearrangement of covalent bonds, high temperatures also are required. The lure of enormous profits resulted in many attempts to perform this transformation. The first bulk production of diamonds was accomplished by the General Electric Company in the 1940s, using high temperatures (about 1600°C) and extremely high pressures (about 5 GPa, that is, about 50 000 times atmospheric pressure). The diamonds produced by this method are small and not of gem quality, although they are ideal for drill bits and as grinding material.

The free energy of diamond is 2.9 $kJ{\cdot}mol^{-1}$ higher than that of graphite. Thus it is only the very slow kinetics of the process that prevents diamonds from crumbling into graphite. For this reason, Western scientists were skeptical when Soviet scientists claimed to have found a method of making layers of diamonds at low temperatures and pressures from a chemical reaction in the gas phase. It was about 10 years before the claims were investigated and shown to be true. We are now aware of the tremendous potential of diamond films as a means of providing very hard coatings—on surgical knives, for example. Diamond films are also promising coatings for computer microprocessor chips. A continuing problem associated with computer chips is their exposure to high temperatures generated by excess heat resulting from electric resistance in the computer's electric circuits. Diamond has a very high thermal conductivity; hence chips with a diamond coating will be undamaged by the heat produced by high-density circuitry. Diamond film technology is predicted to be a major growth industry over the next decade.

Until the nineteenth century, it was thought that graphite and diamond were two different substances. It was Humphry Davy—by borrowing one of his spouse's diamonds and setting fire to it—who showed that carbon dioxide is the only product when diamond burns:

$$C(s) + O_2(g) \rightarrow CO_2(g)$$

Fortunately, Davy's wife was rich enough not to be too upset about the loss of one of her gems to the cause of science. This is one of the more expensive chemical methods of testing whether you really have a diamond.

Graphite

The structure of graphite is quite different from that of diamond. Graphite consists of layers of carbon atoms (Figure 13.3). Within the layers, covalent bonds hold the carbon atoms in six-membered rings. The carbon-carbon bond length in graphite is 141 pm. These bonds are much shorter than those in diamond (154 pm) but very similar to the 140-pm bonds in benzene, C_6H_6, a compound that was mentioned in Chapter 11. This similarity in bond lengths suggests a possible explanation for the short interatomic distance in graphite—there is multiple bonding between the carbon atoms within layers. Like benzene, graphite is assumed to have a delocalized π electron system throughout the plane of the carbon rings, an arrangement that would result in a net $1\frac{1}{3}$ bonds between each pair of carbon atoms. The measured bond length is consistent with this assumption.

The distance between the carbon layers is very large (335 pm) and is more than twice the value of the van der Waals radius of a carbon atom. Hence the attraction between layers is very weak. In the common hexagonal form of graphite (Figure 13.3), alternating layers are aligned to give an *abab* arrangement. Looking at the sequential layers, one-half of the carbon atoms are located in line with carbon atoms in the planes above and below, and the other half are located above and below the centers of the rings.

The layered structure of graphite accounts for one of its most interesting properties—its ability to conduct electricity. More specifically, the conductivity in the plane of the sheets is about 5000 times greater than that at right angles to the sheets. Graphite is also an excellent lubricant by virtue of the ability of sheets of carbon atoms to slide over one another. However, this is not quite the whole story. Graphite also adsorbs gas molecules between its layers. Thus many chemists argue that the graphite sheets are really gliding on molecular "ball bearings," namely, the adsorbed gas molecules.

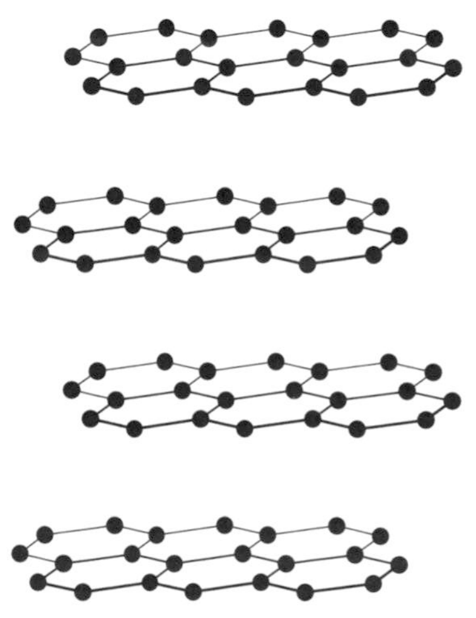

Figure 13.3 Structure of graphite.

Even though thermodynamically more stable than diamond, graphite is kinetically more reactive as a result of the separation of the carbon sheets. Everything from alkali metals through the halogens to metal halide compounds is known to react with graphite. In the resulting products, the graphite structure is essentially preserved, with the intruding atoms or ions fitting between the layers in a fairly stoichiometric ratio. These layer-insertion species are referred to as *intercalation compounds* (in Chapter 10).

Most of the mined graphite comes from the Far East, with China, Siberia, and the two Koreas being the major producers. In North America, Ontario,

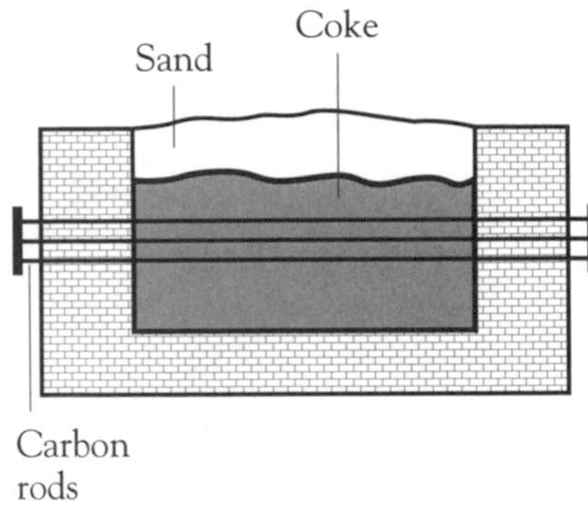

Figure 13.4 Acheson furnace.

Canada, has significant deposits. Graphite is also manufactured from amorphous carbon, the most reliable method being the Acheson process. In this procedure powdered coke (amorphous carbon) is heated at 2500°C for about 30 hours. This temperature is produced in an electric furnace that has carbon rods as heating elements (Figure 13.4). The method is rather similar to sublimation in that pure crystalline material is obtained from an impure powder. The amorphous carbon is covered by a layer of sand to prevent it from oxidizing to carbon dioxide. The process is not very energy efficient, but this equipment has fewer operational problems than other types of furnaces. Thanks to advances in chemical technology, the newer units produce fewer pollutants and are more energy efficient than their predecessors.

Graphite is used in lubricants, as electrodes, and as graphite-clay mixtures in lead pencils. The higher the proportion of clay, the "harder" the pencil. The common mixture is designated "HB." The higher clay (harder) mixtures are designated by various "H" numbers, for example, "2H"; and the higher graphite (softer) mixtures are designated by various "B" numbers. Hence there is no lead in a lead pencil. The term originated from the similarity between the streak left on a surface from a soft lead object and that from graphite.

Fullerenes

Chemistry is full of surprises, and the discovery of a new series of allotropes of carbon must rank as the most unexpected finding of all. The problem with all science is that we are limited by our own imaginations. It has been pointed out that if diamonds did not exist naturally on Earth, it would be very unlikely that any chemist would "waste time" trying to change the structure of graphite by using extremely high pressures. It would be even more unlikely for any agency to advance funding for such a "bizarre" project.

Fullerenes constitute a family of structures in which the carbon atoms are arranged in a spherical or ellipsoidal structure. To make such a structure, the carbon atoms form five- and six-membered rings, similar to the pattern of lines on a soccer ball (the early name for C_{60} was soccerane). The 60-member sphere, C_{60}, buckminsterfullerene, is the easiest to prepare; and it is aesthetically the nicest, being a perfect sphere (Figure 13.5). The 70-member sphere, C_{70}, is the next most commonly available fullerene. The ellipsoidal structure of this allotrope resembles an American football or a rugby ball.

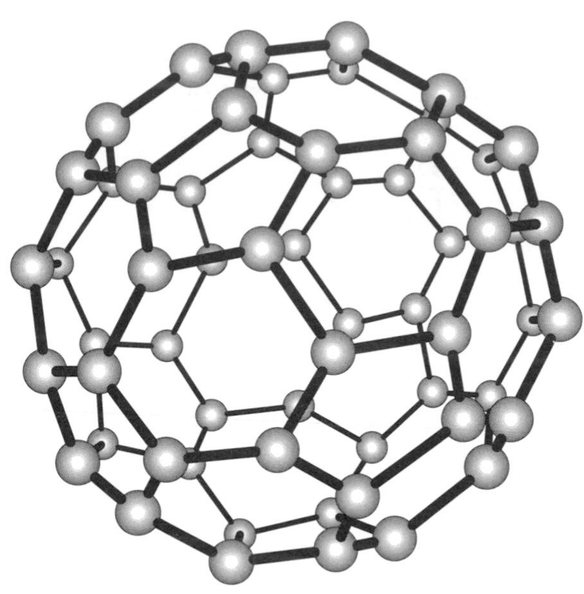

Figure 13.5 Structure of buckminsterfullerene, C_{60}.

This allotrope family was named after R. Buckminster Fuller, a genius of the twentieth century. His name is particularly associated with the geodesic dome, an architectural design that has tremendous strength. Contrary to general belief, he did not invent the dome; this was done by Walter Bauersfield in Germany. However, Buckminster Fuller did make major improvements to the design and popularized it.

One method of manufacturing the fullerenes is to use an intense laser beam to heat graphite to temperatures of over 10 000°C. At these temperatures, sections of the hexagonal planes of carbon atoms peel off the surface and wrap themselves into balls. Although C_{60} and C_{70} are the major products, it is possible to prepare almost all even-number fullerenes by this method. Tubes with the same type of structure—"buckytubes"—also can be made in this way. Now that we know of these molecules, they are turning up everywhere. Common soot contains fullerenes, and they have been found in naturally occurring graphite deposits. Some astrochemists argue that these molecules exist extensively in interstellar space.

Diamond and graphite are insoluble in all solvents because they have network covalent structures. The fullerenes have covalent bonds within the units, but only dispersion forces hold the units together in the solid phase. As a result, they are very soluble in nonpolar solvents such as hexane and toluene. Although black in the solid phase, fullerenes display a wide range of colors in solution: C_{60} gives an intense magenta-purple color; C_{70} is wine red; and C_{76} is bright yellow-green. All the fullerenes sublime when heated, a property providing further evidence of the weak intermolecular forces.

The C_{60} molecules pack together in the same way metal atoms do, forming a face-centered cubic arrangement. The fullerenes have low densities (about 1.5 $g{\cdot}cm^{-3}$), and they are nonconductors of electricity. Molecules of C_{60} (and those of other fullerenes) absorb visible light to produce an unstable excited form, which is represented by the symbol $^{*}C_{60}$. The excited form

The Discovery of Buckminsterfullerene

Discoveries in science are almost always convoluted affairs, not the sudden "Eureka!" of fiction. W. E. Addison had predicted in 1964 that other allotropes of carbon might exist; and David Jones in 1966 actually proposed the existence of "hollow graphitic spheroids."

It was not chemists, however, but two astrophysicists, Donald Huffman of the University of Arizona at Tucson and Wolfgang Krätschmer of the Max Planck Institute for Nuclear Physics at Heidelberg, Germany, who are credited with the first synthesis of fullerenes in 1982. They were interested in the forms of carbon that could exist in interstellar space. They heated graphite rods in a low-pressure atmosphere and obtained a soot. It appeared to have an unusual spectrum, but they attributed that to contamination by oil vapor from the equipment. As a result, they lost interest in the experiment. Two years later, an Australian medical researcher, Bill Burch, at the Australian National University produced a sublimable form of carbon that he patented as "Technogas." This, too, was probably buckminsterfullerene.

Harold Kroto of the University of Sussex, England, and Richard Smalley of Rice University, Texas, performed the crucial experiment. They were also interested in the nature of carbon in space. When Kroto visited Smalley, the former proposed that they use Smalley's high-powered laser to blast fragments off a graphite surface and then identify the products. Between 4 September and 6 September 1985, they found one batch of products that had a very high proportion of a molecule containing 60 carbon atoms. Over the weekend, two research students, Jim Heath and Sean O'Brien, altered the conditions of the experiment until, time after time, they could consistently obtain this unexpected product.

How could the formula C_{60} be explained? Kroto recalled the geodesic dome that housed the U.S. pavilion at Expo 67 in Montreal. However, he thought the structure consisted of hexagonal shapes, like those making up graphite. The chemists were unaware of the work by the eighteenth-century mathematician Leonhard Euler, who had shown that it was impossible to construct closed figures out of hexagons alone. Smalley and Kroto now disagree vehemently over which one of them first realized that a spherical structure could be constructed using 20 hexagons and 12 pentagons. Nevertheless, on 10 September, this was the structure that the group postulated for the mysterious molecule. As a result of the disagreement, relations between the two research groups are acrimonious.

The Kroto-Smalley method produced quantities of buckminsterfullerene that were too small for chemical studies. The discovery cycle was completed in 1988 when Huffman realized that the method that he and Krätschmer had used several years earlier must have been forming large quantities of these molecules. These two physicists resumed their production of the soot and developed methods for producing consistently high yields of the allotropes. From subsequent studies, the chemical evidence proving the structures of C_{60} and C_{70} was independently and almost simultaneously produced by the Kroto and the Smalley groups.

The story of the fullerenes is still developing. In the first seven years of fullerene research, over 3000 research papers have been written about the properties of these compounds.

absorbs light many times more efficiently than normal C_{60}, converting the electromagnetic energy to heat. This is a very important property, because it means that as the intensity of light passed into a solution of C_{60} increases, more $^{*}C_{60}$ will be produced and hence more of the light will be absorbed. The intensity of light leaving the solution will be correspondingly reduced. The solution, therefore, acts as an optical limiter. Coating glasses with this material could prevent eye damage in people working with high-intensity lasers; and in the more common world, such coatings could be used to create instant-response sunglasses.

The chemistry of these novel molecules is, at the time of writing, a field of intense research, and the molecules are now commercially available. The fullerenes are easily reduced to anions; for example, they react with Group 1 and Group 2 metals. Thus rubidium fits within the interstices in the C_{60} lattice to give Rb_3C_{60}. This compound is a superconductor at temperatures below 28 K because its structure is actually $[Rb^+]_3[C_{60}{}^{3-}]$. The extra electrons associated with the fullerenes are free to move throughout the crystal, just like those in a metal. Chemical reaction with the surface of the fullerenes is also possible, one example being the formation of the oxide $C_{60}O$.

Impure Carbon

The major uses of carbon are as an energy source and as a reducing agent. For these purposes, an impure form of carbon (coke) is used. This material is produced by heating coal in the absence of air. In this process, the complex coal structure breaks down, boiling off hydrocarbons and leaving behind a porous, low-density, silvery, almost metallic-looking solid. Essentially, coke is composed of microcrystals of graphite that have small proportions of some other elements, particularly hydrogen, bonded in their structure. Much of the distillate produced by the coking process can be used as raw materials in the chemical industry, but the oily and watery wastes are a nightmare of carcinogens. Coke is utilized in the production of iron from iron ore and in other pyrometallurgical processes. Coke production is considerable, and about 5×10^8 tonnes are used worldwide every year.

Carbon black is a very finely powdered form of carbon. This impure micrographite is produced by incomplete combustion of organic materials. It is used in extremely large quantities—about 1×10^7 tonnes per day. It is mixed with rubber to strengthen tires and reduce wear. About 3 kg are used for the average automobile tire, and it is the carbon content that gives a tire its black color.

Another form of carbon known as activated carbon has a very high surface area—typically 10^3 $m^2{\cdot}g^{-1}$. This material is used for the industrial decolorizing of sugar and in gas filters, as well as for removing impurities from organic reactions in the university laboratory. The physical chemistry of the absorption process is complex, but in part it works by the attraction of polar molecules to the carbon surface.

Blocks of carbon are industrially important as electrodes in electrochemical and thermochemical processes. For example, about 7.5 million tonnes of carbon are used each year just in aluminum smelters. And, of course, the summer season always increases the consumption of carbon in home barbecues.

Isotopes of Carbon

Natural carbon contains three isotopes: carbon-12 (98.89 percent), the most prevalent isotope; a small proportion of carbon-13 (1.11 percent); and a trace of carbon-14. Carbon-14 is a radioactive isotope with a half-life of 5700 years. With such a short half-life, we would expect little sign of this isotope on Earth. Yet it is prevalent in all living tissue, because the isotope is constantly being produced by reactions between cosmic ray neutrons and nitrogen atoms in the upper atmosphere:

$$^{14}_{7}N + ^{1}_{0}n \rightarrow ^{14}_{6}C + ^{1}_{1}H$$

The carbon atoms react with oxygen gas to form radioactive molecules of carbon dioxide. These are absorbed by plants in photosynthesis. All creatures that eat plants and the creatures that eat the creatures that eat the plants will all contain the same proportion of radioactive carbon. After the death of the organism, there is no further intake of carbon, and that already present in the body decays. Thus the age of an object can be determined by measuring the amount of carbon-14 present in a sample. This method provides an absolute scale of dating objects that are between 1000 and 20 000 years old. W. F. Libby was awarded the Nobel prize for chemistry in 1960 for developing the radiocarbon dating technique.

The Extensive Chemistry of Carbon

Carbon has two properties that enable it to form such an extensive range of compounds: *catenation* (the ability to form chains of atoms) and multiple bonding (that is, the ability to form double and triple bonds). The latter property is common to carbon, nitrogen, and oxygen, but catenation is relatively rare. For catenation, three conditions are necessary:

1. A bonding capacity (valence) greater than or equal to 2.
2. An ability of the element to bond with itself; the self-bond must be about as strong as its bonds with other elements.
3. A kinetic inertness of the catenated compound toward other molecules and ions.

We can see why catenation is frequently found in carbon compounds but only rarely in silicon compounds by comparing bond energy data for these two elements (Table 13.2). Notice that the energies of the carbon-carbon and

Table 13.2 Bond energies of various carbon and silicon bonds

Carbon bonds	Bond energy ($kJ \cdot mol^{-1}$)	Silicon bonds	Bond energy ($kJ \cdot mol^{-1}$)
C–C	346	Si–Si	222
C–O	358	Si–O	452

carbon-oxygen bonds are very similar. However, the silicon-oxygen bond is much stronger than that between two silicon atoms. Thus in the presence of oxygen, silicon will form –Si–O–Si–O– chains rather than –Si–Si– linkages. We will see later that the silicon-oxygen chains dominate the chemistry of silicon. There is no such energy "incentive" to break carbon-carbon bonds.

There are two related reasons for this difference in behavior. First, in a *saturated* carbon compound, the *s* and *p* orbitals are all involved in bonding. Attack by an oxygen molecule will be difficult; hence, kinetically, there is no easy pathway for the reaction to occur. In silicon compounds, the empty 3*d* orbitals can be used to bond with the attacking oxygen molecules. Second, the very high strength of the silicon-oxygen bond is explained in terms of a π bond formed between a full *p* orbital of the oxygen and an empty *d* orbital of the silicon (Figure 13.6). It is sobering to realize that two "quirks" of the chemical world make life possible: the hydrogen bond and the catenation of carbon. Without these two phenomena, life of any form (that we can imagine) could not exist.

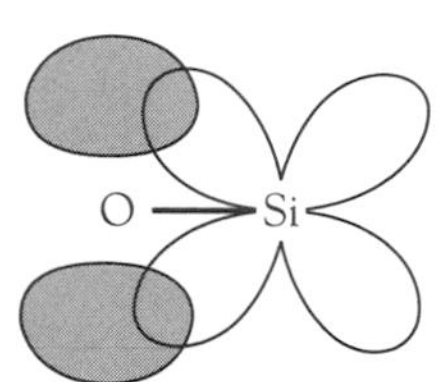

Figure 13.6 Overlap of a full oxygen *p* orbital with an empty silicon *d* orbital.

Carbides

Binary compounds of carbon with less electronegative elements (except hydrogen) are called carbides. There are three types of bonding in carbides: ionic, covalent, and metallic; however, all three categories of carbides are hard solids with high melting points.

Ionic Carbides

Ionic carbides are formed by the most electropositive elements, the alkali and alkaline earth metals and aluminum. Most of these ionic compounds contain the dicarbide(2–) ion that we discussed in the context of calcium dicarbide (Chapter 11). The ionic carbides are the only carbides to show much chemical reactivity. In particular, they react with water to produce ethyne, C_2H_2, formerly called acetylene:

$$Na_2C_2(s) + 2\ H_2O(l) \rightarrow 2\ NaOH(aq) + C_2H_2(g)$$

From their formulas, the red beryllium carbide, Be_2C, and the yellow aluminum carbide, Al_4C_3, appear to contain the C^{4-} ion. The high charge density cations Be^{2+} and Al^{3+} are the only ions that can form stable lattices with such a highly charged anion. However, the cations are so small and highly charged and the anion so large that we assume there must be a large degree of covalency in the bonding. Nevertheless, despite the assumption of covalent bonding, these two carbides do produce methane, CH_4, when they react with water, as would be expected if the C^{4-} ion were present:

$$Al_4C_3(s) + 12\ H_2O(l) \rightarrow 4\ Al(OH)_3(s) + 3\ CH_4(g)$$

Covalent Carbides

Because most nonmetals are more electronegative than carbon, there are few covalent carbides. Silicon carbide, SiC, and boron carbide, B_4C, are the common examples; both are very hard and have high melting points. Silicon carbide is used as a grinding and polishing agent in metallurgical applications, and it is the only nonoxide ceramic product of large-scale industrial importance. Worldwide production of this compound is about 7×10^5 tonnes. Silicon carbide, which is bright green when pure, is produced in an Acheson furnace, like that used to convert coke to graphite. The furnace takes about 18 hours to heat electrically to the reaction temperature of about 2300°C, and it takes another 18 hours for the optimum yield to be formed. The production of this compound is extremely energy intensive, and between 6 and 12 kWh of electricity are needed to produce 1 kg of silicon carbide:

$$SiO_2(s) + 3\,C(s) \rightarrow SiC(s) + 2\,CO(g)$$

There is tremendous interest in silicon carbide as a material for high-temperature turbine blades, which operate at temperatures that cause metals to lose their strength. It is also being used for the backing of high-precision mirrors because it has a very low coefficient of expansion. This property minimizes distortion problems because silicon carbide mirrors undergo only a negligible change in shape as temperatures fluctuate. As you can see, silicon carbide is certainly a promising material for the twenty-first century.

Metallic Carbides

Metallic carbides are compounds in which the carbon atoms fit within the crystal structure of the metal itself, and they are usually formed by the transition metals. To form a metallic carbide, the metal must assume a close-packed structure, and the atoms usually have a metallic radius greater than 130 pm. The carbon atoms can then fit into the octahedral holes (interstices) in the structure; hence metallic carbides are also called *interstitial* carbides. If all the octahedral holes are filled, the stoichiometry of these compounds is 1:1.

Because metallic carbides retain the metallic crystal structure, they look metallic and conduct electricity. They are important because they have very high melting points, show considerable resistance to chemical attack, and are extremely hard. The most important of these compounds is tungsten carbide, WC, of which about 20 000 tonnes are produced annually worldwide. Most of the material is used in cutting tools.

Some metals with a radius below 130 pm form metallic carbides, but their metal lattices are distorted. As a result, such compounds are more reactive than true interstital carbides. The most important of these almost interstitial carbides is Fe_3C, commonly called cementite. It is microcrystals of cementite that cause carbon steel to be harder than pure iron.

Carbon Monoxide

Carbon monoxide is a colorless, odorless gas. It is very poisonous, because it has a 300-fold greater affinity for blood hemoglobin than does oxygen; thus quite low concentrations of carbon monoxide in air are sufficient to prevent oxygen absorption in the lungs. Without a continuous supply of oxygen, the brain loses consciousness and death follows unless the supply of oxygenated hemoglobin is restored. Curiously, there is now evidence that carbon monox-

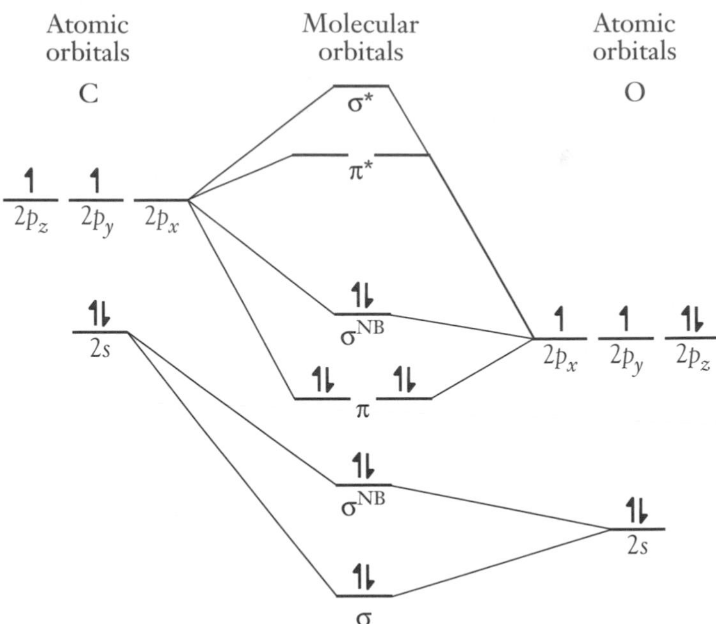

Figure 13.7 Partial simplified molecular orbital diagram for carbon monoxide.

ide is a messenger molecule in some neurons in the brain. Thus what is toxic in large quantities is necessary in tiny amounts for the correct functioning of the brain.

The carbon-oxygen bond in carbon monoxide is very short, about the length that would be expected for a triple bond. A simplified molecular orbital diagram for carbon monoxide is shown in Figure 13.7 for the molecular orbitals derived from 2*s* and 2*p* atomic orbitals. We know the comparative energies of the molecular orbitals from photoelectron spectroscopic analysis (Chapter 3). However, chemists have not agreed on which levels are predominantly bonding, antibonding, and nonbonding. Here, we will follow one of the most popular assignments, which designates two of the filled σ orbitals as nonbonding. This model gives a bond order of three from the filling of one σ bonding orbital and two π bonding orbitals.

Carbon monoxide is produced when any carbon-containing compound, including carbon itself, is burned with an amount of oxygen insufficient for complete combustion:

$$2\ C(s) + O_2(g) \rightarrow 2\ CO(g)$$

As automobile engines have become more efficient, carbon monoxide production has diminished substantially. Thus in one respect the air inhaled on city streets is less harmful than it used to be because its carbon monoxide content is lower.

The pure gas is prepared in the laboratory by warming methanoic (formic) acid with concentrated sulfuric acid. In this decomposition, the sulfuric acid acts as a dehydrating agent:

$$HCOOH(l) + H_2SO_4(l) \rightarrow H_2O(l) + CO(g) + H_2SO_4(aq)$$

Carbon monoxide is quite reactive; for example, it burns with a blue flame to carbon dioxide:

$$2\ CO(g) + O_2(g) \rightarrow 2\ CO_2(g)$$

It reacts with chlorine gas in the presence of light or hot charcoal, which serve as catalysts, to give carbonyl chloride, $COCl_2$, a compound better known as the poison gas phosgene:

$$CO(g) + Cl_2(g) \rightarrow COCl_2(g)$$

When carbon monoxide is passed over heated sulfur, the compound carbonyl sulfide, COS, a promising low-hazard fungicide, is formed:

$$CO(g) + S(s) \rightarrow COS(g)$$

As the Frost diagram in Figure 13.1 shows, carbon monoxide is a strong reducing agent. It is used industrially in this role, for example, in the smelting of iron(III) oxide to iron metal:

$$Fe_2O_3(s) + 3\ CO(g) \rightarrow 2\ Fe(l) + 3\ CO_2(g)$$

It is also an important starting material in industrial organic chemistry. Under high temperatures and pressures, carbon monoxide will combine with hydrogen gas (a mixture known as synthesis gas) to give methanol, CH_3OH. Mixing carbon monoxide with ethene, C_2H_4, and hydrogen gas produces propanal, CH_3CH_2CHO, a reaction known as the *OXO process:*

$$CO(g) + 2\ H_2(g) \rightarrow CH_3OH(g)$$

$$CO(g) + C_2H_4(g) + H_2(g) \rightarrow C_2H_5CHO(g)$$

The active catalytic species in this process is a cobalt compound containing covalent bonds to both hydrogen and carbon monoxide, $HCo(CO)_4$; and it is similar compounds of metals with carbon monoxide that we consider next.

Carbon monoxide forms numerous compounds with transition metals. In these highly toxic, volatile compounds, the metal is considered to have an oxidation number of 0. Among the simple carbonyls are tetracarbonylnickel(0), $Ni(CO)_4$, pentacarbonyliron(0), $Fe(CO)_5$, and hexacarbonylchromium(0), $Cr(CO)_6$. Many of the metal carbonyls can be prepared simply by heating the metal under pressure with carbon monoxide. For example, when heated, nickel reacts with carbon monoxide to give the colorless gas tetracarbonylnickel(0):

$$Ni(s) + 4\ CO(g) \rightarrow Ni(CO)_4(g)$$

These compounds are often used as reagents for the preparation of other low oxidation number compounds of transition metals. The shapes of these three metal carbonyls are shown in Figure 13.8; bonding in these compounds is discussed in Chapter 18.

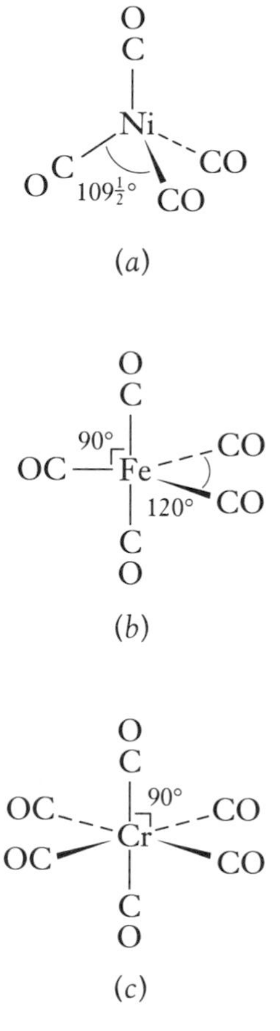

Figure 13.8 Structures of simple metal carbonyls: (*a*) tetracarbonylnickel(0); (*b*) pentacarbonyliron(0); and (*c*) hexacarbonylchromium(0).

Carbon Dioxide

Carbon dioxide is a dense, colorless, odorless gas, which does not burn or, normally, support combustion. The combination of high density and inertness has led to its use for extinguishing fires. Because it is about one and a half times denser than air under the same conditions of temperature and pressure, it flows, almost like a liquid, until air currents mix it with the gases of the atmosphere. Thus it is effective at fighting floor-level fires but almost useless for fighting fires in ceilings. However, carbon dioxide will react with burning metals, such as calcium:

$$2\ Ca(s) + CO_2(g) \rightarrow 2\ CaO(s) + C(s)$$

To extinguish metal fires, a class D fire extinguisher has to be used (classes A, B, and C are used to fight conventional fires). As we mentioned in Chapter 11, this type of extinguisher contains either graphite or sodium chloride, substances that cover the blaze with an inert layer and prevent oxygen from reaching the metal surface.

Carbon dioxide is relatively unique, because it has no liquid phase at normal atmospheric pressures. Instead, the solid sublimes directly to the gas phase. To obtain the liquid phase at room temperature, a pressure of 6.7 MPa (67 times standard atmospheric pressure) must be applied, as shown in the phase diagram in Figure 13.9.

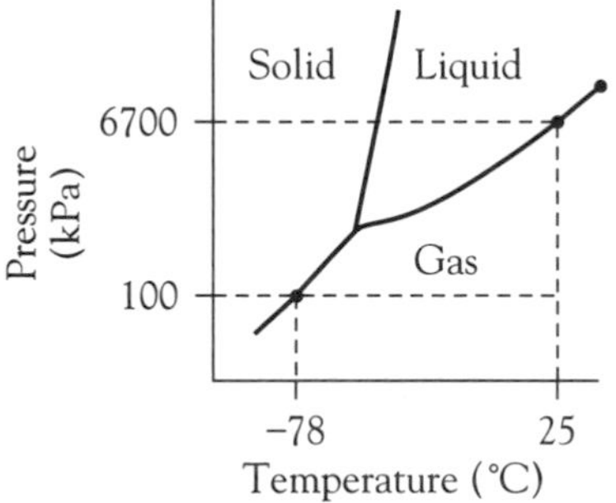

Figure 13.9 Phase diagram for carbon dioxide.

Carbon dioxide is usually conveyed in tank cars and cylinders in the liquid form. When the pressure is released, some of the liquid carbon dioxide vaporizes, but the heat absorbed in the expansion process (overcoming the dispersion forces between molecules) is enough to cool the remaining liquid below its sublimation point, −78°C at atmospheric pressure. By inverting the cylinder and opening the valve, solid carbon dioxide, "dry ice", can be collected in a gauze bag or in a CO_2 "patty maker" at room temperature.

Carbon dioxide is an important industrial chemical. Each year, over 40 million tonnes are used in the United States alone. Half of this quantity is needed as a refrigerant, and another 25 percent is used to carbonate soft drinks. It is also used as a propellant in some aerosol cans, as a pressurized gas to inflate life rafts and life vests, and as a fire-extinguishing material. In its liquid form, it is an excellent nonpolar solvent. In particular, it is used to extract caffeine from coffee. This extraction used to be done with dichloromethane, CH_2Cl_2, but traces of this possibly carcinogenic solvent remained in the coffee. The manufacturers were so concerned that carbon dioxide would be perceived as another nasty "chemical" that the current process was advertised as an extraction technique using "natural effervescence"—a much more reassuring term.

There are a number of sources of industrial carbon dioxide: as a byproduct in the manufacture of ammonia, molten metals, cement, and sugar fermentation products. And, of course, we exhaust carbon dioxide into the atmosphere during the complete combustion of any carbon-containing substance: wood, natural gas, gasoline, and oil.

In the laboratory, carbon dioxide is most conveniently prepared by adding dilute hydrochloric acid to marble chips (chunks of impure calcium carbonate), although any dilute acid with a carbonate or hydrogen carbonate can be used (for example, Alka-Selzer® tablets or baking powder):

$$2\ HCl(aq) + CaCO_3(s) \rightarrow CaCl_2(aq) + H_2O(l) + CO_2(g)$$

To identify carbon dioxide, the limewater test is used. In this test, a gas is bubbled into a saturated solution of calcium hydroxide. When the gas is carbon dioxide, a white precipitate of calcium carbonate forms. Addition of more carbon dioxide results in the disappearance of the precipitate as soluble calcium hydrogen carbonate forms:

$$CO_2(g) + Ca(OH)_2(aq) \rightarrow CaCO_3(s) + H_2O(l)$$

$$CO_2(g) + CaCO_3(s) + H_2O(l) \rightarrow Ca(HCO_3)_2(aq)$$

Bond lengths and bond strengths indicate that there are double bonds between the carbon and oxygen atoms in the carbon dioxide molecule. We would predict this bonding pattern both from an electron-dot representation and from simple hybridization theory. On the basis of hybridization theory, we assume that σ bonds are formed from *sp* hybrid orbitals. The remaining *p* orbitals, which are at right angles to the bond direction, then overlap to form two π molecular orbitals (see Figure 13.10).

Carbon dioxide's multiple bonding can also be discussed in terms of molecular orbital theory. Figure 13.11 shows a simplified molecular orbital diagram for carbon dioxide. Because of the differences in the atomic orbital energies of the two elements, it is the 2*p* orbitals of the oxygen atoms that interact with the 2*s* and 2*p* orbitals of the carbon atom. When this happens, two σ, two π, two π^{NB}, two π^*, and two σ^* orbitals are formed. The 12 available electrons fill all the bonding and nonbonding orbitals. Hence, for the molecule as a whole, the net bonding is two σ and two π, which amounts to one σ and one π bond per pair of atoms—the same result we obtained with the hybridization approach.

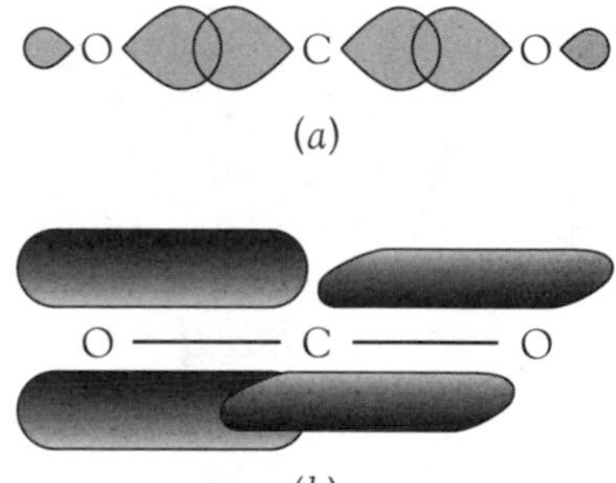

Figure 13.10 (*a*) The σ bonds between the atoms of a carbon dioxide molecule. (*b*) The two π bonds between the same atoms.

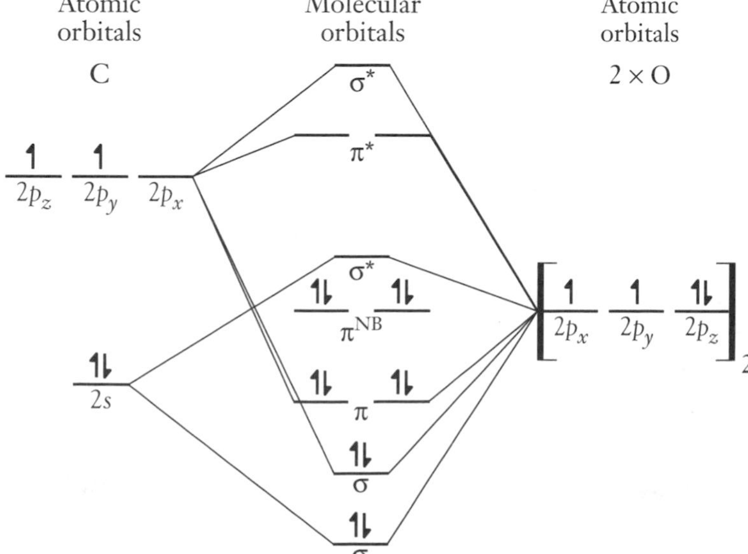

Figure 13.11 A simplified molecular orbital diagram for the carbon dioxide molecule.

Carbon Dioxide, the Killer Gas

We are all aware of carbon dioxide's high solubility in water under pressure. The quick release of a cap on a bottle of carbonated soft drink causes a massive foaming as the compound escapes into the gas phase. Yet this simple phase change is believed to have caused the deaths of about 3000 people and countless animals on one day in 1986.

It was on the evening of 21 August 1986 that every living creature for several kilometers around Lake Nyos in western Cameroon died. How was all the animal life killed without leaving any obvious trace? After years of study, the deaths can now be explained. The following clues led to the solution of this mystery:

The victims were all asphyxiated.

Deaths only occurred in the lake's neighboring valleys, not on the mountain ridges.

The deep lake waters contained extremely high levels of carbon dioxide gas.

The vegetation around this tropical lake showed evidence of frost-burn.

Lake Nyos is in a volcanic crater, and cracks in the crater floor continuously release about 5 million cubic meters of carbon dioxide into the lake bottom waters every year. Because carbon dioxide is very soluble in water, the bottom waters continuously dissolve the gas, just as the solution in a soft-drink bottling plant does. An upper layer of cold water traps this gas-saturated bottom water, in a manner similar to the atmospheric temperature-inversion process that sometimes happens in valleys. In normal circumstances, natural mixing and diffusion processes allow enough of the carbon dioxide to escape into the air to maintain a constant level of dissolved carbon dioxide.

For some reason, on that particular evening—perhaps because of an unusually strong wind—the upper cold water layer began to sink at one end of the lake. Without the cold water "cap," the gas-saturated water below started rising like a fountain, spewing out a frothy mixture. It must have been an incredible sight. As the gas expanded and escaped, it cooled to below freezing, frost-burning the surrounding plants. The carbon dioxide, being much denser than air, must have rolled down the valleys, pushing the air out of its path and asphyxiating all forms of animal life. This process probably continued until the natural mixing of atmospheric gases diluted the carbon dioxide enough to allow life to be sustainable.

What of the future? Chemists are monitoring the renewed buildup of dissolved gas. Because the land around the lake is so fertile, landless people have moved into the area, preferring the possible risk of death by asphyxiation to the certain risk of starvation.

In aqueous solution, almost all the carbon dioxide is present as $CO_2(aq)$; only 0.37 percent is present as carbonic acid, $H_2CO_3(aq)$:

$$CO_2(aq) + H_2O(l) \rightleftharpoons H_2CO_3(aq)$$

It is fortunate for us that the equilibrium lies to the left, not to the right, and that carbonic acid is a weak acid, because it means that carbonation of beverages will not cause them to become unpleasantly acidic.

Carbonic acid only exists in aqueous solution, and any attempt to isolate the acid shifts the above equilibrium to the left, releasing carbon dioxide. Carbonic acid is an extremely weak diprotic acid, as can be seen from the pK_a values corresponding to each of the ionization steps:

$$H_2CO_3(aq) + H_2O(l) \rightleftharpoons H_3O^+(aq) + HCO_3^-(aq) \qquad pK_{a1} = 6.37$$

$$HCO_3^-(aq) + H_2O(l) \rightleftharpoons H_3O^+(aq) + CO_3^{2-}(aq) \qquad pK_{a2} = 10.33$$

Because it is an acid oxide, carbon dioxide reacts with bases to give carbonates. The presence of excess carbon dioxide results in the formation of the hydrogen carbonates of the alkali and alkaline earth elements:

$$2\,KOH(aq) + CO_2(g) \rightarrow K_2CO_3(aq) + H_2O(l)$$

$$K_2CO_3(aq) + CO_2(g) + H_2O(l) \rightarrow 2\,KHCO_3(aq)$$

Hydrogen Carbonates and Carbonates

Hydrogen Carbonates

Only the alkali metals (except lithium) form solid compounds with the hydrogen carbonate ion, HCO_3^-, and even these decompose to the carbonate when heated:

$$2\,NaHCO_3(s) \xrightarrow{\Delta} Na_2CO_3(s) + H_2O(l) + CO_2(g)$$

Solutions of lithium and Group 2 metal hydrogen carbonates can be prepared, but even in solution the hydrogen carbonates decompose to carbonates when heated. As we discussed in Chapter 10, calcium hydrogen carbonate is formed during the dissolution of carbonate rocks; the dissolution process creates caves, and the subsequent decomposition of the hydrogen carbonate produces stalagmites and stalactites within caves:

$$CaCO_3(aq) + CO_2(aq) + H_2O(l) \rightleftharpoons Ca(HCO_3)_2(aq)$$

Household water supplies that come from chalk or limestone regions contain calcium hydrogen carbonate. Heating such water in a hot water tank or a kettle shifts the equilibrium to the right, precipitating out the calcium carbonate as a solid that is often called "scale":

$$Ca(HCO_3)_2(aq) \rightleftharpoons CaCO_3(aq) + CO_2(aq) + H_2O(l)$$

The hydrogen carbonate ion reacts with acids to give carbon dioxide and water, and with bases to give the carbonate ion:

$$HCO_3^-(aq) + H^+(aq) \rightarrow CO_2(g) + H_2O(l)$$

$$HCO_3^-(aq) + OH^-(aq) \rightarrow CO_3^{2-}(aq) + H_2O(l)$$

Carbonates

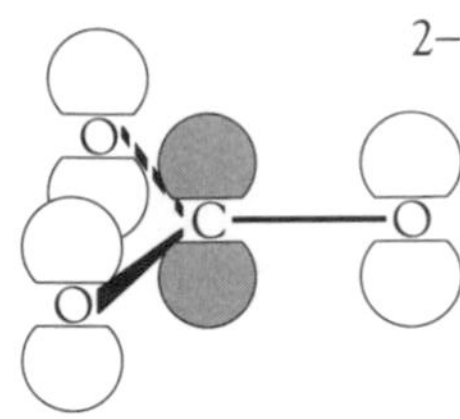

Figure 13.12 Delocalized π bonding in the carbonate ion.

The carbonate ion is very basic in aqueous solution as a result of a hydrolysis reaction that gives hydrogen carbonate and hydroxide ion:

$$CO_3{}^{2-}(aq) + H_2O(l) \rightleftharpoons HCO_3{}^{-}(aq) + OH^{-}(aq)$$

As a result, concentrated solutions of even that "harmless" household substance, sodium carbonate, commonly called washing soda, should be treated with respect (though not fear).

The carbon-oxygen bonds in the carbonate ion are all the same length and are significantly shorter than a single bond. We can consider the bonding in terms of a σ framework that is centered on the carbon atom and uses sp^2 hybrid orbitals. The electron pair in the remaining p orbital of the carbon atom can then form a π bond that is delocalized (shared) over the whole ion (Figure 13.12). With the π bond shared three ways, each carbon-oxygen bond would have a bond order of $1\frac{1}{3}$. The notation used to represent this bond order is shown in Figure 13.13.

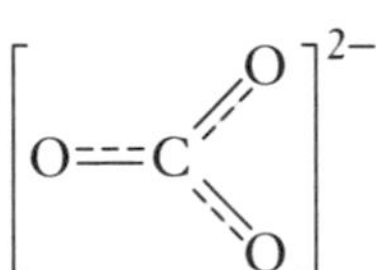

Figure 13.13 Representation of the partial bonds of the carbonate ion.

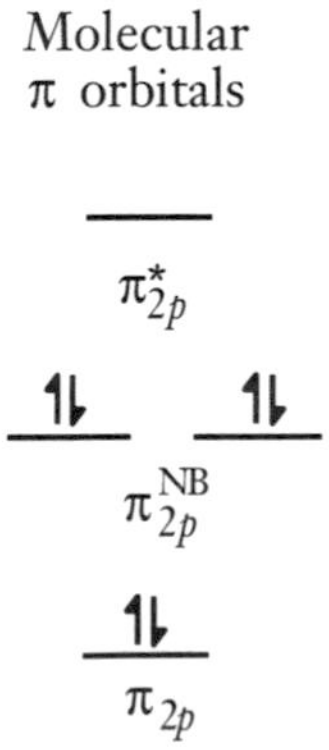

We can also consider the π system in simplified molecular orbital terms. The four p orbitals at right angles to the plane of the molecule interact to form four π molecular orbitals. Appropriate calculations indicate that one will be bonding, two nonbonding, and one antibonding. The three electron pairs that are available to fill these orbitals fill the bonding and nonbonding orbitals (Figure 13.14). Hence, like the hybridization model, molecular orbital theory predicts one π bond shared throughout the four-atom system—or one-third of a π bond for each bonded pair.

Most carbonates are insoluble, the exceptions being ammonium carbonate and the alkali metal carbonates. The alkali metal carbonates (except that of lithium) do not decompose when heated. Lithium carbonate and carbonates of other moderately electropositive metals, such as calcium, give the metal oxide and carbon dioxide when heated:

$$CaCO_3(s) \xrightarrow{\Delta} CaO(s) + CO_2(g)$$

For the carbonates of weakly electropositive metals, such as silver, the metal oxide is itself decomposed by heat. Thus the final products are the metal, carbon dioxide, and oxygen:

$$Ag_2CO_3(s) \xrightarrow{\Delta} Ag_2O(s) + CO_2(g)$$

$$Ag_2O(s) \xrightarrow{\Delta} 2\,Ag(s) + \tfrac{1}{2}\,O_2(g)$$

Molecular π orbitals

π^*_{2p}

π^{NB}_{2p}

π_{2p}

Figure 13.14 Molecular orbitals involved in π bond formation in the carbonate ion.

As is typical of the behavior of the ammonium ion in oxysalts, both the anion and the cation of ammonium carbonate decompose when heated; the products are ammonia, water, and carbon dioxide:

$$(NH_4)_2CO_3(s) \rightarrow 2\,NH_3(g) + H_2O(g) + CO_2(g)$$

Carbon Disulfide

Carbon disulfide is the sulfur analog of carbon dioxide, and it has the same linear geometry. The compound is a colorless, highly flammable, low-boiling liquid with a pleasant smell when pure, but the commercial grade of the compound usually contains very foul-smelling impurities. It is highly toxic, causing damage to the brain and nervous system and, eventually, death. Carbon disulfide is prepared industrially by passing methane gas over molten sulfur at about 700°C, then cooling the products, from which carbon disulfide condenses:

$$CH_4(g) + 4\ S(l) \xrightarrow{\Delta} CS_2(g) + 2\ H_2S(g)$$

Over 1 million tonnes of this reagent are consumed each year, mainly in the production of cellophane and viscose rayon polymers. It is also the starting material for the manufacture of carbon tetrachloride. We tend to forget that industrial chemistry is rarely the conversion of one naturally occurring substance directly to some required product. More often, the product is, itself, only a reagent in the production of many other compounds.

Carbon Tetrahalides

The major divisions of chemistry—inorganic, organic, physical, and analytical—are inventions of chemists attempting to organize this vast and continually growing science. Yet chemistry does not fit into neat little compartments, and the carbon tetrahalides are compounds that belong to the realms of both organic and inorganic chemistry. As a result, they have two sets of names: carbon tetrahalides, according to inorganic nomenclature, and tetrahalomethanes, according to organic nomenclature.

All of the tetrahalides contain the carbon atom tetrahedrally coordinated to the four halogen atoms. The phases of the tetrahalides at room temperature reflect the increasing strength of the intermolecular dispersion forces. Thus carbon tetrafluoride is a colorless gas; carbon tetrachloride is a dense, almost oily liquid; carbon tetrabromide, a pale yellow solid; and carbon tetraiodide, a bright red solid.

Carbon tetrachloride is an excellent nonpolar solvent. In recent years, the discovery of its cancer-causing ability has made it a solvent of last resort. It was formerly used as a fire-extinguishing material, particularly where water could not be used, for example, around electrical wiring and restaurant deep fat fryers. The liquid vaporizes to form a gas that is five times denser than air, effectively blanketing the fire with an inert gas. However, in addition to its carcinogenic nature, it does oxidize in the flames of a fire to give the poison gas carbonyl chloride, $COCl_2$. Carbon tetrachloride is also a "greenhouse" gas and, in the upper atmosphere, a potent ozone destroyer. It is therefore important to minimize emissions of this compound from industrial plants.

The major industrial route for the synthesis of carbon tetrachloride involves the reaction of carbon disulfide with chlorine. In this reaction, iron(III) chloride is the catalyst. In the first step, the products are carbon tetrachloride and disulfur dichloride. Then at a higher temperature, addition of more carbon disulfide produces additional carbon tetrachloride and sulfur. The sulfur can be reused in the production of a new batch of carbon disulfide:

$$CS_2(g) + 3\ Cl_2(g) \rightarrow CCl_4(g) + S_2Cl_2(l)$$

$$CS_2(g) + 2\ S_2Cl_2(l) \xrightarrow{\Delta} CCl_4(g) + 6\ S(s)$$

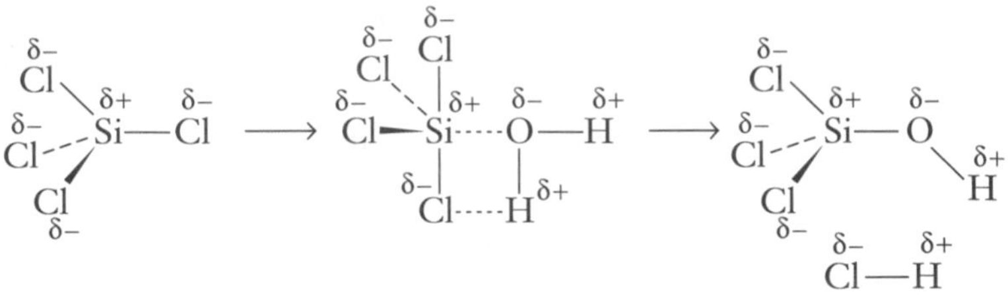

Figure 13.15 First step of the postulated mechanism of hydrolysis of silicon tetrachloride.

The reaction between methane and chlorine is also used to produce carbon tetrachloride:

$$CH_4(g) + 4\ Cl_2(g) \rightarrow CCl_4(l) + 4\ HCl(g)$$

For chemists, one of the interesting properties of carbon tetrachloride is its chemical inertness. For example, it does not react with water, yet silicon tetrachloride does—violently. If we look at the comparative free energies of reaction with water, we see that, if anything, carbon tetrachloride should be more reactive than silicon tetrachloride:

$$CCl_4(l) + 3\ H_2O(l) \rightarrow H_2CO_3(aq) + 4\ HCl(g) \qquad \Delta G = -380\ \text{kJ·mol}^{-1}$$

$$SiCl_4(l) + 3\ H_2O(l) \rightarrow H_2SiO_3(s) + 4\ HCl(g) \qquad \Delta G = -289\ \text{kJ·mol}^{-1}$$

The answer must lie with kinetic factors, that is, the lack of an available pathway for the reaction. We postulate that the reaction between silicon tetrachloride and water occurs in a manner similar to that described for boron trichloride (Figure 12.5): an attack by the polar water molecule in which the partially positive hydrogen bonds to the partially negative chlorine while the partially negative oxygen is bonding to the partially positive silicon (Figure 13.15). This process is repeated three more times, thereby replacing all of the chlorines with hydroxyl groups. The transition states for this process involve a five-coordinate silicon atom. For five bonds, hybridization theory requires the use of at least one *d* orbital of the silicon, as well as the *s* and *p* orbitals. However, carbon, which is a second period element, has no available *d* orbitals. Thus this reaction pathway cannot exist for carbon.

Chlorofluorocarbons

It was Thomas Midgley, Jr., a General Motors chemist, who in 1928 first synthesized dichlorodifluoromethane, CCl_2F_2. This discovery was made as part of a search to find a good, safe refrigerant material. A refrigerant is a compound that, at room temperature, is a gas at low pressures but a liquid at high pressures. Reducing the pressure on the liquid causes it to boil and absorb heat from the surroundings (such as the inside of a refrigerator). The gas is then conveyed outside the cooled container, where it is compressed. Under these conditions, it reliquefies, releasing the enthalpy of vaporization to the surroundings as it does so.

At the time they were discovered, the chlorofluorocarbon family (CFCs) appeared to be a chemist's dream. They were almost completely unreactive and they were nontoxic. As a result, they were soon used in air conditioning systems, as blowing agents for plastic foams, as aerosol propellants, as fire-extinguishing materials, as degreasing agents of electronic circuits, and as anesthetics—to name but a few uses. Annual production amounted to nearly 700 000 tonnes in the peak years. Their lack of reactivity is partly due to the lack of a hydrolysis pathway, but, in addition, the high strength of the carbon-fluorine bond confers extra protection against oxidation.

Chlorofluorocarbons have a specialized and arcane naming system:

1. The first digit represents the number of carbon atoms minus one. For the one-carbon CFCs, the zero is eliminated.

2. The second digit represents the number of hydrogen atoms plus one.

3. The third digit represents the number of fluorine atoms.

4. Structural isomers are distinguished by "a," "b," and so on.

Of the simple CFCs, $CFCl_3$ (CFC-11), and CF_2Cl_2 (CFC-12) were the most widely used.

It was not until the 1970s that the great stability of these compounds—their "best" property—was recognized as a threat to the environment. These compounds were so stable that they remained in the atmosphere for hundreds of years. Some of these molecules were diffusing into the upper atmosphere, where ultraviolet light cleaved a chlorine atom from each of them. The chlorine atom then reacted with ozone molecules in a series of steps that can be represented in a simplified and not wholly accurate form as

$$Cl + O_3 \rightarrow O_2 + ClO$$

$$ClO \rightarrow Cl + O$$

$$Cl + O_3 \rightarrow O_2 + ClO$$

$$ClO + O \rightarrow Cl + O_2$$

The chlorine atom is then free to repeat the cycle time and time again, destroying enormous numbers of ozone molecules. Incidentally, these chemical species are not those that you can find in a chemistry laboratory, but at the low pressures existing in the upper atmosphere, even free chlorine and oxygen atoms can exist for measurable periods of time.

Some nonscientists believe that the CFCs cannot possibly affect the ozone layer, because they are dense gases that will "settle out" of the atmosphere, just as one can "pour" carbon dioxide out of a beaker. The falsity of this claim is shown by personal experience and by an understanding of the properties of matter. Personal experience convinces us not to worry about gases "settling out"; if they did, the Earth would have a blanket of dense gases, such as carbon dioxide and argon, at sea level. Also, we would have to avoid basements because our exhaled carbon dioxide might accumulate down there. In fact, gases mix in any proportion, thus enabling the smogs of our big cities to be dissipated into the global atmosphere. On the basis of the concept of the nature of matter, we assume that gas molecules move at extremely high velocities, of the order of hundreds of meters per second—they are not just "hanging around." Thus without even considering other factors such as bulk atmospheric disturbances, we conclude that CFCs will indeed spread into the upper layers of the atmosphere.

The search was rapidly undertaken to find substitutes for CFCs. This has not been easy, because most potential replacements are flammable, toxic, or suffer from some other major problem. For example, the most promising alternative to the CFC-12 refrigerant is a hydrofluorocarbon (HFC), CF_3–CH_2F (HFC-134a). The absence of a chlorine atom in its structure means that it cannot wreak havoc in the ozone layer. Nevertheless, there are three major problems with its widespread adoption. First, CFC-12 is manufactured in a simple, one-step process from carbon tetrachloride and hydrogen fluoride. The synthesis of HFC-134a, however, requires a costly, multistep procedure. Second, existing refrigeration units have to be altered to operate with the new compound. The major reason is that the pressures used to condense HFC-134a are higher than those used with CFC-12. The costs of building the chemical factories and of modifying refrigerator pumps are a tolerable burden for Western countries but are beyond the means of less developed countries. The final problem is that all the fluorocarbon compounds are also excellent "greenhouse" gases; that is, like carbon dioxide, they absorb infrared radiation and can contribute to global warming. Thus we must make sure that the CFCs and HFCs are used in closed systems, with zero leakage, and with a legal requirement to return the unit to the manufacturer at the end of its operating life. The manufacturer will then recycle the fluorocarbon refrigerant.

The lesson, of course, is clear. Just because a compound is chemically inert in the laboratory does not mean that it is harmless. No product of the chemical industry can be released into the environment without first considering its impact. At the same time, hindsight is a wonderful thing. It is only in recent years that we have become aware of the importance of the ozone layer and of the many chemical reactions that occur in it (discussed in Chapter 15). The more crucial moral is that research into the chemical cycles in nature must continue to be funded. If we do not know how the world works at the chemical level, then it will be impossible to predict the effect of any human perturbation.

Methane

The simplest compound of carbon and hydrogen is methane, CH_4, a colorless, odorless gas. There are enormous quantities of this gas, commonly called natural gas, found in underground deposits and in deposits under the Arctic seabed. It is one of the major sources of thermal energy in use today because it undergoes an exothermic combustion reaction:

$$CH_4(g) + 2\ O_2(g) \rightarrow CO_2(g) + 2\ H_2O(g)$$

Because methane is undetectable by our basic senses (sight and smell), a strong-smelling, organic sulfur-containing compound is added to the gas before it is supplied to customers. We can detect any leakage of the gas by its odor.

For inorganic chemists, this combustion reaction provides quite a challenge. Methane and the myriad other compounds of carbon and hydrogen are kinetically stable to oxidation even though thermodynamically the reaction is very favorable. We need to apply a spark or a flame before reaction will occur. This is in complete contrast to silane, SiH_4, which catches fire the instant it comes in contact with oxygen. Why is there such a difference? Our first impulse is to use the same explanation we gave for the comparative reactivities of carbon and silicon halides with water, that is, the lack of available *d* orbitals on the carbon atom. However, the oxidation mechanism is very different from that of hydrolysis, and it is dangerous simply to accept the first possible explanation.

Let us examine the chemistry of the hydrides more carefully; some react violently with oxygen and some do not. Diborane, B_2H_6, silane, SiH_4, and germane, GeH_4, are hydrides that react violently with oxygen. Methane, CH_4, ammonia, NH_3, and hydrogen sulfide, H_2S, are hydrides that are kinetically stable to oxidation. The key feature seems to be the relative electronegativities of the elements involved. In borane, silane, and germane, hydrogen is the more electronegative element, whereas in methane, ammonia, and hydrogen sulfide, hydrogen is the less electronegative element. Thus the polarity of the bond is reversed. A partially negative hydrogen (a hydridic hydrogen) is much more reactive than a partially positive hydrogen atom.

The Greenhouse Effect

Most of us are aware of the greenhouse effect, yet few really understand the complexity of the problem. A tremendous quantity of research is currently being done by chemists, biologists, physicists, geologists, and geographers to gain information from which we can make more informed judgments on the issue.

The greenhouse effect can be described as follows: The energy from the sun reaches the Earth's surface as radiation that represents most of the electromagnetic spectrum, particularly the ultraviolet, visible, and infrared. This energy is absorbed by the Earth's surface and atmosphere. It is reemitted mainly as infrared radiation ("heat" rays). If all the incoming energy were lost back into space as infrared radiation, the temperature of the Earth's surface would be between –20° and –40°C. Fortunately, there is a reflecting layer in the atmosphere. Many small molecules in the atmosphere, particularly water and carbon dioxide, absorb certain wavelengths of the escaping radiation, the absorbed energy corresponding to the energy of their molecular vibrations. These small molecules act like the glass in a greenhouse. It is the reradiation of their absorbed energy back to Earth that warms the oceans, land, and air. As a result, the average temperature of the Earth's surface is about +14°C. In other words, it is the greenhouse effect that makes this planet habitable.

The wavelengths of the energy absorbed by the molecules in the "greenhouse" layer correspond to certain vibrations of atoms within molecules. All polyatomic molecules, except diatomic molecules, absorb energy in the infrared region. Diatomic molecules only absorb infrared radiation if they contain two different atoms (such as carbon monoxide); and monatomic gases do not absorb infrared radiation. Thus dinitrogen, dioxygen, and argon, the three major constituents of the atmosphere, are essentially transparent to infrared radiation. The vibrations of the water molecule and the carbon dioxide molecule that are responsible for the absorption of infrared radiation are shown in Figure 13.16, together with the corresponding wavelength of the

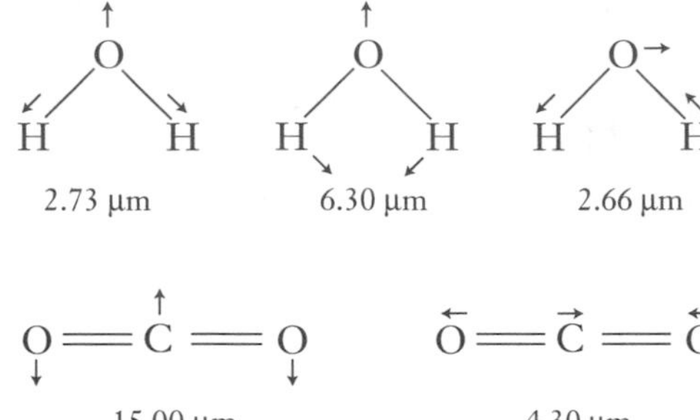

Figure 13.16 The vibrations of water and carbon dioxide and the wavelengths corresponding to the frequencies of these vibrations.

energy absorbed. These molecules also will absorb energy with wavelengths that are multiples of the frequencies of these vibrations, but to a lesser extent.

Water vapor is the predominant greenhouse gas, and its atmospheric concentration of about 1 percent has remained fairly constant over geological time as a result of the large bodies of water on this planet. However, as far as we can tell from the geological record, the levels of carbon dioxide have fluctuated dramatically. It is very clear that in the warm Carboniferous era, from 350 million to 270 million years ago, the carbon dioxide level was over six times higher than it is now. A large proportion of that carbon dioxide is presently "locked away" in the massive coal deposits that exist throughout the world. Another "sink" for the carbon dioxide has been the oceans, which contain large amounts of both dissolved carbon dioxide and calcium carbonate in seashells and coral reefs. The possible limit of absorption of carbon dioxide by the oceans is one of the unknown factors.

There are three major sources of carbon dioxide: natural outpourings of volcanoes, burning of fossil fuels, and burning of vegetation such as the rain forests. Comparison of measurements of carbon dioxide levels from the last century to the present show an increase, and this is a matter of concern. However, the amount of carbon dioxide already in the atmosphere is sufficient to absorb most of the outgoing energy of the appropriate wavelengths. Thus doubling the carbon dioxide level will not double the heat retained. Figure 13.17 shows the infrared spectrum of our atmosphere. The absorptions at wavelengths corresponding to the frequencies of the vibrations of the water and carbon dioxide molecules (Figure 13.16) are the dominant features of the longer wavelengths, whereas at shorter wavelengths, most of the absorption occurs at wavelengths corresponding to the multiples (harmonics) of the frequencies of these vibrations.

Of more concern is the absorption of other infrared wavelengths by molecules whose previous levels in the atmosphere were low or nil; many of these, such as the chlorofluorocarbons, absorb in the present "windows" in the infrared spectrum. Many scientists are particularly concerned by the rising concentration of methane. It is currently the gas whose concentration in the atmosphere is rising most rapidly. The methane molecule absorbs infrared wavelengths different from those currently being absorbed by carbon dioxide and water vapor, particularly those between 3.4 and 3.5 μm. There have always been traces of methane in the atmosphere as a result of vegetation decay in marshes. However, the proportion in the atmosphere has risen drastically over the last century. The rise is due in part to the rapid

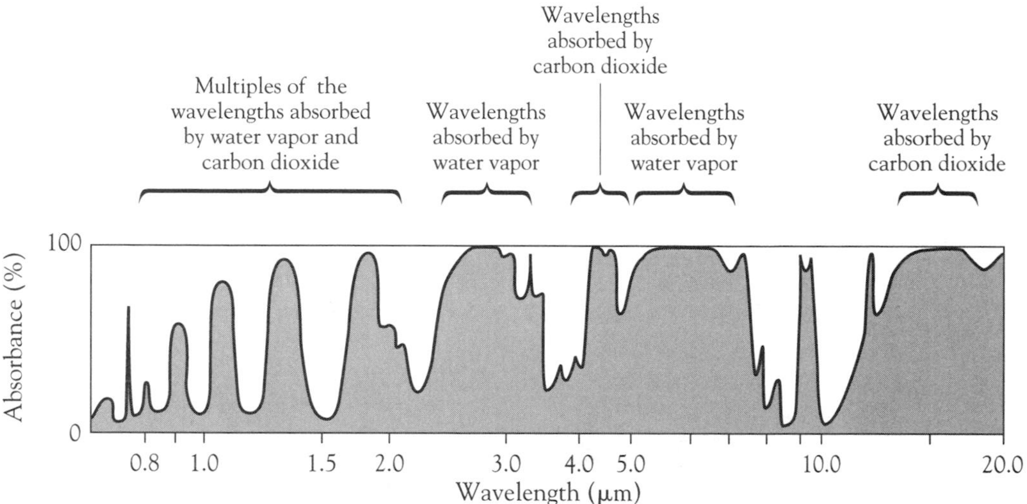

Figure 13.17 Infrared spectrum showing absorption of infrared radiation by various components of the atmosphere.

growth in the number of cattle and sheep (members of the grazing animals, the ruminants). All ruminants produce large quantities of methane in their digestive tracts, and this gas is expelled into the atmosphere. Thus in terms of degree of influence, the rise in methane concentration might be of more concern than that of carbon dioxide. Much research needs to be done before we really understand all of the factors contributing to the greenhouse effect.

It is fortunate that the carbon dioxide levels have dropped over geological time, because the sun has been getting hotter over the same period. Hence the decreasing greenhouse effect has approximately balanced the increasing solar radiation. It should be kept in mind that, if the formation of coal deposits and carbonate rocks continued for many more millions of years, a point would be reached at which photosynthesis—the reduction of CO_2 to sugars—would become impossible. Such a scenario would be in the very distant future, and one of the few positive aspects of combustion of fossil fuels is to delay this time even further.

Silicon

About 27 percent by mass of the Earth's crust is silicon, even though it is never found in nature as the free element but only in compounds containing oxygen-silicon bonds. The element itself is a gray, metallic-looking, crystalline solid. Although it looks metallic, it is not classified as a metal because it has a low electrical conductivity. And although silicon is thermodynamically unstable (particularly toward oxidation), slow kinetics of reaction are used to explain its failure to react with oxygen, water, and other common chemical reagents.

About half a million tonnes per year of silicon is used in the preparation of metal alloys. Although alloy manufacture is the major use, the application having the greatest influence on our lives has been the use of silicon in semiconductor devices for computers and other electronic devices. The purity level of the silicon used in the electronics industry has to be exceedingly high. For example, the presence of only 1 ppb of phosphorus is enough to drop the specific resistance of silicon from 150 to 0.1 kΩ·cm. As a result of the expensive purification process, ultrapure electronic grade silicon sells for over 1000 times the price of metallurgical grade (98 percent pure) silicon.

The element is prepared by heating silicon dioxide (quartz) with coke at over 2000°C in an electrical furnace similar to that used for the Acheson process of calcium carbide synthesis. Liquid silicon (melting point 1400°C) is drained from the furnace:

$$SiO_2(s) + 2\ C(s) \xrightarrow{\Delta} Si(l) + 2\ CO(g)$$

To obtain ultrapure silicon, the crude silicon is heated at 300°C in a current of hydrogen chloride gas. The trichlorosilane product, $SiHCl_3$, can be distilled and redistilled until the impurity levels are below the parts per billion level:

$$Si(s) + 3\ HCl(g) \rightarrow SiHCl_3(g) + H_2(g)$$

The reverse reaction is spontaneous at 1000°C, depositing ultrapure silicon. The hydrogen chloride can be reused in the first part of the process:

$$SiHCl_3(g) + H_2(g) \rightarrow Si(s) + 3\ HCl(g)$$

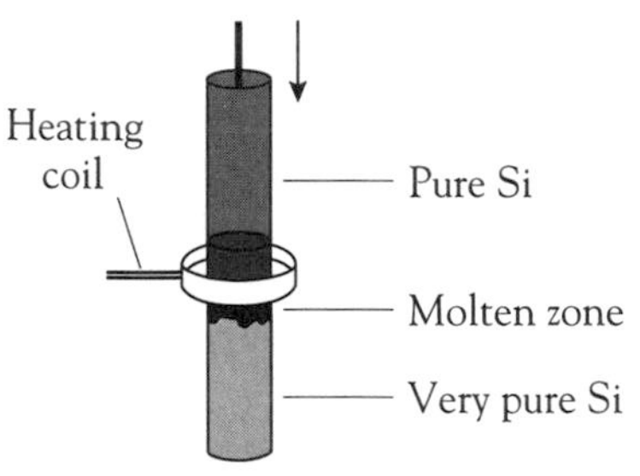

Figure 13.18 Zone refining method for the purification of silicon.

The ultrapure single crystals needed for solar cells are produced by *zone refining* (Figure 13.18). This process depends on the fact that the impurities are more soluble in the liquid phase than in its solid phase. A rod of silicon is moved through a high-temperature electric coil that partially melts the silicon. As part of the rod moves beyond the coil, the silicon resolidifies; during the solidification process, the impurities diffuse into the portion of the rod that is still molten. After the entire rod has passed through the coil, the impurity-rich top portion can be removed. The procedure can be repeated until the desired level of purity (less than 0.1 ppb impurity) is obtained.

Computing devices can function because the silicon chips have been selectively "doped." That is, controlled levels of other elements are introduced. If traces of a Group 15 element, such as phosphorus, are mixed into the silicon, then the extra valence electron of the added element is free to roam throughout the material. Conversely, doping with a Group 3 element will result in electron "holes." These holes act as sinks for electrons. A combination of electron-rich and electron-deficient silicon layers makes up the basic electronic circuit.

Silicon Dioxide

The most common crystalline form of silicon dioxide, SiO_2, commonly called silica, is the mineral quartz. Most sands consist of particles of silica that usually contain impurities such as iron oxides. It is interesting to note that carbon dioxide and silicon dioxide share the same type of formula, yet their properties are very different. Carbon dioxide is a colorless gas at room temperature, whereas solid silicon dioxide melts at 1600°C and boils at 2230°C. The difference in boiling points is due to bonding factors. Carbon dioxide consists of small, triatomic, nonpolar molecular units whose attraction to one another is due to dispersion forces. By contrast, silicon dioxide contains a network of silicon-oxygen covalent bonds in a giant molecular lattice. Each silicon atom is bonded to four oxygen atoms, and each oxygen atom is bonded to two silicon atoms, an arrangement consistent with the SiO_2 stoichiometry of the compound (Figure 13.19).

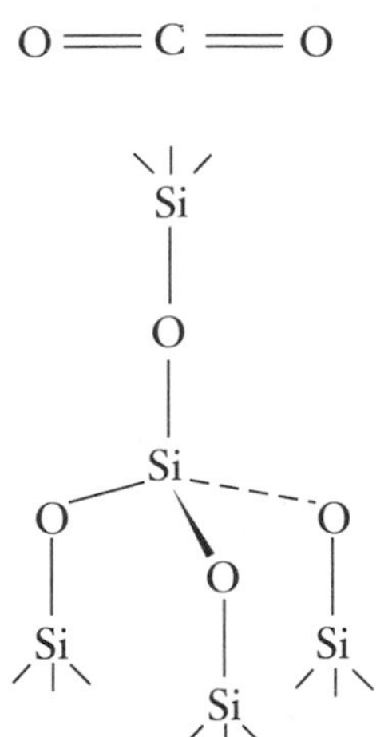

Figure 13.19 Structures of carbon dioxide and silicon dioxide.

How can we explain this difference? First, the carbon-oxygen single bond (C–O bond energy = 358 kJ·mol^{-1}) is much weaker than the carbon-oxygen double bond (C=O bond energy = 799 kJ·mol^{-1}). So formation of a p_π–p_π bond between the carbon and oxygen atoms more than doubles the strength of the bond. Hence it is energetically more favorable to form two C=O double bonds than four C–O single bonds, which would be analogous to the silicon dioxide structure. Conversely, as mentioned earlier in the discussion of catenation, silicon-oxygen single bonds actually have a partial double bond character as a result of the overlap of the empty *d* orbitals on the silicon with the full *p* orbitals of the oxygen (Figure 13.6). Because of this delocalized bonding, the Si–O bond energy is 452 kJ·mol^{-1}. And it is known that multiple bonds in compounds of elements from Period 3 and higher periods have bond energies that are not much greater than those of the corresponding single bond energies. Thus for silicon, four single bonds (with partial multiple bond character) are much preferable to two conventional double bonds.

Silicon dioxide is very unreactive; it reacts only with hydrofluoric acid (or wet fluorine) and molten sodium hydroxide. The reaction with hydrofluoric acid is used to etch designs on glass:

$$SiO_2(s) + 6\ HF(aq) \rightarrow SiF_6^{2-}(aq) + 2\ H^+(aq) + 2\ H_2O(l)$$

$$SiO_2(s) + 2\ NaOH(l) \rightarrow Na_2SiO_3(s) + H_2O(g)$$

Silicon dioxide is mainly used as an optical material. It is hard, strong, and transparent to visible and ultraviolet light; and it has a very low coefficient of expansion. Thus lenses constructed from it do not warp as the temperature changes.

Silica Gel

Silica gel is a hydrated form of silicon dioxide, $SiO_2{\cdot}xH_2O$. It is used as a desiccant (drying agent) in the laboratory and also for keeping electronics and even prescription drugs dry. You may have noticed the packets of a grainy material enclosed with electronic equipment or little cylinders placed in some drug vials by pharmacists. These enclosures keep the product dry even in humid climates. Commercial silica gel contains about 4 percent water by mass, but it will absorb very high numbers of water molecules over the crystal surface. And it has the particular advantage that it can be reused by heating for several hours; the high temperature drives off the water molecules, enabling the gel to function effectively once more.

Aerogels

An American chemist, Samuel Kistler, in the 1930s devised a way of drying wet silica gel without causing it to shrink and crack, like mud on a dried riverbank. At the time, there was little interest in the product. Furthermore, the procedure required extremely high pressures, and one laboratory was destroyed by an explosion during the preparation of this material. Now, about 60 years later, chemists have discovered new and safer synthetic routes to this rediscovered family of materials, called *aerogels*. The basic aerogel is silicon dioxide in which a large number of pores exist—so many, in fact, that 99 percent of an aerogel block consists of air. As a result, the material has an extremely low density, yet is quite strong. The translucent solid is also an excellent thermal insulator, and it promises to be a useful fireproof insulating material. Aerogels also have some unique properties. For example, sound travels through aerogels more slowly than through any other medium. Chemists have now prepared aerogels that incorporate other elements, a technique that enables the chemists to vary the characteristics of the aerogels. What is a laboratory curiosity at present will surely become a commercial product in the near future.

Glasses

Glasses are noncrystalline materials—in fact, they are sometimes described as supercooled liquids. Glass has been used as a material for at least 5000 years. It is difficult to obtain a precise figure of current annual production, but it must be about 100 million tonnes.

Almost all glass is silicate glass; it is based on the three-dimensional network of silicon dioxide. Quartz glass is made simply by heating pure silicon dioxide above 2000°C and then pouring the viscous liquid into molds. The product has great strength and low thermal expansion, and it is highly transparent in the ultraviolet region. However, the high melting point precludes the use of quartz glass for most everyday glassware.

The properties of the glass can be altered by mixing in other oxides. The compositions of three common glasses are shown in Table 13.3. About 90 percent of glass used today is soda-lime glass. It has a low melting point, so it

Table 13.3 Approximate compositions of common glasses

	Composition (%)		
Component	Soda-lime glass	Borosilicate glass	Lead glass
SiO_2	73	81	60
CaO	11	—	—
PbO	—	—	24
Na_2O	13	5	1
K_2O	1	—	15
B_2O_3	—	11	—
Other	2	3	<1

is very easy to form soda-lime glass into containers, such as soft-drink bottles. In the chemistry laboratory, we need a glass that will not crack from thermal stress when heated; borosilicate glass (discussed in Chapter 12) is used for this purpose. Lead glasses have a high refractive index; as a result, cut glass surfaces sparkle like gemstones, and these glasses are used for fine glassware. The element lead is a strong absorber of radiation; hence a very different use for lead glass is in radiation shields, such as those over cathode ray tubes.

Silicates

About 95 percent of the rocks of the Earth's crust are silicates; and there is a tremendous variety of silicate minerals. The simplest silicate ion has the formula SiO_4^{4-}; and zirconium silicate, $ZrSiO_4$, the gemstone zircon, is one of the few minerals to contain this ion. Silicates are generally very insoluble, as one might expect of rocks that have resisted rain for millions of years. The one common exception is sodium silicate, which can be prepared by reacting solid silicon dioxide with molten sodium carbonate:

$$SiO_2(s) + 2\ Na_2CO_3(l) \rightarrow Na_4SiO_4(s) + 2\ CO_2(g)$$

A concentrated solution of sodium (*ortho*)silicate is called water glass, and it is extremely basic as a result of hydrolysis reactions of the silicate anion. Before modern refrigeration became available, the water glass solution was used to preserve eggs, the soft porous shell of calcium carbonate being replaced by a tough, impervious layer of calcium silicate that seals in the egg contents:

$$2\ CaCO_3(s) + SiO_4^{4-}(aq) \rightarrow Ca_2SiO_4(s) + 2\ CO_3^{2-}(aq)$$

This is not the total extent of silicate chemistry. Oxygen atoms can be shared by different silicon atoms. To show these different structures, silicate chemists depict the units in a manner different from that used in conventional molecular geometry. We can illustrate the approach with the silicate ion itself. Most chemists look at an ion from the side, a perspective giving the arrangement depicted in Figure 13.20a. Silicate chemists look down on a silicate ion, sighting along the axis of a Si–O bond (Figure 13.20b). The corner spheres represent three of the oxygen atoms; the central black dot represents the silicon atom; and the circle around the black dot represents the oxygen atom vertically above it. Instead of drawing the covalent bonds, the edges of the tetrahedron are marked with solid lines.

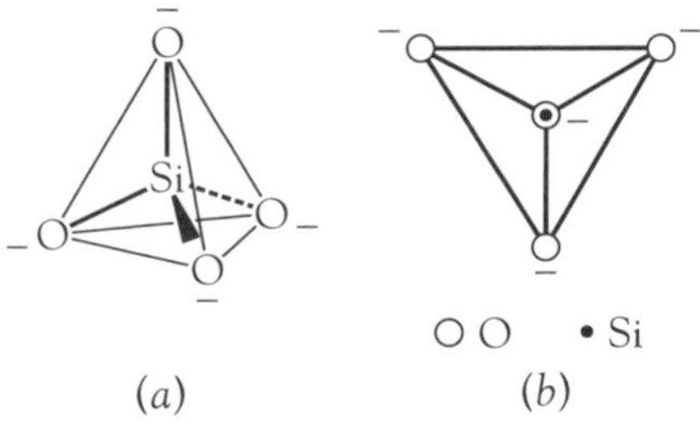

Figure 13.20 Depiction of the tetrahedral shape of the silicate ion in the conventional (*a*) and the silicate chemistry (*b*) forms.

When a small amount of acid is added to the (*ortho*)silicate ion, the pyrosilicate ion, $Si_2O_7^{6-}$, is formed:

$$2\ SiO_4^{4-}(aq) + 2\ H^+(aq) \rightarrow Si_2O_7^{6-}(aq) + H_2O(l)$$

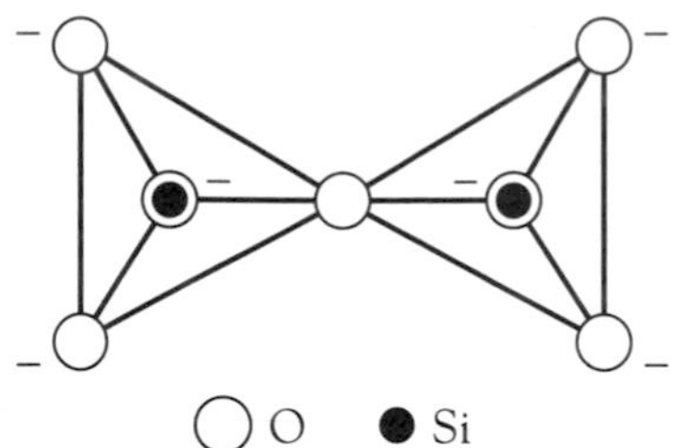

Figure 13.21 Depiction of the $Si_2O_7^{6-}$ ion.

In the pyrosilicate ion, the two silicate ions are linked together by one shared oxygen atom (Figure 13.21). This ion is, itself, not of great importance. However, these silicate units can join to form long chains, and they can cross-link to form a double chain. A polymeric structure with an empirical formula of $Si_4O_{11}^{6-}$ is formed in this way (Figure 13.22). The double chain is an important structure, and it gives rise to a whole family of minerals called the *amphiboles*. The cations that are packed in among these chains determine the identity of the mineral formed. For example, $Na_2Fe_5(Si_4O_{11})_2(OH)_2$ is the mineral crocidolite, more commonly known as blue asbestos.

The double chains of silicate units can link side by side to give sheets of empirical formula $Si_2O_5^{2-}$ (Figure 13.23). One of the sheet silicates is $Mg_3(Si_2O_5)(OH)_4$, chrysotile. This compound, also known as white asbestos, has alternating layers of silicate ions and hydroxide ions, with magnesium ions filling in available holes. Asbestos has been used for thousands of years. For example, the ancient Greeks used it as wicks for their lamps; and the European king Charlemagne astounded his guests in about A.D. 800 by throwing his dirty asbestos tablecloth into a fire and retrieving it unburned and clean. The use of asbestos is declining rapidly, now that we are aware of the health risks from embedded asbestos fibers on the lung surface.

Very few nonchemists realize that there are two common forms of this fibrous mineral and that they have very different chemical structures and different degrees of hazard. In fact, about 95 percent of that currently used is the less harmful white asbestos and only about 5 percent, the more dangerous blue asbestos. Asbestos is a very convenient and inexpensive fireproof material, and it has been extremely difficult for chemists to find hazard-free materials that can substitute for all the 3000 uses of asbestos. In fact, there is still a significant consumption of asbestos for products such as brake linings, engine gaskets, and even as a filter material for wine.

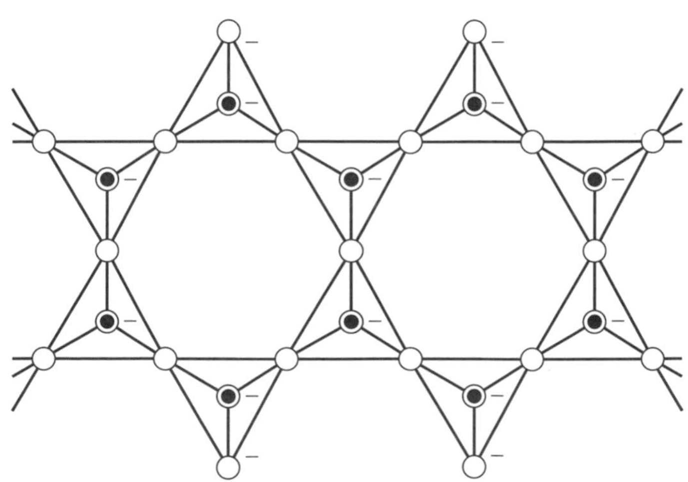

Figure 13.22 Depiction of a section of the $Si_4O_{11}^{6-}$ repeating double chain.

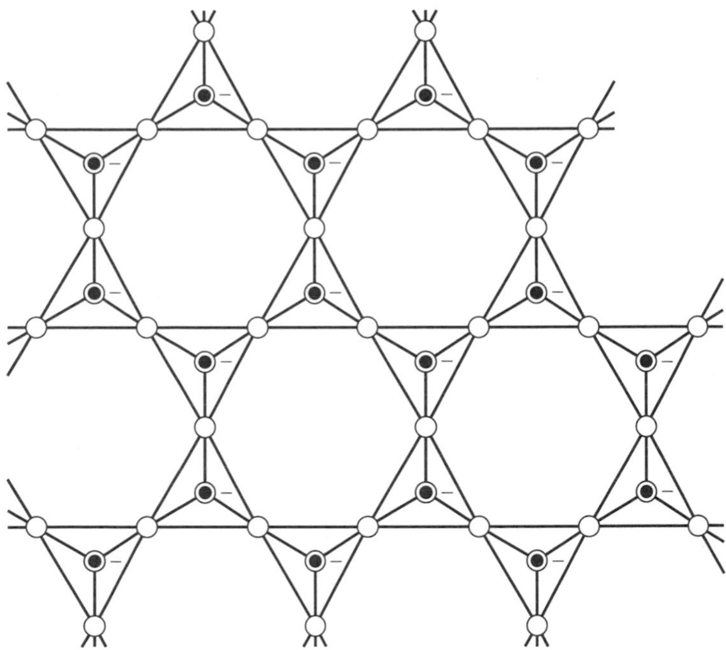

Figure 13.23 Depiction of a section of the sheet silicate, $Si_2O_5^{2-}$.

It is fascinating how minor changes in structure can make major changes in properties. If, instead of alternating layers of magnesium silicate and magnesium hydroxide, we have a layer of hydroxide ions sandwiched between pairs of layers of silicate ions, we get a different formula, $Mg_3(Si_2O_5)_2(OH)_2$, and a different name, talc (see Figure 13.24). Because each sandwich is electrically neutral, it is as almost as slippery as graphite (but white instead of black). Talc is used on a very large scale—about 8 million tonnes worldwide—for ceramics, fine paper, paint, and the cosmetic product talcum powder. Clays are mixtures of these sheet silicates. One particular clay mineral, kaolinite, $Al_2Si_2O_5(OH)_4$, is important for the manufacture of ceramics.

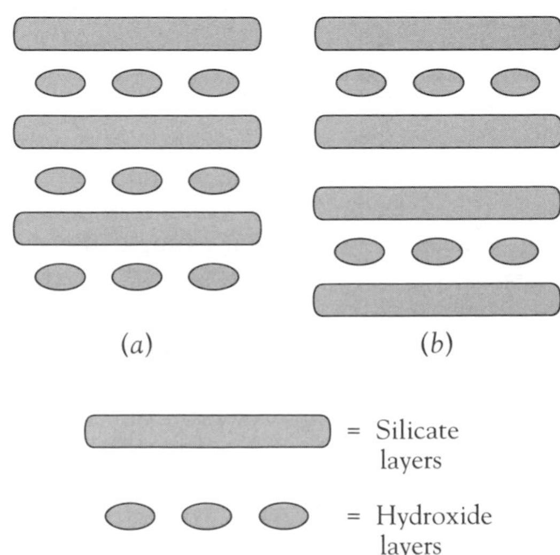

Figure 13.24 The layer structures of (*a*) white asbestos and (*b*) talc.

Besides chains and sheets, three-dimensional structures of linked silicate units can occur; in these structures the basic unit is the neutral silicon dioxide molecule. Variations in the structure can be attained by substituting aluminum for some of the silicon atoms. Replacement of one-fourth of the silicon atoms by aluminum results in an anion of empirical formula $[AlSi_3O_8]^-$; replacement of one-half the silicon atoms gives the formula $[Al_2Si_2O_8]^{2-}$. The charge is counterbalanced by Group 1 or 2 cations. This particular family of minerals comprises the *feldspars*, components of granite. Typical examples are orthoclase, $K[AlSi_3O_8]$, and anorthite, $Ca[Al_2Si_2O_8]$.

Zeolites

One three-dimensional silicate structure has open channels throughout the network. Compounds with this structure are known as *zeolites*, and their industrial importance is skyrocketing. A number of zeolites exist in nature, but chemists have mounted a massive search for zeolites with novel cavities throughout their structures. There are four major uses for zeolites: as ion exchangers, as adsorption agents, for gas separation, and as catalysts.

Zeolites as Ion Exchangers

If "hard" water (water high in calcium and magnesium ion concentration) is passed through a column containing pellets of a sodium ion zeolite, the Group 2 ions push out the sodium ions. The "soft" water emerging from the column requires less detergent for washing purposes and produces less solid deposit (scum) when soap is used. When the cation sites have been fully exchanged, passage of a saturated salt solution through the column pushes out the alkaline earth metal ions by a process based on the Le Châtelier principle.

Zeolites as Adsorption Agents

The pores in a zeolite are just about the right size for holding small covalent molecules, so one major application is the use of a zeolite to dry organic liquids. The water molecule is small enough to fit into a cavity of the zeolite, so it remains in the zeolite, which has effectively "dried" the organic liquid. Strong heating of the "wet" zeolite causes expulsion of the water, so the zeolite can be used again. By choosing a zeolite with a particular pore size, the process can be made quite specific for the removal of particular molecules. These zeolites are called *molecular sieves*. For example, the zeolite of formula $Na_{12}[(AlO_2)_{12}(SiO_2)_{12}]\cdot xH_2O$ has pores that are 400 pm in diameter; pores of this size can accommodate small molecules. The zeolite of formula $Na_{86}[(AlO_2)_{86}(SiO_2)_{106}]\cdot xH_2O$ has holes that are 800 pm in diameter and can accommodate larger molecules.

Zeolites for Gas Separation

Zeolites are very selective in their absorption of gases. In particular, they have a total preference for dinitrogen over dioxygen; 1 L of a typical zeolite contains about 5 L of nitrogen gas. This gas is released when the zeolite is heated. The selective absorption of dinitrogen makes zeolites of great use in the inexpensive separation of the two major components of the atmosphere. For example, a major cost in sewage treatment and steelworks has been the provision of oxygen-enriched air. Traditionally, the only route to oxygen enrichment was liquefying the air and distilling the components. Now, by cycling air through beds of zeolites, the components can be separated inexpensively.

Why is dinitrogen selectively absorbed? After all, both dioxygen and dinitrogen are nonpolar molecules of about the same size. To answer this question, we have to look at the atomic nuclei rather than at the electrons. Nuclei can be spherical or ellipsoidal. If they are ellipsoidal (football-shaped), like nitrogen-14, then the nuclei possess an unevenly distributed nuclear charge, known as an electric quadrupole moment. Oxygen-16, however, contains spherical nuclei and thus does not have an electric quadrupole moment. The interior of a zeolite cavity contains an extremely high electric charge, which attracts nuclei with electric quadrupole moments, such as those in the dinitrogen molecule. The effect is much smaller than the electron dipole moment and, apart from this instance, is of little importance in terms of chemical properties.

Zeolites as Catalysts

The modern oil industry depends on zeolites. Crude oil from the ground does not meet many of our modern requirements. It has a high proportion of long-chain molecules, whereas the fuels we need are short-chain molecules with low boiling points. Furthermore, the long-chain molecules in crude oil have straight chains, which is fine for diesel engines, but the gasoline engine needs branched-chain molecules for optimum performance. Zeolite catalysts can, under specific conditions, convert straight-chain molecules to branched-chain isomers. The cavities in the zeolite structure act as molecular templates, rearranging the molecular structure to match the cavity shape. In addition to the oil industry, several industrial organic syntheses employ zeolite catalysts to convert a starting material to a very specific product. Such "clean" reactions are rare in conventional organic chemistry; in fact, side reactions giving unwanted products are very common.

One of the most important catalysts is $Na_3[(AlO_2)_3(SiO_2)]\cdot xH_2O$, commonly called ZSM-5. This compound does not occur in nature; it was first synthesized by research chemists at Mobil Oil. It is higher in aluminum than most naturally occurring zeolites, and its ability to function depends on the high acidity of water molecules bound to the high charge density aluminum ions (Figure 13.25). In fact, the hydrogen in ZSM-5 is as strong a Brønsted-Lowry acid as that in sulfuric acid.

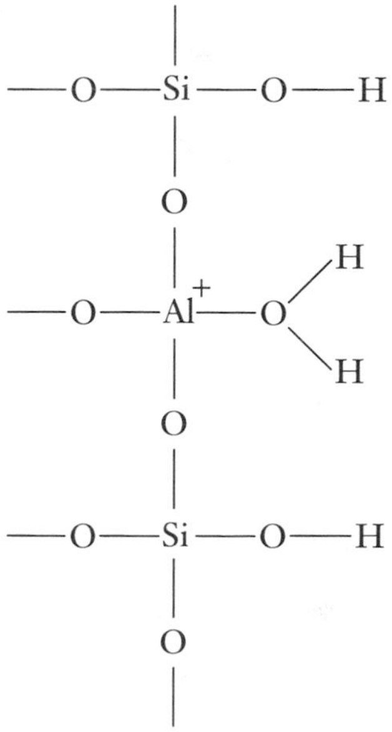

The zeolite ZSM-5 catalyzes reactions by admitting molecules of the appropriate size and shape into its pores and then acting as a strong Brønsted-Lowry acid. This process can be illustrated by the synthesis of ethylbenzene, an important organic reagent, from ethene, C_2H_4, and benzene, C_6H_6. It is believed that the ethene is protonated within the zeolite:

$$H_2C{=}CH_2(g) + H_2O{-}Al^+(s) \rightarrow H_2C{-}CH_3^+(g) + HO{-}Al(s)$$

Figure 13.25 Acidic hydrogen on the surface of zeolite ZSM-5.

The carbocation can then attack a benzene molecule to give ethylbenzene:

$$H_2C{-}CH_3^+(g) + C_6H_6(g) + HO{-}Al(s)$$
$$\rightarrow C_6H_5{-}CH_2{-}CH_3(g) + H_2O{-}Al^+(s)$$

Ceramics

The term *ceramics* describes nonmetallic, inorganic compounds that are prepared by high-temperature treatment. The properties of ceramic materials are a function not only of their chemical composition but also of the conditions of synthesis. Typically, the components are finely ground and mixed to a paste with water. The paste is then formed into the desired shape and heated to about 900°C. At these temperatures, all the water molecules are lost, and numerous high-temperature chemical reactions occur. In particular, long needle crystals of mullite, $Al_6Si_2O_{13}$, are formed. These make a major contribution to the strength of the ceramic material.

Conventional ceramics are made from a combination of quartz with two-dimensional silicates (clays) and three-dimensional silicates (feldspars). Thus a stoneware used for household plates will have a composition of about 45 percent clay, 20 percent feldspar, and 35 percent quartz. By contrast, a dental ceramic for tooth caps is made from about 80 percent feldspar, 15 percent clay, and 5 percent quartz.

The major interest today, however, is in nontraditional ceramics, particularly metal oxides. To form a solid ceramic, the microcrystalline powder is heated to just below its melting point, sometimes under pressure as well. Under these conditions, bonding between the crystal surfaces occurs, a process known as *sintering*. Aluminum oxide is a typical example. Aluminum oxide ceramic is used as an insulator in automobile spark plugs and as a replacement for bone tissue, such as artificial hips. The most widely used non-oxide ceramic, silicon carbide, was discussed earlier in this chapter.

As the search for new materials intensifies, boundaries between compound classifications are disappearing. *Cermets* are materials containing cemented grains of metals and ceramic compounds; *glassy ceramics* are glasses in which a carefully controlled proportion of crystals has been grown. Two examples of compounds that can be formed into glassy ceramics are lithium aluminum silicate, $Li_2Al_2Si_4O_{12}$, and magnesium aluminum silicate, $Mg_2Al_4Si_5O_{18}$. These materials are nonporous and are known for their extreme resistance to thermal shock. That is, they can be heated to red heat and then plunged into cold water without shattering. The major use of this material is in cooking utensils and heat-resistant cooking surfaces. Many of these glassy ceramics are produced by Corning.

Silicones

Silicones constitute an enormous family of polymers, and they all contain a chain of alternating silicon and oxygen atoms. Attached to the silicon atoms are pairs of organic groups, such as the methyl group, CH_3. The structure of this simplest silicone is shown in Figure 13.26, where the number of repeat-

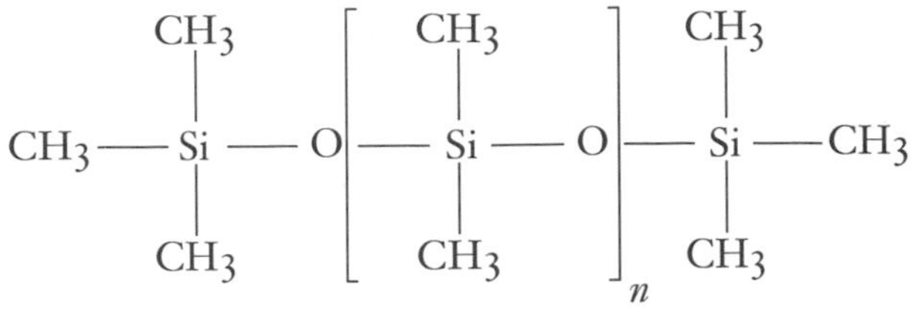

$$\mathrm{CH_3{-}\underset{CH_3}{\overset{CH_3}{Si}}{-}O}\left[\mathrm{{-}\underset{CH_3}{\overset{CH_3}{Si}}{-}O}\right]_n\mathrm{{-}\underset{CH_3}{\overset{CH_3}{Si}}{-}CH_3}$$

Figure 13.26 Structure of the simplest silicon, *catena*-poly-[(dimethylsilicon)-μ-oxo]. The number of repeating units, n, is very large.

ing units, n, is very large. To synthesize this compound, chloromethane, CH_3Cl, is passed over a copper-silicon alloy at 300°C. A mixture of compounds is produced, including $(CH_3)_2SiCl_2$:

$$2\ CH_3Cl(g) + Si(s) \rightarrow (CH_3)_2SiCl_2(l)$$

Water is added, causing hydrolysis:

$$(CH_3)_2SiCl_2(l) + 2\ H_2O(l) \rightarrow (CH_3)_2Si(OH)_2(l) + 2\ HCl(g)$$

The hydroxo compound then polymerizes, with loss of water:

$$n\ (CH_3)_2Si(OH)_2(l) \rightarrow [\text{—O—Si}(CH_3)_2\text{—}]_n(l) + H_2O(l)$$

Silicones are used for a wide variety of purposes. The liquid silicones are more stable than hydrocarbon oils. In addition, their viscosity changes little with temperature, whereas the viscosity of hydrocarbon oils changes dramatically with temperature. Thus silicons are used as lubricants and wherever inert fluids are needed, for example, in hydraulic braking systems. Silicones are very hydrophobic (nonwetting); hence they are used in water-repellent sprays for shoes and other items.

By cross-linking chains, silicone rubbers can be produced. Like the silicone oils, the rubbers show great stability to high temperature and to chemical attack. Their multitudinous uses include the face-fitting edges for snorkel and scuba masks. The rubbers also are very useful in medical applications, such as transfusion tubes. However, silicone gels have attained notoriety in their role as a breast implant material. While sealed in a polymer sack, they are believed to be harmless. The major problem arises when the container walls leak or break. The silicone gel can then diffuse into surrounding tissues. The chemical inertness of silicones turns from a benefit to a problem, because the body has no mechanism for breaking down the polymer molecules. Many medical personnel believe that these alien gel fragments trigger the immune system, thereby causing a number of medical problems.

Tin and Lead

Tin forms two common allotropes: the shiny metallic allotrope, which is thermodynamically stable above 13°C, and the gray, nonmetallic diamond-structure allotrope, which is stable below that temperature. The change at low temperatures to microcrystals of the gray allotrope is slow at first but accelerates rapidly. This transition is a particular problem in poorly heated museums, where priceless historical artifacts can crumble into a pile of tin powder. The effect can spread from one object to another in contact with it, and this lifelike behavior has been referred to as "tin plague" or "museum disease." The soldiers of Napoleon's army had tin buttons fastening their clothes, and they used tin cooking utensils. It is believed by some that, during the bitterly cold winter invasion of Russia, the crumbling of buttons, plates, and pans contributed to the low morale and hence to the ultimate defeat of the Imperial French troops.

The existence of both a metallic and nonmetallic allotrope identifies tin as a real "borderline" or weak metal. Tin is also amphoteric, another of its weak metallic properties. Thus tin(II) oxide reacts with acid to give (covalent) tin(II) salts, and with bases to form the stannite ion, $[Sn(OH)_3]^-$:

$$SnO(s) + 2\ HCl(aq) \rightarrow SnCl_2(aq) + H_2O(l)$$

$$SnO(s) + NaOH(aq) + H_2O(l) \rightarrow Na^+(aq) + [Sn(OH)_3]^-(aq)$$

Lead, the more economically important of the two metals, is a soft, gray-black, dense solid found almost exclusively as lead(II) sulfide, the mineral galena. Lead(II) sulfide cannot be directly reduced with carbon because that reaction has an unfavorable free energy change. This relation is shown in Figure 13.27; note that the lead(II) sulfide line does not cross that of carbon disulfide (the product of oxidation of carbon with sulfur). Instead, the compound is heated with air to oxidize the sulfide ions to sulfur dioxide (which the Ellingham diagram shows to be feasible). The lead(II) oxide can then be reduced with coke to lead metal:

$$2\ PbS(s) + 3\ O_2(g) \rightarrow 2\ PbO(s) + 2\ SO_2(g)$$

$$PbO(s) + C(s) \rightarrow Pb(l) + CO(g)$$

There are two major environmental concerns that arise in connection with this lead extraction process. First, the sulfur dioxide produced contributes to atmospheric pollution unless it is utilized in another process; and second, lead dust must not be permitted to escape during the smelting. Lead is highly toxic, so the best solution is to recycle the metal. At the present time,

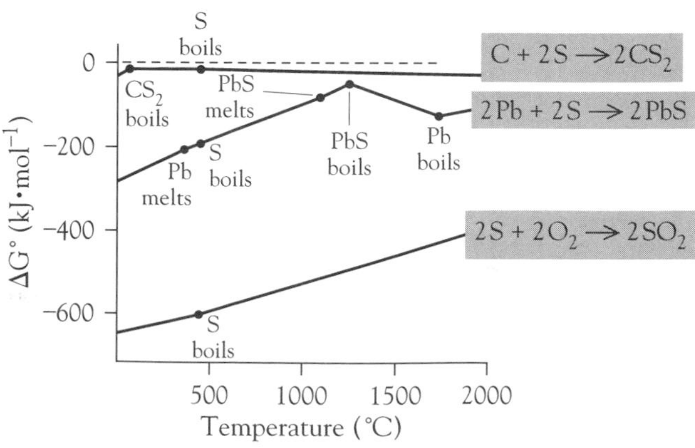

Figure 13.27 Ellingham diagram for lead(II) sulfide, carbon disulfide, and sulfur dioxide.

Table 13.4 Charge densities of lead ions

Ion	Charge density ($C{\cdot}mm^{-3}$)
Pb^{2+}	32
Pb^{4+}	196

close to half of the 6 million tonnes of lead used annually comes from recycling. The aim must be to substantially increase this proportion. In particular, it would help if all defunct lead-acid batteries were returned for disassembly and reuse of the lead contained in them. Of course, such a move would have a negative economic effect as the result of a decline in employment in the lead mining industry. There would, however, be an increase in employment in the labor-intensive recycling and reprocessing sector.

Tin and lead exist in two oxidation states, +4 and +2. It is possible to explain the existence of the +2 oxidation state in terms of the inert-pair effect, as we did for the +1 oxidation state of thallium.

The formation of ions of these metals is rare. Tin and lead compounds in which the metals are in the +4 oxidation state are covalent, except for a few solid-phase compounds. Even when in the +2 oxidation state, tin generally forms covalent bonds, with ionic bonds only being present in compounds in the solid phase. Conversely, lead forms the 2+ ion in solid and in solution. Table 13.4 shows that the charge density for Pb^{2+} is relatively low, whereas that for 4+ ions is extremely high—high enough to cause the formation of covalent bonds with all but the least polarizable anion, fluoride.

Tin and Lead Oxides

The oxides of the heavier members of Group 14 can be regarded as ionic solids. Tin(IV) oxide, SnO_2, is the stable oxide of tin, whereas lead(II) oxide, PbO, is the stable oxide of lead. Lead(II) oxide exists in two crystalline forms, one yellow and the other red. There is also a mixed oxide of lead, Pb_3O_4 (red lead), which behaves chemically as $PbO_2{\cdot}2PbO$; hence its systematic name is lead(II) lead(IV) oxide. The chocolate-brown lead(IV) oxide, PbO_2, is quite stable, and it is a good oxidizing agent. We can see the differences between tin and lead in this respect from the Frost diagram shown in Figure 13.28.

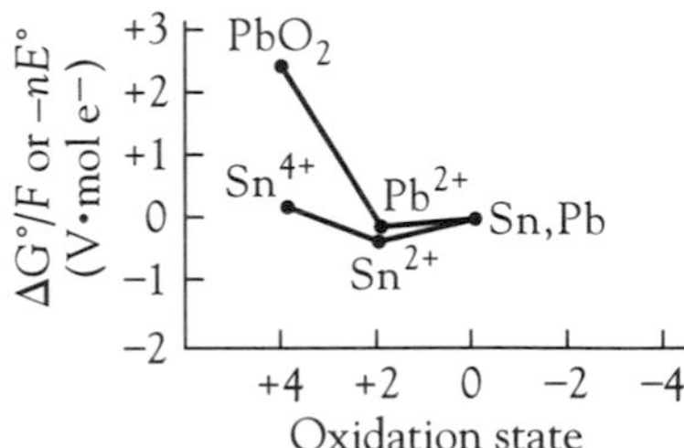

Figure 13.28 Frost diagram for tin and lead.

Tin(IV) oxide is incorporated in glazes used in the ceramics industry. About 3500 tonnes are used annually for this purpose. The consumption of lead(II) oxide is much higher, of the order of 250 000 tonnes annually, because it is used to make lead glass and for the production of the electrode surfaces in lead-acid batteries. In these batteries, both electrodes are formed by pressing lead(II) oxide into a frame of lead metal. The cathode is formed by oxidizing lead(II) oxide to lead(IV) oxide, and the anode is produced by reducing lead(II) oxide to lead metal. The electric current arises when lead(IV) oxide is reduced to insoluble lead(II) sulfate in the sulfuric acid electrolyte while the lead metal is oxidized to lead(II) sulfate on the other electrode:

$$PbO_2(s) + 4\ H^+(aq) + SO_4^{2-}(aq) + 2\ e^- \rightarrow PbSO_4(s) + 2\ H_2O(l)$$

$$Pb(s) + SO_4^{2-}(aq) \rightarrow PbSO_4(s) + 2\ e^-$$

These two half-reactions are reversible. Hence the battery can be recharged by applying an electric current in the reverse direction. In spite of a tremendous quantity of research, it has been very difficult to develop a low-cost, lead-free, heavy-duty battery that can perform as well as the lead-acid battery.

Red lead, Pb_3O_4, has been used on a large scale as a rust-resistant surface coating for iron and steel. Mixed metal oxides, such as calcium lead(IV) oxide, $CaPbO_3$, are now being used as an even more effective protection against salt water for steel structures. The structure of $CaPbO_3$ is discussed in Chapter 15.

As mentioned earlier, the lead(IV) ion is too polarizing to exist in aqueous solution. Oxygen can often be used to stabilize the highest oxidation number of an element, and this phenomenon is true for lead. Lead(IV) oxide is an insoluble solid in which the Pb^{4+} ions are stabilized in the lattice by the high lattice energy. Even then, one can argue that there is considerable covalent character in the structure. Addition of an acid, such as nitric acid, gives immediate reduction to the lead(II) ion and the production of oxygen gas:

$$2\ PbO_2(s) + 4\ HNO_3(aq) \rightarrow 2\ Pb(NO_3)_2(aq) + 2\ H_2O(l) + O_2(g)$$

In the cold, lead(IV) oxide undergoes a double-replacement reaction with concentrated hydrochloric acid to give covalently bonded lead(IV) chloride. When warmed, the unstable lead(IV) chloride decomposes to give lead(II) chloride and chlorine gas:

$$PbO_2(s) + 4\ HCl(aq) \rightarrow PbCl_4(aq) + 2\ H_2O(l)$$

$$PbCl_4(aq) \rightarrow PbCl_2(s) + Cl_2(g)$$

Tin and Lead Chlorides

Tin(IV) chloride is a typical covalent metal chloride. It is an oily liquid that fumes in moist air to give a gelatinous tin(IV) hydroxide, which we represent as $Sn(OH)_4$ (although it is actually more of a hydrated oxide), and hydrogen chloride gas:

$$SnCl_4(l) + 4\ H_2O(l) \rightarrow Sn(OH)_4(s) + 4\ HCl(g)$$

Like so many compounds, tin(IV) chloride has a small but important role in our lives. The vapor of this compound is applied to freshly formed glass, where it reacts with water molecules on the glass surface to form a layer of tin(IV) oxide. This very thin layer substantially improves the strength of the glass, a property particularly important in eyeglasses. A thicker coating of tin(IV) oxide acts as an electrically conducting layer. Aircraft cockpit windows use such a coating. An electric current is applied across the conducting glass surface, and the resistive heat that is generated prevents frost formation when the aircraft descends from the cold upper atmosphere.

Lead(IV) chloride is a yellow oil that, like its tin analog, decomposes in the presence of moisture and explodes when heated. Lead(IV) bromide and iodide do not exist, because the reduction potential of these two halogens is sufficient to reduce lead(IV) to lead(II).

Lead(II) chloride is a white insoluble solid that forms an ionic-type crystal lattice. Tin(II) chloride, however, is a white water-soluble solid that forms a solid-phase structure in which there are covalently bridging chloride ions. Its covalent nature is also apparent from its relatively high solubility in low-polarity organic solvents. In the gas phase, tin(II) chloride is a V-shaped molecule. According to VSEPR theory, this molecular shape is due to the presence of a lone pair (Figure 13.29).

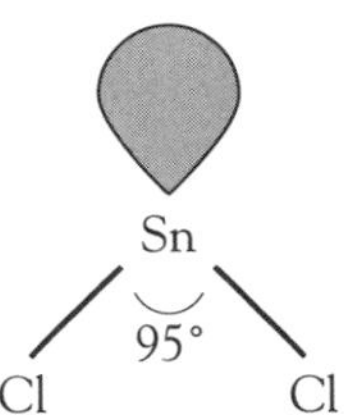

Figure 13.29 The tin(II) chloride molecule in the gas phase.

With a lone pair, tin(II) chloride would be expected to be a Lewis base. But, as I've said before, chemistry is full of surprises: The compound generally behaves like a Lewis acid. Thus the lone pair appears to be unreactive—a truer use of the term *inert-pair effect*. For example, tin(II) chloride reacts with chloride ion to form the trichlorostannate(II) ion, $SnCl_3^-$ (Figure 13.30). The bond angle of about 90° suggests an alternative explanation for the bonding, that is, tin uses pure *p* orbitals. This model would explain the 90° bond angles and the lack of Lewis base behavior. In other words, the electron pair is in a spherical *s* orbital rather than being part of a directional sp^2 or sp^3 hybrid. The tin then uses an empty *p* orbital to bond with the lone pair of the chloride ion.

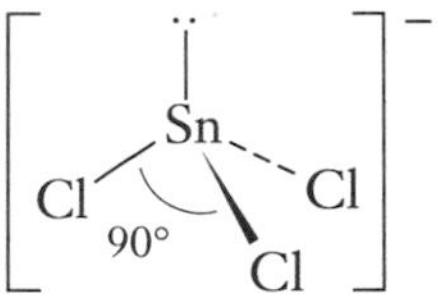

Figure 13.30 The trichlorostannate(II) ion, $SnCl_3^-$.

Tetraethyllead

The less electropositive (more weakly metallic) metals form an extensive range of compounds containing metal-carbon bonds. The metal-carbon compound that has been produced on the largest scale is tetraethyllead, $Pb(C_2H_5)_4$, known as TEL. Tetraethyllead is a stable compound that has a low boiling point and at one time was produced on a vast scale as a gasoline additive.

In a gasoline engine, a spark is used to ignite the mixture of fuel and air. However, straight-chain hydrocarbons will burn simply when compressed with air—the mode of operation of a diesel engine. This reactivity is responsible for the phenomenon of premature ignition (commonly called knocking or pinging), and in addition to making the engine sound as if it is about to fall apart, it can cause severe damage. Branched-chain molecules, however, because of their kinetic inertness, require a spark to initiate combustion (Figure 13.31).

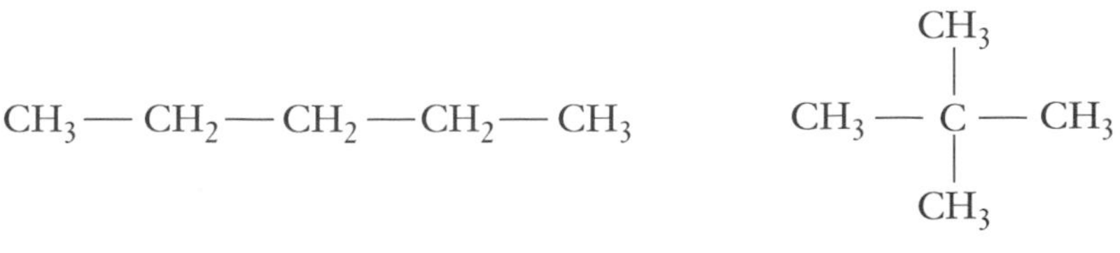

$CH_3—CH_2—CH_2—CH_2—CH_3$

(*a*)

$C(CH_3)_4$: $CH_3—C—CH_3$ with CH_3 above and CH_3 below the central C

(*b*)

Figure 13.31 Two hydrocarbons of the same formula, C_5H_{12}: (*a*) a straight-chain isomer and (*b*) a branched-chain isomer.

The measure of the proportion of branched-chain molecules in gasoline is the *octane rating*; the higher the proportion of branched-chain molecules, the higher the octane rating of the fuel. With the demand for higher performance, higher compression engines, the need for higher octane rated gasoline became acute. The addition of TEL to low octane rated gasoline increases the octane rating, that is, it prevents premature ignition. In the early 1970s, about 500 000 tonnes of TEL were produced annually for addition to gasoline. In fact, the U.S. Environmental Protection Agency allowed up to 3 g of TEL per gallon of gasoline until 1976.

Thomas Midgley discovered both the chlorofluorocarbons and the role of TEL in improving gasolines. The irony is that both discoveries were designed to make life better through progress in chemistry, and yet both have had quite the opposite long-term effect. Tetraethyllead poses both direct and indirect hazards. The direct hazard has been to people working with gasoline, such as gas station attendants. Because it has a low boiling point, the TEL added to gasoline vaporizes readily; hence persons exposed to TEL vapor absorb this neurotoxic lead compound through the lining of their lungs; they develop headaches, tremors, and increasingly severe neurological disorders. Unfortunately, there are some countries in the world where TEL is still a permitted gasoline additive.

By accident, an indirect hazard was also discovered. In studies of the level of trace metals in vegetation, a researcher in Vancouver, Canada, found that some samples collected by students had enormous concentrations of lead. He questioned his students and found that the lead-containing samples, rather than being collected from remote locations, had been collected beside the Trans-Canada Highway. A follow-up study showed that the source of these high levels of lead was vehicle emissions. Up to the 1960s, it had been assumed that the combustion of TEL would produce insoluble lead compounds, such as oxides, that would simply remain inert. However, it is now well established that the fine particles of lead compounds are absorbed by plants and, indeed, by humans.

Once the hazard was recognized, countries started to ban TEL. Germany, Japan, and the former USSR were quick to outlaw TEL; other countries (such as the United States) followed more slowly. One of the problems of eliminating TEL from gasolines was simply that modern vehicles need high octane rated gasoline. Two solutions have been found: the development of the zeolite catalysts that enable oil companies to convert straight-chain molecules to the required branched-chain molecules; and the addition of oxygenated compounds, such as ethanol, to fuels. Thus the need for octane boosters has been eliminated.

Biological Aspects

The Carbon Cycle

There are many biogeochemical cycles on this planet. The largest scale process is the carbon cycle. Of the 2×10^{16} tonnes of carbon, most of it is "locked away" in the Earth's crust as carbonates, coal, and oil. Only about 2.5×10^{12} tonnes are available as carbon dioxide. Every year, about 15 per-

W.H. Freeman and Company

4419 West 1980 South
Salt Lake City, Utah 84104
FAX (801) 977-9712
Telephone (801) 973-4660

ISBN PUB PREFIX 0 - 7167
FID #94-1569566
D.U.N.S. 00-913-5245

WHSE SL

PAGE 1

PACKING LIST

INVOICE NUMBER 6 - B0114997

SHIP TO:
SUSAN JACKELS
WAKE FOREST UNIVERSITY
DEPT OF CHEMISTRY
WINSTON-SALEM NC 27109

BILL TO:
CONNAUGHT COLBERT REP 69
W H FREEMAN AND COMPANY
5620 BRIAR OAKS LANE #602
RALEIGH NC 27612

SHIP VIA	P.O. #	CUST. ACCOUNT NO.	INVOICE DATE
7 LIBRARY	...	SHIP 1*10034028 BILL 1*10006717	10-11-95

QUANTITY	DISCOUNT	ISBN	PRICE	AUTHOR	TITLE	H/P
** BACK ORDER RELEASE **						
ZONE: 8						
1	100.00%	0716728192	0.00	RAYNER	DESCRIPTIVE INORGANIC CHEMISTRY	HAR
	PO#=ES69257828					

TOT BOOKS: 1 TOT WGT: 2.13

W.H. Freeman and Co.
4419 West 1980 South
Salt Lake City, Utah 84104

RETURN REQUESTED

06011499702

7 LIBRARY

SHIP TO:
SUSAN JACKELS
WAKE FOREST UNIVERSITY
DEPT OF CHEMISTRY
WINSTON-SALEM NC 27109

cent of this total is absorbed by plants and algae in the process of photosynthesis, which uses energy from the sun to synthesize complex molecules such as sucrose.

Some plants are eaten by animals (such as humans), and a part of the stored chemical energy is released during their decomposition to carbon dioxide and water. These two products are returned to the atmosphere by the process of respiration. However, the majority of the carbon dioxide incorporated into plants is returned to the atmosphere only after the death and subsequent decomposition of the plant organisms. Another portion of the plant material is buried, thereby contributing to the soil humus or the formation of peat bogs. The carbon cycle is partially balanced by the copious output of carbon dioxide by volcanoes.

In this century the demand for energy has led to the burning of coal and oil, which were formed mainly in the Carboniferous era. This combustion adds about 2.5×10^{10} tonnes of carbon dioxide to the atmosphere each year, in addition to that from natural cycles. Although we are just returning to the atmosphere carbon dioxide that came from there, we are doing so at a very rapid rate, and many scientists are concerned that the rate of return will overwhelm the Earth's absorption mechanisms. This topic is currently being studied in many laboratories.

The Essentiality of Silicon

For several decades, it has been known that silicon is essential to the life of animals (including humans). Up to now, the role of this element has been a mystery. It now appears likely that silicon, in the form of soluble silicate ions, inhibits the toxic effects of the aluminum ion on the body. Scientists assume that the inhibition occurs by the formation of very insoluble aluminum silicates.

Silicon dioxide, in the form of diatomaceous earth, is sometimes used to protect plants against insect infestations. This material is found as large deposits that were formed from the death of countless tiny marine creatures, the diatoms. Instead of calcium carbonate–based skeletons, this family of marine organisms uses silicon dioxide as the structural material of their bodies. It is claimed that a thin layer of diatomaceous earth spread on a garden cuts open the shells (exoskeletons) of insects when they crawl over the sharp, microscopic fragments of silica, thereby causing their death by dehydration.

We have already mentioned the hazards of asbestos. It can cause two serious lung diseases: asbestosis and mesothelioma. The dust of any silicate rock will also cause lung damage, in this case, silicosis. The fundamental cause of the lung problems is due to the total insolubility of the silicates. Once the particles stick in the lungs, they are there for life. The irritation they cause produces scarring and immune responses that lead to the disease state.

The Toxicity of Tin

Although the element and its simple inorganic compounds have a fairly low toxicity, its organometallic compounds are very toxic. Compounds such as hydroxotributyltin, $(C_4H_9)_3SnOH$, are effective against fungal infections in potatoes, grapevines, and rice plants. For many years, organotin compounds were incorporated into the paints used on ships' hulls. The compound would kill the larvae of mollusks, such as barnacles, that tend to attach themselves to a ship's hull, slowing the vessel considerably. However, the organotin compound slowly leaches into the surrounding waters, where, particularly within the confines of a harbor, it destroys other marine organisms. For this reason, its marine use is no longer permitted.

The Severe Hazard of Lead

Lead is one of the most toxic elements to which we are exposed. Although we no longer use "sugar of lead" as a sweetener, in more recent times lead has become a hazard from a number of other sources. Lead-based paints used to be very common; even today, many older housing units contain unacceptable quantities of lead in their peeling paint. Until the recent shift to nonleaded gasolines, exposure to lead from the combustion of leaded gasolines was a major concern. Cigarettes, too, present a hazard, because, with every one consumed, a smoker ingests 1 μg of lead. Finally, lead dust is a continuing problem; it comes from industries that currently use lead or from sites where lead was used in the past, some of which are now used as playgrounds or sites for housing developments.

Lead(II) ion interferes with a wide range of biochemical processes. As a result, it is probably the most dangerous ion of all. Symptoms of lead poisoning depend on the concentration level of the ion. At low concentrations, lead causes anemia and headaches; but at high concentrations, kidney failure, convulsions, brain damage, and then death ensue.

Exercises

13.1. Write balanced chemical equations corresponding to the following chemical reactions:

(a) solid lithium dicarbide(2−) with water
(b) silicon dioxide with carbon
(c) copper(II) oxide heated with carbon monoxide
(d) calcium hydroxide solution with carbon dioxide (two equations)
(e) methane with molten sulfur
(f) silicon dioxide with molten sodium carbonate
(g) lead(IV) oxide with concentrated hydrochloric acid (two equations)

13.2. Write balanced chemical equations corresponding to the following chemical reactions:

(a) solid beryllium carbide with water
(b) carbon monoxide with dichlorine
(c) hot magnesium metal with carbon dioxide
(d) solid sodium carbonate with hydrochloric acid
(e) heating barium carbonate
(f) carbon disulfide gas and chlorine gas
(g) tin(II) oxide with hydrochloric acid

13.3. Define the following terms: (a) catenation; (b) aerogel; (c) ceramic; (d) silicone.

13.4. Define the following terms: (a) synergistic effect; (b) molecular sieves; (c) cermet; (d) galena.

13.5. Contrast the properties of the three main allotropes of carbon: diamond, graphite, and C_{60}.

13.6. Explain why (a) diamond has a very high thermal conductivity; (b) high pressure and temperature are required for the traditional method of diamond synthesis.

13.7. Why are fullerenes soluble in many solvents even though both graphite and diamond are insoluble in all solvents?

13.8. Explain why catenation is common for carbon but not for silicon.

13.9. Write the chemical equation for the reaction used in the commercial production of silicon carbide. Is it enthalpy or entropy driven? Explain your reasoning. Calculate the values of $\Delta H°$ and $\Delta S°$ for the process to confirm your deduction; then calculate $\Delta G°$ at 2000°C.

13.10. Carbon dioxide has a negative enthalpy of formation whereas that of carbon disulfide is positive. Using bond energy data, construct an enthalpy of formation diagram and identify the reason(s) for such different values.

13.11. Contrast the properties of carbon monoxide and carbon dioxide.

13.12. Discuss the bonding in carbon disulfide in terms of hybridization theory.

13.13. From data tables of ΔH_f° and S° values, show that the combustion of methane is a spontaneous process.

13.14. Explain why silane burns in contact with air whereas methane requires a spark before it will combust.

13.15. Describe why the CFCs were once thought to be ideal refrigerants.

13.16. Why is HFC-134a a less than ideal replacement for CFC-12?

13.17. What would be the chemical formula of HFC-134b?

13.18. Why does methane represent a particular concern as a potential greenhouse gas?

13.19. Contrast the properties of carbon dioxide and silicon dioxide and explain these differences in terms of bond types. Explain why the two oxides adopt such dissimilar bonding.

13.20. In crocidolite, $Na_2Fe_5(Si_4O_{11})_2(OH)_2$, how many of the iron ions must have a 2+ charge and how many a 3+ charge?

13.21. Describe the difference in structure between white asbestos and talc.

13.22. Describe the major uses of zeolites.

13.23. If the water in a zeolite is expelled by strong heating, must the absorption of water by the zeolite be an endo- or exothermic process?

13.24. What advantage of silicone polymers becomes a problem when they are used as breast implants?

13.25. Contrast the properties of the oxides of tin and lead.

13.26. Construct the electron-dot structures of tin(IV) chloride and gaseous tin(II) chloride. Draw the corresponding molecular shapes.

13.27. Lead(IV) fluoride melts at 600°C whereas lead(IV) chloride melts at –15°C. Interpret the values in relation to the probable bonding in the compounds.

13.28. To form the electrodes in the lead-acid battery, the cathode is produced by oxidizing lead(II) oxide to lead(IV) oxide and the anode is produced by reducing the lead(II) oxide to lead metal. Write half-equations to represent the two processes.

13.29. Show from standard reduction potentials that lead(IV) iodide cannot exist.

13.30. Our evidence that the Romans ingested high levels of lead(II) comes from examination of skeletons. Suggest why the lead ions would be present in bone tissues.

13.31. What are the main sources of lead in the environment today?

CHAPTER

14

The Group 15 Elements

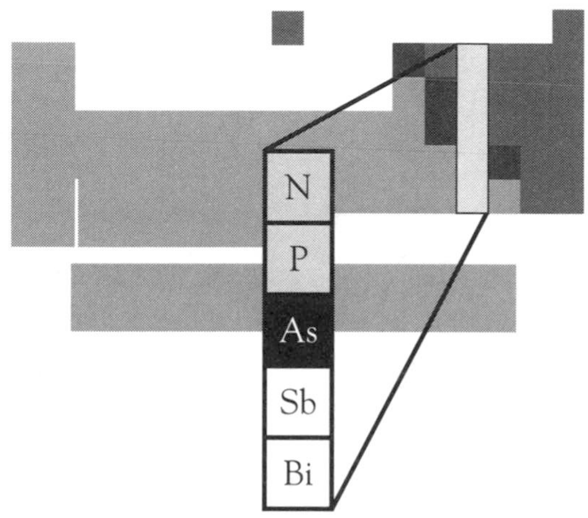

Two of the most dissimilar nonmetallic elements are in the same group: reactive phosphorus and unreactive nitrogen. Of the other members of the group, arsenic is really a semimetal, and the two lower members of the group, antimony and bismuth, exhibit very weakly metallic behavior.

The discovery of phosphorus by the German alchemist Hennig Brand provides the most interesting saga of the members of this group. The discovery occurred by accident during his investigation of urine. Urine was a favorite topic of research in the seventeenth century, for it was believed anything gold-colored, such as urine, had to contain gold! However, when Brand fermented urine and distilled the product, he obtained a white, waxy, flammable solid with a low melting point—white phosphorus. One hundred years later, a route to extract phosphorus from phosphate rock was devised, and chemists no longer needed buckets of urine to synthesize the element.

In these days of pocket butane lighters, we forget how difficult it used to be to generate a flame. So in 1833, people were delighted to find how easily fire could be produced by using white phosphorus matches. This convenience came at a horrendous human cost, because white phosphorus is extremely toxic. The young women who worked in the match factories died in staggering numbers from phosphorus poisoning. This occupational hazard manifested itself as "phossy jaw," a disintegration of the lower jaw, followed by an agonizing death.

In 1845 the air-stable red phosphorus was shown to be chemically identical to white phosphorus. The British industrial chemist Arthur Albright, who had been troubled by the enormous number of deaths in his match factory, learned of this safer allotrope and determined to produce matches bearing red phosphorus. But mixing the inert red phosphorus with an oxidizing agent gave an instant explosion. Prizes were offered for the development of a safe match, and finally in 1848 some now unknown genius proposed to put half the ingredients on the match tip and the remainder on a strip attached to the matchbox. Only when the two surfaces were brought into contact did ignition of the match head occur—so science and technology moved forward together. (For more details, see the section "Matches" later in the chapter.)

Group Trends

The first two members of Group 15, nitrogen and phosphorus, are nonmetals; the remaining three members, arsenic, antimony, and bismuth, have some metallic character. Scientists like to categorize things, but in this group their efforts are frustrated because there is no clear division of properties between nonmetals and metals. Two characteristic properties that we can study are the electrical resistivity of the elements and the acid-base behavior of the oxides (Table 14.1).

Nitrogen and phosphorus are both nonconductors of electricity, and both form acidic oxides, so they are unambiguously classified as nonmetals. The problems start with arsenic. Even though the common allotrope of arsenic looks metallic, subliming and recondensing the solid produce a second allotrope that is a yellow powder. Because it has both metallic-looking and nonmetallic allotropes and forms amphoteric oxides, arsenic can be classified as a semimetal. However, much of its chemistry parallels that of phosphorus, so there is a good case for considering it as a nonmetal.

Table 14.1 Properties of the Group 15 elements

Element	Appearance at SATP	Electrical resistivity (μΩ·cm)	Acid-base properties of oxides
Nitrogen	Colorless gas	—	Acidic and neutral
Phosphorus	White, waxy solid	10^{17}	Acidic
Arsenic	Brittle, metallic solid	33	Amphoteric
Antimony	Brittle, metallic solid	42	Amphoteric
Bismuth	Brittle, metallic solid	120	Basic

Table 14.2 Melting and boiling points of the Group 15 elements

Element	Melting point (°C)		Boiling point (°C)
N_2	–210		–196
P_4	44		281
As		Sublimes at 615	
Sb	631		1387
Bi	271		1564

Antimony and bismuth are almost as borderline as arsenic. Their electrical resistivities are much higher than those of a "true" metal, such as aluminum (2.8 μΩ·cm), and even higher than a typical "weak" metal, such as lead (22 μΩ·cm). Generally, however, these two elements are categorized as metals. All three of these borderline elements form covalent compounds almost exclusively.

If we want to decide where to draw the vague border between metals and semimetals, the melting and boiling points are as good an indicator as any. In Group 15, these parameters increase as we descend the group, except for a small decrease from antimony to bismuth (Table 14.2). As noted for the alkali metals, the melting points of metals tend to decrease down a group, whereas those of nonmetals tend to increase down a group (we will encounter the latter behavior most clearly with the halogens). Thus the increase-decrease pattern shown in Table 14.2 indicates that the lighter members of Group 15 follow the typical nonmetal trend and the shift to the metallic decreasing trend starts at bismuth. In fact, it is only antimony and bismuth that have the fairly characteristic long liquid range of metals. Thus we will refer to arsenic as a semimetal and consider antimony and bismuth to be metals, although very "weak" ones.

Anomalous Nature of Nitrogen

In general, the differences between the chemistry of nitrogen and those of the other members of Group 15 relate to the lack of available *d* orbitals on the nitrogen. The major differences are discussed in the following sections.

High Stability of Multiple Bonds

The dinitrogen molecule is a very stable species. Figure 14.1 shows the molecular orbitals involved in the triple bond of that species. The N≡N bond energy is 942 $kJ \cdot mol^{-1}$, far greater than that for the triple phosphorus-phosphorus bond (481 $kJ \cdot mol^{-1}$) and greater even than that for the triple carbon-carbon bond (835 $kJ \cdot mol^{-1}$). The accepted explanation is that the *p* orbitals involved in forming the two π bonds overlap better in the molecule containing the small nitrogen atoms than they do in the other species.

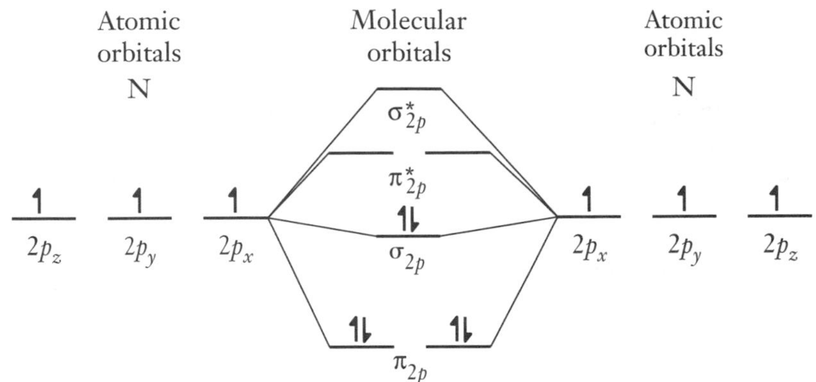

Figure 14.1 Molecular orbital diagram for the $2p$ atomic orbitals of the dinitrogen molecule.

Conversely, nitrogen forms very weak single bonds. The typical nitrogen-nitrogen single bond energy is only 200 kJ·mol^{-1}, considerably lower than the bond energy of 346 kJ·mol^{-1} for a carbon-carbon single bond. The argument is made that as Period 2 is traversed from left to right, the atoms become smaller and smaller. At nitrogen, the atoms become so small that electronic repulsions between the nonbonding electrons "force" the atoms further apart. Thus the nitrogen-nitrogen triple bond is particularly strong, whereas the single bond is comparatively weak. It is this large difference between N≡N and N–N bond strengths (742 kJ·mol^{-1}) that contributes to the preference in nitrogen chemistry for the formation of the dinitrogen molecule in a reaction rather than chains of nitrogen-nitrogen single bonds, as occurs in carbon chemistry. Furthermore, the fact that dinitrogen is a gas means that an entropy factor also favors the formation of the dinitrogen molecule in chemical reactions.

We can see the difference in behavior between nitrogen and carbon by comparing the combustion of hydrazine, N_2H_4, with that of ethene, C_2H_4. The nitrogen compound burns to produce dinitrogen, whereas the carbon compound gives carbon dioxide:

$$N_2H_4(g) + O_2(g) \rightarrow N_2(g) + 2\ H_2O(g)$$

$$C_2H_4(g) + 3\ O_2(g) \rightarrow 2\ CO_2(g) + 2\ H_2O(g)$$

Curiously, in Groups 15 and 16, it is the second members—phosphorus and sulfur—that are prone to catenation.

Lack of Available *d* Orbitals

Nitrogen forms only a trifluoride, NF_3, whereas phosphorus forms two common fluorides, the pentafluoride, PF_5, and the trifluoride, PF_3. To construct a bonding model of phosphorus pentafluoride, we assume that the $3d$ orbitals of the phosphorus are utilized. The nitrogen atom does not have any available d orbitals, and, as a result, it cannot form an analogous compound.

Another example that illustrates the difference in bonding behavior between nitrogen and phosphorus is the pair of compounds, NF_3O and PF_3O. The former contains a weak nitrogen-oxygen bond, whereas the latter

Propellants and Explosives

Propellants and explosives share many common properties. They function by means of a rapid, exothermic reaction that produces a large volume of gas. It is the expulsion of this gas that causes a rocket to be propelled forward (according to Newton's third law of motion), but for the explosive, it is mostly the shock wave from the gas production that causes the damage. There are three factors that make a compound (or a pair of compounds) a potential propellant or explosive:

1. The reaction must be thermodynamically spontaneous and very exothermic, so that a great deal of energy is released in the process.
2. The reaction must be very rapid; in other words, it must be kinetically favorable.
3. The reaction must produce small gaseous molecules, because (according to kinetic theory) small molecules will have high average velocities and hence high momenta.

Although the chemistry of propellants and explosives is a whole science in itself, most of the candidates contain (singly bonded) nitrogen because of the exothermic formation of the dinitrogen molecule. This feature has been of great help in trying to discover terrorist-set explosives in luggage and carry-ons, in that any bags containing abnormally high proportions of nitrogen compounds are suspect.

To illustrate the workings of a propellant, we consider the propellant used in the first rocket-powered aircraft—a mixture of hydrogen peroxide, H_2O_2, and hydrazine, N_2H_4. These combine to give dinitrogen gas and water (as steam):

$$2\ H_2O_2(l) + N_2H_4(l) \rightarrow N_2(g) + 4\ H_2O(g)$$

The bond energies of the reactants are O–H = 460 kJ·mol^{-1}; O–O = 142 kJ·mol^{-1}; N–H = 386 kJ·mol^{-1}; and N–N = 247 kJ·mol^{-1}. Those of the products are N≡N = 942 kJ·mol^{-1} and O–H = 460 kJ·mol^{-1}. Adding the bond energies on each side and finding their difference give the result that 707 kJ·mol^{-1} of heat are released for every 32 g (1 mol) of hydrazine consumed—a very exothermic reaction. And 695 of that 707 kJ·mol^{-1} can be attributed to the conversion of the nitrogen-nitrogen single bond to the nitrogen-nitrogen triple bond.

This mixture clearly satisfies our first criterion for a propellant. Experimentation showed that the reaction is, indeed, very rapid, and it is obvious from the equation and the application of the ideal gas law that very large volumes of gas will be produced from a very small volume of the two liquid reagents. Because these particular reagents are very corrosive and extremely hazardous, safer mixtures have since been devised by using the same criteria of propellant feasibility.

contains a fairly strong phosphorus-oxygen bond. For the nitrogen compound, we assume the oxygen is bonded through a coordinate covalent bond, with the nitrogen donating its lone pair in an sp^3 hybrid orbital to a p orbital of the oxygen atom. The phosphorus bonding is analogous to that of silicon (Chapter 13), so we assume that a second bond is formed with oxygen through the overlap of a full p orbital on the oxygen with an empty d orbital on the phosphorus. Figure 14.2 shows the electron-dot representations for the two compounds; and Figure 14.3 shows an orbital representation of the bonding in the phosphorus compound.

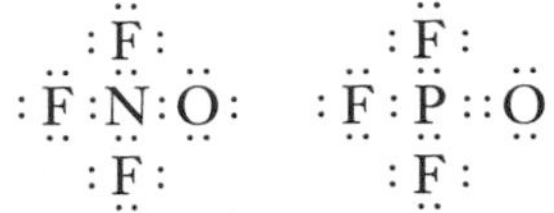

Figure 14.2 Electron-dot representations of the bonding in NF_3O and PF_3O.

Higher Electronegativity

Nitrogen has a much higher electronegativity than the other members of Group 15. As a result, the polarity of the bonds in nitrogen compounds is often the reverse of that in phosphorus and the other heavier members of the group. For example, the different polarities of the N–Cl and P–Cl bonds result in different hydrolysis products of the respective trichlorides:

$$NCl_3(l) + 3\ H_2O(l) \rightarrow NH_3(g) + 3\ HClO(aq)$$

$$PCl_3(l) + 3\ H_2O(l) \rightarrow H_3PO_3(aq) + 3\ HCl(g)$$

Because the nitrogen-hydrogen covalent bond is strongly polar, ammonia is basic, whereas the hydrides of the other Group 15 elements—phosphine, PH_3, arsine, AsH_3, and stibine, SbH_3—are essentially neutral.

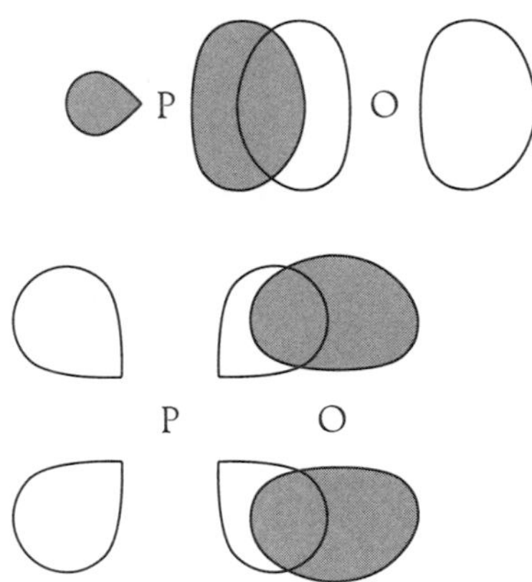

Figure 14.3 Representations of the σ and π bonds between the phosphorus and oxygen atoms in PF_3O.

Nitrogen

The element nitrogen has only one allotrope: the colorless, odorless gas, dinitrogen. Dinitrogen makes up 78 percent of the atmosphere on this planet. Apart from its role in the nitrogen cycle, which we will discuss later, it is very important as an inert diluent for the highly reactive gas in our atmosphere, dioxygen. Without the dinitrogen, every spark in our atmosphere would cause a massive fire. The tragic deaths in 1967 of the astronauts Grissom, White, and Chaffee in an Apollo space capsule were a result of the use of a pure oxygen cabin atmosphere (since discontinued). An accidental electrical spark became a raging inferno within seconds, killing all of the occupants.

Dinitrogen is not very soluble in water, although like most gases, its solubility increases rapidly with increasing pressure. This is a major problem for deep-sea divers. As they dive, additional dinitrogen dissolves in their bloodstream; as they return to the surface, the decreasing pressure brings the dinitrogen out of solution, and it forms tiny bubbles, particularly around the joints. Prevention of this painful and sometimes fatal problem—called the bends—required divers to return to the surface very slowly. In emergency situations, they were placed in decompression chambers, where the pressure was reapplied and then reduced carefully over hours or days. To avoid this hazard, oxygen-helium gas mixtures are now used for deep diving, because helium has a much lower blood solubility than dinitrogen.

Industrially, dinitrogen is prepared by liquefying air and then slowly warming the liquid mixture. The dinitrogen boils at –196°C, leaving behind

The First Dinitrogen Compound

Time and time again, chemists fall into the trap of simplistic thinking. As we have said, dinitrogen is very unreactive, but this does not mean that it is totally unreactive. In Chapter 13 we noted that carbon monoxide could bond to metals. Dinitrogen is isoelectronic with carbon monoxide, although there is the important difference that dinitrogen is nonpolar whereas carbon monoxide is polar. Nevertheless, the isoelectronic concept is useful for predicting the possible formation of a compound.

In early 1964 Caesar Senoff, a Canadian chemistry student at the University of Toronto, was working with compounds of ruthenium. He synthesized a brown compound whose composition he was unable to explain. Time passed, and in May 1965, during a discussion with another chemist, it dawned on him that the only feasible explanation was that the molecule contained the N_2 unit bound to the metal in a manner analogous to the carbon monoxide–metal bond. Excitedly, he told his very skeptical supervisor, Bert Allen. After several months, Allen finally agreed to submit the findings to a journal for publication. The manuscript was rejected—a common occurrence when a discovery contradicts accepted thought. After Allen and Senoff rebutted the criticisms, the journal sent the revised manuscript to 16 other chemists for comment and approval before publishing it. Finally, the article appeared in print, and the world of inorganic chemistry was changed yet again.

Since then, transition metal compounds containing the N_2 unit have become quite well known, and some can be made by simply bubbling dinitrogen gas through the solution of a metal compound. (As a consequence, research chemists no longer use dinitrogen as an inert atmosphere for all their reactions.) Some of the compounds are of interest because they are analogs of compounds soil bacteria produce when they convert dinitrogen to ammonia. None of the compounds, however, has become of great practical significance, although they serve as a reminder to inorganic chemists to never say, "Impossible!"

the dioxygen, b.p. −183°C. On a smaller scale, dinitrogen can be separated from the other atmospheric gases by using a zeolite, as discussed in Chapter 13. In the laboratory, dinitrogen can be prepared by gently warming a solution of ammonium nitrite:

$$NH_4NO_2(aq) \rightarrow N_2(g) + 2\ H_2O(l)$$

Dinitrogen does not burn or support combustion. It is extremely unreactive toward most elements and compounds. Hence it is commonly used to provide an inert atmosphere when highly reactive compounds are being handled or stored. About 60 million tonnes of dinitrogen are used every year, worldwide. A high proportion is used in steel production as an inert atmosphere and in oil refineries to purge the flammable hydrocarbons from the pipes and reactor vessels when they need maintenance. Liquid nitrogen is used as a safe refrigerant where very rapid cooling is required. Finally, a significant proportion is employed in the manufacture of ammonia and other nitrogen-containing compounds.

There are a few chemical reactions in which dinitrogen is a reactant. For example, dinitrogen combines with the most electropositive metals to form an ionic nitride:

$$6\,Li(s) + N_2(g) \rightarrow 2\,Li_3N(s)$$

If a mixture of dinitrogen and dioxygen is sparked, nitrogen dioxide is formed:

$$N_2(g) + 2\,O_2(g) \rightleftharpoons 2\,NO_2(g)$$

On a large scale, this reaction takes place in lightning flashes, where it contributes to the biologically available nitrogen in the biosphere. However, it also occurs under the conditions of high pressure and sparking found in modern high-compression gasoline engines. Local concentrations of nitrogen dioxide may be so high that they become a significant component of urban pollution. The equilibrium position for this reaction actually lies far to the left; or, to express this idea another way, nitrogen dioxide has a positive free energy of formation. Its continued existence depends on its extremely slow decomposition rate. Thus it is kinetically stable. It is one of the roles of the automobile catalytic converter to accelerate the rate of decomposition back to dinitrogen and dioxygen.

Finally, dinitrogen participates in an equilibrium reaction with hydrogen, one that under normal conditions does not occur to any significant extent because of the high activation energy of the reaction (in particular, a single-step reaction cannot occur because it would require a simultaneous four-molecule collision):

$$N_2(g) + 3\,H_2(g) \rightleftharpoons 2\,NH_3(g)$$

We will discuss this reaction in much more detail shortly.

Overview of Nitrogen Chemistry

Nitrogen chemistry is complex. For an overview, consider the oxidation state diagram in Figure 14.4. The first thing we notice is that nitrogen can assume formal oxidation states that range from +5 to −3. Second, because it behaves so differently in acidic and basic conditions, we can conclude that the relative stability of an oxidation state is very dependent on pH. Let us look at some specific features of the chemistry of nitrogen.

1. Molecular dinitrogen is found at a deep minimum on the Frost diagram. Hence it is a thermodynamically very stable species. In acidic solution, ammonium ion, NH_4^+, is slightly lower; thus we might expect that a strong reducing agent would cause dinitrogen to reduce the ammonium ion.

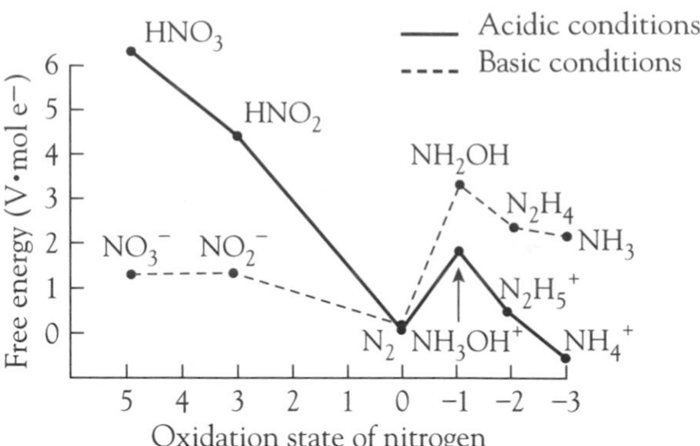

Figure 14.4 Frost diagram for the common nitrogen species in acidic and basic conditions.

However, the diagram does not reveal anything about the kinetics of the process, and it is, in fact, kinetically very slow.

2. Species that are high to the left are strongly oxidizing. Thus nitric acid, HNO_3, is a very strong oxidant, although the nitrate ion, NO_3^-, the conjugate base of nitric acid, is not significantly oxidizing.

3. Species that are high to the right tend to be strong reducing agents. Thus in basic solution, hydroxylamine, NH_2OH, hydrazine, N_2H_4, and ammonia, NH_3, tend to be reducing in their chemical behavior.

4. Both hydroxylamine and its conjugate acid, the hydroxylammonium ion, NH_3OH^+, should readily disproportionate, because they are at convex locations on the diagram. Experimentally, we find that they do disproportionate, but the products are not always those resulting in the greatest decrease in free energy; instead, kinetic factors select the products. Hydroxylamine disproportionates to give dinitrogen and ammonia, whereas the hydroxylammonium ion produces dinitrogen oxide and the ammonium ion:

$$3\ NH_2OH(aq) \rightarrow N_2(g) + NH_3(aq) + 3\ H_2O(l)$$

$$4\ NH_3OH^+(aq) \rightarrow N_2O(g) + 2\ NH_4^+(aq) + 2\ H^+(aq) + 3\ H_2O(l)$$

Ammonia

Ammonia is a colorless, poisonous gas with a very strong characteristic smell. It is the only common gas that is basic. Ammonia dissolves readily in water: At room temperature, over 50 g of ammonia will dissolve in 100 g of water, giving a solution of density 0.880 g·mL^{-1} (known as 880 ammonia). The solution is most accurately called "aqueous ammonia" but is often misleadingly called "ammonium hydroxide." A small proportion does, in fact, react with the water to give ammonium and hydroxide ions:

$$NH_3(aq) + H_2O(l) \rightleftharpoons NH_4^+(aq) + OH^-(aq)$$

This reaction is analogous to the reaction of carbon dioxide with water, and the equilibrium lies to the left. And, like the carbon dioxide and water reaction, evaporating the solution shifts the equilibrium further to the left. Thus there is no such thing as pure "ammonium hydroxide."

Ammonia is prepared in the laboratory by mixing an ammonium salt and a hydroxide, for example, ammonium chloride and calcium hydroxide:

$$2\ NH_4Cl(s) + Ca(OH)_2(s) \xrightarrow{\Delta} CaCl_2(s) + 2\ H_2O(l) + 2\ NH_3(g)$$

It is a reactive gas, burning in air when ignited to give water and nitrogen gas:

$$4\ NH_3(g) + 3\ O_2(g) \rightarrow 2\ N_2(g) + 6\ H_2O(l) \qquad \Delta G^\circ = -1305\ kJ{\cdot}mol^{-1}$$

There is an alternative decomposition route that is thermodynamically less favored, but in the presence of a platinum catalyst, it is kinetically preferred; that is, the (catalyzed) activation energy for this alternative route becomes lower than that for the combustion to nitrogen gas:

$$4\ NH_3(g) + 5\ O_2(g) \xrightarrow{Pt} 4\ NO(g) + 6\ H_2O(l) \qquad \Delta G^\circ = -1132\ kJ{\cdot}mol^{-1}$$

Ammonia acts as a reducing agent in its reactions with chlorine. There are two pathways. With excess ammonia, nitrogen gas is formed and the excess ammonia reacts with the hydrogen chloride gas produced to give clouds of white, solid ammonium chloride:

$$2\ NH_3(g) + 3\ Cl_2(g) \rightarrow N_2(g) + 6\ HCl(g)$$

$$HCl(g) + NH_3(g) \rightarrow NH_4Cl(s)$$

With excess chlorine, a very different reaction occurs. In this case, the product is nitrogen trichloride, a colorless, explosive, oily liquid:

$$NH_3(g) + 3\ Cl_2(g) \rightarrow 3\ HCl(g) + NCl_3(l)$$

As a base, ammonia reacts with acids in solution to give its conjugate acid, the ammonium ion. For example, when ammonia is mixed with sulfuric acid, ammonium sulfate is formed:

$$2\ NH_3(aq) + H_2SO_4(aq) \rightarrow (NH_4)_2SO_4(aq)$$

Ammonia reacts in the gas phase with hydrogen chloride to give a white smoke of solid ammonium chloride:

$$NH_3(g) + HCl(g) \rightarrow NH_4Cl(g)$$

The formation of a white film over glass objects in a chemistry laboratory is usually caused by the reaction of ammonia escaping from reagent bottles with acid vapors, particularly hydrogen chloride.

Ammonia condenses to a liquid at –35°C. This boiling point is much higher than that of phosphine, PH_3 (–134°C), because ammonia molecules form strong hydrogen bonds with their neighbors. Liquid ammonia is a good polar solvent because it autoionizes just as water does to produce the ammonium cation and the amide anion, NH_2^-:

$$2\ NH_3(l) \rightleftharpoons NH_4^+(NH_3) + NH_2^-(NH_3)$$

Recall that the autoionization of water gives

$$2\ H_2O(l) \rightleftharpoons H_3O^+(aq) + OH^-(aq)$$

Thus a whole range of ammonia acid-base chemistry exists in which the ammonium ion is the conjugate acid and the amide ion the conjugate base of ammonia. For example, there is a neutralization reaction,

$$NH_2^-(NH_3) + NH_4^+(NH_3) \rightarrow 2\ NH_3(l)$$

that is analogous to that between the hydroxide ion and the hydronium ion in aqueous solution:

$$OH^-(aq) + H_3O^+(aq) \rightarrow 2\ H_2O(l)$$

With its lone electron pair, ammonia is also a strong Lewis base. One of the "classic" Lewis acid-base reactions involves that between the gaseous electron-deficient boron trifluoride molecule and ammonia to give the white solid compound in which the lone pair on the ammonia is shared with the boron:

$$:NH_3(g) + BF_3(g) \rightarrow F_3B:NH_3(s)$$

Ammonia also acts like a Lewis base when it coordinates to metal ions. For example, it will displace the six water molecules that surround a nickel(II) ion, because it is a stronger Lewis base than water:

$$[Ni(OH_2)_6]^{2+}(aq) + 6\ NH_3(aq) \rightarrow [Ni(NH_3)_6]^{2+}(aq) + 6\ H_2O(l)$$

Nitrogen Fertilizers and the Industrial Synthesis of Ammonia

It has been known for hundreds of years that nitrogen compounds are essential for plant growth. Manure was once the main source of this ingredient for soil enrichment. But the rapidly growing population in Europe during the nineteenth century necessitated a corresponding increase in food production. The solution, at the time, was found in the sodium nitrate (Chile saltpeter) deposits in Chile. This compound was mined in vast quantities and shipped around Cape Horn to Europe. The use of sodium nitrate fertilizer prevented famine in Europe and provided Chile with its main income, turning it into an extremely prosperous nation. However, it was clear that the sodium nitrate deposits would one day be exhausted. Thus chemists rushed to find some method of forming nitrogen compounds from the unlimited resource of unreactive nitrogen gas.

Discovery of the Haber Process

It was Fritz Haber, a German chemist, who showed in 1908 that at about 1000°C, traces of ammonia are formed when nitrogen gas and hydrogen gas are mixed:

$$N_2(g) + 3\ H_2(g) \rightleftharpoons 2\ NH_3(g)$$

A physical chemist, Walther Nernst (of Nernst equation fame), pointed out that a study of the thermodynamics of the system would enable the con-

Haber and Scientific Morality

It has been said that many scientists are amoral because they fail to consider the applications to which their work can be put. The life of Fritz Haber presents a real dilemma: Should we regard him as a hero or as a villain? As discussed earlier, Haber devised the process of ammonia synthesis, which he intended to be used to help feed the world; yet the process was turned into a source of materials to kill millions. He cannot easily be faulted for this, but his other interest is more controversial. Haber argued that it was better to incapacitate the enemy during warfare than to kill them. Thus he worked enthusiastically on poison gas research during the First World War. His wife, Clara Haber, pleaded with him to desist, and when he did not, she committed suicide.

In 1918, Haber was awarded the Nobel prize for his work on ammonia synthesis; but many chemists opposed the award on the basis of his poison gas research. After that war, Haber was a key figure in the rebuilding of Germany's chemical research community. Then in 1933, the National Socialist government took power, and Haber, of Jewish origin himself, was told to fire all of the Jewish workers at his institute. He refused and resigned instead, bravely writing: "For more than 40 years I have selected my collaborators on the basis of their intelligence and their character and not on the basis of their grandmothers, and I am not willing to change this method which I have found so good."

This action infuriated the Nazi leaders, but in view of Haber's international reputation, they did not act against him at that time. In 1934, the year after his death, the German Chemical Society held a memorial service for him. The German government was so angered by this tribute to someone who had stood up against their regime that they threatened arrest of all those chemists who attended. But their threat was hollow. The turnout of so many famous chemists for the event caused the Gestapo to back down.

ditions of maximum yield to be found. In fact, the conversion of dinitrogen and dihydrogen into ammonia is exothermic and results in a decrease in gas volume and a resulting decrease in entropy. To "force" the reaction to the right, the Le Châtelier principle suggests that the maximum yield of ammonia would be at low temperature and high pressure. However, the lower the temperature, the slower the rate at which equilibrium is reached. A catalyst might help, but even then there are limits to the most practical minimum temperature. Furthermore, there are limits to how high the pressure can go, simply in terms of the cost of thick-walled containers and pumping systems.

Haber found that adequate yields could be obtained in reasonable time by using a pressure of 20 MPa (200 atm) and a temperature of 500°C. However, it took five years for a chemical engineer, Carl Bosch, to actually design an industrial-size plant for the chemical company BASF that could work with gases at this pressure and temperature. Unfortunately, completion of the plant coincided with the start of World War I. With Germany blockaded by the Allies, supplies of Chile saltpeter were no longer available; nevertheless, the ammonia produced was used for the synthesis of explosives rather than for crop production. Without the Haber-Bosch process, the German and Austro-Hungarian armies might well have been forced to capitulate earlier than 1918, simply because of a lack of explosives.

The Modern Haber-Bosch Process

To prepare ammonia in the laboratory, we can simply take cylinders of nitrogen gas and hydrogen gas and pass them into a reaction vessel at appropriate conditions of temperature, pressure, and catalyst. But neither dinitrogen nor dihydrogen is a naturally occurring pure reagent. Thus for the industrial chemist, obtaining the reagents inexpensively, on a large scale, with no useless by-products, is a challenge.

The first step is to obtain the dihydrogen gas. This is accomplished by the *steam reforming process*, where a hydrocarbon, such as methane, is mixed with steam at high temperatures (about 750°C) and at high pressures (about 4 MPa). This process is endothermic, so high temperatures would favor product formation on thermodynamic grounds, but high pressure is used for kinetic reasons, to increase the collision frequency (reaction rate). A catalyst, usually nickel, is present for the same reason:

$$CH_4(g) + H_2O(g) \rightarrow CO(g) + 3\ H_2(g)$$

Catalysts are easily inactivated (*poisoned*) by impurities. Thus it is crucial to remove impurities from the reactants (*feedstock*). Sulfur compounds are particularly effective at reacting with the catalyst surface and deactivating it by forming a layer of metal sulfide. Thus, before the methane is used, it is pretreated to convert contaminating sulfur compounds to hydrogen sulfide. The hydrogen sulfide is then removed by passing the impure methane over zinc oxide:

$$ZnO(s) + H_2S(g) \rightarrow ZnS(s) + H_2O(g)$$

Next, air is added to the mixture of carbon monoxide and dihydrogen, which still contains some methane—deliberately. The methane burns to give carbon monoxide; but, by controlling how much methane is present, the amount of dinitrogen left in the deoxygenated area should be that required to achieve the 1:3 stoichiometry of the Haber-Bosch reaction:

$$2\ CH_4(g) + \tfrac{1}{2}\ O_2(g) + 2\ N_2(g) \rightarrow CO(g) + 2\ H_2(g) + 2\ N_2(g)$$

There is no simple way of removing carbon monoxide from the mixture of gases. For this reason, and to produce an additional quantity of hydrogen, the third step involves the oxidation of the carbon monoxide to carbon dioxide by using steam. This *water gas shift process* is performed at fairly low temperatures (350°C) because it is exothermic. Even though a catalyst of iron and chromium oxides is used, the temperature cannot be any lower without reducing the rate of reaction to an unacceptable level:

$$CO(g) + H_2O(g) \rightleftharpoons CO_2(g) + H_2(g)$$

The carbon dioxide can be removed by a number of different methods. Carbon dioxide has a high solubility in water and many other solvents. Alternatively, it can be removed by a chemical process such as the reversible reaction with potassium carbonate:

$$CO_2(g) + K_2CO_3(aq) + H_2O(l) \rightleftharpoons 2\ KHCO_3(aq)$$

The potassium hydrogen carbonate solution is pumped into tanks where it is heated to generate pure carbon dioxide gas and potassium carbonate solution:

$$2\ KHCO_3(aq) \rightleftharpoons CO_2(g) + K_2CO_3(aq) + H_2O(l)$$

The carbon dioxide is liquefied under pressure and sold, and the potassium carbonate is returned to the ammonia processing plant for reuse.

Now that a mixture of the pure reagents of dinitrogen and dihydrogen gas has been obtained, the conditions are appropriate for the simple reaction that gives ammonia:

$$N_2(g) + 3\ H_2(g) \rightarrow 2\ NH_3(g)$$

The practical thermodynamic range of conditions is shown in Figure 14.5. As mentioned earlier, to "force" the reaction to the right, high pressures are used. But the higher the pressure, the thicker the reaction vessels and piping required to prevent an explosion—and the thicker the containers, the higher the cost of construction. Today's ammonia plants utilize pressures between 10 and 100 MPa (100 and 1000 atm). The lower the temperature, the higher the yield but the slower the rate. With current high-performance catalysts, the optimum conditions are 400°C to 500°C. The catalyst is the heart of every ammonia plant. The most common catalyst is specially prepared high surface area iron containing traces of potassium, aluminum, calcium, magnesium, silicon, and oxygen. About 100 tonnes of catalyst are used in a typical reactor vessel, and, provided all potential "poisons" are removed from the incoming gases, the catalyst will have a working life of about 10 years.

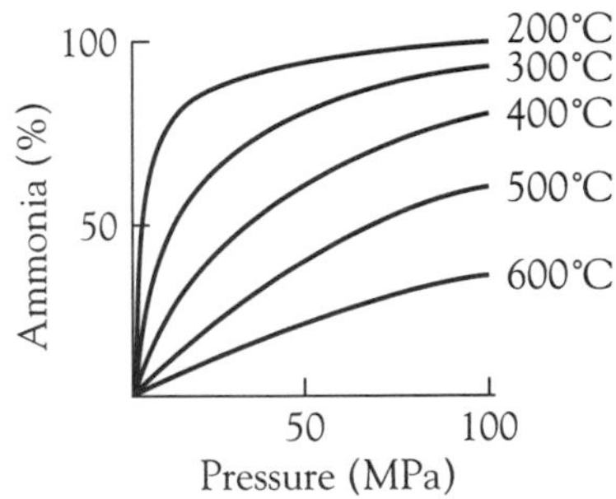

Figure 14.5 Percentage yields of ammonia as a function of pressure, at various temperatures.

After leaving the reactor vessel, the ammonia is condensed. The remaining dinitrogen and dihydrogen are then recycled back through the plant, to be mixed with the fresh incoming gas. A typical ammonia plant produces about 1000 $t \cdot day^{-1}$ (t = tonne) The most crucial concern is to minimize energy consumption. A traditional Haber-Bosch plant consumed about 85 $GJ \cdot t^{-1}$ of ammonia produced, whereas a modern plant, built to facilitate energy recycling, uses only about 30 $GJ \cdot t^{-1}$.

Even today, the most important use of ammonia itself is in the fertilizer industry. The ammonia is often applied to fields as ammonia gas. Ammonium sulfate and ammonium phosphate also are common solid fertilizers. These are simply prepared by passing the ammonia into sulfuric acid and phosphoric acid, respectively:

$$2\ NH_3(g) + H_2SO_4(aq) \rightarrow (NH_4)_2SO_4(aq)$$

$$3\ NH_3(g) + H_3PO_4(aq) \rightarrow (NH_4)_3PO_4(aq)$$

Ammonia is also used in a number of industrial syntheses, particularly that of nitric acid, as we will discuss shortly.

The Ammonium Ion

The colorless ammonium ion is the most common nonmetallic cation used in the chemistry laboratory. This tetrahedral polyatomic ion can be thought of as a pseudo–alkali metal ion, close in size to the potassium ion (see Chapter 9). Not only does it have the same charge, but its salts usually have solubility patterns like those of one of the larger alkali metal ions. Unlike the alkali metal ions, however, the ammonium ion does not always remain intact: It can be hydrolyzed, dissociated, or oxidized.

The ammonium ion is hydrolyzed in water to give its conjugate base, ammonia:

$$NH_4^+(aq) + H_2O(l) \rightleftharpoons H_3O^+(aq) + NH_3(aq)$$

As a result, solutions of ammonium salts of strong acids, such as ammonium chloride, are slightly acidic.

Ammonium salts can volatilize (vaporize) by dissociation. The classic example of this is ammonium chloride:

$$NH_4Cl(s) \rightleftharpoons NH_3(g) + HCl(g)$$

If a sample of ammonium chloride is left open to the atmosphere, it will "disappear." It is this same decomposition reaction that is used in "smelling salts." The pungent ammonia odor, which masks the sharper smell of the hydrogen chloride, has a considerable effect on a semicomatose individual (although it should be noted that the use of smelling salts except by medical personnel is now deemed to be unwise and potentially dangerous).

Finally, the ammonium ion can be oxidized by the anion in the ammonium salt. These are reactions that occur when an ammonium salt is heated, and each one is unique. The three most common examples are the thermal decomposition of ammonium nitrite, ammonium nitrate, and ammonium dichromate. The reaction of ammonium dichromate is often referred to as the "volcano" reaction. A source of heat, such as a lighted match, will cause the orange crystals to decompose, producing sparks and a large volume of dark green chromium(III) oxide. Although this is a very spectacular decomposition reaction, it needs to be performed in a fume hood, because a little ammonium dichromate dust usually is dispersed by the reaction, and this highly carcinogenic material can be absorbed through the lungs:

$$NH_4NO_2(aq) \rightarrow N_2(g) + 2\ H_2O(l)$$

$$NH_4NO_3(s) \rightarrow N_2O(g) + 2\ H_2O(l)$$

$$(NH_4)_2Cr_2O_7(s) \rightarrow N_2(g) + Cr_2O_3(s) + 4\ H_2O(g)$$

Other Hydrides of Nitrogen

Besides ammonia, nitrogen forms three other compounds with hydrogen: hydrazine, N_2H_4; hydroxylamine, NH_2OH; and hydrogen azide, HN_3.

Hydrazine

Hydrazine is a fuming, colorless liquid. It is a weak base, forming two series of salts, in which it is either monoprotonated or diprotonated:

$$N_2H_4(aq) + H_3O^+(aq) \rightleftharpoons N_2H_5^+(aq) + H_2O(l)$$

$$N_2H_5^+(aq) + H_3O^+(aq) \rightleftharpoons N_2H_6^{2+}(aq) + H_2O(l)$$

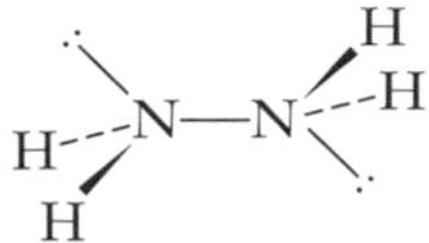

Figure 14.6 The hydrazine molecule.

However, hydrazine is a strong reducing agent, reducing iodine to hydrogen iodide and copper(II) ion to copper metal:

$$N_2H_4(aq) + 2\ I_2(aq) \rightarrow 4\ HI(aq) + N_2(g)$$

$$N_2H_4(aq) + 2\ Cu^{2+}(aq) \rightarrow 2\ Cu(s) + N_2(g) + 4\ H^+(aq)$$

Most of the 20 000 tonnes produced worldwide annually are used as the reducing component of a rocket fuel, usually in the form of dimethyl hydrazine, $(CH_3)_2NNH_2$. Another derivative, dinitrophenylhydrazine, $H_2NNHC_6H_3(NO_2)_2$, is used in organic chemistry to identify carbon compounds containing the C=O grouping. The structure of hydrazine is like that of ethane, except that two ethane hydrogens are replaced by lone pairs of electrons, one pair on each nitrogen atom (Figure 14.6).

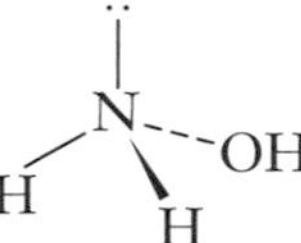

Figure 14.7 The hydroxylamine molecule.

Hydroxylamine

Hydroxylamine is a very poisonous, colorless solid with a low melting point (Figure 14.7). It is thermally unstable and a very weak base. Like hydrazine, hydroxylamine is a strong reducing agent.

Hydrogen Azide

Hydrogen azide, a colorless liquid, is quite different from the other nitrogen hydrides. It is acidic, with a pK_a similar to that of acetic acid:

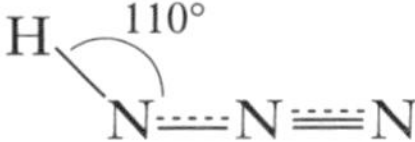

Figure 14.8 The hydrogen azide molecule. The bond orders of the two nitrogen-nitrogen bonds are $1\frac{1}{2}$ and $2\frac{1}{2}$.

$$HN_3(aq) + H_2O(l) \rightleftharpoons H_3O^+(aq) + N_3^-(aq)$$

The compound has a repulsive, irritating odor and is extremely poisonous. Like hydroxylamine, it has a low boiling point. More important, it is highly explosive, producing hydrogen gas and nitrogen gas:

$$2\ HN_3(l) \rightarrow H_2(g) + 3\ N_2(g)$$

The three nitrogen atoms in a hydrogen azide molecule are colinear, with the hydrogen at a 110° angle (see Figure 14.7). The nitrogen-nitrogen bond lengths in hydrogen azide are 124 pm and 113 pm (the end N–N bond is shorter). A typical N=N bond is 120 pm, and the N≡N bond in the dinitrogen molecule is 110 pm. Thus the bond orders in hydrogen azide must be approximately $1\frac{1}{2}$ and $2\frac{1}{2}$, respectively (Figure 14.8). The bonding can be pictured simply as an equal resonance mixture of the two electron-dot structures shown in Figure 14.9, one of which contains two N–N double bonds and the other, an N–N single bond and a triple N–N bond.

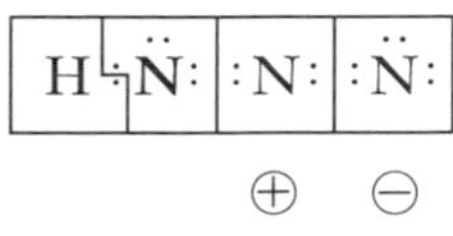

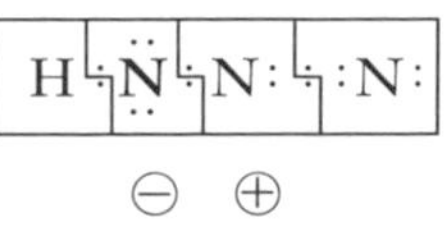

Figure 14.9 The bonding in hydrogen azide can be pictured as a resonance mixture of these two structures.

The azide ion, N_3^-, is isoelectronic with carbon dioxide, and it is presumed to have an identical electronic structure. The nitrogen-nitrogen bonds are of equal length (116 pm), an observation that reinforces the concept that the presence of the hydrogen atom in hydrogen azide causes the neighboring

N–N bond to weaken (and lengthen to 124 pm) and the more distant one to strengthen (and shorten to 113 pm).

Sodium azide is the starting point for most azide chemistry. The most convenient preparation is accomplished by bubbling dinitrogen oxide gas through a solution of sodium metal in liquid ammonia:

$$3\ N_2O(g) + 4\ Na(NH_3) + NH_3(l) \rightarrow NaN_3(s) + 3\ NaOH(NH_3) + 2\ N_2(g)$$

It is interesting how so much of chemistry can be used either destructively or constructively. The azide ion is now used to save lives—in the automobile air bag. It is crucial that an air bag inflate extremely rapidly, before the victim is thrown forward after impact. The only way to produce such a fast response is through a controlled chemical explosion that produces a large volume of gas. For this purpose, sodium azide is preferred: It is about 65 percent nitrogen by mass, can be routinely manufactured to a high purity (at least 99.5 percent), and decomposes cleanly to sodium metal and dinitrogen at 350°C:

$$2\ NaN_3(s) \rightarrow 2\ Na(s) + 3\ N_2(g)$$

In an air bag, this reaction takes place in about 40 ms. Obviously, we would not want the occupants to be saved from a crash and then have to face molten sodium metal. Therefore, iron(III) oxide is included in the reaction container, producing iron and sodium oxide through a single replacement reaction:

$$6\ Na(l) + Fe_2O_3(s) \rightarrow 3\ Na_2O(s) + 2\ Fe(s)$$

Lead(II) azide is important as a detonator: It is a fairly safe compound unless it is impacted, in which case it explosively decomposes. The shock wave produced is usually sufficient to detonate a more stable explosive such as dynamite:

$$Pb(N_3)_2(s) \rightarrow Pb(s) + 3\ N_2(g)$$

Nitrogen Oxides

Nitrogen forms a plethora of common oxides: dinitrogen oxide, N_2O; nitrogen monoxide, NO; dinitrogen trioxide, N_2O_3; nitrogen dioxide, NO_2; dinitrogen tetroxide, N_2O_4; and dinitrogen pentoxide, N_2O_5. Each of the oxides is actually thermodynamically unstable with respect to decomposition to its elements, but all are kinetically stabilized.

Dinitrogen Oxide

The sweet smelling, gaseous dinitrogen oxide is also known as nitrous oxide or, more commonly, laughing gas. This name results from the intoxicating effect of low concentrations. It is sometimes used as an anesthetic, although the high concentrations needed to cause unconsciousness make it unsuitable for more than brief operations such as tooth extraction. Anesthetists have been known to become addicted to the narcotic gas. Because the gas is very soluble in fats, tasteless, and nontoxic, its major use is as a propellant in pressurized cans of whipped cream.

Dinitrogen oxide is a fairly unreactive, neutral gas, although it is the only common gas other than oxygen to support combustion. For example, magnesium burns in dinitrogen oxide to give magnesium oxide and nitrogen gas:

$$N_2O(g) + Mg(s) \rightarrow MgO(s) + N_2(g)$$

The standard method of preparation of dinitrogen oxide involves the thermal decomposition of ammonium nitrate. This reaction can be accomplished by heating the molten solid to about 280°C. An explosion can ensue from strong heating, however, so a safer route is to heat an ammonium nitrate solution that has been acidified with hydrochloric acid:

$$NH_4NO_3(aq) \xrightarrow{H^+} N_2O(g) + 2\ H_2O(l)$$

Dinitrogen oxide is isoelectronic with carbon dioxide and the azide ion. However, contrary to what one might expect, the atoms are arranged asymmetrically, with an N–N bond length of 113 pm and an N–O bond length of 119 pm. These values indicate a nitrogen-nitrogen bond order of close to $2\frac{1}{2}$ and a nitrogen-oxygen bond order close to $1\frac{1}{2}$ (Figure 14.10).

Like hydrogen azide, dinitrogen oxide can be pictured simply as a molecule that resonates between two electron-dot structures, one of which contains an N–O double bond and an N–N double bond and the other, an N–O single bond and an N–N triple bond (Figure 14.11).

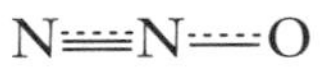

N≡N—O

Figure 14.10 The dinitrogen oxide molecule. The N–N bond order is about $2\frac{1}{2}$, and the N–O bond order is about $1\frac{1}{2}$.

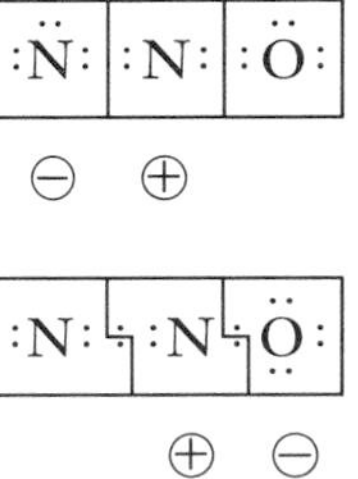

Figure 14.11 The bonding in dinitrogen oxide can be pictured as a resonance mixture of these two structures.

Nitrogen Monoxide

One of the most curious simple molecules is nitrogen monoxide, also called nitric oxide. It is a colorless, neutral, paramagnetic gas. Its molecular orbital diagram resembles that of carbon monoxide, but with one additional electron that occupies an antibonding orbital (Figure 14.12). Hence the predicted net bond order is $2\frac{1}{2}$.

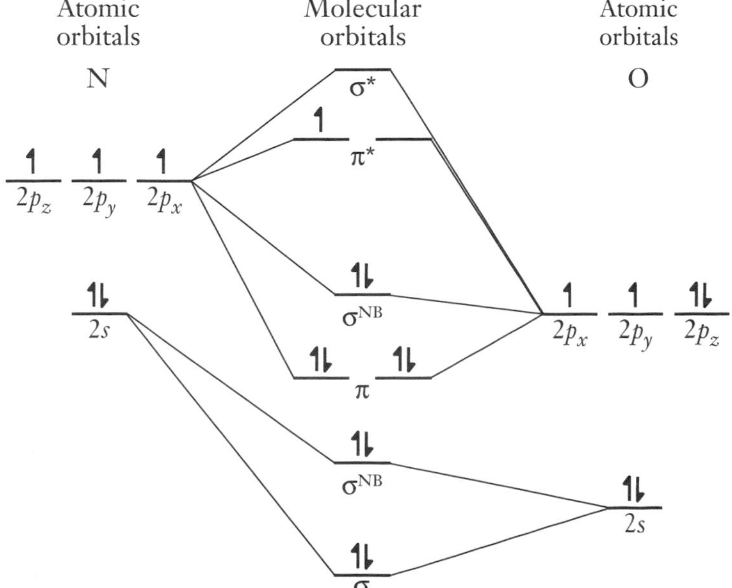

Figure 14.12 Molecular orbital diagram for the *2p* atomic orbitals of the nitrogen monoxide molecule.

Chemists expect molecules containing unpaired electrons to be very reactive. Yet nitrogen monoxide in a sealed container is quite stable. Only when it is cooled to form the colorless liquid or solid does it show a tendency to form a dimer, N_2O_2, in which the two nitrogen atoms are joined by a single bond.

Consistent with the molecular orbital representation, nitrogen monoxide readily loses its electron from the antibonding orbital to form the nitrosyl ion, NO^+, which is diamagnetic and has a shorter N–O bond length (106 pm) than that of the parent molecule (115 pm). This triple-bonded ion is isoelectronic with carbon monoxide, and it forms many analogous metal complexes.

Nitrogen monoxide does show a high reactivity toward dioxygen; and once a sample of colorless nitrogen monoxide is opened to the air, brown clouds of nitrogen dioxide form:

$$2\,NO(g) + O_2(g) \rightleftharpoons 2\,NO_2(g)$$

The molecule is an atmospheric pollutant, commonly formed as a side reaction in high-compression internal combustion engines when dinitrogen and dioxygen are compressed and sparked:

$$N_2(g) + O_2(g) \rightleftharpoons 2\,NO(g)$$

The easiest method for preparing the gas in the laboratory involves the reaction between copper and 50 percent nitric acid:

$$3\,Cu(s) + 8\,HNO_3(aq) \rightarrow 3\,Cu(NO_3)_2(aq) + 4\,H_2O(l) + 2\,NO(g)$$

However, the product is always contaminated by nitrogen dioxide. This contaminant can be removed by bubbling the gas through water, because the nitrogen dioxide reacts rapidly with water.

Until recently, a discussion of simple nitrogen monoxide chemistry would have ended here. Now we realize that this little molecule plays a vital role in our bodies and those of all mammals. In fact, the prestigious journal *Science* called it the 1992 Molecule of the Year. It has been known since 1867 that organic nitro compounds, such as nitroglycerine, can relieve angina, lower blood pressure, and relax smooth muscle tissue. Yet it was not until 1987 that Salvador Moncada and his team of scientists at the Wellcome Research Laboratories in Britain identified the crucial factor in blood vessel dilation as nitrogen monoxide gas. That is, organic nitro compounds were broken down to produce this gas in the organs.

Since this initial work, we have come to realize that nitrogen monoxide is crucial in controlling blood pressure. There is even an enzyme (nitric oxide synthase) whose sole task is the production of nitrogen monoxide. At this point, a tremendous quantity of biochemical research is concerned with the role of this molecule in the body. A lack of nitrogen monoxide is implicated as a cause of high blood pressure, whereas septic shock, a leading cause of death in intensive care wards, is ascribed to an excess of nitrogen monoxide. The gas appears to have a function in memory and in the stomach. Male erections have been proved to depend on production of nitrogen monoxide, and there are claims of important roles for nitrogen monoxide in female uterine contractions. One question still to be answered concerns the life span of these molecules, considering the ease with which they react with oxygen gas.

Dinitrogen Trioxide

Dinitrogen trioxide, the least stable of the common oxides of nitrogen, is a dark blue liquid that decomposes above –30°C. It is prepared by cooling a stoichiometric mixture of nitrogen monoxide and nitrogen dioxide:

$$NO(g) + NO_2(g) \rightleftharpoons N_2O_3(l)$$

Dinitrogen trioxide is the first of the acidic oxides of nitrogen. In fact, it is the acid anhydride of nitrous acid. Thus when dinitrogen trioxide is mixed with water, nitrous acid is formed; but when it is mixed with hydroxide ion, the nitrite ion is produced:

$$N_2O_3(l) + H_2O(l) \rightarrow 2\ HNO_2(aq)$$

$$N_2O_3(l) + 2\ OH^-(aq) \rightarrow 2\ NO_2^-(aq) + H_2O(l)$$

Although, simplistically, dinitrogen trioxide can be considered to contain two nitrogen atoms in the +3 oxidation state, the structure is asymmetric (Figure 14.13), an arrangement that shows it to be a simple combination of the two molecules with unpaired electrons from which it is prepared (nitrogen monoxide and nitrogen dioxide). In fact, the nitrogen-nitrogen bond length in dinitrogen trioxide is abnormally long (186 pm) relative to the length of the single bond in hydrazine (145 pm).

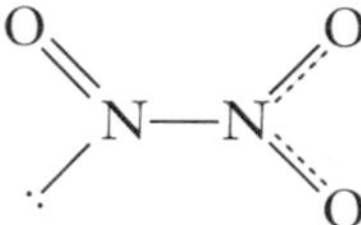

Figure 14.13 The dinitrogen trioxide molecule.

Bond length data indicate that the single oxygen is bonded to the nitrogen with a double bond, whereas the other two oxygen-nitrogen bonds each have a bond order of about $1\frac{1}{2}$. This value is the average of the single and double bond forms that can be constructed with electron-dot formulas.

Nitrogen Dioxide and Dinitrogen Tetroxide

These two toxic oxides coexist in a state of dynamic equilibrium. Low temperatures favor the formation of the colorless dinitrogen tetroxide, whereas high temperatures favor the formation of the dark red-brown nitrogen dioxide:

$$\underset{\text{colorless}}{N_2O_4(g)} \rightleftharpoons \underset{\text{red-brown}}{2\ NO_2(g)}$$

At the boiling point of 21°C, the mixture contains 16 percent nitrogen dioxide; but the proportion of nitrogen dioxide rises to 99 percent at 135°C.

Nitrogen dioxide is prepared by reacting copper metal with concentrated nitric acid:

$$Cu(s) + 4\ HNO_3(l) \rightarrow Cu(NO_3)_2(aq) + 2\ H_2O(l) + 2\ NO_2(g)$$

It is also formed by heating heavy metal nitrates, a reaction that produces a mixture of nitrogen dioxide and oxygen gases:

$$Cu(NO_3)_2(s) \xrightarrow{\Delta} CuO(s) + 2\ NO_2(g) + \tfrac{1}{2}\ O_2(g)$$

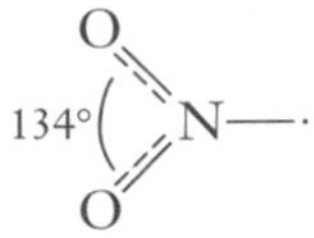

And, of course, it is formed when nitrogen monoxide reacts with dioxygen:

$$2\ NO(g) + O_2(g) \rightarrow 2\ NO_2(g)$$

Figure 14.14 The nitrogen dioxide molecule.

Nitrogen dioxide is an acid oxide, dissolving in water to give nitric acid and nitrous acid:

$$2\ NO_2(g) + H_2O(l) \rightleftharpoons HNO_3(aq) + HNO_2(aq)$$

This potent mixture of corrosive, oxidizing acids is produced when nitrogen dioxide, formed from automobile pollution, reacts with rain. It is a major damaging component of urban precipitation.

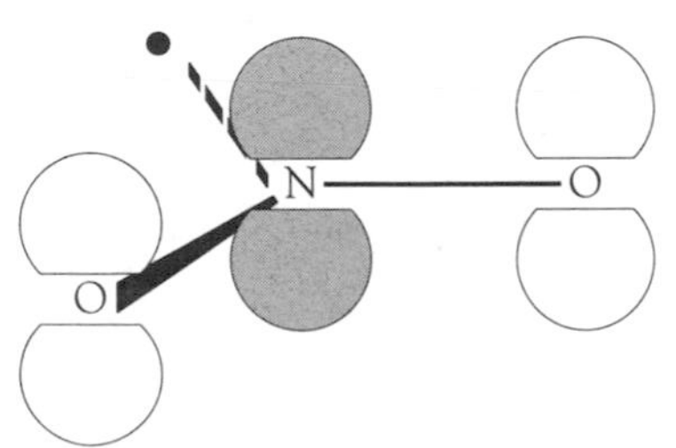

Figure 14.15 Overlap of the *p* orbitals at right angles to the molecular plane of nitrogen dioxide.

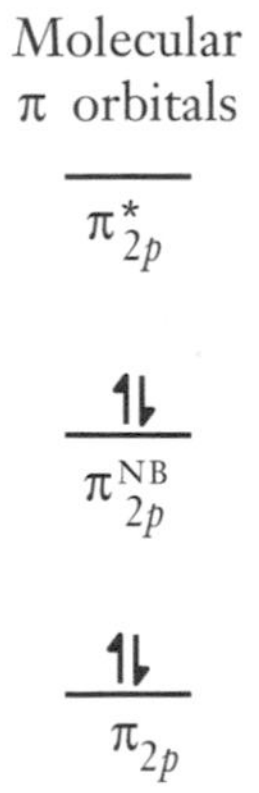

Nitrogen dioxide is a V-shaped molecule with an O–N–O angle of 134°, an angle slightly larger than the true trigonal planar angle of 120°. Because the third bonding site is occupied by a single electron rather than a lone pair, it is not unreasonable for the bonding angle to be opened up (Figure 14.14). The oxygen-nitrogen bond length indicates a $1\frac{1}{2}$ bond order, like that in the NO_2 half of dinitrogen trioxide.

It is useful to compare the π bonding in nitrogen dioxide with that in carbon dioxide. The linear structure of carbon dioxide allows both sets of *p* orbitals that are at right angles to the bonding direction to overlap and participate in π bonding. In the bent nitrogen dioxide molecule, the *p* orbitals are still at right angles to the bonding direction, but, in the plane of the molecule, they are skewed with respect to one another and cannot overlap to form a π system. As a result, the only π bond that can form is at right angles to the plane of the molecule (Figure 14.15). Combining this set of three *p* orbitals results in the formation of three π molecular orbitals: one bonding, one nonbonding, and one antibonding (Figure 14.16). Of the four electrons available, two electrons enter the bonding molecular orbital and two, the nonbonding orbital. This arrangement gives a net single π bond. However, this single π bond is shared between two bonded pairs; hence each pair has one-half a π bond.

Figure 14.16 Molecular orbital diagram of the orbitals involved in π bond formation in nitrogen dioxide.

The "odd" electron is believed to occupy a weakly antibonding σ molecular orbital derived from the nitrogen *p* orbital in the molecular plane. (In hybridization terms, the single electron occupies an sp^2-type orbital on the nitrogen atom, and the other two hybrid orbitals form σ bonds with the oxygen atoms.)

The O–N–O bond angle in the dinitrogen tetroxide molecule is almost identical to that in nitrogen dioxide (Figure 14.17). But dinitrogen tetroxide has an abnormally long (and hence weak) nitrogen-nitrogen bond, although at 175 pm, it is not as weak as the N–N bond in dinitrogen trioxide. The N–N bond is formed by the combination of the weakly antibonding σ orbitals of the two NO_2 units (overlap of the sp^2 hybrid orbitals containing the "odd" electrons, in hybridization terminology). The resulting N–N bonding molecular orbital will have correspondingly weak bonding character. In fact, the N–N bond energy is only about 60 kJ·mol^{-1}.

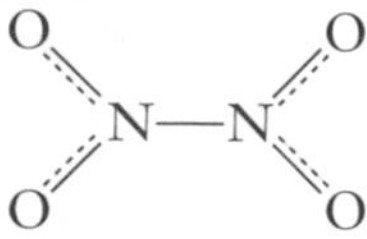

Figure 14.17 The dinitrogen tetroxide molecule.

Dinitrogen Pentoxide

This colorless, solid, deliquescent oxide is the most strongly oxidizing of the nitrogen oxides. It is also strongly acidic, reacting with water to form nitric acid:

$$N_2O_5(s) + H_2O(l) \rightarrow 2\ HNO_3(aq)$$

In the liquid and gas phases, the molecule has a structure related to those of the other dinitrogen oxides, N_2O_3, and N_2O_4, except that an oxygen atom links the two NO_2 units (Figure 14.18). Once again, the two pairs of *p* electrons provide a half π bond to each oxygen-nitrogen pair. Of more interest, however, is the bonding in the solid phase. We have already seen that compounds of metals and nonmetals can be covalently bonded. Here we have a case of a compound of two nonmetals that contains ions! In fact, the crystal structure consists of alternating nitryl cations, NO_2^+, and nitrate anions, NO_3^- (Figure 14.19).

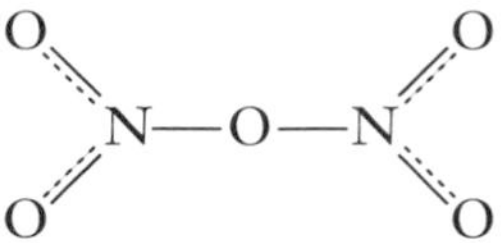

Figure 14.18 The dinitrogen pentoxide molecule.

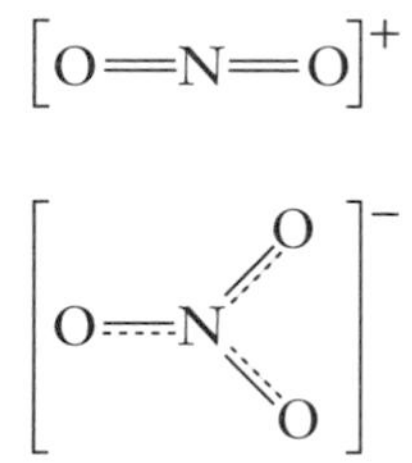

Figure 14.19 The nitryl cation and nitrate anion present in solid-phase dinitrogen pentoxide.

Nitrogen Halides

Nitrogen trichloride is a typical covalent chloride. It is a yellow, oily liquid that reacts with water to form ammonia and hypochlorous acid:

$$NCl_3(aq) + 3\ H_2O(l) \rightarrow NH_3(g) + 3\ HClO(aq)$$

The compound is highly explosive when pure, because it has a positive free energy of formation. However, nitrogen trichloride vapor is used quite extensively (and safely) to bleach flour.

By contrast, nitrogen trifluoride is a thermodynamically stable, colorless, odorless gas of low chemical reactivity. For example, it does not react with water at all. Such stability and low reactivity are quite common among covalent fluorides. In spite of having a lone pair like ammonia (Figure 14.20), it is a weak Lewis base. The F–N–F bond angle in nitrogen trifluoride (102°) is significantly less than the tetrahedral angle. One explanation for the weak Lewis base behavior and the decrease in bond angle from $109\frac{1}{2}°$ is that the nitrogen-fluorine bond has predominantly *p* orbital character (for which 90° would be the optimum angle), and the lone pair is in the nitrogen *s* orbital rather than in a more directional sp^3 hybrid.

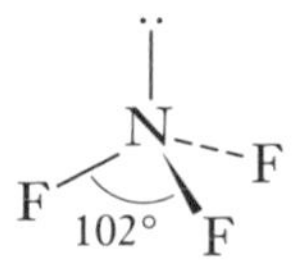

Figure 14.20 The nitrogen trifluoride molecule.

There is one unusual reaction in which nitrogen trifluoride does act as a Lewis base: It forms the stable compound nitrogen oxide trifluoride, NF_3O, when an electric discharge provides the energy for its reaction with oxygen gas at very low temperature:

$$2\ NF_3(g) + O_2(g) \rightarrow 2\ NF_3O(g)$$

Nitrogen oxide trifluoride is often used as the classic example of a compound with a coordinate covalent bond between the nitrogen and oxygen atoms.

Nitrous Acid

Nitrous acid is a weak acid that is unstable, except in solution. It can be prepared by mixing a metal nitrite and a solution of a dilute acid at 0°C in a double replacement reaction. Barium nitrite and sulfuric acid give a pure solution of nitrous acid, because the barium sulfate that is formed has a very low solubility:

$$Ba(NO_2)_2(aq) + H_2SO_4(aq) \rightarrow 2\ HNO_2(aq) + BaSO_4(s)$$

The shape of the nitrous acid molecule is shown in Figure 14.21.

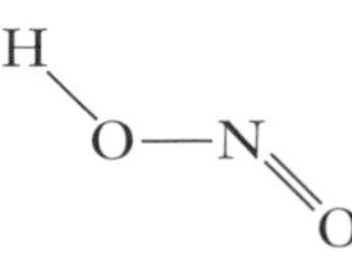

Figure 14.21 The nitrous acid molecule.

Even at room temperature, disproportionation of aqueous nitrous acid occurs to give nitric acid and bubbles of nitrogen monoxide. The latter reacts rapidly with the oxygen gas in the air to produce brown fumes of nitrogen dioxide:

$$3\ HNO_2(aq) \rightarrow HNO_3(aq) + 2\ NO(g) + H_2O(l)$$

$$2\ NO(g) + O_2(g) \rightarrow 2\ NO_2(g)$$

Nitrous acid is used as a reagent in organic chemistry; for example, diazonium salts are produced when nitrous acid is mixed with an organic amine (in this case, aniline, $C_6H_5NH_2$):

$$C_6H_5NH_2(aq) + HNO_2(aq) + HCl(aq) \rightarrow C_6H_5N_2^+Cl^-(s) + 2\ H_2O(l)$$

The diazonium salts are used, in turn, to synthesize a wide range of organic compounds.

Nitric Acid

A colorless, oily liquid when pure, nitric acid is extremely hazardous. It is obviously dangerous as an acid, but, as can be seen from the Frost diagram (Figure 14.4), it is a very strong oxidizing agent, making it a potential danger in the presence of any oxidizable material. The acid, which melts at −42°C and boils at +83°C, is usually slightly yellow as a result of a light-induced decomposition reaction:

$$4\ HNO_3(aq) \rightarrow 4\ NO_2(g) + O_2(g) + 2\ H_2O(l)$$

When pure, liquid nitric acid is almost completely nonconducting. A small proportion ionizes as follows (all species exist in nitric acid solvent):

$$2\ HNO_3(l) \rightleftharpoons H_2NO_3^+ + NO_3^-$$

$$H_2NO_3^+ \rightleftharpoons H_2O + NO_2^+$$

$$H_2O + HNO_3 \rightleftharpoons H_3O^+ + NO_3^-$$

giving an overall reaction of

$$3\ HNO_3 \rightleftharpoons NO_2^+ + H_3O^+ + 2\ NO_3^-$$

The nitryl cation is important in the nitration of organic molecules, for example, the conversion of benzene, C_6H_6, to nitrobenzene, $C_6H_5NO_2$, an important step in numerous organic industrial processes.

Concentrated nitric acid is actually a 68 percent solution in water, whereas "fuming nitric acid," an extremely powerful oxidant, is a red solution of nitrogen dioxide in pure nitric acid. Even when dilute, it is such a strong oxidizing agent that the acid rarely evolves hydrogen when mixed with metals; instead, a mixture of nitrogen oxides is produced and the metal is oxidized to its cation.

The terminal O–N bonds are much shorter (121 pm) than the O–N bond attached to the hydrogen atom (141 pm). This bond length indicates multiple bonding between the nitrogen and the two terminal oxygen atoms.

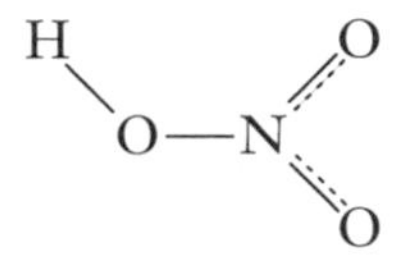

In addition to the electrons in the σ system, there are four electrons involved in the O–N–O π system, two in a bonding orbital and two in a nonbonding orbital, a system giving a bond order of $1\frac{1}{2}$ for each of those nitrogen-oxygen bonds (Figure 14.22).

Figure 14.22 The nitric acid molecule.

Much of the ammonia produced industrially is used in nitric acid synthesis. The process is performed in three steps. First, a mixture of ammonia and dioxygen (or air) is passed through a platinum metal gauze. This is a very efficient, highly exothermic process that causes the gauze to glow red-hot. The step is performed at low pressures to take advantage of the entropy effect, that is, the formation of 10 gas moles from 9 gas moles (an application of the Le Châtelier principle) to shift the equilibrium to the right:

$$4\ NH_3(g) + 5\ O_2(g) \rightarrow 4\ NO(g) + 6\ H_2O(g)$$

Additional oxygen is added to oxidize the nitrogen monoxide to nitrogen dioxide. To improve the yield of this exothermic reaction, heat is removed from the gases, and the mixture is placed under pressure:

$$2\ NO(g) + O_2(g) \rightarrow 2\ NO_2(g)$$

Finally, the nitrogen dioxide is mixed with water to give a solution of nitric acid:

$$3\ NO_2(g) + H_2O(l) \rightarrow 2\ HNO_3(l) + NO(g)$$

This reaction also is exothermic. Again, cooling and high pressures are used to maximize yield. The nitrogen monoxide is returned to the second stage for reoxidation.

Pollution used to be a major problem for nitric acid plants. The older plants were quite identifiable by the plume of yellow-brown gas—escaping nitrogen dioxide. State-of-the-art plants have little trouble in meeting the current emission standards of less than 200 ppm nitrogen oxides in their flue gases. Older plants now mix stoichiometric quantities of ammonia into the nitrogen oxides, a mixture producing harmless dinitrogen and water vapor:

$$NO(g) + NO_2(g) + 2\ NH_3(g) \rightarrow 2\ N_2(g) + 3\ H_2O(g)$$

Worldwide, about 80 percent of the nitric acid is used in fertilizer production. This proportion is only about 65 percent in the United States, because about 20 percent is required for explosives production.

Nitrites

The colorless nitrite ion reacts with acids to form nitrous acid. At room temperature, the nitrous acid rapidly decomposes, as noted earlier in this chapter:

$$NO_2^-(aq) + H^+(aq) \rightarrow HNO_2(aq)$$

$$3\ HNO_2(aq) \rightarrow HNO_3(aq) + 2\ NO(g) + H_2O(l)$$

$$2\ NO(g) + O_2(g) \rightarrow 2\ NO_2(g)$$

The nitrite ion is a weak oxidizing agent; hence nitrites of metals in their lower oxidation states cannot be prepared. For example, nitrite will oxidize iron(II) ion to iron(III) ion and is simultaneously reduced to lower oxides of nitrogen.

Sodium nitrite is a commonly used meat preservative, particularly in cured meats such as ham, hot dogs, sausages, and bacon. The nitrite ion inhibits the growth of bacteria, particularly *Clostridium botulinum*, an organism that produces the deadly botulism toxin. Sodium nitrite is also used to treat packages of red meat, such as beef. Blood exposed to the air rapidly produces a brown color, but shoppers much prefer their meat purchases to look bright red. Thus the meat is treated with sodium nitrite; the nitrite ion is reduced to nitrogen monoxide, which then reacts with the hemoglobin to form a very stable bright red compound. It is true that the nitrite will prevent bacterial growth in this circumstance as well, but these days, the meat is kept at temperatures low enough to inhibit bacteria. To persuade shoppers to prefer brownish rather than red meat will require a lot of reeducation. Now that all meats are treated with sodium nitrite, there is concern that the cooking process will cause the nitrite ion to react with amines in the meat to produce nitrosamines, compounds containing the –NNO functional group. These compounds are known to be carcinogenic. However, as long as preserved meats are consumed in moderation, it is generally believed that the cancer risk is minimal.

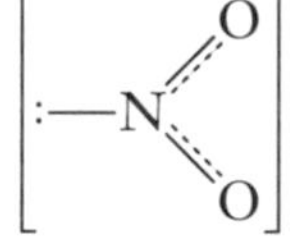

Figure 14.23 The nitrite ion.

The nitrite ion is V-shaped, as a result of the lone pair on the central nitrogen (Figure 14.23). The N–O bond length is 124 pm, longer than that in nitrogen dioxide (120 pm), but still much shorter than the N–O single bond (143 pm). As mentioned earlier, the lone electron in nitrogen dioxide is believed to occupy a weakly antibonding orbital. With analogous orbitals, the additional electron in the nitrite ion should also enter the antibonding orbital, thus leading to a bond weaker than that in nitrogen dioxide.

Nitrates

Nitrates of almost every metal in their common oxidation states are known; and, of particular note, all are water-soluble. For this reason, nitrates tend to be used whenever a solution of a cation is required. Although nitric acid is strongly oxidizing, the colorless nitrate ion is not, under normal conditions (Figure 14.4). Hence one can obtain nitrates of metals in their lower oxidation states, such as iron(II).

The most important nitrate is ammonium nitrate; in fact, this one chemical accounts for the major use of nitric acid. About 1.5×10^7 tonnes are produced annually, worldwide. It is prepared simply by the reaction of ammonia with nitric acid:

$$NH_3(g) + HNO_3(aq) \rightarrow NH_4NO_3(aq)$$

Ammonium nitrate is a convenient and concentrated source of nitrogen fertilizer, although it has to be handled with care. At low temperatures, it decomposes to dinitrogen oxide; but at higher temperatures, explosive decomposition to dinitrogen, dioxygen, and water vapor occurs:

$$2\ NH_4NO_3(s) \rightarrow 2\ N_2(g) + O_2(g) + 4\ H_2O(g)$$

The hazards of this particular reaction were discussed in Chapter 9.

Other nitrates decompose by different routes when heated. Sodium nitrate melts; and then, when strongly heated, bubbles of oxygen gas are produced, leaving behind sodium nitrite:

$$2\ NaNO_3(l) \rightarrow 2\ NaNO_2(s) + O_2(g)$$

Most other metal nitrates give the metal oxide, nitrogen dioxide, and oxygen. For example, heating blue crystals of copper(II) nitrate heptahydrate first yields a green liquid, as the water of hydration is released and dissolves the copper(II) nitrate itself. Continued heating boils off the water, and the green solid then starts to release dioxygen and the brown fumes of nitrogen dioxide and leaves the black residue of copper(II) oxide:

$$2\ Cu(NO_3)_2(s) \rightarrow 2\ CuO(s) + 4\ NO_2(g) + O_2(g)$$

Both nitrates and nitrites can be reduced to ammonia in basic solution with zinc or Devarda's alloy (a combination of aluminum, zinc, and copper). This reaction is a common test for nitrates and nitrites, because there is no characteristic precipitation test for this ion. The ammonia is usually detected by odor or with damp red litmus paper (which will turn blue):

$$3\ NO_3^-(aq) + 8\ Al(s) + 5\ OH^-(aq) + 2\ H_2O(l) \rightarrow 3\ NH_3(g) + 8\ AlO_2^-(aq)$$

The "brown ring test" for nitrate involves the reduction of nitrate with iron(II) in very acidic solution, followed by replacement of one of the coordinated water molecules of the iron(II) to give a brown complex ion:

$$4\ H^+(aq) + 3\ Fe^{2+}(aq) + NO_3^-(aq) \rightarrow 3\ Fe^{3+}(aq) + NO(g) + 2\ H_2O(l)$$

$$[Fe(OH_2)_6]^{2+}(aq) + NO(g) \rightarrow [Fe(OH_2)_5NO]^{2+}(aq) + H_2O(l)$$

The nitrate ion is trigonal planar and has short nitrogen-oxygen bonds (122 pm)—bonds slightly shorter than those in the nitrite ion. Figure 14.24 shows an electron-dot diagram of one of the three possible resonance structures. If each contributes equally, the bond order will be $1\frac{1}{3}$. The partial bonds of this ion are shown in Figure 14.25.

In molecular orbital terms, we envisage a *p* electron system involving one *p* orbital of the nitrogen atom and one *p* orbital of each of the three oxygen atoms. Molecular orbital calculations show that combining the atomic orbitals of four planar atoms gives four π molecular orbitals: one bonding, two nonbonding, and one antibonding. Six electrons are available; thus two electrons enter the bonding molecular orbital and four, the nonbonding orbital. This arrangement gives a net single π bond (Figure 14.26). However, this π system is shared over three bonds; hence each bond possesses one-third of a π bond—the same result as that derived from the simple resonance structures.

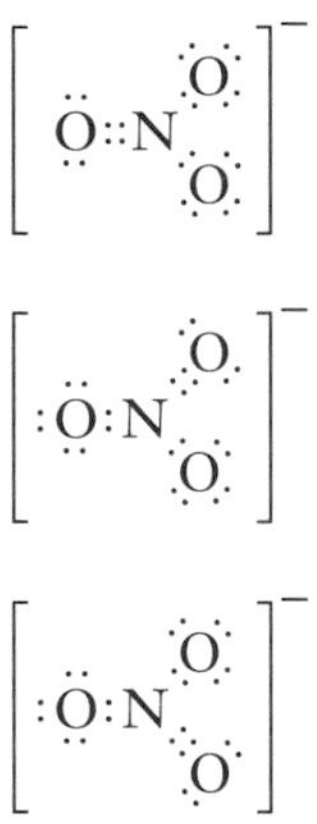

Figure 14.24 The three resonance forms of the nitrate ion.

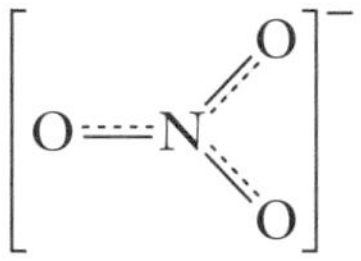

Figure 14.25 The nitrate ion.

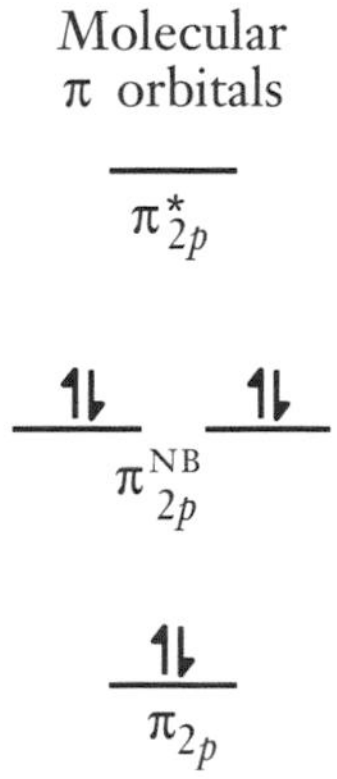

Molecular π orbitals

π^*_{2p}

π^{NB}_{2p}

π_{2p}

Figure 14.26 The molecular orbitals involved in the π bond in the nitrate ion.

Overview of Phosphorus Chemistry

Although they are neighbors in the periodic table, the redox behavior of nitrogen and phosphorus could not be more different (Figure 14.27). Whereas the higher oxidation states of nitrogen are strongly oxidizing in acidic solution, those of phosphorus are quite stable. In fact, the highest

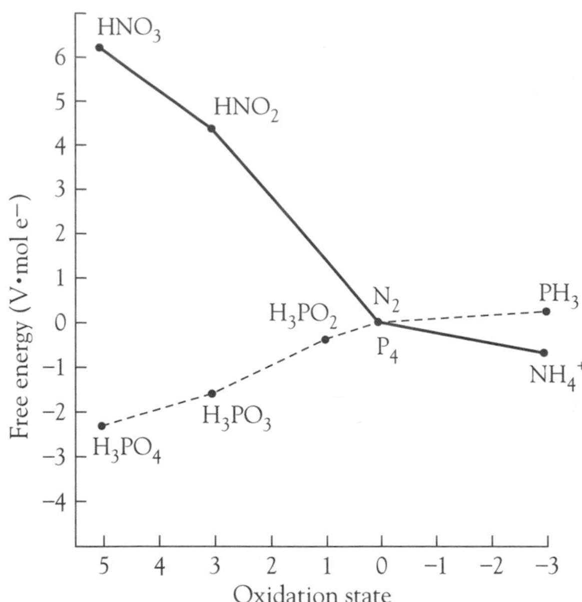

Figure 14.27 Frost diagram comparing the stability of the oxidation states of phosphorus and nitrogen in acidic solution.

oxidation state of phosphorus is the most thermodynamically stable and the lowest oxidation state the least stable—the converse of nitrogen chemistry.

Allotropes of Phosphorus

Phosphorus has several allotropes. The most common is white phosphorus (sometimes called yellow phosphorus); the other common one is red phosphorus. White phosphorus is a very poisonous, white, waxy-looking substance. It is a tetratomic molecule with the phosphorus atoms at the corners of a tetrahedron (Figure 14.28). White phosphorus is an extremely reactive substance, possibly because of its highly strained bond structure. It burns vigorously in air to give tetraphosphorus decaoxide:

$$P_4(s) + 5\ O_2(g) \rightarrow P_4O_{10}(s)$$

The oxide is formed in an electronically excited state; and as the electrons fall to the lowest energy state, visible light is released. In fact, the name *phosphorus* is derived from the phosphorescent glow when white phosphorus is exposed to air in the dark.

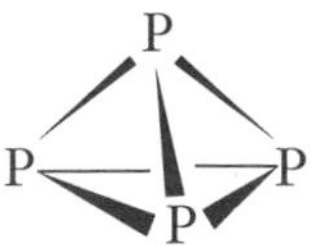

Figure 14.28 The white phosphorus molecule.

Because it is so reactive toward oxygen, white phosphorus has to be stored under water. This allotrope, having only weak dispersion forces between neighboring molecules, melts at 44°C. Although insoluble in hydrogen bonding solvents such as water, it is extremely soluble in nonpolar organic solvents such as carbon disulfide.

Even though white phosphorus is formed when liquid phosphorus solidifies, it is the least thermodynamically stable of the allotropes. When exposed to ultraviolet radiation (for example, from fluorescent lights), the white phosphorus slowly turns to red phosphorus. In this allotrope, one of the bonds in the tetrahedral structure of white phosphorus has broken open and joined to

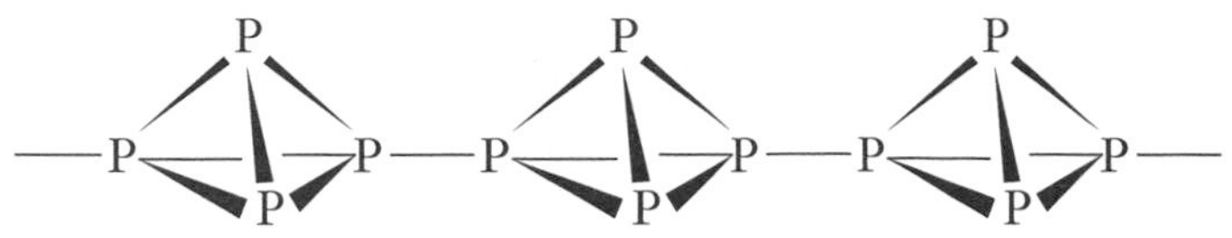

Figure 14.29 The arrangement of atoms in red phosphorus.

a neighboring unit (Figure 14.29). Thus red phosphorus is a polymer with bonds less strained than those of the white allotrope.

The more thermodynamically stable red phosphorus has properties that are completely different from those of the white allotrope. It is stable in air, reacting with the dioxygen in air only above about 400°C. The melting point of the red allotrope is about 600°C, at which temperature the polymer chains break to give the same P_4 units contained in white phosphorus. As we would expect for a covalently bonded polymer, red phosphorus is insoluble in all solvents.

Curiously, the most thermodynamically stable form of phosphorus, black phosphorus, is the hardest to prepare. To make black phosphorus, white phosphorus is heated under pressures of about 1.2 GPa! This densest allotrope (as might be expected from the preferred reaction conditions) has a complex polymeric structure.

Industrial Extraction of Phosphorus

Phosphorus is such a reactive element that we have to resort to quite extreme methods to extract it from its compounds. Calcium phosphate is used as the raw material. This compound is found in large deposits in central Florida, the Morocco-Sahara region, and the Pacific island of Nauru. The origins of these deposits are not well understood, although they may have been the result of the interaction of the calcium carbonate of coral reefs with the phosphate-rich droppings of seabirds over a period of hundreds of thousands of years.

Processing of phosphate rock is highly dependent on electric energy. As a result, the ore is usually shipped to countries where electric power is abundant and inexpensive, such as North America and Europe. The conversion of phosphate rock to the element is accomplished in a very large electric furnace containing 60-tonne carbon electrodes. In this electrothermal process, the furnace is filled with a mixture of ore, sand, and coke, and a current of about 180,000 A (at 500 V) is applied across the electrodes. At the 1500°C operating temperature of the furnace, the calcium phosphate reacts with carbon monoxide to give calcium oxide, carbon dioxide, and gaseous tetraphosphorus:

$$2\ Ca_3(PO_4)_2(s) + 10\ CO(g) \rightarrow 6\ CaO(s) + 10\ CO_2(g) + P_4(g)$$

The carbon dioxide is then reduced back to carbon monoxide by the coke:

$$5\ CO_2(g) + 5\ C(s) \rightarrow 10\ CO(g)$$

Some of the gas is reused, but the remainder escapes from the furnace. The calcium oxide reacts with silicon dioxide (sand) to give calcium silicate (slag):

$$CaO(s) + SiO_2(s) \rightarrow CaSiO_3(l)$$

Nauru, the World's Richest Island

In our studies of industrial chemistry, or any branch of science, it is important to consider the human element. The extraction of phosphate rock from the island of Nauru is an illustrative case history.

The Republic of Nauru, located in the Pacific Ocean, has an area of 21 km^2, yet it is one of the world's major suppliers of calcium phosphate. Between 1 and 2 million tonnes of phosphate rock are mined there each year, thus providing that small nation with a gross national product of about \$200 million per year. For the approximately 5000 native Nauruans, this income provides an opulent lifestyle, with all the conveniences of modern living from washing machines to VCRs. In addition, they hire servants and maids from Asia and from other island countries to perform the work around the house and garden.

The downside of this "idyllic" life is that there is little incentive to work or study. Obesity, heart disease, and alcohol abuse have suddenly become major problems. The long-term effects on the environment have also been catastrophic. Extraction of the phosphate rock is like large-scale dentistry. The ore is scooped up from between enormous toothlike stalks of coral limestone, some of which are 25 m high. These barren pinnacles of limestone will be all that is left of 80 percent of the island when the deposits are exhausted. Furthermore, silt runoff from the mining operation has damaged the offshore coral reefs that once provided abundant fishing resources.

To provide a future for the island and the islanders, the majority of the mining royalties are now placed in the Phosphate Royalties Trust. The money is intended to give the islanders a long-term benefit. This may involve the importation of millions of tonnes of soil to resurface the island. Alternatively, some planners envisage the purchase of some sparsely populated island for relocation of the entire population and an abandonment of Nauru. Whatever the future, phosphate mining has changed the lives of these islanders forever.

The escaping carbon monoxide is burned and the heat is used to dry the three raw materials:

$$2\ CO(g) + O_2(g) \rightarrow 2\ CO_2(g)$$

To condense the gaseous tetraphosphorus, it is pumped into a tower and sprayed with water. The liquefied phosphorus collects at the bottom of the tower and is drained into holding tanks. The average furnace produces about 5 tonnes of tetraphosphorus per hour.

There are two common impurities in the phosphate ore. First, there are traces of fluorapatite, $Ca_5(PO_4)_3F$, that react at the high temperatures to produce toxic and corrosive silicon tetrafluoride. This contaminant is removed from the effluent gases by treating them with sodium carbonate solution. The process produces sodium hexafluorosilicate, Na_2SiF_6, which is a commercially useful product. The second impurity is iron(III) oxide, which reacts with

Table 14.3 Materials consumed and produced in the extraction of 1 tonne of phosphorus

Required	Produced
10 t calcium phosphate (phosphate rock)	1 t white phosphorus
3 t silicon dioxide (sand)	8 t calcium silicate (slag)
$1\frac{1}{2}$ t carbon (coke)	$\frac{1}{4}$ t iron phosphides
14 MWh electrical energy	0.1 t filter dust
	2500 m^3 flue gas

the tetraphosphorus to form ferrophosphorus (mainly Fe_2P, one of the several interstitial iron phosphides), a dense liquid that can be tapped from the bottom of the furnace below the liquid slag layer. Ferrophosphorus can be used in specialty steel products such as railroad brake shoes. The other by-product from the process, calcium silicate (slag), has little value apart from road fill. The cost of this whole process is staggering, not only for its energy consumption but also for the total mass of materials. These are listed in Table 14.3.

The major pollutants from the process are dusts, flue gases, phosphorus-containing sludge, and process water from the cooling towers. Older plants had very bad environmental records. In fact, the technology has changed to such an extent that it is now more economical to abandon an old plant and build a new one that will produce as little pollution as is possible using modern technology. However, the abandoned plant may become a severe environmental problem for the community in which it is located as a result of leaching from the waste material dumps.

The need for pure phosphorus is in decline because the energy costs of its production are too high to make it an economical source for most phosphorus compounds. Furthermore, demand for phosphate-based detergents has dropped because of ecological concerns. Nevertheless, elemental phosphorus is still the preferred route for the preparation of high-purity phosphorus compounds, such as phosphorus-based insecticides and match materials.

Matches

In spite of the prevalence of cheap butane lighters, match consumption is still between 10^{12} and 10^{13} per year. As mentioned at the beginning of this chapter, the modern safety match depends on a chemical reaction between the match head and the strip on the matchbox. The head of the match is mostly potassium chlorate, $KClO_3$, an oxidizing agent, whereas the strip contains red phosphorus and antimony sulfide, Sb_2S_3, both of which oxidize very exothermically when brought in contact with the potassium chlorate.

In addition to the safety match, there is also the "strike-anywhere" match. In this case, the two chemical components, the oxidizing agent (potassium chlorate) and the reducing agent (tetraphosphorus trisulfide, P_4S_3), are mixed in the match head. Any source of friction, such as the glass-paper strip on the matchbox or a brick wall, can provide the activation energy necessary to start the reaction.

Phosphine

The analog of ammonia—phosphine, PH_3—is a colorless, highly poisonous gas. The two hydrides differ substantially because the P–H bond is much less polar than the N–H bond. Thus phosphine is a very weak base, and it does not form hydrogen bonds. In fact, the phosphonium ion, PH_4^+, the equivalent of the ammonium ion, is difficult to prepare. Phosphine itself can be prepared by mixing a phosphide of a very electropositive metal with water:

$$Ca_3P_2(s) + 6\ H_2O(l) \rightarrow 2\ PH_3(g) + 3\ Ca(OH)_2(aq)$$

In phosphine, the P–H bond angle is only 93° rather than 107°, the angle of the N–H bond in ammonia. The phosphine angle suggests that the phosphorus atom is using *p* orbitals rather than sp^3 hybrids for bonding.

Phosphorus Oxides

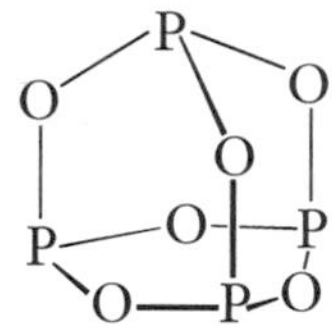

Figure 14.30 The tetraphosphorus hexaoxide molecule.

Phosphorus forms two oxides: tetraphosphorus hexaoxide, P_4O_6, and tetraphosphorus decaoxide, P_4O_{10}. They are both white solids at room temperature. Tetraphosphorus hexaoxide is formed by heating white phosphorus in an environment with a shortage of oxygen:

$$P_4(s) + 3\ O_2(g) \rightarrow P_4O_6(s)$$

Conversely, tetraphosphorus decaoxide, the more common and more important oxide, is formed by heating white phosphorus in the presence of an excess of oxygen:

$$P_4(s) + 5\ O_2(g) \rightarrow P_4O_{10}(s)$$

Tetraphosphorus decaoxide can be used as a dehydrating agent because it reacts vigorously with water in a number of steps to give, ultimately, phosphoric acid:

$$P_4O_{10}(s) + 6\ H_2O(l) \rightarrow 4\ H_3PO_4(l)$$

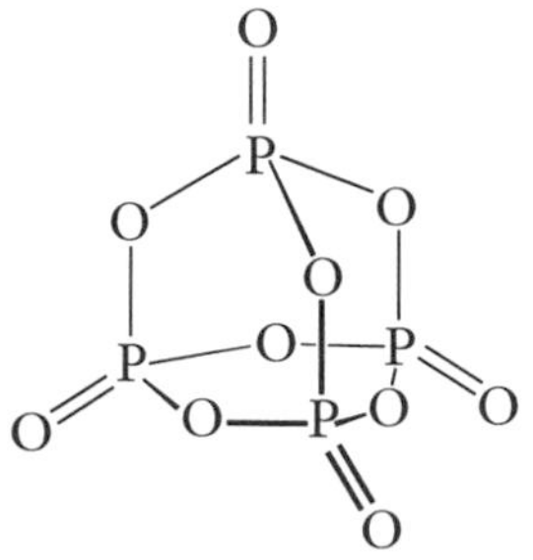

Tetraphosphorus decaoxide will dehydrate many compounds, for example, nitric acid to dinitrogen pentoxide and organic amides, $RCONH_2$, to nitriles, RCN.

Figure 14.31 The tetraphosphorus decaoxide molecule.

The structure of both these oxides is based on the tetrahedron of white phosphorus (tetraphosphorus) itself. In tetraphosphorus hexaoxide, oxygen atoms have inserted themselves in all the phosphorus-phosphorus bonds (Figure 14.30). In tetraphosphorus decaoxide, four additional oxygen atoms form coordinate covalent bonds to the phosphorus atoms, extending out from the corners of the tetrahedron (Figure 14.31). These bonds are close to being double bonds, because, like most of the third (and subsequent) period elements, empty *d* orbitals can be used to form a π bond with a full *p* orbital on an oxygen atom (or other Period 2 element).

Phosphorus Chlorides

In parallel with the oxides, there are two chlorides: phosphorus trichloride, PCl_3, a colorless liquid, and phosphorus pentachloride, PCl_5, a white solid. Phosphorus trichloride is produced when chlorine gas reacts with an excess of phosphorus:

$$P_4(s) + 6\ Cl_2(g) \rightarrow 4\ PCl_3(l)$$

An excess of chlorine results in phosphorus pentachloride:

$$P_4(s) + 10\ Cl_2(g) \rightarrow 4\ PCl_5(l)$$

Phosphorus trichloride reacts with water to give phosphonic acid, H_3PO_3, and hydrogen chloride gas (Figure 14.32):

$$PCl_3(l) + 3\ H_2O(l) \rightarrow H_3PO_3(l) + 3\ HCl(g)$$

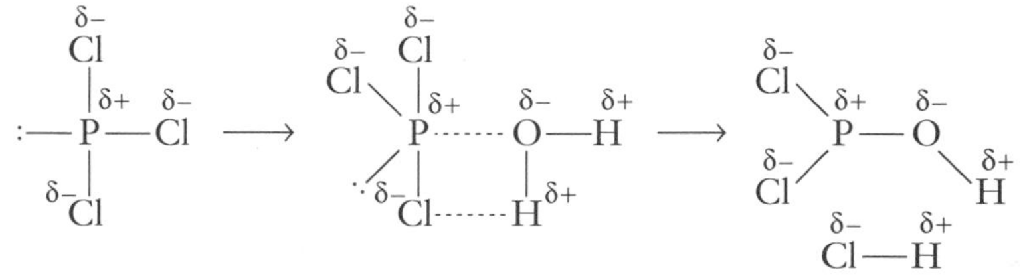

Figure 14.32 The proposed mechanism for the first step in the reaction between phosphorus trichloride and water.

This behavior contrasts with that of nitrogen trichloride, which, as mentioned earlier, hydrolyzes to give ammonia and hypochlorous acid (Figure 14.33):

$$NCl_3(l) + 3\ H_2O(l) \rightarrow NH_3(g) + 3\ HClO(aq)$$

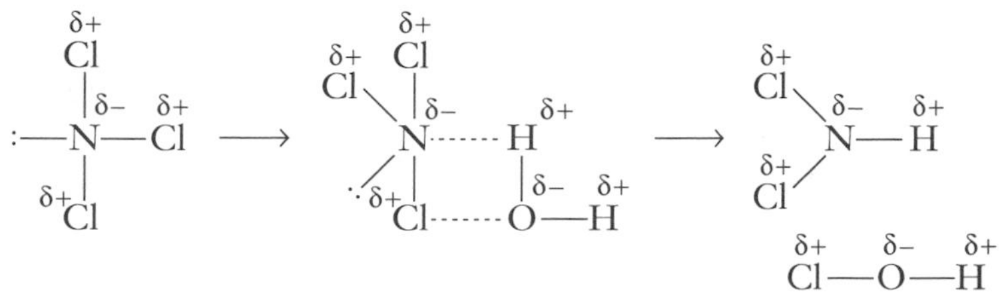

Figure 14.33 The proposed mechanism for the first step in the reaction between nitrogen trichloride and water.

Phosphorus trichloride is an important reagent in organic chemistry, and its worldwide production amounts to about 250 000 tonnes. For example, it can be used to convert alcohols to chloro compounds. Thus 1-propanol is converted to 1-chloropropane by phosphorus trichloride:

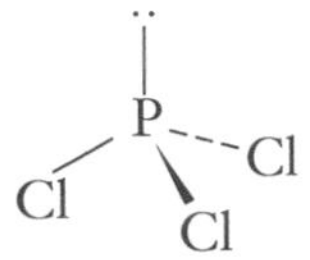

$$PCl_3(l) + 3\ C_3H_7OH(l) \rightarrow 3\ C_3H_7Cl(g) + H_3PO_3(l)$$

Phosphorus trichloride has a trigonal shape that is explained by the lone pair on the phosphorus atom (Figure 14.34).

Phosphorus pentachloride is also used as an organic reagent, but it is less important, annual production being only about 20 000 tonnes worldwide. Like phosphorus trichloride, it reacts with water, but in a two-step process, the first step yielding phosphoryl chloride, $POCl_3$:

$$PCl_5(s) + H_2O(l) \rightarrow POCl_3(l) + 2\ HCl(g)$$

$$POCl_3(l) + 3\ H_2O(l) \rightarrow H_3PO_4(l) + 3\ HCl(g)$$

In the gas phase, the phosphorus pentachloride is a trigonal bipyramidal covalent molecule (Figure 14.35). To illustrate again that these compounds are on the border of ionic-covalent stability, in the solid phase, phosphorus pentachloride adopts the structure $PCl_4^+PCl_6^-$ (Figure 14.36).

Figure 14.34 The phosphorus trichloride molecule.

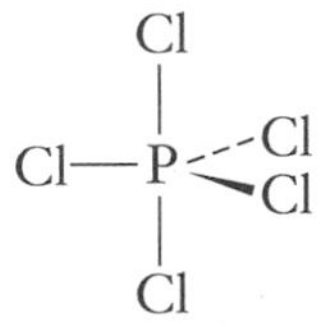

Figure 14.35 The shapes of the phosphorus pentachloride molecule in liquid and gas phases.

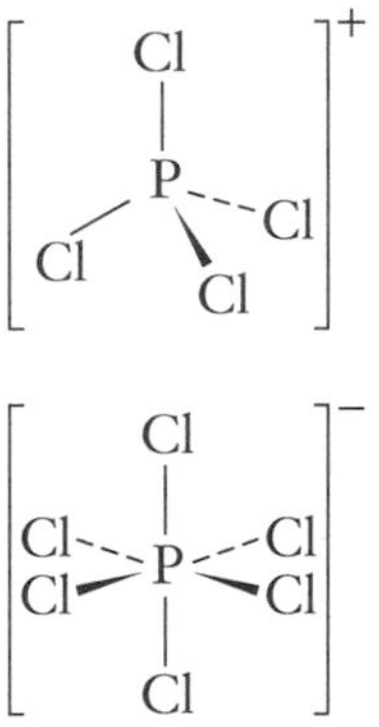

Figure 14.36 The two ions present in solid-phase phosphorus pentachloride.

Common Oxyacids of Phosphorus

There are three oxyacids of phosphorus that we will mention here: phosphoric acid, H_3PO_4; phosphonic acid, H_3PO_3 (commonly called phosphorous acid); and phosphinic acid, H_3PO_2 (commonly called hypophosphorous acid). The first, phosphoric acid, is really the only oxyacid of phosphorus that is important. However, the other two acids are useful for making a point about the character of oxyacids.

In an oxyacid, for the hydrogen to be significantly acidic, it must be attached to an oxygen atom—and this is normally the case. In general, as we progress through a series of oxyacids—for example, nitric acid, $(HO)NO_2$, and nitrous acid, (HO)NO—it is one of the terminal oxygen atoms that is lost as the oxidation state of the central element is reduced. Phosphorus is almost unique in that the oxygens linking a hydrogen to the phosphorus are the ones that are lost. Thus phosphoric acid possesses three ionizable hydrogen atoms; phosphonic acid, only two; and phosphinic acid, only one (Figure 14.37).

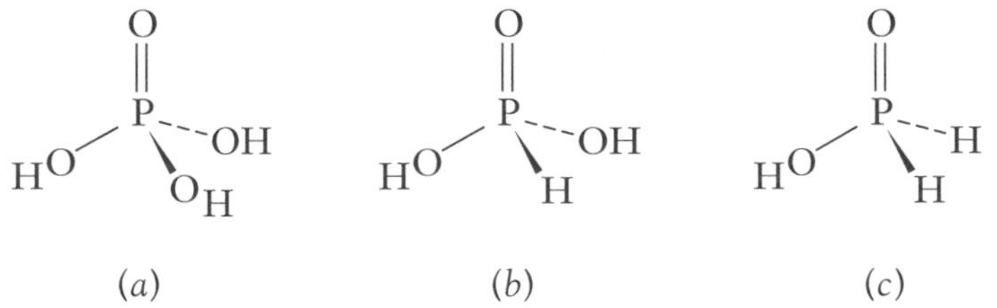

(a) (b) (c)

Figure 14.37 The bonding in the three common oxyacids of phosphorus: (*a*) phosphoric acid; (*b*) phosphonic acid; (*c*) phosphinic acid.

Phosphoric Acid

Pure (*ortho*)phosphoric acid is a colorless solid, melting at 42°C. A concentrated aqueous solution of the acid (85 percent by mass) is called "syrupy" phosphoric acid, its viscous nature being caused by extensive hydrogen bonding. As discussed earlier, the acid is essentially nonoxidizing. In solution, phosphoric acid is a weak acid, undergoing three ionization steps:

$$H_3PO_4(aq) + H_2O(l) \rightleftharpoons H_3O^+(aq) + H_2PO_4^-(aq)$$

$$H_2PO_4^-(aq) + H_2O(l) \rightleftharpoons H_3O^+(aq) + HPO_4^{2-}(aq)$$

$$HPO_4^{2-}(aq) + H_2O(l) \rightleftharpoons H_3O^+(aq) + PO_4^{3-}(aq)$$

The pure acid is prepared by burning white phosphorus to give tetraphosphorus decaoxide, then treating the oxide with water:

$$P_4(s) + 5\ O_2(g) \rightarrow P_4O_{10}(s)$$

$$P_4O_{10}(s) + 6\ H_2O(l) \rightarrow 4\ H_3PO_4(l)$$

Such high purity is not required for most purposes, so where trace impurities can be tolerated, it is much more energy efficient to treat calcium phosphate with sulfuric acid to give a solution of phosphoric acid and a precipitate of calcium sulfate:

$$Ca_3(PO_4)_2(s) + 3\ H_2SO_4(aq) \rightarrow 3\ CaSO_4(s) + 2\ H_3PO_4(aq)$$

The only problem associated with this process is the disposal of the calcium sulfate. Some of this product is used in the building industry, but production of calcium sulfate exceeds the uses; hence most of it must be dumped. Furthermore, when the phosphoric acid is concentrated at the end of the process, many of the impurities precipitate out. This "slime" must be disposed of in an environmentally safe manner.

Heating phosphoric acid causes a stepwise loss of water; in other words, the phosphoric acid molecules undergo *condensation*. The first product is pyrophosphoric acid, $H_4P_2O_7$. Like phosphoric acid, each phosphorus atom is tetrahedrally coordinated (Figure 14.38). The next product is tripolyphosphoric acid, $H_5P_3O_{10}$:

$$2\ H_3PO_4(l) \rightarrow H_4P_2O_7(l) + H_2O(l)$$

$$3\ H_4P_2O_7(l) \rightarrow 2\ H_5P_3O_{10}(l) + H_2O(l)$$

Subsequent condensations give products with even greater degrees of polymerization.

Most of the phosphoric acid is used for fertilizer production. Phosphoric acid also is a common additive to soft drinks, its weak acidity preventing bacterial growth in the bottled solutions. It often serves a second purpose in

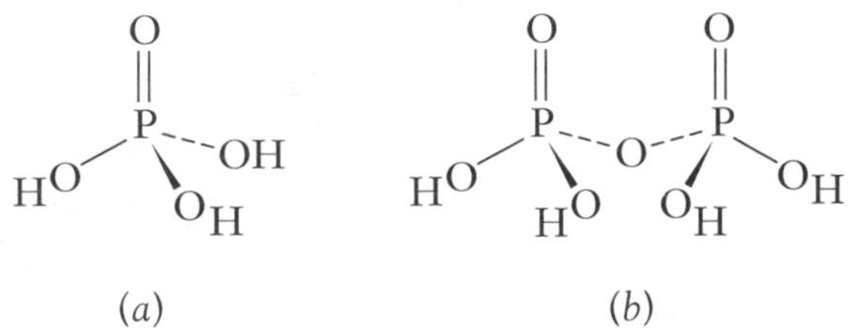

Figure 14.38 The shape of (*a*) (*ortho*)phosphoric acid and (*b*) pyrophosphoric acid.

metal containers. Metal ions may be leached from the container walls, but the phosphate ions will react with the metal ions to give an inert phosphate compound, thus preventing any potential metal poisoning. Phosphoric acid is also used on steel surfaces as a rust remover, both in industry and for home automobile repairs.

Phosphates

With high lattice energies resulting from the high anion charge, most phosphates are insoluble. The alkali metals and ammonium phosphates are the only common exceptions to this rule. There are the three series of phosphate salts: the phosphates, containing the PO_4^{3-} ion; the hydrogen phosphates, containing the HPO_4^{2-} ion; and the dihydrogen phosphates, containing the $H_2PO_4^-$ ion. In solution, there is an equilibrium between the three species and phosphoric acid itself. For example, a solution of phosphate ion will hydrolyze:

$$PO_4^{3-}(aq) + H_2O(l) \rightleftharpoons HPO_4^{2-}(aq) + OH^-(aq)$$

$$HPO_4^{2-}(aq) + H_2O(l) \rightleftharpoons H_2PO_4^-(aq) + OH^-(aq)$$

$$H_2PO_4^-(aq) + H_2O(l) \rightleftharpoons H_3PO_4(aq) + OH^-(aq)$$

Each successive equilibrium lies more and more to the left. Thus the concentration of actual phosphoric acid is minuscule. A solution of sodium phosphate, then, will be quite basic, almost entirely as a result of the first equilibrium. It is this basicity (good for reaction with greases) and the complexing ability of the phosphate ion that make a solution of sodium phosphate a common kitchen cleaning solution (known as TSP, trisodium phosphate).

Figure 14.39 shows how the proportions of the phosphate species depend on pH. A solution of sodium hydrogen phosphate, $Na_2(HPO_4)$, will be basic as a result of the second step of the preceding sequence:

$$HPO_4^{2-}(aq) + H_2O(l) \rightleftharpoons H_2PO_4^-(aq) + OH^-(aq)$$

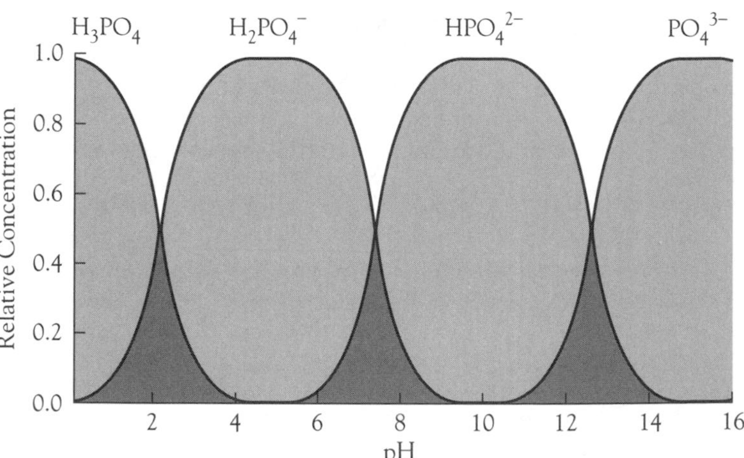

Figure 14.39 The relative concentrations of phosphate species at different values of pH.

A solution of sodium dihydrogen phosphate, $Na(H_2PO_4)$, however, is slightly acidic as a result of the reaction:

$$H_2PO_4^-(aq) + H_2O(l) \rightleftharpoons H_3O^+(aq) + HPO_4^{2-}(aq)$$

Solid hydrogen phosphates and dihydrogen phosphates are only known for most of the monopositive cations (the alkali metals ions and the ammonium ion) and a few of the dipositive cations, such as calcium ion. As we have seen before, to stabilize a large low-charge anion, a low charge density cation is required. For most dipositive and all tripositive metal ions, the metal ion precipitates the small proportion of phosphate ion from a solution of one of the acid ions. The Le Châtelier principle then drives the equilibria to produce more phosphate ion, causing additional metal phosphate to precipitate.

The phosphates have a tremendous range of uses. As mentioned earlier, trisodium phosphate is used as a household cleaner. Other sodium phosphates, such as sodium pyrophosphate, $Na_4P_2O_7$, and sodium tripolyphosphate, $Na_5P_3O_{10}$, are often added to detergents, because they react with calcium and magnesium ions in the tap water to form soluble compounds, preventing the deposition of scum in the washing. However, when the phosphate-rich wastewater reaches lakes, it can cause a rapid growth of algae and other simple plant life. This formation of green, murky lakes is called *eutrophication*. Phosphates are also added to detergents as fillers. Fillers are required because we are used to pouring cupfuls of solid detergent into a washing machine. However, only small volumes of cleaning agents are actually needed, so most of the detergent is simply inert materials. In more frugal societies, such as Japan, people are used to adding spoonfuls of detergent—hence less wasteful filler is required.

Disodium hydrogen phosphate is used in the preparation of pasteurized processed cheese, although even today, the reason why this ion aids in the cheese-making process is not well understood. The ammonium salts, diammonium hydrogen phosphate and ammonium dihydrogen phosphate, are useful nitrogen-phosphorus combination fertilizers. Ammonium phosphates also make excellent flame retardants for drapes, theater scenery, and disposable paper clothing and costumes.

The calcium phosphates are used in many circumstances. For example, "combination baking powder" relies on the reaction between calcium dihydrogen phosphate and sodium hydrogen carbonate to produce the carbon dioxide gas so essential for baking. The reaction can be simplistically represented as

$$Ca(H_2PO_4)_2(aq) + NaHCO_3(aq) \rightarrow CaHPO_4(aq) + NaH_2PO_4(aq) + CO_2(g) + H_2O(g)$$

Other calcium phosphates are used as mild abrasives and polishing agents in toothpaste. Finally, calcium dihydrogen phosphate is used as a fertilizer. The calcium phosphate rock is too insoluble to provide phosphate for plant growth, so it is treated with sulfuric acid to produce calcium dihydrogen phosphate:

$$Ca_3(PO_4)_2(s) + 2\ H_2SO_4(aq) \rightarrow Ca(H_2PO_4)_2(s) + 2\ CaSO_4(s)$$

This compound is only slightly soluble in water, but soluble enough to release a steady flow of phosphate ions into the surrounding soil, where they can be absorbed by plant roots.

Biological Aspects

Nitrogen

Just as there is a carbon cycle, there is also a nitrogen cycle, for all plant life requires nitrogen for growth and survival. Between 10^8 and 10^9 tonnes of nitrogen are cycled between the atmosphere and the lithosphere in a one-year period. The dinitrogen in the atmosphere is converted by bacteria to compounds of nitrogen. Some of the bacteria exist free in the soil, but members of the most important group, the *Rhizobium*, form nodules on the roots of pea, bean, alder, and clover plants. This is a symbiotic relationship, with the bacteria providing the nitrogen compounds to the plants, and the plants providing a stream of nutrients to the bacteria. To do this at a high rate at normal soil temperatures, the bacteria use enzymes such as nitrogenase.

Nitrogenase contains two components: a large protein containing two metals, iron and molybdenum, and a smaller one containing iron. The bioinorganic chemistry involved in this process is still not well understood, but it is believed that one of the crucial steps involves the formation of a molybdenum bond with the dinitrogen molecule. It is hoped that an understanding of the route used by bacteria will enable us to produce ammonia for fertilizers by a room-temperature process rather than the energy-intensive Haber-Bosch procedure.

Phosphorus

Phosphorus is another element essential for life. For example, the free hydrogen phosphate and dihydrogen phosphate ions are involved in the blood buffering system. More important, phosphate is the linking unit in the sugar esters of DNA and RNA; and phosphate units make up part of ATP, the essential energy storage unit in living organisms. Finally, bone is a phosphate mineral, calcium hydroxide phosphate, $Ca_5(OH)(PO_4)_3$, commonly called apatite.

Arsenic

Amazing as it may seem, arsenic also is an element essential to life. But we only need trace amounts of this element, whose role is still unknown. Anything more than a tiny amount causes arsenic poisoning. Arsenic, in measurable quantities, causes death through enzyme inhibition.

The most famous case of arsenic poisoning is believed to be that of Napoleon. Modern chemical analysis has shown very high levels of arsenic in Napoleon's hair. His British captors, or perhaps French rivals, were the prime suspects; but chemical research has turned up the most likely source of poisoning—his wallpaper. At that time, copper(II) hydrogen arsenite, $CuHAsO_3$, was used as a pigment to give a beautiful green color in wall-

Paul Erhlich and His "Magic Bullet"

Arsenic has also been used as a lifesaver. In the nineteenth century, physicians had no means of combating infections, and patients usually died. The whole nature of medicine changed in 1863 when a French scientist, Béchamps, noticed that an arsenic compound was toxic to some microorganisms. A German, Paul Erhlich, decided to synthesize new arsenic compounds, testing each one for its organism-killing ability. With his 606th compound of arsenic, he found a substance that selectively killed the syphilis organism. At the time, syphilis was a feared and widespread disease for which there was no cure, only suffering, dementia, and death. Erhlich's arsenic compound, what he dubbed a "magic bullet," provided miraculous cures; and a search for other chemical compounds that could be used in the treatment of disease was launched.

This field, chemotherapy, has produced one of the most effective tools of controlling bacterial infections and those of many other microorganisms. Chemotherapy also provides one of the lines of attack against cancerous tissues. And it all started with an arsenic compound.

papers. In a dry climate, this pigment was quite safe; but in the chronically damp house in which Napoleon was held on the island of St. Helena, molds grew on the walls. Many of these molds metabolize arsenic compounds to trimethylarsenic, $(CH_3)_3As$, a gas. Thus Napoleon probably inhaled this toxic gas while in bed, and the sicker he became, the more time he spent in his bedroom, thus hastening his death from this toxic element.

Exercises

14.1. Write balanced chemical equations for the following chemical reactions:

(a) arsenic trichloride with water
(b) magnesium with dinitrogen
(c) ammonia with excess chlorine
(d) methane with steam
(e) hydrazine and oxygen
(f) heating a solution of ammonium nitrate
(g) sodium hydroxide solution with dinitrogen trioxide
(h) heating sodium nitrate
(i) heating tetraphosphorus decaoxide with carbon

14.2. Write balanced chemical equations for the following chemical reactions:

(a) heating a solution of ammonium nitrite
(b) ammonium sulfate with sodium hydroxide
(c) ammonia with phosphoric acid
(d) decomposition of silver azide
(e) nitrogen monoxide and nitrogen dioxide
(f) heating solid lead nitrate
(g) tetraphosphorus with an excess of dioxygen
(h) calcium phosphide with water
(i) hydroxylamine solution and dilute hydrochloric acid

14.3. Why is it hard to categorize arsenic as either a metal or a nonmetal?

14.4. What are the factors that distinguish the chemistry of nitrogen from that of the other members of Group 15?

14.5. Contrast the behavior of nitrogen and carbon by comparing the properties of (a) methane and ammonia; (b) ethene and hydrazine.

14.6. Contrast the bonding to oxygen in the two compounds NF_3O and PF_3O.

14.7. (a) Why is dinitrogen very stable? (b) Yet why is dinitrogen not always the product during redox reactions involving nitrogen compounds?

14.8. When ammonia is dissolved in water, the solution is often referred to as "ammonium hydroxide." Discuss whether this terminology is appropriate.

14.9. In the Haber process for ammonia synthesis, the recycled gases contain increasing proportions of argon gas. Where does the argon come from? Suggest how it might be removed.

14.10. Why is it surprising that high pressure is used in the steam reforming process during ammonia synthesis?

14.11. Discuss the differences between the ammonium ion and those of the alkali metals.

14.12. Using bond energies, calculate the heat released when gaseous hydrazine burns in air (oxygen) to give water vapor and nitrogen gas.

14.13. Construct a possible electron-dot structure for the azide ion. Identify the location of the formal charges.

14.14. Construct three possible electron-dot structures for the theoretical molecule, N–O–N. By assignment of formal charges, suggest why the actual dinitrogen oxide molecule has its asymmetrical structure.

14.15. Calculate the volume of nitrogen gas produced at 25°C and 102 kPa pressure when 5.0 g of sodium azide is detonated.

14.16. Nitrogen monoxide can form a cation, NO^+, and an anion, NO^-. Calculate the bond order in each of these species.

14.17. Draw the shape of each of the following molecules: (a) hydroxylamine; (b) dinitrogen trioxide; (c) dinitrogen pentoxide (solid and gas phases); (d) phosphorus pentafluoride.

14.18. Draw the shape of each of the following molecules: (a) dinitrogen oxide; (b) dinitrogen tetroxide; (c) phosphorus trifluoride; (d) phosphonic acid.

14.19. Describe the physical properties of (a) nitric acid; (b) ammonia.

14.20. Explain why, in the synthesis of nitric acid, the reaction of nitrogen monoxide with dioxygen is performed at high pressure and with cooling.

14.21. Write balanced equations for the following reactions:

(a) the reduction of nitric acid to the ammonium ion by zinc metal

(b) the reaction of solid silver sulfide with nitric acid to give silver ion solution, elemental sulfur, and nitrogen monoxide

14.22. Contrast the properties of the two common allotropes of phosphorus.

14.23. Contrast the properties of ammonia and phosphine.

14.24. Phosphine, PH_3, dissolves in liquid ammonia to give $NH_4^+PH_2^-$. What does this tell you about the relative acid-base strengths of the two Group 15 hydrides?

14.25. In the "strike-anywhere" match, assume that the potassium chlorate is reduced to potassium chloride and the tetraphosphorus trisulfide is oxidized to tetraphosphorus decaoxide and sulfur dioxide. Write a balanced chemical equation for the process and identify the oxidation number changes that have occurred.

14.26. A compound is known to have the formula NOCl (nitrosyl chloride). Construct an electron-dot diagram for the molecule and identify the oxidation number of nitrogen. What is the anticipated nitrogen-oxygen bond order? From the $\Delta H_f°$ value for this compound of $+52.6\ kJ\cdot mol^{-1}$ and appropriate bond energy data, calculate the N–O bond energy in this compound. Compare it with values for N–O single and double bonds.

14.27. Construct an electron-dot structure for $POCl_3$ (assume it to be similar to PF_3O) and then draw its molecular shape.

14.28. In the solid phase, PCl_5 forms $PCl_4^+PCl_6^-$. However, PBr_5 forms $PBr_4^+Br^-$. Suggest a reason why the bromine compound has a different structure.

14.29. Suggest why acidification promotes the decomposition of ammonium nitrate to dinitrogen oxide and water. (*Hint:* Consult Figure 14.4.)

14.30. A solution of the hydrogen phosphate ion is basic, whereas a solution of the dihydrogen phosphate ion is acidic. Write chemical equilibria for the predominant reactions that account for this difference in behavior.

14.31. Explain the terms (a) eutrophication; (b) symbiotic relationship; (c) chemotherapy; (d) apatite.

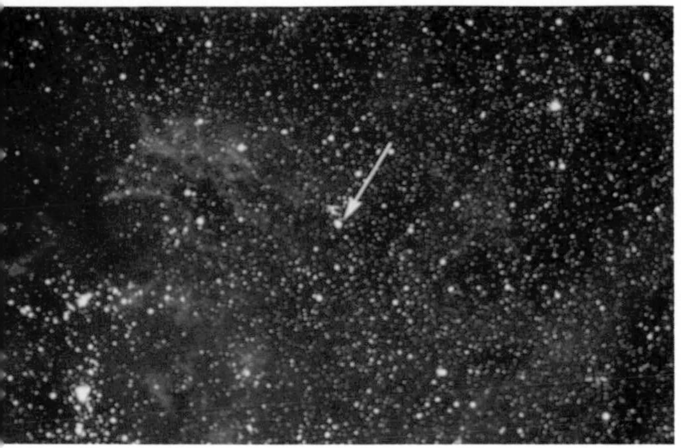

The great supernova of 1987. As the star Sanduleak exploded, it increased 2500 times in brightness. During the death of stars like this one, heavier elements are created from lighter ones. (*Source*: Anglo Australian Telescope Board)

The Atomium, Brussels, Belgium. This enormous representation of a face-centered cubic structure was built for the 1958 World's Fair. (*Source*: Belgian Tourist Office)

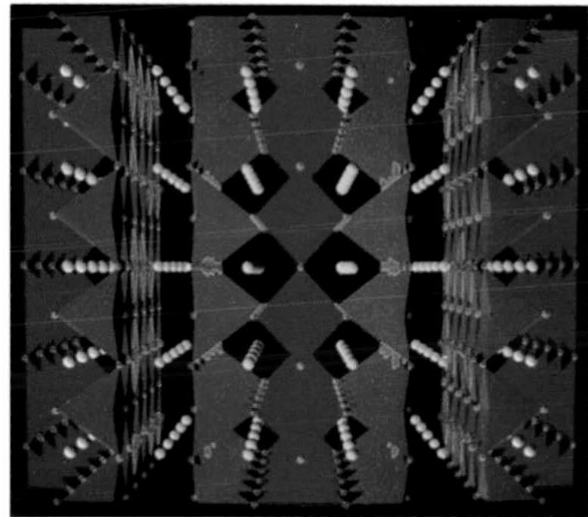

A computer representation of one of the superconductor structures. The green pyramids represent silicate sites. (*Source*: Photo Researchers)

The wreck of the *Titanic*. As the steel oxidizes, a cascade of colorful oxidation products is produced. In time, these chemical reactions will completely consume the remains of the ship. (*Source*: Woods Hole Oceanographic Institution)

Although photosynthesis is the key redox reaction for life on the surface of Earth, it is chemosynthesis, the oxidation of sulfides, that is the energy source for the marine organisms that live around deep sea vents. (*Source*: Photo Researchers)

The Taj Mahal, India. The facing of this beautiful tomb consists of white marble, a form of calcium carbonate. Carbonates react rapidly with acidic gases, so the survival of this building in the industrial city of Agra is increasingly unlikely. (*Source*: Photo Researchers)

A very large scale neutralization reaction. To neutralize a spill of concentrated acid in Denver, Colorado, sodium carbonate was sprayed over the spill using a snowblower. (*Source*: Bill Wunsch, *The Denver Post*)

Lithium floating on a layer of oil, which is less dense than water. Lithium is the least dense of metals. (*Source*: Chip Clark)

An aerial view of the mine in Boron, California, the largest borax mine in the world. Borax, $Na_2B_4O_7 \cdot 10\ H_2O$, is still extracted on a large scale for use in glassmaking and detergents. (*Source*: U.S. Borax)

Preparing an advanced material, boron-tungsten monofilaments. Elecrically heated glowing tungsten microwires are passed down a tube containing boron trichloride. The compound is thermally decomposed, depositing a boron layer over the tungsten. (*Source*: Textron, Inc.)

The industrial extraction of aluminum is accomplished in these series of electrochemical cells. The liquid aluminum metal is siphoned from each cell in turn into a large crucible. (*Source*: Alcan Aluminum Ltd.)

An aerogel. This new material is an ultra-low-density, translucent form of silica that has great potential as a thermal insulator. (*Source*: Quesada Burke)

The platinum gauze that is used as the catalyst in the industrial oxidation of ammonia to nitrogen monoxide. (*Source*: Photo Researchers)

Roots of a pea plant showing the nodules that contain the *Rhizobium* bacteria. These bacteria convert the dinitrogen in the air to compounds of nitrogen. (*Source*: Photo Researchers)

White phosphorus and red phosphorus. These two common allotropes of phosphorus differ considerably in physical and chemical properties. For example, red phosphorus is air-stable, whereas white phosphorus burns in contact with the air. Hence the white allotrope is stored under water. (*Source*: Chip Clark)

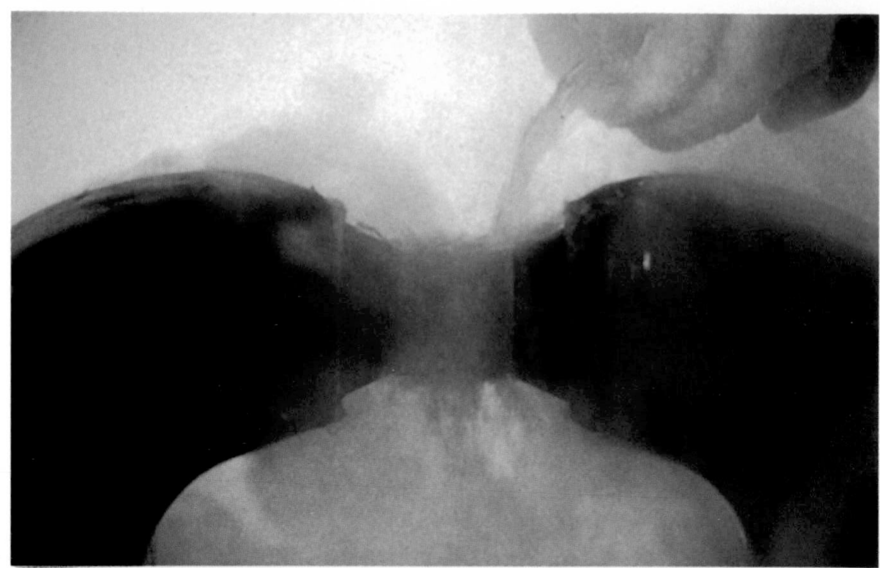

When liquid dioxygen is poured over a magnet, it is held between the poles. This behavior is an indication of the paramagnetism of the molecule. (*Source*: Donald Clegg)

Heating red mercury(II) oxide causes it to decompose to mercury and dioxygen. The metallic mercury condenses on the cooler upper walls of the tube. (*Source*: Chip Clark)

Io, one of the moons of Jupiter, has sulfur volcanoes, which give the moon its uniquely bright coloration. (*Source*: NASA)

These metallic-looking crystals are not metals but metal sulfides: lead(II) sulfide, PbS (galena); mercury(II) sulfide, HgS (cinnabar); iron(II) disulfide, FeS_2 (iron pyrite); and zinc sulfide, ZnS (sphalerite). (*Source*: Chip Clark)

Containers of powdered transition metal oxides, such as chromium(III) oxide, for use as pigments. (*Source*: Photo Researchers)

A variety of gemstones. These silicate minerals contain trace proportions of transition metal ions, and it is the electron transitions within these ions that give the colors that are so highly valued. Shown here are (1) kunzite, (2) garnet, (3) zircon, (4) aquamarine, (5) amethyst, (6) peridot, (7) morganite, (8) topaz, (9) ruby, (10) tourmaline (indicolite), (11) chrome tourmaline, (12) rose quartz, (13) rubellite tourmaline, (14) kyanite, (15) citrine, and (16) green tourmaline (*Source*: Chip Clark)

This metallic-looking crystal is calcium titanate, $CaTiO_3$. This mineral, perovskite, gives its name to a large number of compounds that share the same crystal structure. (*Source*: Quesada Burke)

The Statue of Liberty. The pleasant green color of the statue, and of other copper objects such as copper roofs, is a result of the oxidation of copper to copper(II) hydroxide carbonate, $Cu_2(OH)_2CO_3$. (*Source*: National Parks Service)

Horseshoe crabs acting as "blood donors." The blue blood of these crabs, the color caused by the copper-containing protein hemocyanin, is used in a test for bacterial contamination of drugs. The crabs are subsequently released and returned to the sea. (*Source*: Associates of Cape Cod)

CHAPTER

15

The Group 16 Elements

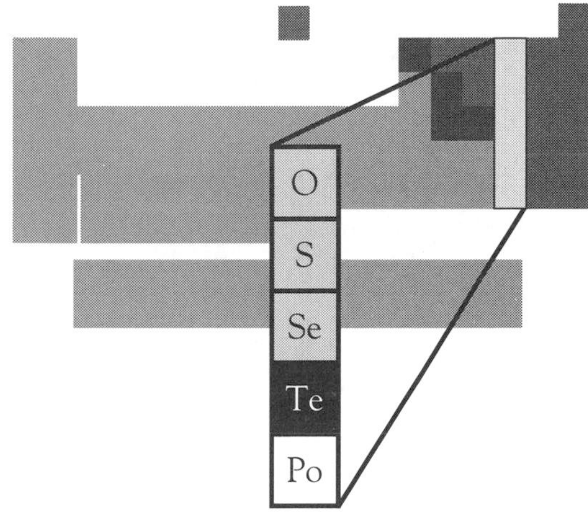

Again, it is the first two members of the group that have the most significant chemistry. In Group 16, the first two, oxygen and sulfur, are nonmetals; selenium and tellurium both possess some semimetallic behavior; and it is only the heaviest member, the radioactive element polonium, that can be said to possess anything like metallic character. The differences between the first and second members that we saw for the Group 15 elements (nitrogen and phosphorus) are repeated with the first and second Group 16 elements (oxygen and sulfur).

It is a common fallacy to link a particular discovery with a specific name. Our modern perception of the progress of science shows that discoveries usually involve the work of many individuals. For example, the discovery of oxygen is credited to the eighteenth-century chemist Joseph Priestley, when, in fact, it was a long-forgotten Dutch inventor, Cornelius Drebble, who first reported the preparation of the gas about 150 years earlier. Nevertheless, Priestley does deserve the bulk of the credit, for he made extensive studies of pure oxygen gas and, very bravely, breathed this gas, then known as "dephlogisticated air." Priestley performed these experiments in Birmingham,

England, where he was a Nonconformist minister. He was known for his "leftist" views on politics and religion—for example, he supported the French and American revolutions. A mob burned his church, home, and library. He fled to the United States, where he dedicated one of his books to Vice-President John Adams, noting, "It is happy that, in this country, religion has no connection with civil power."

The discovery of oxygen marked the end of the phlogiston theory of combustion. According to that theory, burning involved the loss of phlogiston. However, the French scientist Guyton de Morveau (see Chapter 9) showed that burning a metal gave a product that showed a gain in weight. His colleague Antoine Lavoisier realized that something had to be added in the combustion process. It was oxygen. But revolutionary concepts are often slow to be accepted in science, and this was true of the idea that combustion is linked to the addition of oxygen. In fact, many chemists of the time, including Joseph Priestley, never did accept this idea.

Group Trends

Oxygen, sulfur, and selenium are the nonmetals of the group; tellurium is generally regarded as a semimetal; and polonium is considered to be the only true metal in Group 16. Certainly the melting and boiling points show the rising trend characteristic of nonmetals, followed by the falling trend at polonium, a trend characteristic of metals (Table 15.1). Our categorization of polonium as a metal is supported by its low electrical resistivity of 43 μΩ·cm. On the basis of electrical resistivity, the common allotrope of selenium is a nonmetal (10^{16} μΩ·cm), whereas tellurium is usually classified as a semimetal (10^{6} μΩ·cm).

Except for oxygen, of course, there are patterns in the oxidation states of the Group 16 elements. We find all of the even-numbered oxidation states from +6, through +4 and +2, to −2. The stability of the −2 and +6 oxidation states decreases down the group, whereas that of the +4 state increases. As happens in many groups, the trends are not as regular as we would like. For example, the acids containing atoms in the +6 oxidation state are sulfuric acid, which we can represent as $(HO)_2SO_2$ to indicate the bonding, and selenic acid, $(HO)_2SeO_2$; but telluric acid has the formula $(HO)_6Te$, or H_6TeO_6.

Table 15.1 Melting and boiling points of the Group 16 elements

Element	Melting point (°C)	Boiling point (°C)
O_2	−219	−183
S_8	119	445
Se_8	221	685
Te	452	987
Po	254	962

Oxygen Isotopes in Geology

Although we usually consider oxygen atoms to have eight neutrons (oxygen-16), there are in fact two other stable isotopes of the element. The isotopes and their abundances are

Isotope	Abundance (%)
Oxygen-16	99.763
Oxygen-17	0.037
Oxygen-18	0.200

Thus one oxygen atom in every 500 has a mass that is 12 percent greater than the other 499. This "heavy" oxygen will have slightly different physical properties, both as the element and in its compounds. In particular, $H_2{}^{18}O$ has a vapor pressure significantly lower than that of $H_2{}^{16}O$. Hence, in an equilibrium between liquid and gaseous water, the gas phase will be deficient in oxygen-18. Because the most evaporation occurs in tropical waters, it is those waters that will have a higher concentration of oxygen-18. This increased proportion of oxygen-18 will be found in all the marine equilibria that involve oxygen.

We can use the ratio of the two isotopes of oxygen to determine the temperature of the seas in which shells were formed millions of years ago simply by determining the oxygen isotopic ratio in the calcium carbonate of the shells. The higher the proportion of oxygen-18, the warmer the waters of those ancient seas.

Anomalous Nature of Oxygen

The anomalies of oxygen chemistry are similar to those of nitrogen, that is, the formation of strong π bonds using the $2p$ atomic orbitals and a lack of any d orbitals, a deficiency precluding the possibility of more than four covalent bonds.

High Stability of Multiple Bonds

Like nitrogen, the oxygen-oxygen double bond (494 $kJ{\cdot}mol^{-1}$) is much stronger than the oxygen-oxygen single bond (142 $kJ{\cdot}mol^{-1}$). The oxygen-oxygen single bond is particularly weak; the carbon-carbon single bond energy is 335 $kJ{\cdot}mol^{-1}$.

If we consider a double bond energy to consist of the single bond (σ bond) energy plus the energy of the second (π) bond, we can see from Table 15.2 that double bond formation results in a considerable energy gain for oxygen but very little for sulfur and selenium. This difference accounts for the lack of multiple bond formation by the other members of the group.

Table 15.2 Bond energies for the Group 16 elements

Bond	σ bond energy ($kJ{\cdot}mol^{-1}$)	π bond energy ($kJ{\cdot}mol^{-1}$)
Oxygen-oxygen	142	350
Sulfur-sulfur	270	155
Selenium-selenium	210	125

Lack of Catenated Compounds

In Group 14, the ability to catenate decreases down the group. However, in Group 16, sulfur forms the longest chains. In fact, compounds containing two oxygen atoms bonded together are usually strong oxidizing agents, and compounds containing three oxygen atoms bonded together are virtually unknown. Such behavior can be explained by postulating that the oxygen-oxygen bond is weaker than its bonds to other elements. For example, the oxygen-sulfur bond energy of 275 $kJ{\cdot}mol^{-1}$ is almost twice as strong as the oxygen-oxygen single bond. Thus oxygen will endeavor to bond to other elements rather than to itself. Conversely, the sulfur-sulfur single bond energy of 270 $kJ{\cdot}mol^{-1}$ is only slightly lower than that of its bonds to other elements, thereby stabilizing catenation in sulfur compounds.

Lack of Available *d* Orbitals

Oxygen forms only one oxide with fluorine, OF_2, whereas sulfur forms several compounds with fluorine, including SF_6. To covalently bond to six fluorine atoms, the sulfur must use its *d* orbitals. Thus the lack of any oxygen equivalent can be attributed to a lack of available *d* orbitals in Period 2 elements.

Oxygen

Oxygen exists in two allotropic forms: the common dioxygen and the less common trioxygen, commonly called ozone.

Dioxygen

Dioxygen is a colorless, odorless gas that condenses to a pale blue liquid. Because it has a low molar mass and forms a nonpolar molecule, it has very low melting and boiling points. The gas does not burn, but it does support combustion. In fact, almost all elements will react with oxygen at room temperature or when heated. The main exceptions are the "noble" metals, such as platinum, and the noble gases. For a reaction to occur, the state of division of the reactant is often important. For example, very finely powdered metals such as iron, zinc, and even lead will catch fire in air at room temperature.

These finely divided forms of metals are sometimes called *pyrophoric*, a term reflecting their ability to catch fire. For example, zinc dust will inflame to give white zinc oxide:

$$2\ Zn(s) + O_2(g) \rightarrow 2\ ZnO(s)$$

Dioxygen is the reactive gas that makes up 21 percent of Earth's atmosphere. This oxidizing gas is not naturally occurring in planetary atmospheres. The "normal" atmosphere of a planet is reducing, containing hydrogen, methane, ammonia, and carbon dioxide. It was the process of photosynthesis that started to convert the carbon dioxide component of Earth's early atmosphere to dioxygen about 2.5×10^9 years ago; its present oxygen-rich state was attained about 5×10^7 years ago. Thus we can look for signs of life similar to our own on planets around other stars just by sending dioxygen detectors.

Dioxygen is not very soluble in water, about 5 g per 100 mL at 0°C, compared with 170 g per 100 mL for carbon dioxide. Nevertheless, the concentration of oxygen in natural waters is high enough to support marine organisms. The solubility of dioxygen decreases with increasing temperature; hence it is the cold waters, such as the Labrador and Humboldt currents, that are capable of supporting the largest fish stocks—and have been the focus of the most severe overfishing. Even though the solubility of dioxygen is low, it is twice that of dinitrogen. Hence the gas mixture released by heating air-saturated water will actually be enriched in dioxygen.

The measurement of dissolved oxygen (sometimes called DO) is one of the crucial determinants of the health of a river or lake. Low levels of dissolved oxygen can be caused by eutrophication (excessive algae and plant growth) or an input of high-temperature water from an industrial cooling system. As a temporary expedient, air-bubbling river barges can be used to increase dissolved dioxygen levels. This has been done in London, England, to help bring back game fish to the river Thames. Almost the opposite of DO is BOD—biological oxygen demand; this measure indexes the potential for oxygen consumption by aquatic organisms. Thus a high BOD can indicate potential problems in a lake or river.

Dioxygen is a major industrial reagent; about 10^9 tonnes are used worldwide every year, most in the steel industry. Dioxygen is also used in the synthesis of nitric acid from ammonia (Chapter 14). Almost all the oxygen is obtained by fractional distillation of liquid air. Dioxygen is also consumed in large quantities by hospital facilities. In that context, it is mostly used to raise the dioxygen partial pressure in gas mixtures given to people with respiratory problems, making absorption of oxygen gas easier for poorly functioning lungs.

In the laboratory, there are a number of ways of making dioxygen gas. For example, strong heating of potassium chlorate in the presence of manganese(IV) oxide gives potassium chloride and oxygen gas:

$$2\ KClO_3(l) \rightarrow 2\ KCl(s) + 3\ O_2(g)$$

However, a much safer route is the catalytic decomposition of aqueous hydrogen peroxide. Again, manganese(IV) oxide can be used as the catalyst:

$$2\ H_2O_2(aq) \rightarrow 2\ H_2O(l) + O_2(g)$$

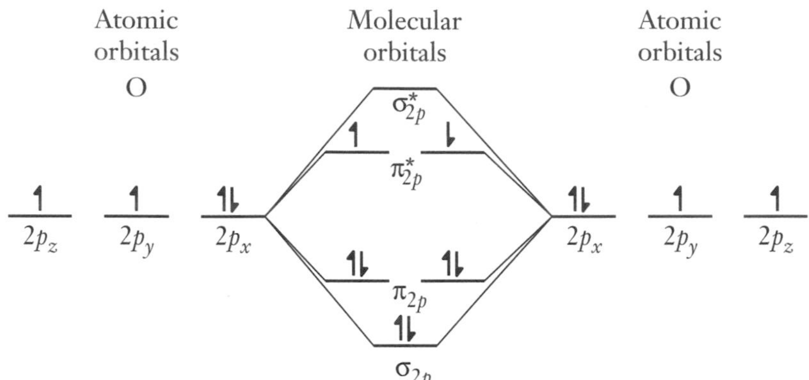

Figure 15.1 Molecular orbital diagram showing the combination of the *2p* atomic orbitals in the dioxygen molecule.

As discussed in Chapter 2, the only representation of the bonding in dioxygen that fits with experimental evidence is the molecular orbital model. Figure 15.1 indicates that the net bond order is 2 (six bonding electrons and two antibonding electrons), with the two antibonding electrons having parallel spins. The molecule, therefore, is paramagnetic.

However, an energy input of only 95 $kJ{\cdot}mol^{-1}$ is required to cause one of the antibonding electrons to "flip" and pair with the other antibonding electron (Figure 15.2). This spin-paired (diamagnetic) form of dioxygen reverts to the paramagnetic form in seconds or minutes, depending on the concentration and environment of the molecule. The diamagnetic form can be prepared by the reaction of hydrogen peroxide with sodium hypochlorite:

$$H_2O_2(aq) + ClO^-(aq) \rightarrow O_2(g)[\text{diamagnetic}] + H_2O(l) + Cl^-(aq)$$

or, alternatively, by irradiating paramagnetic oxygen with ultraviolet radiation in the presence of a sensitizing dye.

Diamagnetic dioxygen is an important reagent in organic chemistry and gives products different from those of the paramagnetic form. Furthermore, diamagnetic oxygen, which is very reactive and formed by ultraviolet radia-

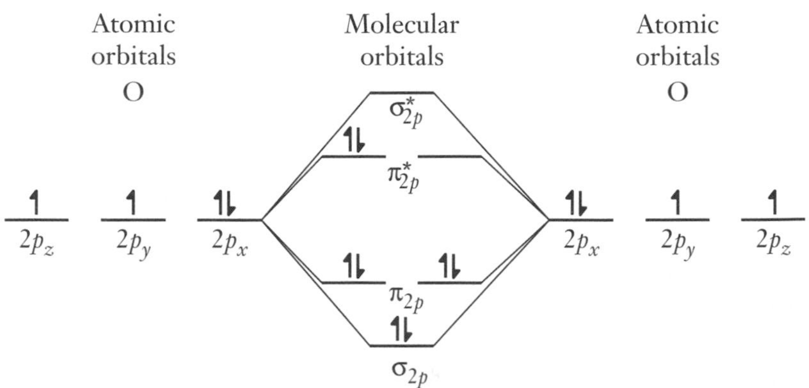

Figure 15.2 Molecular orbital diagram showing the combination of the *2p* atomic orbitals in the more common of the two diamagnetic forms of the dioxygen molecule.

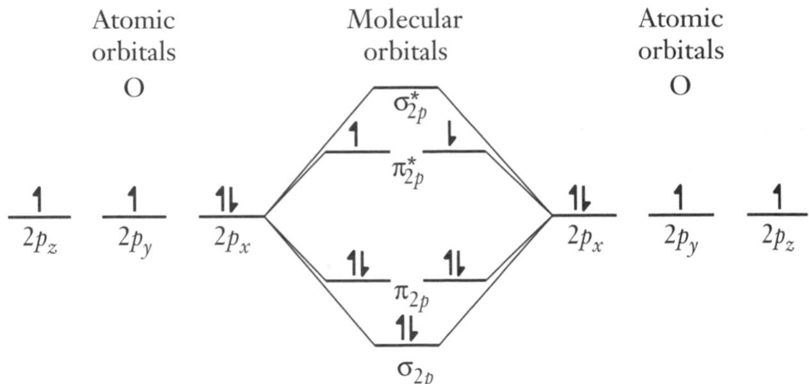

Figure 15.3 Molecular orbital diagram showing the combination of the $2p$ atomic orbitals in the less common of the two diamagnetic forms of the dioxygen molecule.

tion, has been implicated in skin cancer induction. Diamagnetic oxygen is often referred to as singlet oxygen, and the paramagnetic form is called triplet oxygen.

It requires 95 kJ·mol^{-1} to pair up the electrons in the antibonding orbital. There is a second singlet form of dioxygen in which the spin of one electron is simply flipped over, so the resulting unpaired electrons have opposite spins (Figure 15.3). Surprisingly, this arrangement requires much more energy to attain, about 158 kJ·mol^{-1}. As a result, this other singlet form is of little laboratory importance. In Chapter 13, we showed that most of the absorption features in the infrared spectrum of the atmosphere could be explained in terms of vibrations of the water and carbon dioxide molecules. However, there was an absorption at the precise wavelength of 0.76 μm that we did not explain (Figure 15.4). This wavelength, in fact, represents the electronic absorption of energy corresponding to the production of the higher energy singlet form of dioxygen from the normal triplet oxygen. The absorption corresponding to the formation of the lower energy singlet form is hidden under a massive carbon dioxide absorption at 1.27 μm (shown in Figure 13.17).

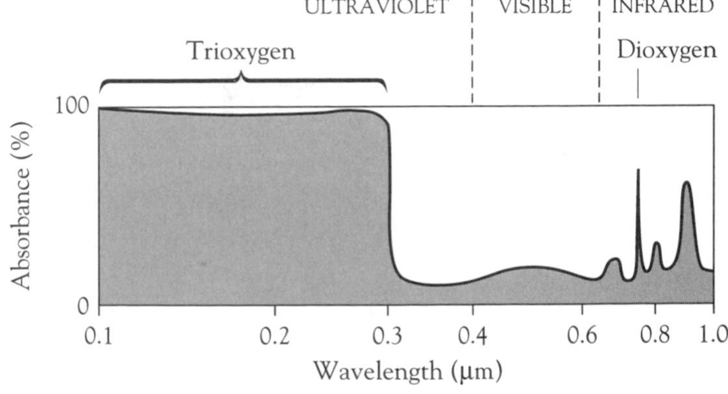

Figure 15.4 The infrared, visible, and ultraviolet portions of the electromagnetic spectrum of the atmosphere.

Trioxygen (Ozone)

This thermodynamically unstable allotrope of oxygen is a diamagnetic gas with a strong odor. In fact, the "metallic" smell of ozone can be detected in concentrations as low as 0.01 ppm. The gas is extremely toxic; the maximum permitted concentration for extended exposure is 0.1 ppm. The gas is produced in regions of high voltages; thus photocopying machines and laser printers have been responsible for high levels of ozone in many office environments. The ozone generated may well have been one cause of headaches and other complaints by office workers. Some machines have carbon filters on the air exhaust to minimize the trioxygen emissions, and these need to be replaced periodically according to the manufacturer's recommendations. However, technological advances have enabled the development of duplicators and printers that produce very low levels of trioxygen.

A convenient way to generate trioxygen is to pass a stream of dioxygen through a 10- to 20-kV electric field. This field provides the energy necessary for the reaction:

$$3\ O_2(g) \rightarrow 2\ O_3(g) \qquad \Delta H_f^\circ = +143\ \text{kJ·mol}^{-1}$$

At equilibrium, the concentration of trioxygen is about 10 percent. The trioxygen slowly decomposes to dioxygen, although the rate of conversion depends on the phase (gas or aqueous solution).

Trioxygen is a very powerful oxidizing agent, much more powerful than dioxygen, as can be seen from a comparison of reduction potentials in acid solution:

$$O_3(g) + 2\ H^+(aq) + 2\ e^- \rightarrow O_2(g) + H_2O(l) \qquad E^\circ = +2.07\ \text{V}$$

$$O_2(g) + 4\ H^+(aq) + 4\ e^- \rightarrow 2\ H_2O(l) \qquad E^\circ = +1.23\ \text{V}$$

In fact, in acid solution, fluorine and the perxenate ion, XeO_6^{4-}, are the only common oxidizing agents that are stronger than trioxygen. Its great oxidizing ability is illustrated by the fact that trioxygen will oxidize iodide to iodine and sulfide to sulfate:

$$O_3(g) + 2\ I^-(aq) + H_2O(l) \rightarrow O_2(g) + I_2(aq) + 2\ OH^-(aq)$$

$$PbS(s) + 4\ O_3(g) \rightarrow PbSO_4(s) + 4\ O_2(g)$$

It is the strongly oxidizing nature of trioxygen that enables it to be used as a bactericide. For example, it is used to kill bacteria in bottled waters; and in France particularly, it is used to kill organisms in municipal water supplies and in public swimming pools. Conversely, water purification experts in North America have preferred the use of chlorine gas for water purification. There are advantages and disadvantages of both bactericides. Ozone changes to dioxygen over a fairly short period of time; thus its bactericidal action is not long lasting. However, ozone is chemically innocuous in water supplies. Dichlorine remains in the water supply to ensure bactericidal action, but it reacts with any organic contaminants in the water supply to form hazardous organochlorine compounds.

On the surface of Earth, ozone is a dangerous compound, a major atmospheric pollutant in urban areas. In addition to its damaging effect on lung tissue and even on exposed skin surfaces, ozone reacts with the rubber of tires, causing them to become brittle and crack. However, as most people now know, it is a different story in the upper atmosphere, where the ozone in the stratosphere provides a vital protective layer for life on Earth. The absorption region that results from atmospheric ozone absorption is shown in Figure 15.4.

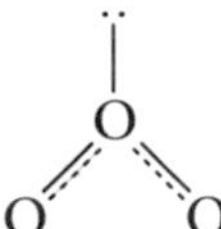

Figure 15.5 The ozone molecule.

The process is quite complex, but the main steps are as follows: First, the shorter wavelength ultraviolet radiation reacts with dioxygen to produce atomic oxygen:

$$O_2(g) \xrightarrow{UV} 2\ O(g)$$

The atomic oxygen reacts with dioxygen to give trioxygen:

$$O(g) + O_2(g) \rightarrow O_3(g)$$

The trioxygen absorbs longer wavelength ultraviolet radiation and decomposes back to dioxygen:

$$O_3(g) \xrightarrow{UV} O_2(g) + O(g)$$

$$O_3(g) + O(g) \rightarrow 2\ O_2(g)$$

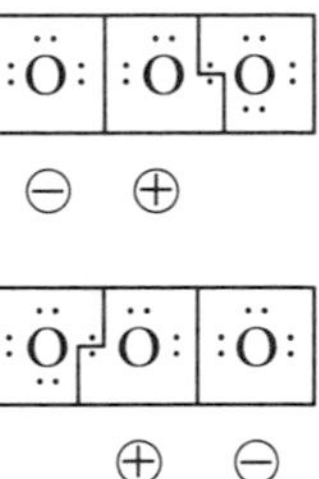

Figure 15.6 The bonding of trioxygen can be interpreted as the mean of these two resonance structures.

Trioxygen is a V-shaped molecule with a bond angle of 117°. Its oxygen-oxygen bonds are of equal length and have a bond order of about $1\frac{1}{2}$ (Figure 15.5). The bond angles and bond lengths of trioxygen are very similar to those of the isoelectronic nitrite ion.

We can use resonance structures to account for the bond order (Figure 15.6). However, a molecular orbital approach provides a better representation (Figure 15.7). Like the nitrogen dioxide arrangement, we envisage the 2*p* orbitals to be those responsible for the bonding. A set of σ bonds is formed by overlap of the four *p* orbitals along the bonding axes: two of the *p* orbitals on the central oxygen plus one each on the terminal oxygen atoms. This arrangement results in the formation of four molecular orbitals: two bonding and two antibonding. The three *p* orbitals at right angles to the molecular plane provide three π molecular orbitals: one bonding, one nonbonding, and one antibonding. Finally, the remaining *p* orbital (in the molecular plane but at right angles to the bonding direction) on each of the terminal oxygen atoms will be nonbonding. Filling these orbitals with the twelve 2*p* electrons results in two net σ bonds and one π bond, the latter being shared by the three atoms. Like the resonance approach, the $1\frac{1}{2}$ bond order matches the experimental bond information.

Molecular orbitals

σ^* σ^*

π^*

σ^{NB} π^{NB} σ^{NB}

π

σ σ

Figure 15.7 The molecular orbitals derived from the 2*p* atomic orbitals in trioxygen.

Ozone forms compounds with the alkali and alkaline earth metals. These compounds contain the trioxide(1−) ion, O_3^-. As we would expect from lattice stability arguments, it is the larger cations, such as cesium, that form the most stable trioxides. It has been shown that the trioxide(1−) is also V-shaped, so the molecular orbital diagram should be similar to that of trioxygen itself (Figure 15.8). Thus the additional electron in the anion should enter the π antibonding orbital. This arrangement would reduce the π bonding to one-half, or one-fourth per oxygen-oxygen bond. Experimental measurements have shown this to be the case. The oxygen-oxygen bond length is 135 pm in the trioxide(1–) ion, slightly longer than the 128-pm bond in trioxygen itself.

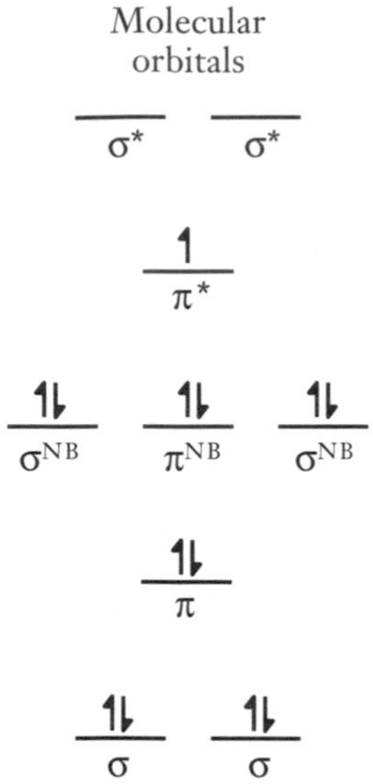

Molecular orbitals

Figure 15.8 The molecular orbitals derived from the 2*p* atomic orbitals in the trioxide(1–) ion.

Bonding in Covalent Oxygen Compounds

Oxygen atoms usually form two single covalent bonds or one multiple bond, ordinarily a double bond. When they form two covalent bonds, the angle between the bonds can be significantly different from the $109\frac{1}{2}°$ tetrahedral angle. The traditional explanation for the bond angle of $104\frac{1}{2}°$ in water asserts that the lone pairs occupy more space than bonding pairs, thus "squashing" the bond angle in the water molecule.

However, when we compare the two halogen-oxygen compounds—oxygen difluoride, OF_2 (with a bond angle of 103°), and dichlorine oxide, Cl_2O (with a bond angle of 111°)—we have to look for a different explanation. The best explanation relates to the degree of orbital mixing. In Chapter 3, we considered the hybridization model of bonding, where orbital characters mixed. At the time, we considered integral values of mixing: for example, one *s* orbital with three *p* orbitals to form four sp^3 hybrid orbitals. However, there is no reason why the mixing cannot be fractional. Thus some covalent bonds can have more *s* character and others, more *p* character. Also, beyond Period 2, we have to keep in mind that *d* orbitals might be mixed in as well. This approach of partial orbital mixing is, in fact, moving toward the more realistic molecular orbital representation of bonding.

It was Henry A. Bent who proposed an empirical rule to explain, among other things, the variation in bond angles of oxygen compounds. The rule states: More electronegative substituents "prefer" hybrid orbitals with less *s* character, and more electropositive substituents "prefer" hybrid orbitals with more *s* character. Thus, with fluorine (more electronegative), the bond angle tends toward the 90° angle of two "pure" *p* orbitals on the oxygen atom. Conversely, the angle for chlorine (less electronegative) is greater than that for an sp^3 hybrid orbital, somewhere between the $109\frac{1}{2}°$ angle of sp^3 hybridization and the 120° angle of sp^2 hybridization. An alternative explanation for the larger angle in dichlorine oxide is simply that there is a steric repulsion between the two large chlorine atoms, thus increasing the angle.

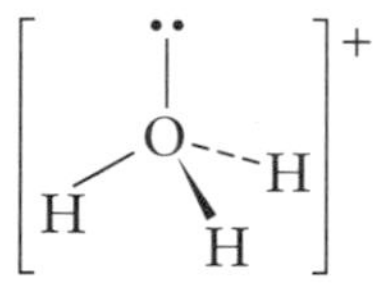

Figure 15.9 The hydronium ion.

Oxygen can form coordinate covalent bonds, either as a Lewis acid or as a Lewis base. The former are very rare; the compound NF_3O (mentioned in Chapter 14) is one such case. However, oxygen readily acts as a Lewis base, for example, in the bonding of water molecules to transition metal ions through a lone pair on the oxygen. There are also numerous examples of double bonded oxygen, such as that in PF_3O. In these cases, there is a π bond that involves a lone pair on the oxygen and an empty *d* orbital on the other element.

There are also other bonding modes of oxygen. In particular, oxygen can form three equivalent covalent bonds. The classic example is the hydronium ion, in which each bond angle is close to that of the tetrahedral value of $109\frac{1}{2}°$ (Figure 15.9). However, such oxygen molecules are not always tetrahedral. In the unusual cation $[O(HgCl)_3]^+$, the atoms are all coplanar and each bond angle is 120° (Figure 15.10). To explain this, we must assume that the lone

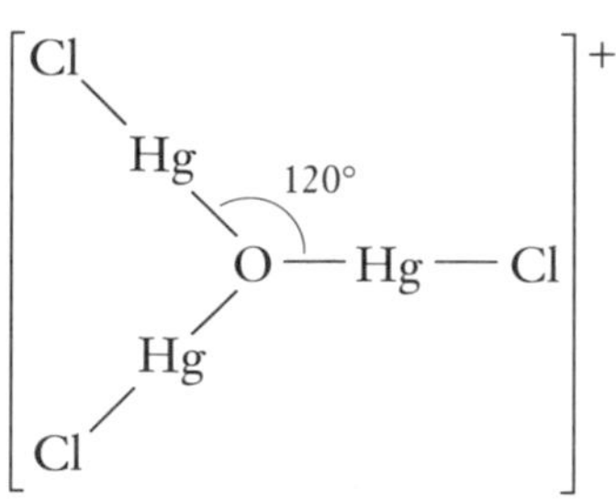

Figure 15.10 The $[O(HgCl)_3]^+$ ion.

pair on the oxygen atom is not in its usual sp^3 hybrid orbital but in a *p* orbital, where it can form a π bond with empty orbitals of the mercury atom.

Trends in Oxygen Compounds

The oxides of electropositive metals are ionic and basic. Some of these, such as barium oxide, react directly with water to give a base:

$$BaO(s) + H_2O(l) \rightarrow Ba(OH)_2(aq)$$

Other basic oxides, such as copper(II) oxide, are water-insoluble, but they will react with dilute acids:

$$CuO(s) + 2\ H^+(aq) \rightarrow Cu^{2+}(aq) + H_2O(l)$$

The oxides of weakly electropositive metals such as aluminum, zinc, and tin are amphoteric, that is, they react with both acids and bases. For example, zinc oxide reacts with acids to give the hexaaquazinc(II) cation, $[Zn(OH_2)_6]^{2+}$, which we can simply represent as $Zn^{2+}(aq)$; and it reacts with bases to give the tetrahydroxozincate(II), $[Zn(OH)_4]^{2-}$, anion:

$$ZnO(s) + 2\ H^+(aq) \rightarrow Zn^{2+}(aq) + H_2O(l)$$

$$ZnO(s) + 2\ OH^-(aq) + H_2O(l) \rightarrow [Zn(OH)_4]^{2-}(aq)$$

If a metal forms more than one oxide, then the lower oxide is always basic (and sometimes reducing), whereas the higher oxide is often acidic and oxidizing. For example, chromium(III) oxide is ionic, and it dissolves in acid to give the chromium(III) ion, whereas chromium(VI) oxide is covalently bonded, and it reacts with water to give chromic acid:

$$Cr_2O_3(s) + 6\ H^+(aq) \rightarrow 2\ Cr^{3+}(aq) + 3\ H_2O(l)$$

$$CrO_3(s) + H_2O(l) \rightarrow H_2CrO_4(aq)$$

Oxygen often "brings out" a higher oxidation state in the elements of Periods 3–6 than does fluorine, probably as a result of the ability of oxygen to form a π bond, using one of its own full *p* orbitals and an empty *d* orbital on the other element (Table 15.3).

The oxides of nonmetals are always covalently bonded. Those with the element in a low oxidation state tend to be neutral, whereas those with the element in the higher oxidation states tend to be acidic. For example,

Table 15.3 Highest oxidation states of oxides and fluorides of three elements

Element	Highest oxide	Highest fluoride
Chromium	CrO_3 (+6)	CrF_5 (+5)
Xenon	XeO_4 (+8)	XeF_6 (+6)
Osmium	OsO_4 (+8)	OsF_7 (+7)

dinitrogen oxide, N_2O, is neutral, whereas dinitrogen pentoxide dissolves in water to give nitric acid:

$$N_2O_5(g) + H_2O(l) \rightarrow 2\ HNO_3(l)$$

As we have seen, there are oxygen ions in which oxygen itself has an abnormal oxidation state. These include the dioxide(2–) ion, O_2^{2-}, the dioxide(1–) ion, O_2^-, and the trioxide(1–) ion, O_3^-. These exist only in solid-phase compounds, specifically those in which the metal cation has a charge density low enough to stabilize these large, low-charge anions.

Mixed Metal Oxides

In Chapter 11 we discussed a family of mixed metal oxides, the spinels. These compounds have empirical formulas of AB_2X_4, where A is usually a dipositive metal ion; B, usually a tripositive metal ion; and X, a dinegative anion, usually oxygen. The crystal lattice consists of a framework of oxide ions with metal ions in octahedral and tetrahedral sites. This is not the only possible structure of mixed metal oxides: There are many more, one of which is the *perovskite* structure; and, like the spinels, perovskites are of great interest to materials scientists.

Perovskites have the general formula ABO_3, where A is usually a large dipositive metal ion and B is generally a small tetrapositive metal ion. It is important to distinguish these mixed metal oxides from the oxyanion salts that we generally study in inorganic chemistry. The metal salts of oxyanions can have a formula parallel to that of a perovskite: AXO_3, involving a metal (A), a nonmetal (X), and oxygen. In these compounds, the XO_3 consists of a covalently bonded polyatomic ion. For example, sodium nitrate consists of the Na^+ and NO_3^- ions arrayed in a sodium chloride structure, in which each nitrate ion occupies the site equivalent to the chloride ion site. However, in perovskites, such as the parent compound, calcium titanate, $CaTiO_3$, there is no such thing as a "titanate ion." Instead, the crystal lattice consists of independent Ca^{2+}, Ti^{4+}, and O^{2-} ions.

The packing in the perovskite unit cell is shown in Figure 15.11. The large calcium ion occupies the center of the cube; it is surrounded by 12 oxide ions. Eight titanium(IV) ions are located at the cube corners; each has six oxide neighbors (three being in adjacent unit cells). Many perovskites are ferroelectric materials (although they contain no iron). Such compounds can convert a mechanical pulse into an electrical signal (and vice versa), a property that is important for many electronic devices. In Chapter 13, another perovskite, $CaPbO_3$, was mentioned as a rust-resistant coating for metal surfaces.

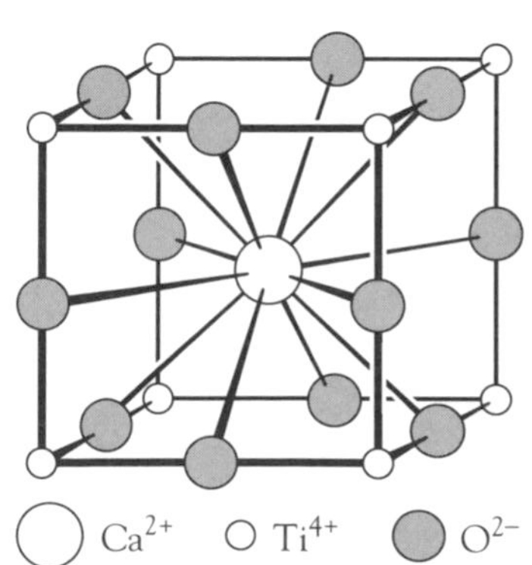

Figure 15.11 The perovskite unit cell.

The perovskite sodium tungstate, $NaWO_3$, has very unusual properties. This compound can be prepared with less than stoichiometric proportions of sodium ion; that is, Na_xWO_3, where $x < 1$. The stoichiometric tungstate is white, but as the mole proportion of sodium drops to 0.9, it becomes a metallic golden yellow. As the proportion drops from 0.9 to 0.3, colors from metallic orange to red to bluish black are obtained. This material and its relatives, called the *tungsten bronzes*, are often used for metallic paints. Not only do the compounds look metallic, their electrical conductivity approaches that of a metal. In the crystal, increasing proportions of the cell centers, where the large alkali metal would be found, are vacant. As a result, the conduction band, which in the stoichiometric compound would be full, is now partially empty. Under these circumstances, electrons can move through the π system along the cell edges by using tungsten ion d orbitals and oxide ion p orbitals. It is this electron mobility that produces the color and the electrical conductivity.

Water

Water is the only common liquid on this planet. Without water as a solvent for our chemical and biochemical reactions, life would be impossible. Yet a comparison of water with the other hydrides of Group 16 would lead us to expect it to be a gas at the common range of temperatures found on Earth. In fact, on the basis of a comparison with the other Group 16 hydrides, we would expect water to melt at about –100°C and boil at about –90°C (Figure 15.12). As discussed in Chapter 7, the cause of its abnormally high melting and boiling points is the strength of the hydrogen bonding between water molecules. To break these bonds, more energy (and hence a higher temperature) is necessary for the phase change.

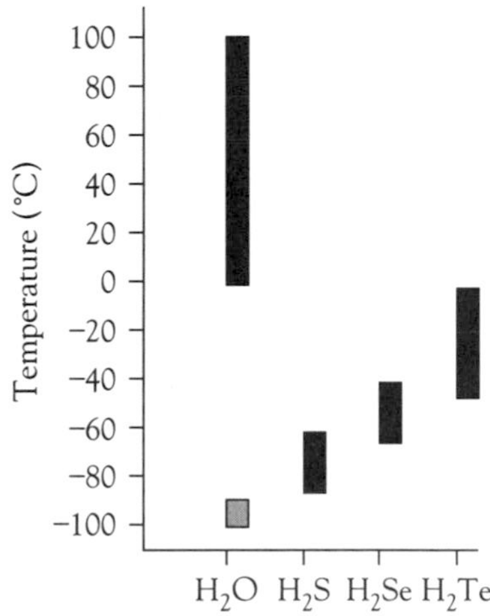

Figure 15.12 The liquid ranges of the Group 16 hydrides. If water molecules were not bound to one another by hydrogen bonds, then its liquid range would be between –90°C and –100°C (shaded rectangle).

Liquid water has formed and re-formed Earth's surface over geological time. It has been able to do this because it has the ability to dissolve ionic substances, particularly the alkali and alkaline earth metals and the common anions such as chloride and sulfate. Thus the composition of seawater reflects the leaching of ions from minerals since the time that Earth cooled enough to form liquid water. The current composition of this solution—the oceans, which make up 97 percent of the water on Earth—is shown in Table 15.4.

Many of our mineral deposits were formed by aqueous processes. We assume that the massive deposits of the alkali and alkaline earth minerals were formed by deposition from ancient seas and lakes. Less obvious is the mechanism of formation of heavy metal sulfide deposits, such as lead(II) sulfide. In fact, they also are the result of aqueous solution processes. Even though these minerals are extremely insoluble at common temperatures and pressures, this is not the case under the extremely high pressures and very high temperatures that exist deep under Earth's surface. Under those conditions, many ions dissolve and are transported to near Earth's surface, where reductions in temperature and pressure cause precipitation to occur.

Table 15.4 Ion proportions of the principal ionic constituents in seawater

Cation	Concentration (%)	Anion	Concentration (%)
Sodium	86.3	Chloride	94.5
Magnesium	10.0	Sulfate	4.9
Calcium	1.9	Hydrogen carbonate	0.4
Potassium	1.8	Bromide	0.1

Water acts as a solvent for these ionic compounds by the interaction of ions in the crystal lattice with the dipoles of the water molecules. Thus the partially negative oxygen atoms of water molecules are attracted to the cations and the partially positive hydrogen atoms of water molecules are attracted toward the anions. As discussed in Chapter 5, it is the balance of the enthalpy and entropy terms for the breakup of the crystal lattice and those terms for the formation of the hydrated ions that determines solubility. It is generally true that the interaction between the anion and water is simply that of ion-dipole electrostatic forces. For the cation, the picture is murkier. Certainly for the low charge density alkali metals, the interaction can still be attributed to ion-dipole electrostatic forces. When alkali metal salts crystallize, either the water molecules are lost in the process (as in sodium chloride) or the water molecules simply fill holes in the crystal lattice (as in sodium sulfate decahydrate).

When we consider higher charge density cations, particularly those of the transition metals, the interaction is much stronger and more closely resembles that of a covalent bond. For example, the iron(III) ion crystallizes as the hexaaquairon(III) ion, $[Fe(H_2O)_6]^{3+}$, with the oxygen atoms of the water molecules organized in a precise octahedral arrangement around the metal ion. In these hydrates we consider the water molecules to be essentially bonded to the cations by using an oxygen lone pair to form a coordinate covalent bond. In this way, the metal ion is acting as a Lewis acid and the water molecule is the Lewis base. To indicate that it is the oxygen end of the water molecule that is bonded to the metal ion, we conventionally write such a hydrated ion as $[Fe(OH_2)_6]^{3+}$, reversing the order of the hydrogen and oxygen. The interaction between the water molecules and transition metal ions is discussed in detail in Chapter 18.

The hydration process is usually exothermic. Anhydrous calcium chloride, for example, reacts with water to form the hexahydrate:

$$CaCl_2(s) + 6\,H_2O(l) \rightarrow [Ca(OH_2)_6]^{2+}(aq) + 2\,Cl^-(aq)$$

This exothermic reaction is used in some varieties of “instant heat” packets. Anhydrous calcium chloride is also used to melt ice, where part of its benefits result from the exothermic hydration as well as its effect of lowering the freezing point of water.

Water is also the basis of our acid-base system, autoionizing to give the hydronium and hydroxide ions, the strongest acid and base, respectively, in our aqueous world:

$$2\,H_2O(l) \rightleftharpoons H_3O^+(aq) + OH^-(aq)$$

Although in chemical equations we often write the hydronium ion as $H^+(aq)$, such a tiny high charge density ion could not exist in solution. There is good evidence of the existence of the hydronium ion: When hydrochloric acid freezes, it gives crystals of hydronium chloride, $H_3O^+Cl^-$. The hydronium ion resembles the ammonium ion in some respects. For example, perchloric acid forms a solid, $H_3O^+ClO_4^-$, whose crystals are the same shape as ammonium perchlorate, $NH_4^+ClO_4^-$. Actually, even the hydronium ion is only an approximation of the actual hydrogen ion environment, because there is evidence that four other water molecules hydrogen bond to the ion to give an ion of formula $H_9O_4^+$.

Hydrogen Peroxide

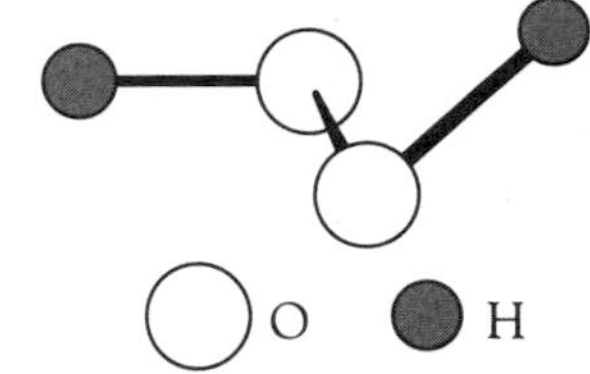

Figure 15.13 The hydrogen peroxide molecule.

Pure hydrogen peroxide is an almost colorless (slightly bluish), viscous liquid; its high viscosity is a result of the high degree of hydrogen bonding. It is an extremely corrosive substance that should always be handled with great care. The shape of the molecule is unexpected; the H–O–O bond angle in the gas phase is only $94\frac{1}{2}°$ (about 10° less than the H–O–H bond angle in water), and the two H–O units form a dihedral angle of 111° with each other (Figure 15.13).

Hydrogen peroxide is thermodynamically very unstable with respect to disproportionation:

$$H_2O_2(l) \rightarrow H_2O(l) + \tfrac{1}{2}\,O_2(g) \qquad \Delta G = -119.2\ \text{kJ·mol}^{-1}$$

However, when pure, it is slow to decompose because of kinetic factors (the reaction pathway must have a high activation energy). Almost anything—transition metal ions, metals, blood, dust—will catalyze the decomposition.

A solution of hydrogen peroxide can be prepared in the laboratory by the reaction of sodium peroxide with water:

$$Na_2O_2(s) + H_2O(l) \rightarrow 2\ NaOH(aq) + H_2O_2(aq)$$

It is advisable to handle even dilute solutions of hydrogen peroxide with gloves and eye protection, because it attacks the skin. Hydrogen peroxide can act as an oxidizing or reducing agent in both acidic and basic solutions. Oxidations are usually performed in acidic solution and reductions in basic solution:

$$H_2O_2(aq) + 2\ H^+(aq) + 2\ e^- \rightarrow 2\ H_2O(l) \qquad E° = +1.77\ V$$

$$HO_2^-(aq) + OH^-(aq) \rightarrow O_2(g) + H_2O(l) + 2\ e^- \qquad E° = +0.08\ V$$

Hydrogen peroxide will oxidize iodide ion to iodine and reduce permanganate ion in acid solution to manganese(II) ion. Hydrogen peroxide has an important application to the restoration of antique paintings. One of the favored white pigments was white lead, a mixed carbonate-hydroxide, $Pb_3(OH)_2(CO_3)_2$. Traces of hydrogen sulfide cause the conversion of this white compound to black lead(II) sulfide, which discolors the painting. Application of hydrogen peroxide oxidizes the lead(II) sulfide to white lead(II) sulfate, thereby restoring the correct color of the painting:

$$PbS(s) + 4\ H_2O_2(aq) \rightarrow PbSO_4(s) + 4\ H_2O(l)$$

Apart from being a convenient redox reagent, hydrogen peroxide is used to test for the presence of chromium ions. Addition of hydrogen peroxide to dichromate ion solutions results in the formation of blue chromium(VI) oxide peroxide, $CrO(O_2)_2$. This covalent compound can be extracted into a low-polarity organic solvent, such as ethoxyethane (diethyl ether).

Hydrogen peroxide is a major industrial chemical; about 10^6 tonnes are produced worldwide every year. Almost all industrial syntheses use a complex series of organic reactions known as the anthoquinone process. Its uses are highly varied, from paper bleaching to household products, particularly hair bleaches. Hydrogen peroxide is also used as an industrial reagent, for example, in the synthesis of sodium peroxoborate (Chapter 12).

Hydroxides

Almost every metallic element forms a hydroxide. The colorless hydroxide ion is the strongest base in aqueous solution. It is very hazardous, because it reacts with the proteins of the skin, producing a white opaque layer. For this reason, it is particularly hazardous to the eyes. In spite of its dangerous nature, many household products, particularly oven and drain cleaners, utilize solid or concentrated solutions of sodium hydroxide. It is also important to realize that, through hydrolysis, very high levels of hydroxide ion are present in products that do not appear to contain them. For example, the phosphate ion, used in sodium phosphate–containing cleansers, reacts with water to give hydroxide ion and the hydrogen phosphate ion:

$$PO_4^{3-}(aq) + H_2O(l) \rightarrow HPO_4^{2-}(aq) + OH^-(aq)$$

Sodium hydroxide is prepared by electrolysis of aqueous brine (Chapter 10). Calcium hydroxide is obtained by heating calcium carbonate to give calcium oxide, which is then mixed with water to give calcium hydroxide:

$$CaCO_3(s) \xrightarrow{\Delta} CaO(s) + CO_2(g)$$

$$CaO(s) + H_2O(l) \rightarrow Ca(OH)_2(s)$$

Calcium hydroxide is actually somewhat soluble in water—soluble enough to give a significantly basic solution. A mixture of the saturated solution with a suspension of excess solid calcium hydroxide was referred to as "whitewash," and it was used as a low-cost white coating for household painting.

Many metal hydroxides can be prepared by adding a metal ion solution to a hydroxide ion solution. Thus green-blue copper(II) hydroxide can be prepared by mixing solutions of copper(II) chloride with sodium hydroxide:

$$CuCl_2(aq) + 2\ NaOH(aq) \rightarrow Cu(OH)_2(s) + 2\ NaCl(aq)$$

Many of the insoluble hydroxides precipitate out of solution as a *gelatinous* (jellylike) solid, making them difficult to filter. Furthermore, some of the metal hydroxides are very unstable; they lose water to form the oxide, which, with its higher charge, forms a more stable lattice. For example, even gentle warming of the green-blue copper(II) hydroxide gel produces the black solid copper(II) oxide:

$$Cu(OH)_2(s) \xrightarrow{\Delta} CuO(s) + H_2O(l)$$

Solutions of the soluble hydroxides (the alkali metals, barium, and ammonium) react with the acidic oxide, carbon dioxide, present in the air to give the metal carbonate. For example, sodium hydroxide reacts with carbon dioxide to give sodium carbonate solution:

$$2\ NaOH(aq) + CO_2(g) \rightarrow Na_2CO_3(aq) + H_2O(l)$$

For this reason, solutions of hydroxides should be kept sealed except while being used. It is also one of the reasons why sodium hydroxide contained in glass bottles should be sealed with a rubber stopper rather than a glass stopper. Some of the solution in the neck of the bottle will react to form crystals of sodium carbonate—enough to effectively "glue" the glass stopper into the neck of the bottle.

Alkali and alkaline earth metal hydroxides react with carbon dioxide, even when they are in the solid phase. In fact, "whitewashing" involves the penetration of the partially soluble calcium hydroxide into the wood or plaster surface (often the hydroxide ion acts additionally as a degreasing agent). Over the following hours and days, it reacts with the carbon dioxide in the air to give microcrystalline, very insoluble, intensely white calcium carbonate:

$$Ca(OH)_2(aq) + CO_2(g) \rightarrow CaCO_3(s) + H_2O(l)$$

This process, performed by many of our ancestors, involved some very practical chemistry! As discussed in Chapter 11, this reaction was also employed as a simple adhesive for brick and stone structures before the invention of cement.

Sulfur

Sulfur (the other nonmetal in Group 16) has a range of even oxidation states from +6, through +4 and +2, to −2. The oxidation state diagram for sulfur in acidic and basic solutions is shown in Figure 15.14. The comparatively low free energy of the sulfate ion in acidic solution indicates that the ion is only weakly oxidizing. In basic solution, the sulfate ion is completely nonoxidizing, and it is the most thermodynamically stable sulfur species. Although on a convex curve, the +4 oxidation state is actually quite stable. The Frost diagram does show that in acidic solution the +4 state tends to be reduced; whereas in basic solution it tends to be oxidized. The element itself is usually reduced in acidic environments but oxidized in base. Figure 15.14 also shows that the sulfide ion (basic solution) is a fairly strong reducing agent but that hydrogen sulfide is a thermodynamically stable species.

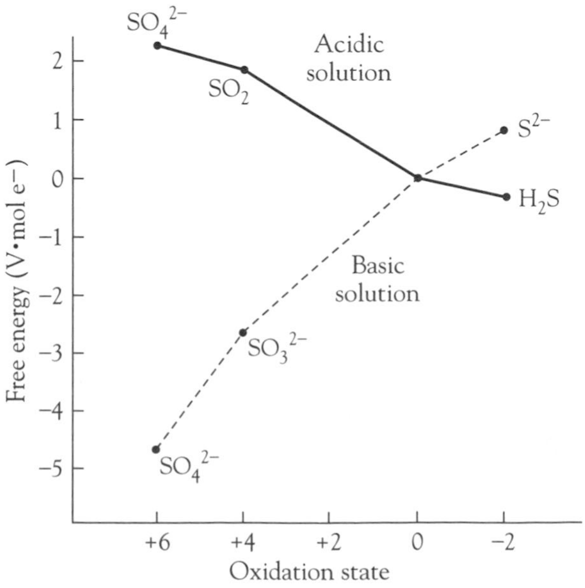

Figure 15.14 Frost diagram for sulfur in acidic and basic solutions.

After carbon, sulfur is the element most prone to catenate. However, there are only two available bonds. Thus the structures are typically chains of sulfur atoms with some other element or group of elements at each end: Dihydrogen polysulfides have the formula, HS–S_n–SH, and the polysulfur dichlorides, ClS–S_n–SCl, where n can have any value between 0 and 20.

Allotropes of Sulfur

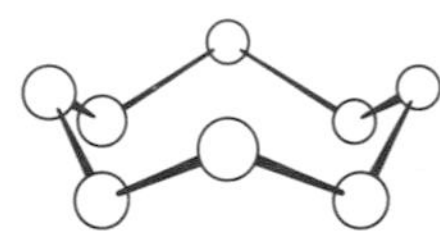

Figure 15.15 The cyclooctasulfur molecule.

Although elemental sulfur has been known from the earliest times, it is only in the last 20 years that the allotropy of sulfur has been clarified. The most common naturally occurring allotrope, S_8, cyclooctasulfur, has a zigzag arrangement of the atoms around the ring (Figure 15.15). This allotrope crystallizes to form needle crystals above 95°C; but below that temperature, "chunky" crystals are formed. The crystals, which are referred to as monoclinic and rhombic forms, differ simply in the way in which the molecules pack. These two forms are *polymorphs* of each other, not allotropes. Polymorphs are defined as different crystal forms in which the identical units of the same compound are packed differently.

In 1891 a sulfur allotrope with a ring size other than eight was first synthesized. This allotrope, S_6, cyclohexasulfur, was the second of many true allotropes of sulfur to be discovered. To distinguish allotropes and polymorphs, we can more correctly define *allotropes* as forms of the same element that contain different molecular units. Sulfur allotropes with ring sizes that range from 6 to 20 have been definitely synthesized, and there is evidence that allotropes with much larger rings exist. The most stable, apart from cyclooctasulfur, is S_{12}, cyclododecasulfur. The structures of cyclohexasulfur and cyclododecasulfur are shown in Figure 15.16.

Cyclohexasulfur can be synthesized by mixing sodium thiosulfate, $Na_2S_2O_3$, and concentrated hydrochloric acid:

$$6\ Na_2S_2O_3(aq) + 12\ HCl(aq) \rightarrow S_6(s) + 6\ SO_2(g) + 12\ NaCl(aq) + 6\ H_2O(l)$$

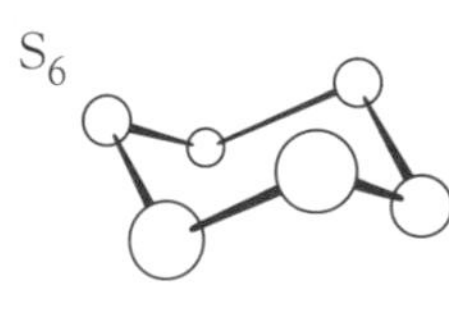

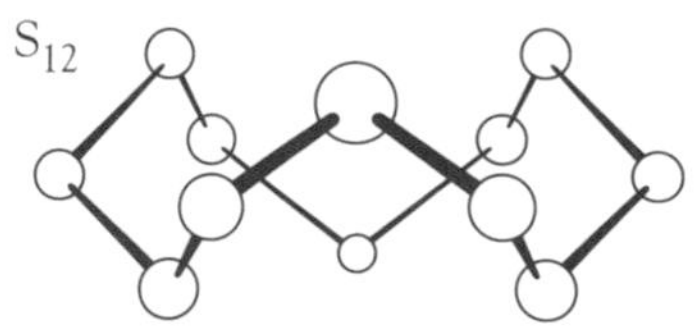

Figure 15.16 The cyclohexasulfur and cyclododecasulfur molecules.

However, there is now a fairly logical synthesis of the even-numbered rings (which are more stable than the odd-numbered rings). The method involves the reaction of the appropriate hydrogen polysulfide, H_2S_x, with the appropriate polysulfur dichloride, S_yCl_2, such that $(x + y)$ equals the desired ring size. Thus cyclododecasulfur can be prepared by mixing dihydrogen octasulfide, H_2S_8, and tetrasulfur dichloride, S_4Cl_2, in ethoxyethane, $(C_2H_5)_2O$, a solvent:

$$H_2S_8(eth.) + S_4Cl_2(eth.) \rightarrow S_{12}(s) + 2\ HCl(g)$$

However, cyclooctasulfur is the allotrope found almost exclusively in nature and as the product of almost all chemical reactions, so we will concentrate on the properties of this allotrope. At its melting point, cyclooctasulfur forms a low-viscosity, straw-colored liquid. But when the liquid is heated, there is an abrupt change in properties at 159°C. The most dramatic transformation is a 10^4-fold increase in viscosity. The liquid also darkens considerably. We can explain these changes in terms of a rupture of the rings. The octasulfur chains then link one to another to form polymers containing as many as 20,000 sulfur atoms. The rise in viscosity, then, is explained by a replacement of the free-moving S_8 molecules by these intertwined chains, which have strong dispersion force interactions.

As the temperature increases toward the boiling point of sulfur (444°C), the viscosity slowly drops as the polymer units start to fragment as a result of the greater thermal motion. If this liquid is poured into cold water, a brown transparent rubbery solid—plastic sulfur—is formed. This material slowly changes to microcrystals of rhombic sulfur.

Boiling sulfur produces a green gaseous phase, most of which consists of cyclooctasulfur. Raising the temperature even more causes the rings to fragment; and by 700°C, a violet gas is observed. This gas contains disulfur molecules, S_2, analogous to dioxygen.

Industrial Extraction of Sulfur

Elemental sulfur is found in large underground deposits in the United States and Poland. It is believed they were formed by the action of anaerobic bacteria on lake-bottom deposits of sulfate minerals. A discovery causing great excitement among planetary scientists is the evidence for large deposits of sulfur on Jupiter's moon, Io.

The method of extraction was devised by a Canadian scientist, Herman Frasch, after whom the process is named (Figure 15.17). The sulfur deposits are between 150 and 750 m underground and are typically about 30 m thick. A pipe 20 cm in diameter is sunk almost to the bottom of the deposit. Then a 10-cm pipe is inserted inside the larger one; this pipe is a little shorter than the outer pipe. Finally, a 2.5-cm pipe is inserted into the middle pipe, but ends about halfway down the length of the outer pipes.

Water at 165°C is initially pumped down both outer pipes; this water melts the surrounding sulfur. The flow of superheated water down the 10-cm pipe is discontinued, and liquid pressure starts to force the dense liquid sulfur up that pipe. Compressed air is pumped down the 2.5-cm pipe, producing a low-density froth that flows freely up the 10-cm pipe to the surface. At the surface, the sulfur-water-air mixture is pumped into gigantic tanks, where it cools and the violet sulfur liquid crystallizes to a solid yellow block. The

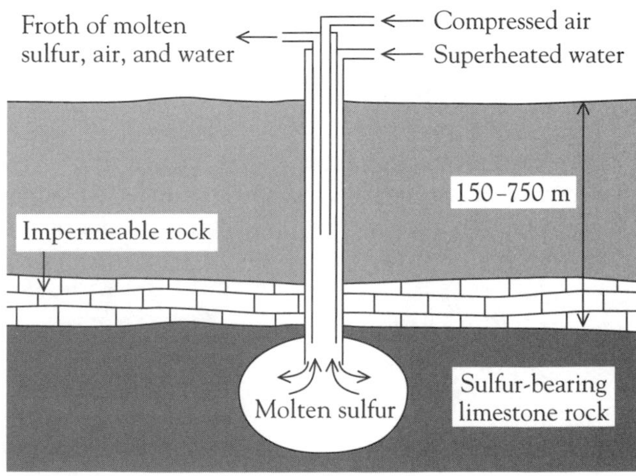

Figure 15.17 The Frasch method of extraction of sulfur.

retaining walls of the tank are then removed, and dynamite is used to break up the blocks to a size that can be transported by railcar.

The United States and Poland are the only countries fortunate enough to have large underground deposits of elemental sulfur. Other nations must resort to sources such as natural gas deposits. Many of these deposits contain high levels of hydrogen sulfide, a toxic and very odorous gas. Deposits that are low in hydrogen sulfide are known as "sweet gas," and those containing high levels, between 15 and 20 percent typically, are known as "sour gas." Gas producers are obviously only too pleased to find a market for this contaminant of the hydrocarbon mixtures.

The production of elemental sulfur from the hydrogen sulfide in natural gas is accomplished by using the Claus process. The hydrogen sulfide is first extracted from the sour natural gas by bubbling the gas through a basic organic solvent, $HOCH_2CH_2NH_2$, the hydrogen sulfide acting as a Brønsted-Lowry acid:

$$HOCH_2CH_2NH_2(l) + H_2S(g) \rightleftharpoons HOCH_2CH_2NH_3^+(solvent) + HS^-(solvent)$$

The solution is removed and warmed, causing the hydrogen sulfide gas to be released. The hydrogen sulfide is then mixed with dioxygen in a 2:1 mole ratio rather than the 2:3 mole ratio that would be needed to oxidize all the hydrogen sulfide to water and sulfur dioxide. One-third of the hydrogen sulfide burns to give sulfur dioxide gas, and the sulfur dioxide produced then reacts with the remaining two-thirds of the hydrogen sulfide to give elemental sulfur: To meet modern emissions standards, the process has been improved; in modern plants, 99 percent conversion occurs, an extraction much better than the 95 percent conversion achieved in older plants.

$$2\ H_2S(g) + 3\ O_2(g) \rightarrow 2\ SO_2(g) + 2\ H_2O(g)$$

$$4\ H_2S(g) + 2\ SO_2(g) \rightarrow 6\ S(s) + 4\ H_2O(g)$$

$$6\ H_2S(g) + 3\ O_2(g) \rightarrow 6\ S(s) + 6\ H_2O(g)$$

About 53 percent of world sulfur production comes from by-product sulfur produced in the Claus (or a related) process; about 23 percent comes from the Frasch process; and about 18 percent is obtained by heating the mineral iron pyrite, FeS_2 (iron(II) disulfide). Heating this compound in the absence of air decomposes the S_2^{2-} ion to elemental sulfur and iron(II) sulfide:

$$FeS_2(s) \xrightarrow{\Delta} FeS(s) + S(g)$$

Most of the world's sulfur production is needed for the synthesis of sulfuric acid, a process discussed later in this chapter. The remainder is used to synthesize sulfur chemicals such as carbon disulfide. Some of the elemental sulfur is added to asphalt mixes to make more frost-resistant highway surfaces.

Hydrogen Sulfide

Most people have heard of the gas that smells like "rotten eggs," although not as many could identify which gas it is. In fact, the obnoxious odor of hydrogen sulfide is almost unique. More important, this colorless gas is extremely toxic—more toxic than hydrogen cyanide. Because it is much more common, hydrogen sulfide presents a much greater hazard. As mentioned earlier, it sometimes is a component of the natural gas that issues from the ground; thus gas leaks from natural gas wellheads can be dangerous.

Hydrogen sulfide is used in enormous quantities in the separation of "heavy water" from regular water. Communities located near such plants usually have quick-response evacuation procedures to minimize the inherent dangers of this industry. The odor can be detected at levels as low as 0.02 ppm; headaches and nausea occur at about 10 ppm, and death at 100 ppm. Using smell to detect the gas is not completely effective, because it kills by affecting the central nervous system, including the sense of smell.

Hydrogen sulfide is produced naturally by anaerobic bacteria. In fact, this process, which occurs in rotting vegetation and bogs and elsewhere, is the source of most of the natural-origin sulfur in the atmosphere. The gas can be prepared in the laboratory by reacting a metal sulfide with a dilute acid, such as iron(II) sulfide with dilute hydrochloric acid:

$$FeS(s) + 2\,HCl(aq) \rightarrow FeCl_2(aq) + H_2S(g)$$

Hydrogen sulfide burns in air to give sulfur or sulfur dioxide, depending on the gas to air ratio:

$$2\,H_2S(g) + O_2(g) \rightarrow 2\,H_2O(l) + 2\,S(s)$$

$$2\,H_2S(g) + 3\,O_2(g) \rightarrow 2\,H_2O(l) + 2\,SO_2(g)$$

In solution, it is oxidized to sulfur by almost any oxidizing agent:

$$H_2S(aq) \rightarrow 2\,H^+(aq) + S(s) + 2\,e^- \qquad E^\circ = +0.141\text{ V}$$

The common test for the presence of significant concentrations of hydrogen sulfide utilizes lead(II) acetate paper (or a filter paper soaked in any soluble lead(II) salt, such as the nitrate). In the presence of hydrogen sulfide, the colorless lead(II) acetate is converted to black lead(II) sulfide:

$$Pb(CH_3CO_2)_2(s) + H_2S(g) \rightarrow PbS(s) + 2\,CH_3CO_2H(g)$$

In an analogous reaction, the blackening of silver tableware is usually attributed to the formation of black silver(I) sulfide.

The hydrogen sulfide molecule has a V-shaped structure, as would be expected for an analog of the water molecule. However, as we descend the group, the bond angles in their hydrides decrease (see Table 15.5). The variation of bond angle can be explained in terms of a decreasing use of hybrid orbitals by elements beyond Period 2. Hence it can be argued that the

Table 15.5 Bond angles of three Group 16 hydrides

Hydride	Bond angle
H_2O	$104\frac{1}{2}°$
H_2S	$92\frac{1}{2}°$
H_2Se	90°

bonding in hydrogen selenide involves *p* orbitals only. This reasoning is the most commonly accepted explanation, because it is consistent with observed bond angles in other sets of compounds.

Sulfides

Only the Groups 1 and 2 metals and aluminum form soluble sulfides. These readily hydrolyze in water, and, as a result, solutions of sulfides are very basic:

$$S^{2-}(aq) + H_2O(l) \rightarrow HS^-(aq) + OH^-(aq)$$

There is enough hydrolysis of the hydrogen sulfide ion, in turn, to give the solution a strong odor of hydrogen sulfide:

$$HS^-(aq) + H_2O(l) \rightarrow H_2S(g) + OH^-(aq)$$

The sodium-sulfur system provides the basis for a high-performance battery. In most batteries, the electrodes are solids and the electrolyte a liquid. In this battery, however, the two electrodes, sodium and sulfur, are liquids and the electrolyte, $NaAl_{11}O_{17}$, is a solid. The electrode processes are

$$Na(l) \rightarrow Na^+(NaAl_{11}O_{17}) + e^-$$

$$n\ S(l) + 2\ e^- \rightarrow S_n^{2-}(NaAl_{11}O_{17})$$

The battery is extremely powerful and it can be recharged readily. It shows great promise for industrial use, particularly for commercial electricity-driven delivery vehicles. However, adoption of this battery for household purposes is unlikely, because it operates at about 300°C. Of course, it has to remain sealed to prevent reaction of the sodium and sulfur with the oxygen or water vapor in the air.

All other metal sulfides are very insoluble. Many minerals are sulfide ores. The most common of these are listed in Table 15.6. Sulfides tend to be used for specialized purposes. The intense black diantimony trisulfide was one of the first cosmetics, being used as eye shadow from earliest recorded

Table 15.6 Common sulfide minerals

Common name	Formula	Systematic name
Cinnabar	HgS	Mercury(II) sulfide
Galena	PbS	Lead(II) sulfide
Pyrite	FeS_2	Iron(II) disulfide
Sphalerite	ZnS	Zinc sulfide
Orpiment	As_2S_3	Diarsenic trisulfide
Stibnite	Sb_2S_3	Diantimony trisulfide
Chalcopyrite	$CuFeS_2$	Copper(II) iron(II) sulfide

Disulfide Bonds and Hair

Hair consists of amino acid polymers (proteins) cross-linked by disulfide units. In about 1930, it was shown by researchers at the Rockefeller Institute that these links could be broken by sulfides or molecules containing –SH groups in slightly basic solution. This discovery proved to be the key to the present-day method for "permanently" changing the shape of hair, from curly to straight or vice versa.

In the process, a solution of the thioglycollate ion, $HSCH_2CO_2^-$, is poured on the hair, reducing the –S–S– cross-links to –SH groups:

$$2\ HSCH_2CO_2^-(aq) + \text{–S–S–}(hair) \rightarrow [SCH_2CO_2^-]_2(aq) + 2\ \text{–S–H}(hair)$$

By using curlers or straighteners, the protein chains can then be mechanically shifted with respect to their neighbors. Application of a solution of hydrogen peroxide then reoxidizes the –SH groups to re-form new cross-links of –S–S–, thus holding the hair in the new orientation:

$$2\ \text{–S–H}(hair) + H_2O_2(aq) \rightarrow \text{–S–S–}(hair) + 2\ H_2O(l)$$

times. In a different context, religious statues made from realgar, diarsenic disulfide, were popular among devotees of the Chinese Taoist religion. Handling the statues was believed to restore health. In this particular case, chemistry rather than faith might have contributed, for many people in tropical areas suffer from internal parasites and handling the statues would result in arsenic absorption through the skin, enough to kill the parasites but not enough to kill the devotee.

Today, sodium sulfide is the sulfide in highest demand. Between 10^5 and 10^6 tonnes are produced every year by the high-temperature reduction of sodium sulfate with coke:

$$Na_2SO_4(s) + 2\ C(s) \rightarrow Na_2S(l) + 2\ CO_2(g)$$

Sodium sulfide is used to remove hair from hides in the tanning of leather. It is also used in ore separation by flotation, for the manufacture of sulfur-containing dyes, and in the chemical industry, such as the precipitation of toxic metal ions, particularly lead.

Other sulfides in commercial use are selenium disulfide, SeS_2, a common additive to antidandruff hair shampoos, and molybdenum(IV) sulfide, MoS_2, an excellent lubricant for metal surfaces, either on its own or suspended in oil. Metal sulfides tend to be dense, opaque solids, and it is this property that makes the bright yellow cadmium sulfide, CdS, a popular pigment for oil painting.

The formation of insoluble metal sulfides used to be common in inorganic qualitative analysis. Hydrogen sulfide is bubbled through an acid solution containing unknown metal ions. The presence of the high hydrogen ion concentration reduces the sulfide ion concentration to extremely low levels:

$$H_2S(aq) + 2\ H_2O(l) \rightleftharpoons 2\ H_3O^+(aq) + S^{2-}(aq)$$

This very low level of sulfide ion is still enough to precipitate the most insoluble metal sulfides, those with a solubility product, K_{sp}, smaller than 10^{-30}. These metal sulfides are separated by filtration. The pH of the filtrate is increased by adding base. This increase shifts the sulfide equilibrium to the right, thereby raising the concentration of sulfide ions to the point where those metal sulfides with a solubility product between 10^{-20} and 10^{-30} (mainly the transition metals of Period 4) precipitate. Specific tests could then be used to identify which metal ions are present. More recently, thioacetamide, a reagent that hydrolyzes to hydrogen sulfide, has been used for such tests.

In addition to forming conventional sulfides, some elements form disulfides, S_2^{2-}, ions analogous to dioxide(2–). Thus FeS_2 does not contain iron in a high oxidation state, but the disulfide ion and Fe^{2+}. Also, the alkali and alkaline earth metals form polysulfides, which contain the S_n^{2-} ion, where n has values between 2 and 6.

Sulfur Oxides

Sulfur Dioxide

The common oxide of sulfur, sulfur dioxide, is a colorless, dense, toxic gas with an acid "taste." The maximum tolerable levels for humans is about 5 ppm, but plants begin to suffer in concentrations as low as 1 ppm. The taste is a result of the reaction of sulfur dioxide with water on the tongue to give the weak acid, sulfurous acid:

$$SO_2(g) + H_2O(l) \rightleftharpoons H_2SO_3(aq)$$

Sulfur dioxide is very water-soluble, but like ammonia and carbon dioxide, almost all the dissolved gas is present as the sulfur dioxide molecule; only a very small proportion forms sulfurous acid. To prepare the gas in the laboratory, a dilute acid is added to a solution of a sulfite or a hydrogen sulfite:

$$SO_3^{2-}(aq) + 2\ H^+(aq) \rightarrow H_2O(l) + SO_2(g)$$

$$HSO_3^-(aq) + H^+(aq) \rightarrow H_2O(l) + SO_2(g)$$

Sulfur dioxide is one of the few common gases that is a reducing agent, itself being easily oxidized to the sulfate ion:

$$SO_2(aq) + 2\ H_2O(l) \rightarrow SO_4^{2-}(aq) + 4\ H^+(aq) + 2\ e^-$$

To test for reducing gases, such as sulfur dioxide, we can use an oxidizing agent that undergoes a color change, the most convenient one being the dichromate ion. A filter paper soaked in acidified orange dichromate ion will turn green as a result of the formation of the chromium(III) ion:

$$Cr_2O_7^{2-}(aq) + 14\ H^+(aq) + 6\ e^- \rightarrow 2\ Cr^{3+}(aq) + 7\ H_2O(l)$$

Since Earth first solidified, sulfur dioxide has been produced by volcanoes in large quantities. However, we are now adding additional, enormous quantities of this gas to the atmosphere. Combustion of coal is the worst offender, because most coals contain significant levels of sulfur compounds. In London, the yellow smog of the 1950s caused by home coal fires led to thousands of premature deaths. Currently, the coal-fired electric power stations are the major sources of unnatural sulfur dioxide in the atmosphere. Oil, too, contributes to the atmospheric burden of sulfur dioxide, for the lowest cost heating oil is sulfur rich. Thus many schools and hospitals, in their need to conserve finances, become the cause of poorer air quality when they choose the lowest cost oil for their heating purposes. Finally, many metals are extracted from sulfide ores, and the traditional smelting process involves the oxidation of sulfide to sulfur dioxide, thereby providing an additional source of the gas.

In the past, the easiest solution to industrial air pollution problems was to provide ever taller smokestacks so that the sulfur dioxide would travel appreciable distances from the source. But, during its time in the upper atmosphere, the sulfur dioxide is oxidized and hydrated to give droplets of sulfuric acid, a much stronger acid than sulfurous acid:

$$SO_2(g) + H_2O(g) + \text{“O”} \rightarrow H_2SO_4(aq)$$

This product precipitates as acid rain, many hundreds of thousands of kilometers away, making the problem “someone else’s.” Currently, researchers are studying methods to minimize sulfur dioxide emissions. One of these involves the conversion of sulfur dioxide to solid calcium sulfate. In a modern coal-burning power plant, powdered limestone (calcium carbonate) is mixed with the powdered coal. The coal burns, producing a flame at about 1000°C, a temperature high enough to decompose the calcium carbonate:

$$CaCO_3(s) \xrightarrow{\Delta} CaO(s) + CO_2(g)$$

Then the calcium oxide reacts with sulfur dioxide and oxygen gas to give calcium sulfate:

$$2\ CaO(s) + 2\ SO_2(g) + O_2(g) \xrightarrow{\Delta} 2\ CaSO_4(s)$$

Because the second step is about as exothermic as the first step is endothermic, no heat is lost in the overall process. The fine dust of calcium sulfate is captured by electrostatic precipitators.

The solid calcium sulfate can be used for fireproof insulation and for roadbed cement. However, as this process becomes more and more widely used, the supply of calcium sulfate will outstrip demand and using it for landfill will become more and more common. Thus we have replaced a gaseous waste by a less harmful solid waste, but we have not eliminated the waste problem completely.

Sulfur dioxide does have some positive uses. It is used as a bleach and as a preservative, particularly for fruits. In this latter role, it is very effective at killing molds and other fruit-destroying organisms. Unfortunately, some people are sensitive to traces of the substance.

The sulfur dioxide molecule is V-shaped, with an S–O bond length of 143 pm and a O–S–O bond angle of 119°. The bond length is much shorter than that of a sulfur-oxygen single bond (163 pm) and very close to that of a typical sulfur-oxygen double bond (140 pm). The similarity of the sulfur dioxide bond angle to the trigonal angle of 120° (sp^2 hybridization) can be explained in terms of a σ bond between each sulfur-oxygen pair and a lone pair of electrons on the sulfur atom. We might expect the π bond system to resemble that of the nitrite ion. However, sulfur has empty $3d$ orbitals that can be involved in σ bonding. By using these and the full p orbitals of the oxygens, we can construct a conventional π system over the molecule and envisage a double bond between each pair of atoms (see Figure 15.18).

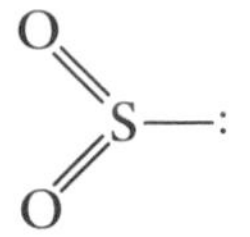

Figure 15.18 The sulfur dioxide molecule.

Sulfur Trioxide

Most people have heard of sulfur dioxide, but few have heard of the other important oxide, sulfur trioxide, a colorless liquid at room temperature. The liquid and gas phases contain a mixture of sulfur trioxide, SO_3, and a trimer, S_3O_9 (Figure 15.19). The liquid freezes at 16°C to give crystals of trisulfur nonaoxide. In the presence of moisture, long-chain solid polymers are formed and have the structure $HO(SO_3)_nOH$, where n is about 10^5. Sulfur trioxide is a very acidic, deliquescent oxide, reacting with water to form sulfuric acid:

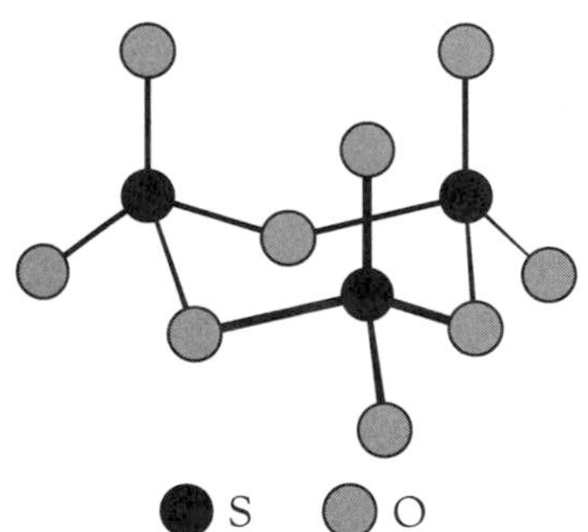

Figure 15.19 The trisulfur nonaoxide molecule, a solid allotrope of sulfur trioxide.

$$SO_3(s) + H_2O(l) \rightarrow H_2SO_4(l)$$

This oxide is so little known because oxidation of sulfur almost always gives sulfur dioxide, not sulfur trioxide. Even though the formation of sulfur trioxide is even more thermodynamically favored than that of sulfur dioxide (–370 kJ·mol^{-1} for sulfur trioxide, –300 kJ·mol^{-1} for sulfur dioxide), the oxidation has a high activation energy. Thus the pathway from sulfur dioxide to sulfur trioxide is kinetically controlled:

$$2\ SO_2(g) + O_2(g) \rightarrow 2\ SO_3(g)$$

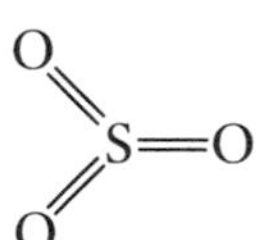

Figure 15.20 The sulfur trioxide molecule in the gas phase.

When liquid sulfur trioxide boils, the gaseous molecules formed are planar SO_3 (Figure 15.20). Like sulfur dioxide, all the sulfur-oxygen bond lengths are equally short (142 pm) and very close to the typical double bond value. Again, this bonding is best interpreted by means of a π system involving the sulfur $3d$ orbitals.

Sulfuric Acid

Sulfuric acid is an oily, dense liquid that freezes at 10.4°C. It mixes with water very exothermically. For this reason, the concentrated acid should be slowly added to water, not the reverse process, and the mixture should be stirred continuously. The molecule contains a tetrahedral arrangement of oxygen atoms around the central sulfur atom (Figure 15.21). The short bond lengths and the high bond energies indicate that there must be double bonds to each terminal oxygen atom.

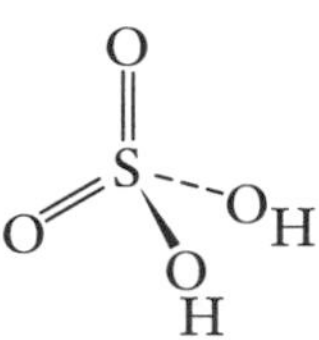

Figure 15.21 The sulfuric acid molecule.

The pure liquid has a significant electrical conductivity as a result of self-ionization reactions:

$$2\ H_2SO_4(l) \rightleftharpoons H_2O(H_2SO_4) + H_2S_2O_7(H_2SO_4)$$

$$H_2O(H_2SO_4) + H_2SO_4(l) \rightleftharpoons H_3O^+(H_2SO_4) + HSO_4^-(H_2SO_4)$$

$$H_2S_2O_7(H_2SO_4) + H_2SO_4(l) \rightleftharpoons H_3SO_4^+(H_2SO_4) + HS_2O_7^-(H_2SO_4)$$

Concentrated sulfuric acid is a water mixture with an acid concentration of 18 mol·L^{-1}. We usually think of sulfuric acid as just an acid, but in fact it can react in five different ways.

Sulfuric Acid as an Acid

Dilute sulfuric acid is used most often as an acid. It is a strong, diprotic acid, forming two ions, the hydrogen sulfate ion and the sulfate ion:

$$H_2SO_4(aq) + H_2O(l) \rightleftharpoons H_3O^+(aq) + HSO_4^-(aq)$$

$$HSO_4^-(aq) + H_2O(l) \rightleftharpoons H_3O^+(aq) + SO_4^{2-}(aq)$$

The first equilibrium lies far to the right, but the second one, less so. Thus the predominant species in a solution of sulfuric acid are the hydronium ion and the hydrogen sulfate ion.

Sulfuric Acid as a Dehydrating Agent

The concentrated acid will remove the elements of water from a number of compounds. For example, sugar is converted to carbon and water. This exothermic reaction is spectacular:

$$C_{12}H_{22}O_{11}(s) + H_2SO_4(l) \rightarrow 12\ C(s) + 11\ H_2O(g) + H_2SO_4(aq)$$

The acid serves this function in a number of important organic reactions. For example, addition of concentrated sulfuric acid to ethanol produces ethene, C_2H_4, or ethoxyethane, $(C_2H_5)_2O$, depending on the reaction conditions:

$$C_2H_5OH(l) + H_2SO_4(l) \rightarrow C_2H_5OSO_3H(aq) + H_2O(l)$$

$$C_2H_5OSO_3H(aq) \rightarrow C_2H_4(g) + H_2SO_4(aq) \qquad \text{[excess acid]}$$

$$C_2H_5OSO_3H(aq) + C_2H_5OH(l) \rightarrow (C_2H_5)_2O(l) + H_2SO_4(aq) \qquad \text{[excess ethanol]}$$

Sulfuric Acid as an Oxidizing Agent

Although sulfuric acid is not as strongly oxidizing as nitric acid, if it is hot and concentrated, it will function as an oxidizing agent. For example, hot concentrated sulfuric acid reacts with copper metal to give the copper(II) ion, and the sulfuric acid itself is reduced to sulfur dioxide and water:

$$Cu(s) \rightarrow Cu^{2+}(aq) + 2\ e^-$$

$$2\ H_2SO_4(l) + 2\ e^- \rightarrow SO_2(g) + 2\ H_2O(l) + SO_4^{2-}(aq)$$

Sulfuric Acid as a Sulfonating Agent

The concentrated acid is used in organic chemistry to replace a hydrogen atom by the sulfonic acid group ($-SO_3H$):

$$H_2SO_4(l) + CH_3C_6H_5(l) \rightarrow CH_3C_6H_4SO_3H(s) + H_2O(l)$$

Sulfuric Acid as a Base

A Brønsted-Lowry acid can only act as a base if it is added to a stronger proton donor. Sulfuric acid is a very strong acid; hence only extremely strong acids such as fluorosulfonic acid (Chapter 8) can cause it to behave as a base:

$$H_2SO_4(l) + HSO_3F(l) \rightleftharpoons H_3SO_4^+(H_2SO_4) + SO_3F^-(H_2SO_4)$$

Industrial Synthesis of Sulfuric Acid

Sulfuric acid is synthesized in larger quantities than any other chemical. All synthetic routes use sulfur dioxide, and in some plants this reactant is obtained directly from the flue gases of smelting processes. However, in North America most of the sulfur dioxide is produced by burning molten sulfur in dry air:

$$S(s) + O_2(g) \rightarrow SO_2(g)$$

It is more difficult to oxidize sulfur further. As we mentioned earlier, there is a kinetic barrier to the formation of sulfur trioxide. Thus an effective catalyst must be used to obtain commercially acceptable rates of reaction. We also need to ensure that the position of equilibrium is to the right side of the equation. To accomplish this, we invoke the Le Châtelier principle, which predicts that an increase in pressure will favor the side of the equation with the fewer moles of gas—in this case, the product side. This reaction is also exothermic; so the choice of temperature must be high enough to produce a reasonable rate of reaction, even though these conditions will result in a decreased yield:

$$2\ SO_2(g) + O_2(g) \xrightarrow{V_2O_5} 2\ SO_3(g)$$

In the process, pure, dry sulfur dioxide and dry air are passed through a catalyst of vanadium(V) oxide on an inert support. The gas mixture is heated to 400°–500°C, which is the optimum temperature for conversion to sulfur trioxide with a reasonable yield at an acceptable rate.

Sulfur trioxide reacts violently with water—not a process that is acceptable in industry. However, it does react more controllably with concentrated sulfuric acid itself to give pyrosulfuric acid, $H_2S_2O_7$ (Figure 15.22):

$$SO_3(g) + H_2SO_4(l) \rightarrow H_2S_2O_7(l)$$

The pyrosulfuric acid is then diluted with water to produce an additional mole of sulfuric acid:

$$H_2S_2O_7(l) + H_2O(l) \rightarrow 2\ H_2SO_4(l)$$

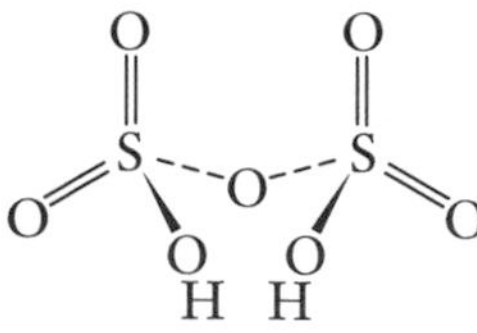

Figure 15.22 The pyrosulfuric acid molecule.

All the steps in the process are exothermic. In fact, the entire process of converting elemental sulfur to sulfuric acid produces 535 $kJ{\cdot}mol^{-1}$ of heat. An essential feature of any sulfuric acid plant is effective utilization of this waste heat, either as direct heating for some other industrial process or in the production of electricity.

This process is associated with two potential pollution problems. First, some of the sulfur dioxide escapes. Legislation in most pollution-conscious countries limits emissions to less than 0.5 percent of the processed gas. Second, in spite of using the pyrosulfate route, some of the sulfuric acid escapes as a fine mist. Newer plants have mist eliminators to reduce this problem.

Usage of the sulfuric acid varies from country to country. In the United States the vast majority of acid is employed in the manufacture of fertilizers, such as the conversion of the insoluble calcium phosphate to the more soluble calcium dihydrogen phosphate:

$$Ca_3(PO_4)_2(s) + 2\ H_2SO_4(aq) \rightarrow Ca(H_2PO_4)_2(s) + 2\ CaSO_4(s)$$

or the production of ammonium sulfate fertilizer:

$$2\ NH_3(g) + H_2SO_4(aq) \rightarrow (NH_4)_2SO_4(aq)$$

In Europe, however, a higher proportion of the acid is used for manufacturing other products such as paints, pigments, and sulfonate detergents.

There is an increasing interest in trying to reclaim waste sulfuric acid. At the present time, the cost of removing contaminants and concentrating the dilute acid is greater than the cost of preparing the acid from sulfur. However, recovery is now preferred over dumping. If the acid is pure but too dilute (in other words, the only contaminant is water), then pyrosulfuric acid is added to increase the concentration of acid to usable levels. For contaminated acid, high-temperature decomposition produces gaseous sulfur dioxide, which can be removed and used to synthesize fresh acid:

$$2\ H_2SO_4(aq) \xrightarrow{\Delta} 2\ SO_2(g) + 2\ H_2O(l) + O_2(g)$$

Sulfites

Although sulfurous acid is mostly an aqueous solution of sulfur dioxide, the sulfite and hydrogen sulfite ions are real entities. In fact, sodium sulfite is a major industrial chemical; about 10^6 tonnes are produced annually. It is most commonly prepared by bubbling sulfur dioxide into sodium hydroxide solution:

$$2\ NaOH(aq) + SO_2(g) \rightarrow Na_2SO_3(aq) + H_2O(l)$$

In the laboratory and in industry, sodium sulfite is used as a reducing agent, itself being oxidized to sodium sulfate:

$$SO_3^{2-}(aq) + H_2O(l) \rightarrow SO_4^{2-}(aq) + 2\ H^+(aq) + 2\ e^-$$

The main use of sodium sulfite is as a bleach in the Kraft process for the production of paper. In the process, the sulfite ion attacks the polymeric material (lignin) that binds the cellulose fibers together (the loose cellulose fibers make up the paper structure). A secondary use, as we will see shortly, is in the manufacture of sodium thiosulfate. Like sulfur dioxide, sodium sulfite can be added to fruit as a preservative.

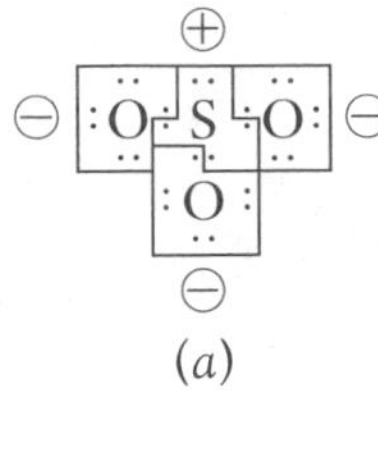
(*a*)

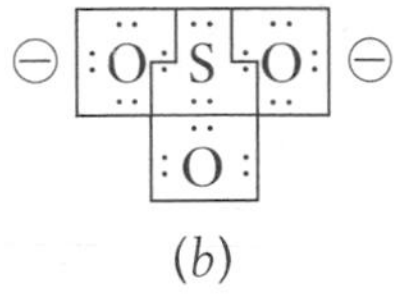
(*b*)

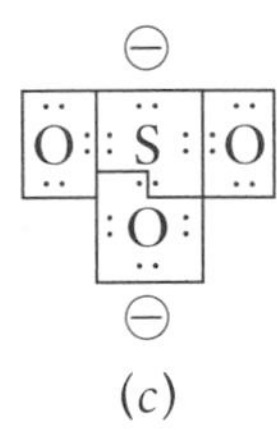
(*c*)

Figure 15.23 Three formal charge representations for the sulfite ion.

The sulfur-oxygen bond lengths in the sulfite ion are 151 pm, slightly longer than the 140-pm S–O double bond. Although it is possible to draw electron-dot structures with all single bonds, we can use formal charge representations to see why multiple bonds are preferred (Figure 15.23).

In Figure 15.23a, the single bond representation has formal charges on each atom, which makes this bonding arrangement unlikely. Figure 15.23c shows two double bonds and negative charges on neighboring atoms, again an unlikely scenario. It is the structure in Figure 15.23b, with one double bond, that has the minimum formal charge arrangement. However, it should be kept in mind that formal charge is a very simplistic method of approaching bonding and that a molecular orbital study of the overlap of sulfur 3*d* orbitals with oxygen 2*p* orbitals provides a much more valid picture. If we take Figure 15.23b to represent one of three possible resonance structures, then each sulfur-oxygen bond can be assigned an average bond order of $1\frac{1}{3}$.

Sulfates

Sulfates and nitrates are the most commonly encountered metal salts. There are several reasons for the use of sulfates:

1. Most sulfates are water-soluble, making them a useful source of the metal cation. Two important exceptions are lead(II) sulfate, which plays an important role in the lead-acid battery, and barium sulfate, used in X-rays of soft tissues such as the stomach.

2. The sulfate ion is not oxidizing or reducing. Hence the sulfate ion can form salts with metals in both their higher and lower common oxidation states, for example, iron(II) sulfate and iron(III) sulfate. Furthermore, when dissolved in water, the sulfate ion will not initiate a redox reaction with any other ion present.

3. The sulfate ion is the conjugate base of a moderately strong acid (the hydrogen sulfate ion), so the anion will not significantly alter the pH of a solution.

4. The sulfates tend to be thermally stable, at least more stable than the equivalent nitrate salt.

Sulfates can be prepared by the reaction between a base, such as sodium hydroxide, and the stoichiometric quantity of dilute sulfuric acid:

$$2\,NaOH(aq) + H_2SO_4(aq) \rightarrow Na_2SO_4(aq) + 2\,H_2O(l)$$

or by the reaction between an electropositive metal, such as zinc, and dilute sulfuric acid:

$$Zn(s) + H_2SO_4(aq) \rightarrow ZnSO_4(aq) + H_2(g)$$

or by the reaction between a metal carbonate, such as copper(II) carbonate, and dilute sulfuric acid:

$$CuCO_3(s) + H_2SO_4(aq) \rightarrow CuSO_4(aq) + CO_2(g) + H_2O(l)$$

The common test for the presence of sulfate ion is the addition of barium ion, which reacts with the anion to give a dense white precipitate, barium sulfate:

$$Ba^{2+}(aq) + SO_4^{2-}(aq) \rightarrow BaSO_4(s)$$

Like the sulfite ion, the sulfate ion has a short sulfur-oxygen bond, a characteristic indicating considerable multiple bond character. In fact, at 149 pm, its length is about the same as that in the sulfite ion, within experimental error.

Hydrogen Sulfates

Like the hydrogen carbonates, only the alkali and alkaline earth metals have charge densities low enough to stabilize these large low-charge cations in the solid phase. The value of the second ionization of sulfuric acid is quite large, so the hydrogen sulfates give an acidic solution:

$$HSO_4^{-}(aq) + H_2O(l) \rightleftharpoons H_3O^{+}(aq) + SO_4^{2-}(aq)$$

It is the high acidity of the solid sodium hydrogen sulfate that makes it useful as a household cleaning agent, such as Saniflush®.

Hydrogen sulfates can be prepared by mixing the stoichiometric quantities of sodium hydroxide and sulfuric acid and evaporating the solution:

$$NaOH(aq) + H_2SO_4(aq) \rightarrow NaHSO_4(aq) + H_2O(l)$$

They decompose in an unexpected manner when heated, forming the metal pyrosulfate:

$$2\ NaHSO_4(s) \rightarrow Na_2S_2O_7(s) + H_2O(l)$$

Thiosulfates

The thiosulfate ion resembles the sulfate ion, except that one oxygen atom has been replaced by a sulfur atom (*thio-* is a prefix meaning "sulfur"). These two sulfur atoms are in completely different environments; the additional sulfur behaves more like a sulfide ion. In fact, a formal assignment of oxidation numbers gives the central sulfur a value of +5 and the other one a value of −1, as discussed in Chapter 9. The shape of the thiosulfate ion is shown in Figure 15.24. Although the ion is depicted as having two double bonds and two single bonds, the multiple bond character is actually spread more evenly over all the bonds.

Sodium thiosulfate pentahydrate, commonly called "hypo," can easily be prepared by boiling sulfur in a solution of sodium sulfite:

$$SO_3^{2-}(aq) + S(s) \rightarrow S_2O_3^{2-}(aq)$$

The thiosulfate ion is not very thermally stable, first dissolving in the water of crystallization:

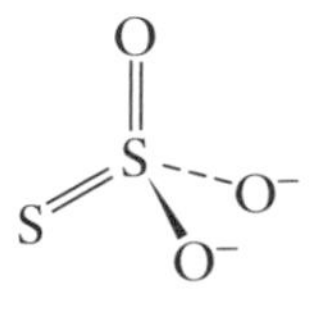

Figure 15.24 The thiosulfate ion.

The Chemistry of Photography

The thiosulfate ion is particularly important in photography. Film is coated with silver bromide. During picture-taking with black-and-white film, the light reduces a few of the silver ions to silver metal. In a silver bromide microcrystal, from 10 to 100 atoms of silver metal are usually produced. This number would be far too few to see; thus the first processing step is to add a developer, usually hydroquinone, $C_6H_6O_2$. A developer selectively reduces all the silver ions in any crystal that already contains silver atoms. By this means, we enhance the image by a factor of about 10^{10} times.

Next, we must remove the remaining insoluble silver bromide, otherwise the film would all turn black when it is exposed to light. To accomplish this, thiosulfate ion is added. This reagent reacts with the remaining silver ion to form a soluble complex, the tris(thiosulfato)silver(I) ion, $Ag(S_2O_3)_3{}^{5-}$, which is washed away. The black particles of silver metal are left behind to form the image. The process is essentially the same for color photography, except that organic dyes are involved as well:

$$AgBr(s) + 3\ S_2O_3{}^{2-}(aq) \rightarrow [Ag(S_2O_3)_3]^{5-}(aq) + Br^-(aq)$$

Although sodium thiosulfate used to be the preferred salt, ammonium thiosulfate is becoming more widely used because it allows faster processing and silver can be recovered from the waste solution more easily.

$$Na_2S_2O_3{\cdot}5H_2O(s) \xrightarrow{\Delta} Na_2S_2O_3(aq) + 5\ H_2O(l)$$

then, when heated more strongly, disproportionating to give three different oxidation states of sulfur: sodium sulfate, sodium sulfide, and sulfur:

$$4\ Na_2S_2O_3(s) \xrightarrow{\Delta} 3\ Na_2SO_4(s) + Na_2S(s) + 4\ S(s)$$

The gentle warming that causes the sodium thiosulfate pentahydrate to lose the water of crystallization is a reversible endothermic process:

$$Na_2S_2O_3{\cdot}5H_2O(s) \rightleftharpoons Na_2S_2O_3(aq) + 5\ H_2O(l) \qquad \Delta H^\circ = +55\ kJ{\cdot}mol^{-1}$$

The equilibrium has generated considerable interest as a heat storage system. In this process, heat from the sun is absorbed by solar panels and transferred to an underground tank of the hydrated compound. This input of heat causes the hydrate to decompose and dissolve in the water produced. Then, in the cool of the night, heat released as the compound crystallizes can be used to heat the dwelling.

When handling solutions of the thiosulfate ion, it is important to avoid the presence of acid. The hydrogen (hydronium) ion first reacts to form thiosulfuric acid, which decomposes rapidly to give a white suspension of sulfur and the characteristic odor-taste of sulfur dioxide. This particular disproportionation is further evidence that the two sulfur atoms are in different oxidation states. Presumably, it is the central sulfur that provides the higher oxidation state sulfur in sulfur dioxide:

$$S_2O_3^{2-}(aq) + 2\ H^+(aq) \rightarrow H_2S_2O_3(aq)$$

$$H_2S_2O_3(aq) \rightarrow H_2O(l) + S(s) + SO_2(g)$$

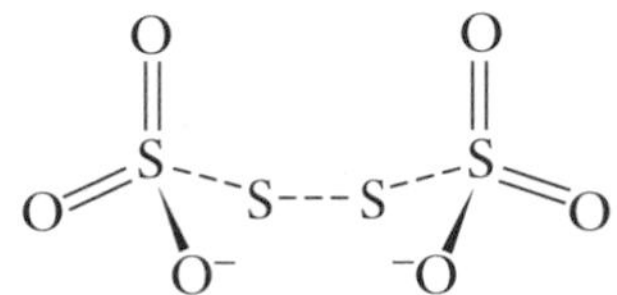

Figure 15.25 The tetrathionate ion.

Sodium thiosulfate is also used in redox titrations. For example, it is used to determine the concentration of iodine in aqueous solutions. During the assay, the iodine is reduced to iodide, and the thiosulfate ion of known concentration is oxidized to the tetrathionate ion:

$$2\ S_2O_3^{2-}(aq) \rightarrow S_4O_6^{2-}(aq) + 2\ e^-$$

$$I_2(aq) + 2\ e^- \rightarrow 2\ I^-(aq)$$

The tetrathionate ion (Figure 15.25) contains bridging sulfur atoms analogous to the oxygen atoms of the peroxodisulfate ion (Figure 15.26).

Mixing cold solutions of thiosulfate ion and iron(III) ion gives a characteristic deep purple complex ion:

$$Fe^{3+}(aq) + 2\ S_2O_3^{2-}(aq) \rightarrow Fe(S_2O_3)_2^-(aq)$$

When warmed, this bis(thiosulfato)ferrate(III) ion, $[Fe(S_2O_3)_2]^-$, undergoes a redox reaction to give the iron(II) ion and the tetrathionate ion:

$$[Fe(S_2O_3)_2]^-(aq) + Fe^{3+}(aq) \rightarrow 2\ Fe^{2+}(aq) + S_4O_6^{2-}(aq)$$

Peroxodisulfates

Although the sulfate ion contains sulfur in its highest possible oxidation state of +6, it can be oxidized electrolytically to the peroxodisulfate ion by using smooth platinum electrodes, acidic solution, and high current densities. These conditions favor oxidations that do not produce gases such as the competing oxidation of water to dioxygen:

$$2\ HSO_4^-(aq) \rightarrow S_2O_8^{2-}(aq) + 2\ H^+(aq) + 2\ e^-$$

This ion contains a dioxo bridge (see Figure 15.26). Hence the two sulfur atoms still have formal oxidation states of +6, but the bridging oxygen atoms have been oxidized from –2 to –1. The terminal S–O bond lengths are all equivalent at 150 pm; once again, there must be considerable multiple bonding. The parent acid, peroxodisulfuric acid, is a white solid, but it is the two salts, potassium peroxodisulfate and ammonium peroxodisulfate, that are important as powerful, stable oxidizing agents:

$$S_2O_8^{2-}(aq) + 2\ e^- \rightarrow 2\ SO_4^{2-}(aq) \qquad E^\circ = +2.01\ V$$

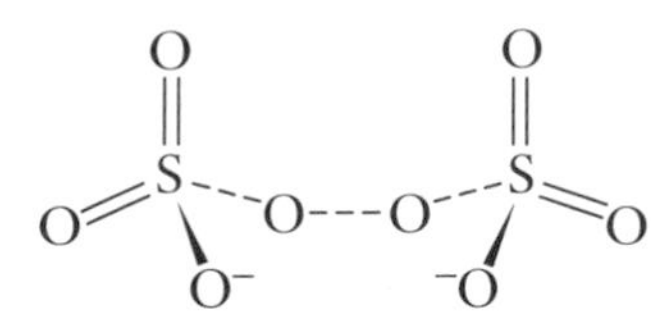

Figure 15.26 The peroxodisulfate ion.

Sulfur Halides

The most important compound of sulfur and fluorine is sulfur hexafluoride, SF_6. This compound is a colorless, odorless, unreactive gas. As a result of its stability, low toxicity, and inertness, sulfur hexafluoride is used as an insulating gas in high-voltage electrical systems. At a pressure of about 250 kPa, it will prevent a discharge across a 1 MV potential difference that is separated by only 5 cm. The gas is also used in double- and triple-glazed windows, where it acts as both a thermal and noise insulator.

The very high molar mass of this gas makes it useful for several scientific applications. For example, air pollution can be tracked for thousands of kilometers by releasing a small amount of sulfur hexafluoride at the pollution source. The extremely high molar mass is so unique that the contaminated air mass can be identified days later by its tiny concentration of sulfur hexafluoride molecules. Similarly, deep ocean currents are being identified by bubbling sulfur hexafluoride into deep-water layers and then tracking the movement of the gas.

The compound, which is produced in the thousand-tonne range, can be synthesized by simply burning molten sulfur in fluorine gas:

$$S(l) + 3\ F_2(g) \rightarrow SF_6(g)$$

As would be expected from simple VSEPR theory, the molecule is octahedral (Figure 15.27).

Figure 15.27 The sulfur hexafluoride molecule.

It is interesting that the other common sulfur-fluorine compound, sulfur tetrafluoride, is extremely reactive. It decomposes in the presence of moisture to hydrogen fluoride and sulfur dioxide:

$$SF_4(g) + 2\ H_2O(l) \rightarrow SO_2(g) + 4\ HF(g)$$

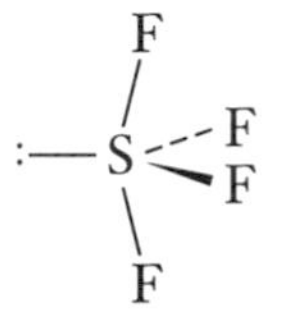

Its high reactivity might be due to the "exposed" lone pair site where reaction can take place. The compound is a convenient reagent for the fluorination of organic compounds. For example, it converts ethanol to fluoroethane. As simple VSEPR theory predicts, it has a slightly distorted seesaw shape (Figure 15.28).

Figure 15.28 The sulfur tetrafluoride molecule.

Whereas sulfur forms high oxidation state compounds with fluorine, it forms stable low oxidation state compounds with chlorine. Bubbling chlorine through molten sulfur produces disulfur dichloride, S_2Cl_2, a toxic, yellow liquid with a revolting odor:

$$2\ S(l) + Cl_2(g) \rightarrow S_2Cl_2(g)$$

The compound is used in the *vulcanization* of rubber, that is, the formation of disulfur cross-links between the carbon chains that make the rubber stronger. The shape of the molecule resembles that of hydrogen peroxide (Figure 15.29).

Surprisingly, no compound of sulfur and chlorine containing a higher sulfur oxidation state than +2 is stable at room temperature. If chlorine is bubbled through disulfur dichloride in the presence of catalytic diiodine, sulfur dichloride, SCl_2, is formed:

$$S_2Cl_2(l) + Cl_2(g) \rightarrow 2\ SCl_2(l)$$

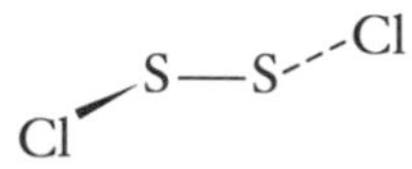

Figure 15.29 The disulfur dichloride molecule.

This foul-smelling red liquid is used in the manufacture of a number of sulfur-containing compounds, including the notorious mustard gas, $S(CH_2CH_2Cl)_2$. Mustard gas was used in the First World War and, more recently, by the Iraqi government against some of its citizens. Liquid droplets containing this gas cause severe blistering of the skin, followed by death. As predicted by VSEPR theory, the sulfur dichloride molecule is V-shaped (Figure 15.30).

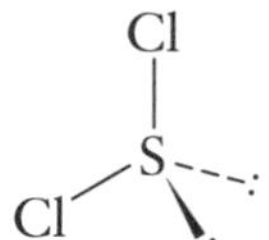

Figure 15.30 The sulfur dichloride molecule.

Sulfur-Nitrogen Compounds

There are several sulfur-nitrogen compounds. Some of these are of interest because their shapes and bond lengths cannot be explained in terms of simple bonding theory. The classic example is tetrasulfur tetranitride, S_4N_4. Unlike the crown structure of octasulfur, it has a closed, basketlike shape, with multiple bonding around the ring and weak bonds cross-linking the pairs of sulfur atoms (Figure 15.31).

Of much more interest, however, is the polymer $(SN)_x$, commonly called polythiazyl. This bronze-colored, metallic-looking compound was first synthesized in 1910, yet it was not until 50 years later that an investigation of its properties showed it to be an excellent electrical conductor. In fact, at very low temperatures (0.26 K), it becomes a superconductor. There is an intense interest in making related nonmetallic compounds that have metallic properties, both because of their potential for use in our everyday lives and because they may help us develop the theory of metals and superconductivity.

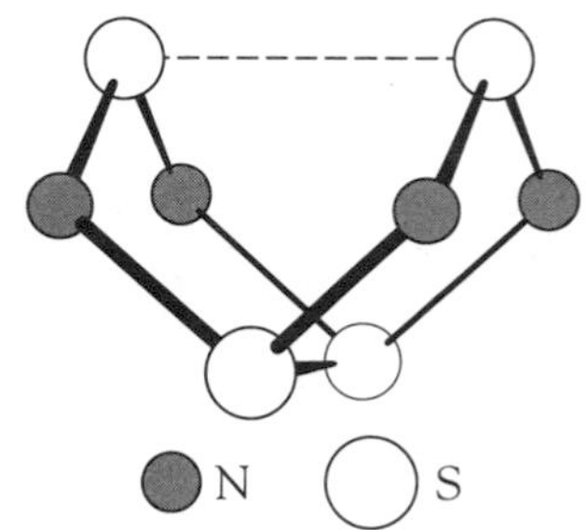

Figure 15.31 The tetrasulfur tetranitride molecule.

Selenium

Until the 1960s, the only major use of selenium was as a glass additive. Addition of cadmium selenide, CdSe, to a glass mixture results in a pure ruby-red color that is much valued by glass artisans. Cadmium selenide is a semiconductor compound used in photocells because its electrical conductivity varies as a function of the light intensity to which it is exposed.

It was the invention of xerography (from the Greek *xero*, "dry," *graphy*, "writing") as a means for duplicating documents that turned an element of little interest to one that affects everyone's life. Xerography is made possible by the unique photoconducting properties of selenium. The heart of a photocopier (and a laser printer) is a drum coated with selenium. The surface is charged in an electric field of about 10^5 V·cm^{-1}. The areas exposed to a high light intensity (the white areas of the image) lose their charge as a result of photoconductivity. Toner powder then adheres to the charged areas of the drum (corresponding to the black parts of the image). In the next step, the toner is transferred to the paper, where a heat source melts the particles, bonding them to the paper fibers and producing the photocopy. Tellurium is used in color photocopiers to alter the color sensitivity of the drum.

Biological Aspects

Oxygen

We can live without food for days, without water for hours or days (depending on the temperature), but without dioxygen, life ceases in a very short time. We inhale about 10 000 L of air per day, from which we absorb about 500 L of oxygen gas. The dioxygen molecule bonds at the lung surface to a hemoglobin molecule; in fact, an oxygen molecule covalently bonds to each of the four iron atoms in a hemoglobin molecule. The process of uptake is amazing in that once the first dioxygen molecule is bonded to an iron atom, the ease of bonding of the second dioxygen is increased, as is the third and fourth, in turn. This *cooperative effect* contrasts strikingly with the normal chemical equilibria, in which successive steps are usually less favored.

The hemoglobin transports the dioxygen to the muscles and other energy-utilizing tissues, where it is transferred to myoglobin molecules. The myoglobin molecule (similar to one of the units in hemoglobin) contains one iron ion, and it bonds the dioxygen molecule even more strongly than the hemoglobin molecule does. Once the first dioxygen molecule is removed from the hemoglobin, the cooperative effect operates again, this time making it easier and easier to remove the remaining dioxygen molecules. The myoglobin molecules store the dioxygen until it is needed in the energy-producing redox reaction with sugars that provides the energy our bodies require to survive and function.

Sulfur

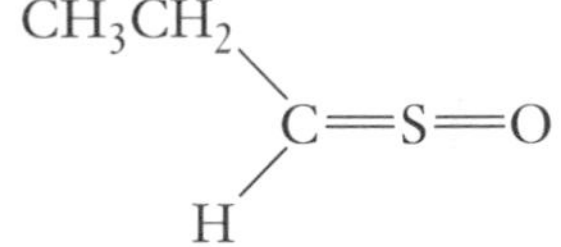

Figure 15.32 The molecule responsible for the eye irritation that accompanies the task of peeling onions.

In the –2 oxidation state, the majority of the simple sulfur-containing compounds have obnoxious odors. For example, the odorous molecules from onion, garlic, and skunk all contain sulfur in this oxidation state. Ethanethiol, CH_3CH_2SH, which is listed by the *Guinness Book of World Records* as the world's most evil-smelling substance, is actually used to save lives. It is added to odorless natural gas supplies so that we can detect leaks. Concentrations as low as 50 ppb are detectable by the human nose.

The important amino acids cysteine and methionine contain sulfur, as does vitamin B_1 (thiamine) and the coenzyme biotin (which, in spite of its name, does not contain tin). Furthermore, many of our antibiotics, such as penicillin, cephalosporin, and sulfanilamide are sulfur-containing substances. Many of the naturally occurring sulfur-containing molecules involve rather bizarre chemical structures. For example, the lachrymatory (tear-inducing) factor in onions is the molecule depicted in Figure 15.32, containing the unusual C=S=O group.

Selenium

Selenium is essential to health. It is utilized in enzymes and in amino acids such as selenomethionine. Among other roles, selenium compounds break down peroxides that will damage the cytoplasm in cells. Unfortunately, this element shows one of the narrowest ranges of tolerance. Clinical deficiency sets in below levels of about 0.05 ppm in food intake, but concentrations over 5 ppm cause chronic poisoning. Severe muscular problems found among residents of north central China were traced to food grown in selenium-deficient soils. In the United States, the Pacific Northwest, the Northeast, and Florida have soils that are very low in selenium. By contrast, the western-central states have adequate levels of selenium in the soil. It is in this latter region where "blind staggers" and "alkali disease" occur in animals that graze on wild plants that accumulate selenium. To reassure the reader, food crops grown in high-selenium regions do not accumulate this element. Conversely, for those readers who live in selenium-deficient regions, the continent-wide movement of foodstuffs ensures an adequate level of selenium in a balanced diet.

Exercises

15.1. Write balanced chemical equations for the following chemical reactions:
(a) finely divided iron with dioxygen
(b) solid barium sulfide with trioxygen
(c) solid barium dioxide(2–) and water
(d) potassium hydroxide solution with carbon dioxide
(e) sodium sulfide solution with dilute sulfuric acid
(f) sodium sulfite solution and sulfuric acid
(g) sodium sulfite solution with cyclooctasulfur.

15.2. Write balanced chemical equations for the following chemical reactions:
(a) heating potassium chlorate
(b) solid iron(II) oxide with dilute hydrochloric acid
(c) iron(II) chloride solution with sodium hydroxide solution
(d) dihydrogen octasulfide with octasulfur dichloride in ethoxyethane
(e) heating sodium sulfate with carbon
(f) sulfur trioxide gas and liquid sulfuric acid
(g) peroxodisulfate ion with sulfide ion.

15.3. Why is polonium the only element in this group to be classified as a metal?

15.4. Discuss the essential differences between oxygen and the other members of Group 16.

15.5. Define the following terms: (a) pyrophoric; (b) polymorphs; (c) cooperative effect.

15.6. Define the following terms: (a) mixed metal oxide; (b) vulcanization; (c) the Claus process.

15.7. Why is Earth's atmosphere so chemically different from that of Venus?

15.8. River and lake water are commonly used by electrical generating plants for cooling purposes. Why is this a potential problem for fish?

15.9. Predict the bond order in the trioxygen cation, O_3^+. Explain your reasoning. Is the ion paramagnetic or diamagnetic?

15.10. As we have seen, dioxygen forms two anions, O_2^- and O_2^{2-}, with bond lengths of 133 pm and 149 pm, respectively; the length of the bond in the dioxygen molecule itself is 121 pm. In addition, dioxygen can form a cation, O_2^+. The bond length in this ion is 112 pm. Use a molecular orbital diagram to deduce the bond order and the number of unpaired electrons in the dioxygen cation. Is the bond order what you would expect for the bond length?

15.11. Dibromine oxide decomposes above –40°C. Would you expect the Br–O–Br bond angle to be larger or smaller than the Cl–O–Cl bond angle in dichlorine oxide? Explain your reasoning.

15.12. Osmium forms osmium(VIII) oxide, OsO_4, but the fluoride with the highest oxidation number of osmium is osmium(VII) fluoride, OsF_7. Suggest an explanation.

15.13. Suggest a structure for the O_2F_2 molecule, explaining your reasoning. Determine the oxidation number of oxygen in this compound and comment on it.

15.14. The mineral thortveitite, $Sc_2Si_2O_7$, contains the $[O_3Si–O–SiO_3]^{6-}$ ion. The Si–O–Si bond angle in this ion has the unusual value of 180°. Use hybridization concepts to account for this.

15.15. The compound $F_3C–O–O–O–CF_3$ is unusual for oxygen chemistry. Explain why.

15.16. Barium forms a sulfide of formula BaS_2. Use an oxidation number approach to account for the structure of this compound.

15.17. Draw structures of the following molecules and ions: (a) sulfuric acid; (b) the SF_5^- ion; (c) sulfur tetrafluoride; (d) the SOF_4 molecule. (*Hint:* The oxygen is in the equatorial plane.)

15.18. Draw structures of the following molecules and ions: (a) thiosulfate ion; (b) pyrosulfuric acid; (c) peroxodisulfuric acid; (d) the SO_2Cl_2 molecule.

15.19. Disulfur difluoride, S_2F_2, rapidly converts to thiothionyfluoride, SSF_2. Construct electron-dot diagrams for these two molecules. Use oxidation numbers to explain why this rearrangement would occur.

15.20. The unstable molecule, SO_4, contains a three-membered ring of the sulfur atom and two oxygen atoms. The other two oxygen atoms are doubly bonded to the sulfur atom. Draw an electron-dot formula for the compound. Then derive the oxidation states of each atom in this molecule and show that no abnormal oxidation states are involved.

15.21. Describe the hazards of (a) trioxygen; (b) hydroxide ion; (c) hydrogen sulfide.

15.22. Describe, using a chemical equation, why "whitewash" was such an effective and inexpensive painting material.

15.23. Although sulfur catenates, it does not have the extensive chemistry that is found for carbon. Explain briefly.

15.24. Describe the changes in cyclooctasulfur as it is heated. Explain the observations in terms of changes in molecular structure.

15.25. Describe the essential features of the Frasch and the Claus processes.

15.26. The bond angle in hydrogen telluride, H_2Te, is $89\frac{1}{2}°$; that in water is $104\frac{1}{2}°$. Suggest an explanation.

15.27. Explain why a solution of sodium sulfide has an odor of hydrogen sulfide.

15.28. Describe the five ways in which sulfuric acid can behave in chemical reactions.

15.29. Why must the formation of sulfur trioxide from sulfur dioxide be an exothermic reaction?

15.30. Suggest two alternative explanations why telluric acid has the formula H_6TeO_6 rather than H_2TeO_4, analogous to sulfuric and selenic acids.

15.31. Construct formal charge representations for (a) sulfate ion; (b) sulfurous acid.

15.32. S_2F_{10} is an unusual fluoride of sulfur. It consists of two SF_5 units joined by a sulfur-sulfur bond. Calculate the oxidation number for the sulfur atoms. This molecule disproportionates. Suggest the products and write a balanced equation. Use oxidation numbers to explain why the products would be those that you have selected.

15.33. Calculate the enthalpy of formation of (theoretical) gaseous sulfur hexachloride and compare it with the tabulated value for sulfur hexafluoride. Suggest why the values are so different.

15.34. Why is the sulfate anion commonly used in chemistry?

15.35. What are the chemical tests used to identify (a) hydrogen sulfide? (b) sulfate ion?

15.36. What are the major uses for (a) sulfur hexafluoride? (b) sodium thiosulfate?

15.37. Ammonium thioglycollate is used for hair straightening or curling, whereas calcium thioglycollate is used for hair removal. The ammonium glycollate does not act so drastically on hair because the solution is less basic than that of the calcium salt. Explain the reason for the difference in solution pH.

15.38. Construct an electron-dot structure for the NSF_3 molecule in which the nitrogen-sulfur bond is (a) double; (b) triple. Decide which structure is most likely the major contributor to the bonding on the basis of formal charge.

15.39. What would happen on this planet if hydrogen bonding ceased to exist between water molecules?

15.40. "Selenium is beneficial and toxic to life." Discuss this statement.

CHAPTER

16

The Group 17 Elements: The Halogens

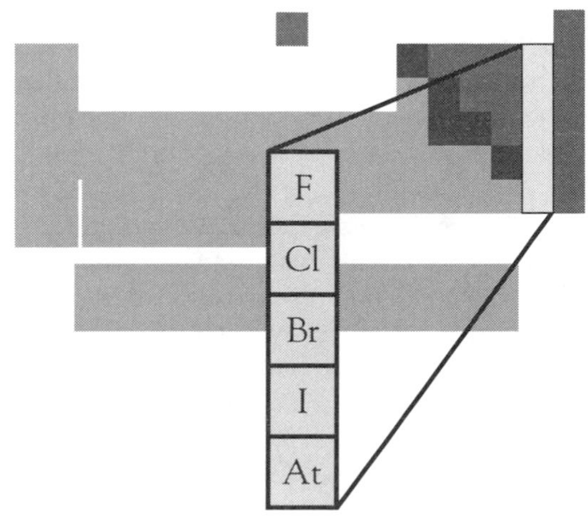

The halogens provide a complete contrast to the alkali metals. We started with the group containing the most reactive metals, and now we have reached the group containing the most reactive nonmetals. Whereas the reactivity of the alkali metals increases down the group, the most reactive halogen is at the top of the group.

Each discovery of a new halogen provided a major advance in our knowledge of chemistry. For example, in the late 1700s, chemists believed that all acids contained oxygen. By this reasoning, hydrochloric acid (then known as muriatic acid) had to contain oxygen. When Scheele, in 1774, prepared a new green gas—dephlogisticated muriatic acid—from hydrochloric acid, it was argued by Lavoisier and most chemists that the substance (chlorine gas) was simply a new compound, containing more oxygen than the muriatic acid itself. This misconception lasted until 1810, when Davy showed that the gas was indeed a new element, chlorine, and in the process overthrew the definition of an acid.

The discovery of iodine involved the field of natural products chemistry, a most important research area today. It had been known for possibly thousands of years that ingestion of burnt sponge was an effective treatment for goiter. Physicians at the time wanted to know what it was in the sponge that actually provided the cure, particularly because ingestion of the entire sponge could also cause the side effect of severe stomach cramps. In 1819, the French chemist J. F. Coindet showed that the beneficial ingredient was iodine and that potassium iodide would produce the same benefits without the side effects. Even today, we consume "iodized salt" as a goiter preventive.

Bromine was the next halogen to be discovered, and the significance of this discovery lies in the fact that three elements—chlorine, bromine, and iodine—were thereby shown to have similar properties. It was the first indication that there were patterns in the properties of elements. The proposal by the German chemist Döbereiner of groups of three elements, or "triads," was one of the first steps toward the discovery of the periodicity of chemical elements.

Fluorine proved to be the most elusive. Many, many attempts were made in the nineteenth century to obtain this reactive element from its compounds. Hydrogen fluoride, an incredibly poisonous substance, was often used as the starting material. At least two chemists died from inhaling the fumes, and many more suffered lifelong pain from damaged lungs. It was the French chemist Henri Moissan, together with the laboratory assistance of his spouse, Léonie Lugan, who devised an electrolytic apparatus for its synthesis in 1886. Moissan, who received the Nobel prize for his discovery of fluorine, died prematurely, himself a victim of hydrogen fluoride poisoning.

Group Trends

Under normal conditions (SATP), fluorine is a colorless gas (although many texts erroneously refer to it as pale green); chlorine is a pale green gas; bromine an oily, red-brown liquid; and iodine a black, metallic-looking solid. Bromine is the only nonmetallic element that is liquid at room temperature. The vapor pressures of bromine and iodine are quite high. Thus toxic, red-brown bromine vapor is apparent when a container of bromine is opened; and toxic violet vapors are produced whenever iodine crystals are gently heated. Although iodine looks metallic, it behaves like a typical nonmetal in most of its chemistry.

As before, we will ignore the chemistry of the radioactive member of the group, in this case, astatine. All of astatine's isotopes have very short half-lives; hence they emit high-intensity radiation. Nevertheless, the chemistry of this element has been shown to follow the trends seen in the other members of this group. Astatine is formed as a rare decay product from isotopes of uranium. Astatine is probably the rarest element on Earth; the top 1 km of Earth's entire crust is estimated to contain no more than 44 mg of astatine. In spite of this, one of the long-overlooked pioneers of radiochemistry, the Austrian scientist Berta Karlik and her student Trudy Beinert actually managed to prove the existence of this element in nature.

Table 16.1 shows the smooth increase in the melting and boiling points of these nonmetallic elements. Because the diatomic halogens possess only dispersion forces between molecules, their melting and boiling points depend on the polarizability of the molecules, a property that, in turn, is dependent on the total number of electrons.

Table 16.1 Melting and boiling points of the Group 17 elements

Element	Melting point (°C)	Boiling point (°C)	Number of electrons
F_2	–219	–188	18
Cl_2	–101	–34	34
Br_2	–7	+60	70
I_2	+114	+185	106

The oxidation state of fluorine is always –1, whereas the other halogens have common oxidation numbers of –1, +1, +3, +5, and +7. As the oxidation state diagram in Figure 16.1 shows, the higher the oxidation state, the stronger the oxidizing ability. Whatever its positive oxidation state, a halogen atom is more oxidizing in an acidic solution than in a basic solution. The chloride ion is the most stable halogen species, for the dichlorine molecule can be reduced to the chloride ion in both acidic and basic solutions. In basic solution, the convex point on which the dichlorine molecule is located indicates that it will undergo disproportionation to the chloride and hypochlorite ions.

All the halogens have odd atomic numbers; hence, as discussed in Chapter 1, they are expected to have few naturally occurring isotopes. In fact, fluorine and iodine each have only one isotope; chlorine has two (76 percent chlorine-35; 24 percent chlorine-37); and bromine also has two (51 percent bromine-79; 49 percent bromine-81).

The formulas of the oxyacids in which the halogen atom is in its highest oxidation state parallel those of the Group 16 acids, with the Period 5 member (iodine) having a structure different from those of the lighter elements of the group. Thus the structure of perchloric acid can be represented as $(HO)ClO_3$, and perbromic acid as $(HO)BrO_3$. Periodic acid, however, has the structure of $(HO)_5IO$ (or H_5IO_6), similar to that of telluric acid, H_6TeO_6.

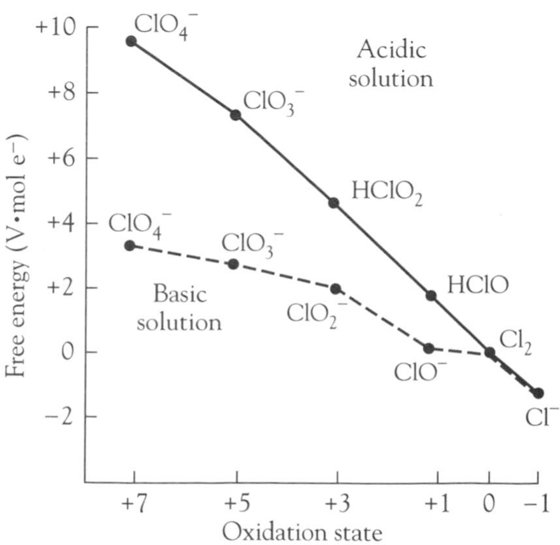

Figure 16.1 Frost diagram for chlorine in acidic and basic solutions.

Table 16.2 Bond energies of the Group 17 elements

Element	Bond energy ($kJ{\cdot}mol^{-1}$)
F_2	155
Cl_2	240
Br_2	190
I_2	149

Anomalous Nature of Fluorine

As previously noted for nitrogen and oxygen, the Period 2 members owe much of their distinctiveness to a lack of available *d* orbitals for bonding. However, fluorine has some additional unique features.

The Weakness of the Fluorine-Fluorine Bond

The bond energies of the halogens from chlorine to iodine show a systematic decrease, but the bond energy of fluorine does not fit the pattern (Table 16.2). To fit the trend, we would expect a fluorine-fluorine bond energy of about 300 $kJ{\cdot}mol^{-1}$ rather than the actual value of 155 $kJ{\cdot}mol^{-1}$. Although a number of reasons have been suggested, most chemists believe that the weak fluorine-fluorine bond is a result of repulsions between the nonbonded electrons of the two atoms of the molecule. It accounts in part for the high reactivity of fluorine gas.

Lack of Available *d* Orbitals

Like the other Period 2 elements, fluorine lacks available *d* orbitals, so it is limited to an octet of electrons in its covalent compounds. Thus fluorine almost always forms just one covalent bond, one of the few exceptions being the H_2F^+ ion.

The High Electronegativity of Fluorine

With its very high electronegativity, the fluorine atom forms the strongest hydrogen bonds of any element. Apart from its major effect on the melting and boiling points of hydrogen fluoride, the hydrogen bonding results in the formation of a very stable polyatomic anion, HF_2^-.

The Ionic Nature of Many Fluorides

Metals in "normal" oxidation states form fluorides that often are ionic, whereas their equivalent compounds with chlorine, bromine, and iodine are covalent. This difference is due to the low polarizability of the small fluoride ion. For example, aluminum fluoride is ionic in its behavior, whereas the other aluminum halides show appreciable covalent behavior.

The High Oxidation States Found in Fluorides

Fluorine, being a very strong oxidizing agent, often "brings out" a higher oxidation state in a metal than the other halogens do. For example, vanadium forms vanadium(V) fluoride, VF_5 (oxidation state of vanadium, +5), but the highest oxidation state that vanadium achieves in compounds with chlorine is that in vanadium(IV) chloride, VCl_4 (oxidation state of vanadium, +4).

Differences in Fluoride Solubilities

Because the fluoride ion is much smaller than the other halide ions, the solubilities of metal fluorides differ from those of the other halides. For example, silver fluoride is soluble, whereas the other silver halides are not. Conversely, calcium fluoride is insoluble, whereas the other calcium halides are soluble. This pattern is a result of the differences in lattice energy between the metal fluorides and the other metal halides. Thus the large silver ion has a relatively low lattice energy with the small fluoride ion. Conversely, the lattice energy is maximized when the small, high charge density calcium ion is matched with the small fluoride ion.

Fluorine

Difluorine is the most reactive element in the periodic table—in fact, it has been called the "Tyrannosaurus rex" of the elements. Fluorine gas is known to react with every other element in the periodic table except helium, neon, and argon. In the formation of fluorides, it is the enthalpy factor that is usually the predominant thermodynamic driving force.

For covalent fluorides, the negative enthalpy of formation is partially due to the weak fluorine-fluorine bond, which is readily broken, and partially to the extremely strong element-fluoride bond that is formed. For example, in Table 16.3 the bond energies of fluorine and chlorine, for bonds with themselves and with carbon, are compared. For ionic compounds, the crucial enthalpy factors are, again, the weak fluorine-fluorine bond that is broken and the high lattice energy due to the small, high charge density fluoride ion. The significant effect of the fluoride ion on lattice energies can be seen in a comparison of lattice energies for the sodium halides (Table 16.4), all of which adopt the NaCl structure.

The synthesis of fluorides usually produces compounds whose other element has the highest possible oxidation state. Thus powdered iron burns

Table 16.3 Comparison of fluorine and chlorine bond energies

Bond	Bond energy ($kJ\cdot mol^{-1}$)	Bond	Bond energy ($kJ\cdot mol^{-1}$)
F–F	155	C–F	485
Cl–Cl	240	C–Cl	327

Table 16.4 The lattice energies of sodium halides

Halide	Lattice energy ($kJ\cdot mol^{-1}$)
NaF	915
NaCl	781
NaBr	743
NaI	699

spectacularly in fluorine gas to form iron(III) fluoride, not iron(II) fluoride:

$$2\ Fe(s) + 3\ F_2(g) \rightarrow 2\ FeF_3(s)$$

Similarly, sulfur burns brightly to give sulfur hexafluoride:

$$S(s) + 3\ F_2(g) \rightarrow SF_6(g)$$

Fluorine oxidizes water to oxygen gas while simultaneously being reduced to fluoride ion:

$$F_2(g) + 2\ e^- \rightarrow 2\ F^-(aq) \qquad E^\circ = +2.87\ V$$

$$2\ H_2O(l) \rightarrow 4\ H^+(aq) + O_2(g) + 4\ e^- \qquad E^\circ = -1.23\ V$$

We can explain the reason for such a high fluorine reduction potential by comparing the free energy cycles for the formation of the hydrated fluoride and chloride ions from their respective gaseous elements (Figure 16.2). The first step is (half of) the bond dissociation free energy, for which chlorine has the slightly higher value. In the next step, the energy required is the electron affinity, and the value for chlorine is again slightly higher, thus almost canceling out the fluorine advantage in the first step. It is the third step, the hydration of the respective ions, for which the free energy change of the flu-

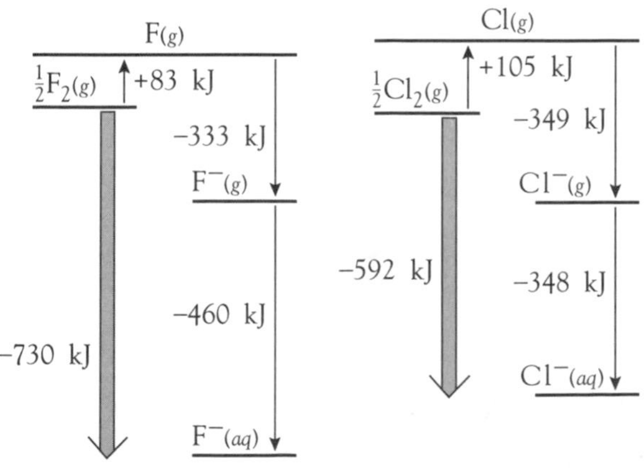

Figure 16.2 The free energy terms in the reduction of difluorine and dichlorine to the aqueous fluoride and chloride ions, respectively.

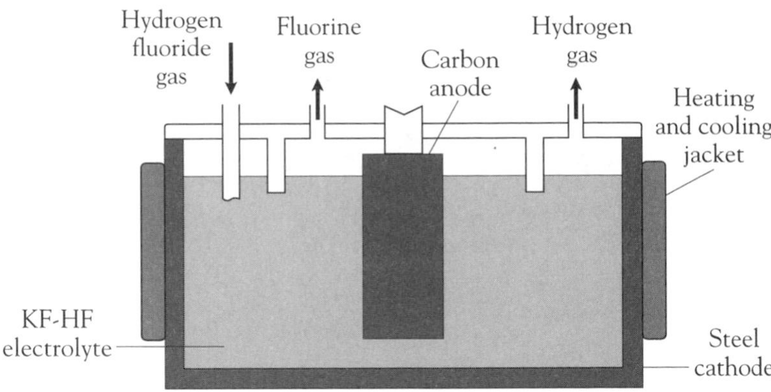

Figure 16.3 The cell used in the production of difluorine.

oride ion is very much greater than that of the chloride ion. This large free energy change results from the strong interactions of the high charge density fluoride ion with surrounding water molecules. Because $\Delta G° = -nFE°$, the large free energy of reduction directly translates to a very positive standard reduction potential, thus accounting for the great strength of difluorine as an oxidizing agent.

Difluorine is still produced by the Moisson electrochemical method, a process devised 100 years ago. The cells can be laboratory-size, running at currents between 10 and 50 A, or industrial-size, with currents up to 15 000 A. The cell contains a molten mixture of potassium fluoride and hydrogen fluoride in a 1:2 ratio and operates at about 90°C. The jacket of the apparatus is used to heat up the cell initially and then to cool it as the exothermic electrolysis occurs. At the central carbon anode, fluoride ion is oxidized to fluorine; and at the steel cathode walls of the container, hydrogen gas is produced (Figure 16.3):

$$2\ F^- \rightarrow F_2(g) + 2\ e^-$$

$$2\ H^+ + 2\ e^- \rightarrow H_2(g)$$

Hydrogen fluoride gas must be bubbled into the cell continuously to replace that used in the process. Annual production is at least 10^4 tonnes. About 55 percent of the fluorine is used to prepare uranium(VI) fluoride. This low boiling point uranium(VI) halide is used to separate the isotopes of uranium; uranium-235 is used in the manufacture of bombs and also in certain types of nuclear reactors. Uranium(VI) fluoride is prepared in two steps. Uranium(IV) oxide, UO_2, reacts with hydrogen fluoride to give uranium(IV) fluoride, UF_4:

$$UO_2(s) + 4\ HF(g) \rightarrow UF_4(s) + 2\ H_2O(g)$$

As expected, this lower oxidation state fluoride is an ionic compound and thus a solid. The uranium(IV) fluoride is then treated with fluorine gas to oxidize the uranium to the +6 oxidation state:

$$UF_4(s) + F_2(g) \rightarrow UF_6(g)$$

In such a high oxidation state, even this fluorine compound exhibits covalent properties, such as a sublimation temperature of about 60°C.

About 40 percent of the industrially produced fluorine gas is used to prepare sulfur hexafluoride, by burning molten sulfur in fluorine gas (as discussed in Chapter 15):

$$S(l) + 3\ F_2(g) \rightarrow SF_6(g)$$

Chlorine

This pale green, dense, toxic gas is also very reactive, although not as reactive as fluorine. It reacts with many elements, usually to give the higher common oxidation state of the element. For example, iron burns to give iron(III) chloride, not iron(II) chloride; and phosphorus burns in excess chlorine to give phosphorus pentachloride:

$$2\ Fe(s) + 3\ Cl_2(g) \rightarrow 2\ FeCl_3(s)$$

$$2\ P(s) + 5\ Cl_2(g) \rightarrow 2\ PCl_5(s)$$

However, particularly with nonmetals, the highest oxidation state of the nonmetal in chlorides is usually much lower than its oxidation state in the equivalent fluoride. Thus, as discussed in Chapter 15, sulfur dichloride, SCl_2 (oxidation number of sulfur, +2), is the chloride with the highest oxidation state of the sulfur atom at room temperature, whereas sulfur hexafluoride, SF_6 (oxidation number of sulfur, +6), is obtainable with the strongly oxidizing fluorine.

Chlorine gas is most easily prepared in the laboratory by adding concentrated hydrochloric acid to solid potassium permanganate. The chloride ion is oxidized to dichlorine and the permanganate ion is reduced to the manganese(II) ion:

$$2\ HCl(aq) \rightarrow 2\ H^+(aq) + Cl_2(g) + 2\ e^-$$

$$MnO_4^-(aq) + 8\ H^+(aq) + 5\ e^- \rightarrow Mn^{2+}(aq) + 4\ H_2O(l)$$

Dichlorine can act as a chlorinating agent. For example, mixing ethene, C_2H_4, with dichlorine gives 1,2-dichloroethane, $C_2H_4Cl_2$:

$$CH_2{=}CH_2(g) + Cl_2(g) \rightarrow CH_2Cl{-}CH_2Cl(g)$$

Chlorine also is a strong oxidizing agent, having a very positive standard reduction potential (although much less than that of difluorine):

$$Cl_2(aq) + 2\ e^- \rightarrow 2\ Cl^-(aq) \qquad E° = +1.36\ V$$

Dichlorine reacts with water to give a mixture of hydrochloric and hypochlorous acids:

$$Cl_2(aq) + H_2O(l) \rightleftharpoons H^+(aq) + Cl^-(aq) + HClO(aq)$$

At room temperature, a saturated solution of chlorine in water contains about two-thirds hydrated dichlorine molecules and one-third of the acid mixture.

It is the hypochlorite ion in equilibrium with the hypochlorous acid, rather than chlorine itself, that is used as an active oxidizing (bleaching) agent:

$$HClO(aq) + H_2O(l) \rightleftharpoons H_3O^+(aq) + ClO^-(aq)$$

Prior to the discovery of the bleaching effect of aqueous chlorine, the only way to produce white linen was to leave it in the sun for weeks at a time. The cloth was placed over wooden frames in vast arrays, and these were known as the bleach-fields or bleachers (hence the name for the bench seats at ball games). These days, however, most linen is bleached with hydrogen peroxide solution.

Chlorine gas is very poisonous; a concentration of over 30 ppm is lethal after a 30-minute exposure. It is the dense, toxic nature of dichlorine that led it to be used as the first wartime poison gas. In 1915, as a result of a German chlorine gas attack, 20 000 Allied soldiers were incapacitated and 5000 of them died. The toxicity of low concentrations of chlorine toward microorganisms has saved millions of human lives, however. It is through chlorination that waterborne disease-causing organisms have been virtually eradicated from domestic water supplies in Western countries. Curiously, there used to be a great enthusiasm for the benefits of chlorine gas. President Calvin Coolidge, among others, used sojourns in a "chlorine chamber" as a means of alleviating his colds. It is probable that many who tried this "cure" finished up with long-term lung damage instead.

The industrial preparation of chlorine is accomplished by electrolysis of aqueous sodium chloride solution (brine); the other product is sodium hydroxide (this process was discussed in Chapter 10). Chlorine is produced on a vast scale, about 10^8 tonnes annually, worldwide. The majority of the product is used for the synthesis of organochlorine compounds. And appreciable quantities are used in the pulp and paper industry to bleach paper, in water treatment, and in the production of titanium(IV) chloride, $TiCl_4$, an intermediate step in the extraction of titanium from its ores. There is increasing concern about the use of chlorine for both pulp and paper production and for water purification. In the former case, the dichlorine reacts with organic compounds in the wood pulp to produce toxic chloro compounds. These by-products end up in the wastewater and are pumped into rivers and seas. Because of increasingly stringent limits on the organochlorine levels in effluents from pulp mills, companies must now reduce such emissions to near-zero levels.

The problem of organochlorines in municipal water supplies is much more localized and certainly a controversial subject. The problem is greatest in communities that take their water supplies from sources that are already relatively high in organic pollutants, such as the Great Lakes. The chlorination process kills the bacteria but also chlorinates the organic compounds. Thus the water supplies contain very low but measurable quantities of carcinogenic substances such as carbon tetrachloride. Now that chemists can measure substances at incredibly low concentrations, the controversy revolves around the level at which any substance becomes a health hazard. To establish an international standard, the World Health Organization has set a maximum acceptable limit of 30 $\mu g{\cdot}L^{-1}$ for trichloromethane, $CHCl_3$, in drinking water.

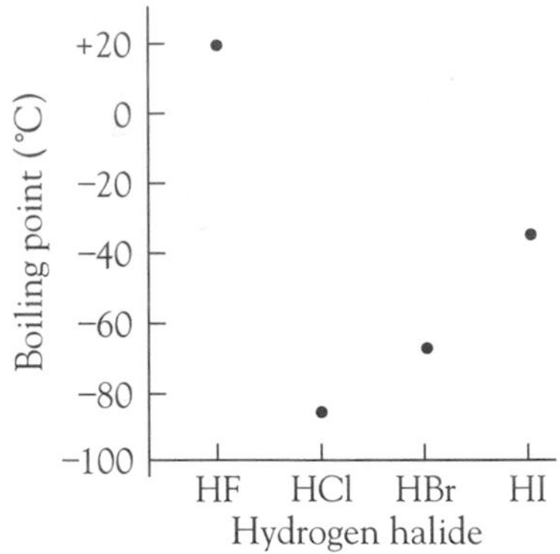

Figure 16.4 Boiling points of the hydrogen halides.

The long-term solution is to reduce substantially the levels of organic pollutants in domestic water supplies. For the Great Lakes, this effort will require continued close liaison between U.S. state and federal officials and Canadian provincial and federal officials. Even then, there will be organochlorines produced from naturally occurring organic compounds, the humic acids. One solution is to use trioxygen (see Chapter 15) or chlorine dioxide (see below) for primary water sterilization. However, because trioxygen decomposes within a short time, a small concentration of chlorine needs to be added to water supplies to maintain potability as it travels many kilometers through the aging, leaky pipes of most of our cities to its destination in your home.

Hydrogen Fluoride and Hydrofluoric Acid

Hydrogen fluoride is a colorless, fuming liquid with a boiling point of 20°C. This is much higher than the boiling points of the other hydrogen halides, as can be seen from Figure 16.4. The high boiling point of hydrogen fluoride is a result of the very strong hydrogen bonding between neighboring hydrogen fluoride molecules. Fluorine has the highest electronegativity of all the elements, so the hydrogen bond formed with fluorine is the strongest of all. The hydrogen bonds are linear with respect to the hydrogen atoms but are oriented at 120° with respect to the fluorine atoms. Thus the molecules adopt a zigzag arrangement (Figure 16.5).

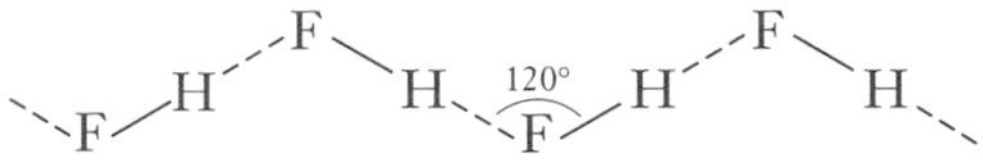

Figure 16.5 Hydrogen bonding in hydrogen fluoride.

Hydrogen fluoride is miscible with water and forms hydrofluoric acid:

$$HF(aq) + H_2O(l) \rightleftharpoons H_3O^+(aq) + F^-(aq)$$

It is a weak acid with a pK_a of 3.2, unlike the other hydrohalic acids, which are very strong, all having negative pK_a values.

As discussed in Chapter 8, the relative weakness of hydrofluoric acid can be ascribed to the fact that the hydrogen-fluorine bond is much stronger than the other hydrogen-halide bonds. Thus the dissociation into ions is less energetically favorable.

In more concentrated solutions, hydrofluoric acid ionizes to an even greater degree, the converse of the behavior of other acids. The cause of this behavior is well understood: a second equilibrium step that becomes more important at higher hydrogen fluoride concentrations and gives the linear hydrogen difluoride ion:

$$F^-(aq) + HF(aq) \rightleftharpoons HF_2^-(aq)$$

The hydrogen difluoride ion is so stable that alkali metal salts, such as potassium hydrogen difluoride, KHF_2, can be crystallized from solution.

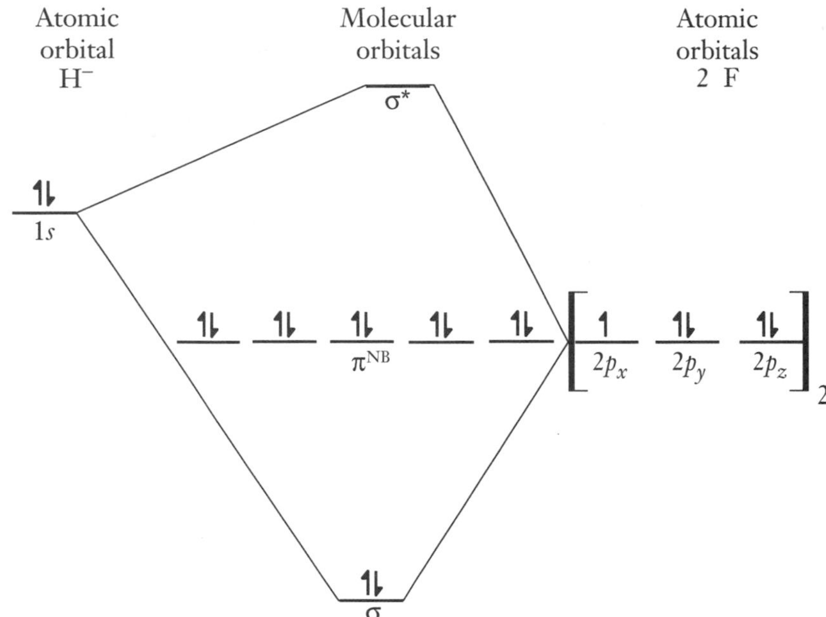

Figure 16.6 Simplified molecular orbital diagram for the hydrogen difluoride ion.

This acid ion is unique, for it involves a bridging hydrogen atom. It used to be regarded as a hydrogen fluoride molecule with a fluoride ion hydrogen bonded to it. However, recent studies have shown that the hydrogen is centrally located between the two fluorine atoms. This molecular arrangement can be explained in molecular orbital terms as a three-center two-electron bond (Figure 16.6).

Hydrofluoric acid is very corrosive, even though it is a weak acid. It is one of the few substances to attack glass, forming the hexafluorosilicate ion, SiF_6^{2-}:

$$SiO_2(s) + 6\,HF(aq) \rightarrow SiF_6^{2-}(aq) + 2\,H^+(aq) + 2\,H_2O(l)$$

This property is used in the etching of glass. An object to be etched is dipped in molten wax, and the wax is allowed to harden. The required pattern is then cut through the wax layer. Dipping the glass in hydrofluoric acid enables the hydrofluoric acid to react with only those parts of the glass surface that are exposed. After a suitable depth of glass has been dissolved, the object is removed from the acid bath, rinsed with water, and the wax melted off, leaving the glass with the desired etched pattern.

About 10^6 tonnes of hydrofluoric acid are produced each year, worldwide. Hydrogen fluoride is produced by heating calcium fluoride, the mineral fluorspar, with concentrated sulfuric acid:

$$CaF_2(s) + H_2SO_4(l) \xrightarrow{\Delta} 2\,HF(g) + CaSO_4(s)$$

The product is either liquefied by refrigeration or added to water to give hydrofluoric acid. To lower the cost of this endothermic reaction, plants have been constructed next to sulfuric acid production facilities. The heat from the exothermic reactions in the sulfuric acid plant is then used for the hydrogen fluoride process.

Obviously, in the production of a substance as toxic as hydrofluoric acid, the flue gases have to be carefully "scrubbed" to prevent traces of hydrogen fluoride from escaping into the environment. The other product in the reaction is the ubiquitous calcium sulfate. A simple stoichiometric calculation shows that, for every tonne of hydrogen fluoride, nearly 4 tonnes of calcium sulfate are produced. As with the other industrial processes that produce this by-product, some is utilized, but much of it is used as landfill.

We have discussed the use of hydrofluoric acid for glass etching, but, in fact, almost all the hydrofluoric acid produced commercially is used as the starting material for the synthesis of other fluorine-containing chemicals. For example, sodium fluoride is produced by mixing hydrofluoric acid with sodium hydroxide solution:

$$\mathrm{NaOH}(aq) + \mathrm{HF}(aq) \rightarrow \mathrm{NaF}(aq) + \mathrm{H_2O}(l)$$

Evaporation gives sodium fluoride crystals, the compound usually used for water fluoridation. The reaction between hydrofluoric acid and potassium hydroxide solution, in a 2:1 mole ratio, gives the acid salt, potassium hydrogen fluoride, which is used in the manufacture of fluorine gas:

$$\mathrm{KOH}(aq) + 2\ \mathrm{HF}(aq) \rightarrow \mathrm{KHF_2}(aq) + \mathrm{H_2O}(l)$$

Hydrochloric Acid

Hydrogen chloride is extremely soluble in water; in fact, concentrated hydrochloric acid contains about 38 percent by mass of hydrogen chloride, a concentration of 12 $\mathrm{mol{\cdot}L^{-1}}$. This acid is a colorless liquid with a pronounced acidic odor, which is due to the equilibrium between gaseous and aqueous hydrogen chloride:

$$\mathrm{HCl}(aq) \rightleftharpoons \mathrm{HCl}(g)$$

The technical grade reagent often has a yellowish color from an iron(III) ion impurity.

In contrast to hydrofluoric acid, hydrochloric acid is a strong acid ($pK_a = -7$), ionizing almost completely:

$$\mathrm{HCl}(aq) + \mathrm{H_2O}(l) \rightarrow \mathrm{H_3O^+}(aq) + \mathrm{Cl^-}(aq)$$

As the oxidation state diagram shows (Figure 16.1), the chloride ion is a very stable species. Hence dilute hydrochloric acid is often the acid of choice over the oxidizing nitric acid and, to a lesser extent, sulfuric acid. For example, zinc metal will react with hydrochloric acid to give the zinc ion and hydrogen gas:

$$\mathrm{Zn}(s) + 2\ \mathrm{HCl}(aq) \rightarrow \mathrm{ZnCl_2}(aq) + \mathrm{H_2}(g)$$

When zinc reacts with nitric acid, there is often some reduction of the nitrate ion to give nitrogen dioxide.

The traditional method of producing hydrochloric acid in the laboratory is to generate hydrogen chloride gas from the reaction of sodium chloride with sulfuric acid. In the first step, at 150°C, sodium hydrogen sulfate is formed:

$$NaCl(s) + H_2SO_4(l) \xrightarrow{\Delta} NaHSO_4(s) + HCl(g)$$

The mixture is heated to 550°C, at which temperature the sodium hydrogen sulfate reacts with an excess of sodium chloride to form the sodium sulfate and additional hydrogen chloride gas. The gas is then dissolved in water to form the acid:

$$NaHSO_4(s) + NaCl(s) \xrightarrow{\Delta} Na_2SO_4(s) + HCl(g)$$

This method is used in industry, too; and hydrogen chloride is also produced by the direct combination of dichlorine and dihydrogen gas:

$$H_2(g) + Cl_2(g) \rightarrow 2\ HCl(g)$$

The largest proportion of hydrogen chloride produced actually is a by-product from other industrial processes, such as the synthesis of carbon tetrachloride:

$$CH_4(g) + 4\ Cl_2(g) \rightarrow CCl_4(l) + 4\ HCl(g)$$

About 10^7 tonnes of hydrochloric acid are used worldwide every year. It has a wide range of uses: as a common acid; for removing rust from steel surfaces (a process called "pickling"); in the purification of glucose and corn syrup; in the acid treatment of oil and gas wells; and in the manufacture of chlorine-containing chemicals. The acid is available in many hardware stores under the archaic name of muriatic acid, its main home uses being for the cleaning of concrete surfaces and for rust removal. A very small number of persons are unable to synthesize enough stomach acid, and these individuals must ingest capsules of dilute hydrochloric acid with every meal.

Halides

Ionic Halides

Most ionic chlorides, bromides, and iodides are soluble in water, dissolving to give the metal ion and the halide ion. However, many metal fluorides are insoluble. For example, as mentioned earlier, calcium chloride is very water-soluble, whereas calcium fluoride is insoluble. We explain these observations in terms of the greater lattice energy in crystals containing the small, high charge density anion and the high charge density cation.

Solutions of soluble fluorides are basic because the fluoride ion is the conjugate base of the weak acid hydrofluoric acid:

$$F^-(aq) + H_2O(l) \rightleftharpoons HF(aq) + OH^-(aq)$$

There are two possible ways to form metal halides: combine metal and halogen to give a metal ion with the higher oxidation state; and combine metal and hydrogen halide to give a metal ion with the lower oxidation state. The preparations of iron(III) chloride and iron(II) chloride illustrate this point:

$$2\ Fe(s) + 3\ Cl_2(g) \rightarrow 2\ FeCl_3(s)$$

$$Fe(s) + 2\ HCl(g) \rightarrow FeCl_2(s) + H_2(g)$$

In the first case, dichlorine is acting as a strong oxidizing agent, whereas in the second case the hydrogen ion is a weak oxidizing agent.

Hydrated metal halides can be prepared from the metal oxide, carbonate, or hydroxide and the appropriate hydrohalic acid. For example, magnesium chloride hexahydrate can be prepared from magnesium oxide and hydrochloric acid, followed by crystallization of the solution:

$$MgO(s) + 2\ HCl(aq) + 5\ H_2O(l) \rightarrow MgCl_2{\cdot}6H_2O(s)$$

In most cases, the anhydrous salt cannot be prepared by heating the hydrate, because decomposition occurs instead. Thus magnesium chloride hexahydrate gives magnesium hydroxide chloride, Mg(OH)Cl, when heated:

$$MgCl_2{\cdot}H_2O(s) \xrightarrow{\Delta} Mg(OH)Cl(s) + HCl(g) + 5\ H_2O(l)$$

To obtain the anhydrous chloride from the hydrate, we have to chemically remove the water. This can be done by using thionyl chloride, $SOCl_2$; the reaction gives sulfur dioxide and hydrogen chloride gases:

$$MgCl_2{\cdot}6H_2O(s) + 6\ SOCl_2(l) \rightarrow MgCl_2(s) + 6\ SO_2(g) + 12\ HCl(g)$$

This is a common way to dehydrate metal chlorides.

Not all metal iodides in which the metal ion takes its higher oxidation state can be prepared, because the iodide ion itself is a reducing agent. For example, iodide ion will reduce copper(II) ion to copper(I):

$$2\ Cu^{2+}(aq) + 4\ I^-(aq) \rightarrow 2\ CuI(s) + I_2(s)$$

As a result, copper(II) iodide does not exist.

The common test for distinguishing chloride, bromide, and iodide ions involves the addition of silver nitrate solution to give a precipitate. Using X^- to represent the halide ion, we can write a general equation:

$$X^-(aq) + Ag^+(aq) \rightarrow AgX(s)$$

Silver chloride is white; silver bromide is cream colored; and silver iodide is yellow. Like most silver compounds, they are light sensitive; and, over a period of hours, the solids change to shades of gray as metallic silver forms.

To confirm the identity of the halogen, ammonia solution is added to the silver halide. A precipitate of silver chloride reacts with dilute ammonia to form the soluble diamminesilver(I) ion, $[Ag(NH_3)_2]^+$:

$$AgCl(s) + 2\ NH_3(aq) \rightarrow [Ag(NH_3)_2]^+(aq) + Cl^-(aq)$$

The other two silver halides are not attacked. Silver bromide does react with concentrated ammonia, but silver iodide remains unreactive even under these conditions.

There also are specific tests for each halide ion. The chlorine test is quite hazardous, because it involves the reaction of the suspected chloride with a mixture of potassium dichromate and concentrated sulfuric acid. When warmed gently, the volatile, red, toxic compound chromyl chloride, CrO_2Cl_2, is produced:

$$K_2Cr_2O_7(s) + 4\ NaCl(s) + 6\ H_2SO_4(l)$$
$$\rightarrow 2\ CrO_2Cl_2(l) + 2\ KHSO_4(s) + 4\ NaHSO_4(s) + 3\ H_2O(l)$$

The vapor can be bubbled into water, where it forms a yellow solution of chromic acid, H_2CrO_4:

$$CrO_2Cl_2(g) + 2\ H_2O(l) \rightarrow H_2CrO_4(aq) + 2\ HCl(aq)$$

To test for bromide ion and iodide ion, a solution of dichlorine in water (aqueous chlorine) is added to the halide ion solution. The appearance of a yellow to brown color suggests the presence of either of these ions:

$$Cl_2(aq) + 2\ Br^-(aq) \rightarrow Br_2(aq) + 2\ Cl^-(aq)$$
$$Cl_2(aq) + 2\ I^-(aq) \rightarrow I_2(aq) + 2\ Cl^-(aq)$$

To distinguish them, we rely on the fact that the halogens themselves are nonpolar molecules. Hence they will "prefer" to dissolve in nonpolar or low-polarity solvents, such as carbon tetrachloride. If the brownish aqueous solution is shaken with such a solvent, the halogen should transfer to the low-polarity, nonaqueous layer. If the unknown is dibromine, the color will be brown, whereas that of diiodine will be bright purple.

There is another, very sensitive test for iodine: It reacts with starch to give a blue color (blue-black when concentrated solutions are used). In this unusual interaction, the starch polymer molecules wrap themselves around the iodine molecules. There is no actual chemical bond involved. The equilibrium is employed qualitatively in starch-iodide paper. When the paper is exposed to an oxidizing agent, the iodide is oxidized to iodine. The starch in the paper forms the starch-iodine complex, and the blue-black color is readily observed. Quantitatively, starch is used as the indicator in redox titrations involving the iodide-iodine redox reaction.

Diiodine, as already mentioned, is a nonpolar molecule. Thus its solubility in water is extremely low. It will, however, "dissolve" in a solution of iodide ion. This is, in fact, a chemical reaction producing the triiodide ion, I_3^- (discussed later):

$$I_2(s) + I^-(aq) \rightleftharpoons I_3^-(aq)$$

This large, low-charge anion will actually form solid compounds with low charge density cations such as rubidium, with which it forms rubidium triiodide, RbI_3.

Iodide ion will also undergo a redox reaction with iodate ion, IO_3^-, in acid solution to give diiodine:

$$IO_3^-(aq) + 6\ H^+(aq) + 5\ I^-(aq) \rightarrow 3\ I_2(s) + 3\ H_2O(l)$$

This reaction is often used in titrimetric analysis of iodide solutions. The diiodine can then be titrated with thiosulfate ion of known concentration:

$$I_2(s) + 2\ S_2O_3^{2-}(aq) \rightarrow 2\ I^-(aq) + S_4O_6^{2-}(aq)$$

Covalent Halides

As a result of weak intermolecular forces, most covalent halides are gases or liquids with low boiling points. The boiling points of these nonpolar molecules are directly related to the strength of the dispersion forces between the molecules. This intermolecular force, in turn, is dependent on the number of electrons in the molecule. A typical series is that of the boron halides (Table 16.5), which illustrates the relationship between boiling point and number of electrons.

Many covalent halides can be prepared by treating the element with the appropriate halogen. When more than one compound can be formed, the mole ratio can be altered to favor one product over the other. For example, in the presence of excess chlorine, phosphorus forms phosphorus pentachloride, whereas in the presence of excess phosphorus, phosphorus trichloride is formed:

$$2\ P(s) + 5\ Cl_2(g) \rightarrow 2\ PCl_5(s)$$

$$2\ P(s) + 3\ Cl_2(g) \rightarrow 2\ PCl_3(l)$$

If the element itself is inert, such as dinitrogen, an alternative route must be used. For nitrogen, the preferred method is to mix ammonia and chlorine gas:

$$NH_3(g) + 3\ Cl_2(g) \rightarrow NCl_3(l) + 3\ HCl(g)$$

Most covalent halides react vigorously with water. For example, phosphorus trichloride reacts with water to give phosphonic acid and hydrogen chloride:

$$PCl_3(l) + 3\ H_2O(l) \rightarrow H_3PO_3(l) + 3\ HCl(g)$$

Some covalent halides are kinetically inert, however, particularly the fluorides, such as carbon tetrafluoride and sulfur hexafluoride.

It is important to remember that metal halides can contain covalent bonds even when the metal is in a high oxidation state. For example, tin(IV) chloride behaves like a typical covalent halide. It is a liquid at room temperature, and it reacts violently with water:

$$SnCl_4(l) + 2\ H_2O(l) \rightarrow SnO_2(s) + 4\ HCl(g)$$

If a nonmetallic element exists in a number of possible oxidation states, then the highest oxidation state is usually stabilized by fluorine and the low-

Table 16.5 Boiling points of the boron halides

Compound	Boiling point (°C)	Number of electrons
BF_3	−100	32
BCl_3	+12	56
BBr_3	+91	110
BI_3	+210	164

est by iodine. This pattern reflects the decreasing oxidizing ability of elements in Group 17 as the group is descended. However, we must always be careful with our application of simplistic arguments. For example, the nonexistence of phosphorus pentaiodide, PI_5, is more likely to be due to the fact that the size of the iodine atom limits the number of iodine atoms that will fit around the phosphorus atom rather than to the spontaneous reduction of phosphorus from the +5 to the +3 oxidation state.

Halogen Oxides

The only stable compound of oxygen and fluorine is the pale yellow gas oxygen difluoride, OF_2. The main claim to fame of this V-shaped molecule is that it is the only well-known compound in which oxygen has a formal oxidation state of +2.

Chlorine forms several oxides, all of which have positive values of free energy of formation. As a result of this and a low activation energy of decomposition, they are very unstable and have a tendency to explode. Chlorine dioxide, ClO_2, is the only oxide of major interest. It is a yellow gas that condenses to a deep red liquid at 11°C. The compound is quite soluble in water, giving a fairly stable, green solution.

Chlorine dioxide is paramagnetic, like nitrogen dioxide. Yet it shows no tendency to dimerize. The chlorine-oxygen bond length is only 140 pm, much shorter than the 170 pm that is typical for a single bond length, and it is very close to that of a typical chlorine-oxygen double bond. A possible electron-dot structure reflecting this bond order is shown in Figure 16.7.

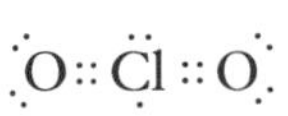

O :: Cl :: O

Figure 16.7 A possible electron-dot representation of the bonding in chlorine dioxide.

The bond angle in chlorine dioxide is 118°, a value suggesting that the σ bond framework involves sp^2 hybrid orbitals on the chlorine. If this were the case, two of these orbitals would be used in bonding to the oxygen atoms and the third would hold the lone pair (accounting for the approximately 120° bond angle). The odd electron would then be in the unhybridized chlorine *p* orbital. The technique of electron spin resonance (ESR) can be used to investigate the properties of unpaired electrons. In a magnetic field, unpaired electrons have a preference for one spin direction ($-\frac{1}{2}$) over the other ($+\frac{1}{2}$). By applying energy, in this case, in the microwave region of the spectrum, we can cause the electron to reverse its spin. The energy (or energies) needed to do this gives us information about the electron environment. When we study chlorine dioxide by this means, we find that the odd electron is indeed in a pure *p* orbital of the chlorine atom at right angles to the molecular plane. In addition, it spends about 15 percent of its time in a *p* orbital of each oxygen atom.

These findings provide an explanation for why nitrogen dioxide dimerizes and chlorine dioxide does not. The ESR results confirm that the unpaired electron of nitrogen dioxide is largely in an *sp*-type hybrid orbital of nitrogen (Chapter 14), although it spends more time delocalized on the oxygen atoms (about 25 percent on each) than it does in chlorine dioxide. However, in chlorine dioxide, the unhybridized *p* orbital has two lobes and is not oriented in a single direction like that of a hybrid orbital. Hence it can be argued that the lone electron is less available for pairing with another lone electron in a second chlorine dioxide molecule. This explanation of the bonding is still a simplistic one, for the chlorine *d* orbitals must be involved in the π bonding to account for the substantial double bond character.

Chlorine dioxide, usually diluted with dinitrogen or carbon dioxide for safety, is a very powerful oxidizing agent. For example, to bleach flour to make white bread, chlorine dioxide is 30 times more effective than dichlorine. Large quantities of chlorine dioxide also are used as dilute aqueous solutions for bleaching wood pulp to make white paper. In this role, it is preferred over dichlorine, because chlorine dioxide bleaches without forming the hazardous chlorinated wastes. Similarly, chlorine dioxide is being used increasingly for domestic water treatment, because, in this context too, it does not chlorinate hydrocarbon pollutants that are present in the water. Hence use of this reagent avoids the problems discussed earlier.

Thus, even though pure chlorine dioxide is explosive, it is of major industrial importance. About 10^6 tonnes are produced every year, worldwide. It is difficult to determine the exact production total, because the gas is so hazardous that it is generally produced in comparatively small quantities at the sites where it is to be used. The synthetic reaction involves the reduction of chlorine in the +5 (ClO_3^-) oxidation state by chlorine in the −1 (Cl^-) oxidation state in very acid conditions to give chlorine in the +4 (ClO_2) and 0 (Cl_2) oxidation states:

$$2\ ClO_3^-(aq) + 4\ H^+(aq) + 2\ Cl^-(aq) \rightarrow 2\ ClO_2(g) + Cl_2(g) + 2\ H_2O(l)$$

In North America, sulfur dioxide is added to reduce (and remove) the dichlorine gas to chloride ion; the sulfur dioxide is simultaneously oxidized to sulfate:

$$SO_2(g) + 2\ H_2O(l) \rightarrow SO_4^{2-}(aq) + 4\ H^+(aq) + 2\ e^-$$

$$Cl_2(g) + 2\ e^- \rightarrow 2\ Cl^-(aq)$$

However, this process produces sodium sulfate waste. A German process separates the dichlorine from the chlorine dioxide, then reacts the dichlorine with hydrogen gas to produce hydrochloric acid. The acid can then be reused in the synthesis.

The bromine oxides are of little interest and importance, but one of the iodine oxides, diiodine pentoxide, I_2O_5, is of use. It is a white, thermodynamically stable solid with a structure similar to that of dinitrogen pentoxide—one oxygen atom links two IO_2 units. The compound reacts with reducing gases, such as carbon monoxide:

$$I_2O_5(s) + 5\ CO(g) \rightarrow I_2(s) + 5\ CO_2(g)$$

The production of iodine can be used as a qualitative or quantitative method for determining the presence of carbon monoxide in auto exhausts or flue gases.

Chlorine Oxyacids and Oxyanions

Chlorine forms a series of oxyacids and oxyanions for each of its positive odd oxidation states from +1 to +7. The shapes of the ions (and related acids) are based on a tetrahedral arrangement around the chlorine atom (Figure 16.8). The short chlorine-oxygen bonds in each of the ions indicate that multiple bonding must be present. We envisage that this π bonding utilizes the full *p*

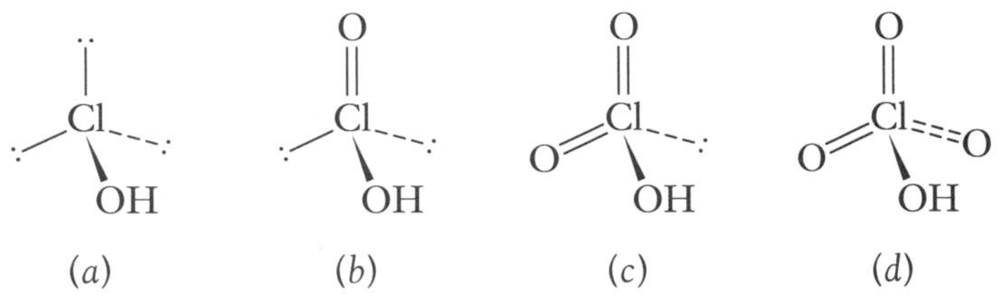

Figure 16.8 (*a*) Hypochlorous acid; (*b*) chlorous acid; (*c*) chloric acid; and (*d*) perchloric acid.

orbitals on the oxygen atoms and empty *d* orbitals on the chlorine atom. As discussed in Chapter 8, acid strength increases as the number of oxygen atoms increases. Thus hypochlorous acid is very weak; chlorous acid, weak; chloric acid, strong; and perchloric acid, very strong.

Hypochlorous Acid and the Hypochlorite Ion

Hypochlorous acid is a very weak acid; thus solutions of hypochlorites are very basic as a result of the hydrolysis reaction:

$$ClO^-(aq) + H_2O(l) \rightleftharpoons HClO(aq) + OH^-(aq)$$

Hypochlorous acid is a strong oxidizing agent and, in the process, is reduced to chlorine gas:

$$2\ HClO(aq) + 2\ H^+(aq) + 2\ e^- \rightarrow Cl_2(g) + 2\ H_2O(l) \qquad E^\circ = +1.64\ V$$

The hypochlorite ion, however, is a weaker oxidizing agent that is usually reduced to the chloride ion:

$$ClO^-(aq) + H_2O(l) + 2\ e^- \rightarrow Cl^-(aq) + 2\ OH^-(aq) \qquad E^\circ = +0.89\ V$$

It is this oxidizing (bleaching and bactericidal) power that renders the hypochlorite ion useful.

Hypochlorous acid and hydrochloric acid are formed when dichlorine is dissolved in cold water:

$$Cl_2(aq) + H_2O(l) \rightleftharpoons H^+(aq) + Cl^-(aq) + HClO(aq)$$

The two compounds of industrial importance are sodium hypochlorite and calcium hypochlorite. The former is prepared by the electrolysis of brine, in a manner similar to that used for the preparation of sodium hydroxide (Chapter 10). However, to produce sodium hypochlorite, the electrolysis is performed in a single chamber with stirrers to mix the hydroxide ion produced at the cathode with the chlorine produced at the anode:

$$2\ Cl^-(aq) \rightarrow Cl_2(aq) + 2\ e^- \qquad \text{(anode)}$$

$$2\ H_2O(l) + 2\ e^- \rightarrow H_2(g) + 2\ OH^-(aq) \qquad \text{(cathode)}$$

$$Cl_2(aq) + 2\ OH^-(aq) \rightarrow Cl^-(aq) + ClO^-(aq) + H_2O(l) \qquad \text{(mixing reaction)}$$

The reaction vessel must be cooled, for as we will see in our discussion of the chlorate ion, a different reaction occurs in warm solutions. Commercial

Swimming Pool Chemistry

In North America, we usually rely on dichlorine or chlorine-based compounds, such as calcium hypochlorite, to destroy the microorganisms in our swimming pools. In fact, the most potent disinfectant is hypochlorous acid. In many public pools, this compound is formed when dichlorine reacts with water:

$$Cl_2(aq) + H_2O(l) \rightleftharpoons H^+(aq) + Cl^-(aq) + HClO(aq)$$

To neutralize the hydronium ion, sodium carbonate (soda ash) is added:

$$CO_3^{2-}(aq) + H^+(aq) \rightarrow HCO_3^-(aq)$$

As a secondary result of this addition, the chlorine equilibrium shifts to the right, thus providing more hypochlorous acid.

In smaller pools, the hydrolysis of the hypochlorite ion provides the hypochlorous acid:

$$ClO^-(aq) + H_2O(l) \rightarrow HClO(aq) + OH^-(aq)$$

Acid must then be added to reduce the pH:

$$H^+(aq) + OH^-(aq) \rightarrow H_2O(l)$$

It is important to keep the hypochlorous acid concentration at levels high enough to protect against bacteria and other pool organisms. This is a particularly difficult task in outdoor pools because hypochlorous acid decomposes in bright light and at high temperatures:

$$2\ HClO(aq) \rightarrow 2\ HCl(aq) + O_2(g)$$

The production of stinging eyes in a swimming pool is usually blamed on "too much chlorine." In fact, it is the converse problem, for the irritated eyes are caused by the presence of chloramines in the water, such as NH_2Cl. These nasty compounds are formed through the reaction of hypochlorous acid with ammonia-related compounds, such as urea from urine, provided by the bathers:

$$NH_3(aq) + HClO(aq) \rightarrow NH_2Cl + H_2O(l)$$

To destroy them, we need to add more dichlorine, a process known as *superchlorination*. This additional dichlorine will react with the chloramines, decomposing them to give hydrochloric acid and dinitrogen:

$$2\ NH_2Cl(aq) + Cl_2(aq) \rightarrow N_2(g) + 4\ HCl(aq)$$

An increasingly popular approach is to use trioxygen (ozone) as the primary bactericide. This disinfectant causes much less eye irritation. However, because trioxygen slowly decomposes into dioxygen, a low level of chlorine-based compounds has to be added to the water to maintain safe conditions.

hypochlorite solutions such as those in Clorox® or Javex® contain an approximately equimolar mixture of sodium hypochlorite and sodium chloride.

Sodium hypochlorite is not stable in the solid phase; thus calcium hypochlorite is often used as a solid source of hypochlorite ion. There are several processes for its synthesis, but the most elegant is the reaction between a suspension of calcium hydroxide and chlorine gas:

$$2\ Ca(OH)_2(s) + 2\ Cl_2(g) \rightarrow Ca(ClO)_2{\cdot}2H_2O(s) + CaCl_2(aq)$$

The calcium chloride produced is soluble, but calcium hypochlorite dihydrate is insoluble. Hence the latter can be separated by filtration to provide a high-purity product.

Sodium hypochlorite solution is used for bleaching and decolorization of wood pulp and textiles; both sodium and calcium hypochlorites are used in disinfection. The latter process has its hazards. Although the labels on sodium hypochlorite solution containers warn about the hazards of mixing cleansers, a knowledge of chemistry is required to understand the problem. As we mentioned, commercial sodium hypochlorite solution contains chloride ion. In the presence of hydrogen (hydronium) ion such as that in a sodium hydrogen sulfate–based cleanser, the hypochlorous acid reacts with the chloride ion to produce chlorine gas:

$$ClO^-(aq) + H^+(aq) \rightarrow HClO(aq)$$

$$HClO(aq) + Cl^-(aq) + H^+(aq) \rightleftharpoons Cl_2(g) + H_2O(l)$$

Several injuries and deaths have been caused by this simple redox reaction.

The Chlorate Ion

Although chlorites are of little interest, chlorates have several uses. Sodium chlorate can be prepared by bubbling dichlorine into a hot solution of sodium hydroxide. The sodium chloride, which is less soluble than sodium chlorate, precipitates:

$$3\ Cl_2(aq) + 6\ NaOH(aq) \xrightarrow{\Delta} NaClO_3(aq) + 5\ NaCl(s) + 3\ H_2O(l)$$

Industrially, the process is performed in a cell like that used for the production of sodium hypochlorite, except hot solution conditions are employed.

Potassium chlorate is used in large quantities to make matches and fireworks. Like all chlorates, it is a strong oxidizing agent that can explode unpredictably when mixed with reducing agents. Considerable amounts of sodium chlorate are consumed in the production of chlorine dioxide.

Chlorates decompose when heated, although in an unusual manner. The route for producing potassium chlorate has been studied in the most detail. When heated to temperatures below 370°C, disproportionation occurs to give potassium chloride and potassium perchlorate:

$$4\ KClO_3(l) \xrightarrow{\Delta} KCl(s) + 3\ KClO_4(s)$$

This is a synthetic route for the perchlorate. When potassium chlorate is heated above 370°C, the perchlorate that is formed by disproportionation decomposes:

$$KClO_4(s) \xrightarrow{\Delta} KCl(s) + 2\ O_2(g)$$

The pathway for the slow, uncatalyzed reaction is different from that for the reaction catalyzed by manganese(IV) oxide. When catalyzed, the pathway giving potassium chloride and dioxygen involves potassium permanganate (which produces a purple color) and potassium manganate, K_2MnO_4. The mechanism is a nice illustration of the chemical participation of a catalyst.

$$2\ KClO_3(s) + 2\ MnO_2(s) \rightarrow 2\ KMnO_4(s) + Cl_2(g) + O_2(g)$$

$$2\ KMnO_4(s) \rightarrow K_2MnO_4(s) + MnO_2(s) + O_2(g)$$

$$K_2MnO_4(s) + Cl_2(g) \rightarrow 2\ KCl(s) + MnO_2(s) + O_2(g)$$

$$2\ KClO_3(s) \rightarrow 2\ KCl(s) + 3\ O_2(g) \qquad \text{[net equation]}$$

Thus the oxygen is oxidized from the oxidation state of –2 to 0; the manganese cycles from +4 through +7 and +6 back to +4; and the chlorine is reduced from +5 to 0 to –1. Additional evidence for this mechanism is a faint odor of dichlorine, which is released during the first step.

Perchloric Acid and the Perchlorate Ion

The strongest simple acid of all is perchloric acid. The pure acid is a colorless liquid that can explode unpredictably. As a result of its oxidizing nature and high oxygen content, contact with organic materials such as wood or paper causes an immediate fire. Concentrated perchloric acid, usually a 60 percent aqueous solution, is rarely used as an acid but is far more often used as a very powerful oxidizing agent, for example, to oxidize metal alloys to the metal ions so that they can be analyzed. Special perchloric acid fume hoods should be used when these oxidations are performed. Cold dilute solutions of perchloric acid are reasonably safe, however.

Sodium perchlorate is prepared industrially by the electrolytic oxidation of sodium chlorate:

$$ClO_3^-(aq) + H_2O(l) \rightarrow ClO_4^-(aq) + 2\ H^+(aq) + 2\ e^- \qquad \text{(anode)}$$

$$2\ H^+(aq) + 2\ e^- \rightarrow H_2(g) \qquad \text{(cathode)}$$

The solubility of an alkali metal salt decreases with increasing cation size. That is, the larger cations have a higher lattice energy with the large, low-charge anion. Thus potassium perchlorate is only slightly soluble (2 g per 100 mL of water). By contrast, silver perchlorate is amazingly soluble, to the extent of 500 g per 100 mL of water. The high solubility of silver perchlorate in low-polarity organic solvents as well as water suggests that its bonding in the solid phase is essentially covalent rather than ionic. That is, only dipole attractions need to be overcome to solubilize the compound rather than the much stronger electrostatic attractions in an ionic crystal lattice, which can be overcome only by a very polar solvent.

Potassium perchlorate is used in fireworks and flares, but about half the commercially produced perchlorate is used in the manufacture of ammonium perchlorate. This compound is used as a component, along with the reducing agent, aluminum, in solid booster rockets:

The Discovery of the Perbromate Ion

The perchlorate ion, ClO_4^-, and the periodate ion, IO_6^{5-}, have been known since the nineteenth century, yet the perbromate ion could never be prepared. Many scientists, including Linus Pauling, devised theories to account for its nonexistence. For example, it was argued that the stability of the perchlorate ion was due to the strong π bonds that involved the chlorine 3*d* orbitals. The claim was made that, for bromine, a very poor overlap of the 4*d* bromine orbitals with the 2*p* orbitals of the oxygen destabilized the theoretical perbromate ion.

The theories had to be rewritten in 1968 when the American chemist E. H. Appelman discovered synthetic routes to this elusive perbromate ion. One of the routes involved another new discovery, a compound of xenon. In this process, xenon difluoride was used as an oxidizing agent:

$$XeF_2(aq) + BrO_3^-(aq) + H_2O(l) \rightarrow Xe(g) + 2\ HF(aq) + BrO_4^-(aq)$$

The second route, which is now used to produce the perbromate ion on a large scale, involves the use of difluorine as the oxidizing agent in basic solution:

$$BrO_3^-(aq) + F_2(g) + 2\ OH^-(aq) \rightarrow BrO_4^-(aq) + 2\ F^-(aq) + H_2O(l)$$

So why is the ion so elusive even though it is thermodynamically stable? The answer lies in the high oxidation potential of the ion:

$$BrO_4^-(aq) + 2\ H^+(aq) + 2\ e^- \rightarrow BrO_3^-(aq) + H_2O(l) \qquad E^\circ = +1.74\ V$$

By contrast, the oxidation potential for perchlorate is +1.23 V and for periodate, +1.64 V. Hence only extremely strong oxidizing agents such as xenon difluoride and difluorine are capable of oxidizing bromate to perbromate. Thus, before dismissing any conjectured compound as impossible to synthesize, we must always be sure to explore all the possible preparative routes and conditions.

$$6\ NH_4ClO_4(s) + 8\ Al(s) \rightarrow 4\ Al_2O_3(s) + 3\ N_2(g) + 3\ Cl_2(g) + 12\ H_2O(g)$$

Each shuttle launch uses 850 tonnes of the compound, and total U.S. consumption is about 30 000 tonnes. Until recently, the only two U.S. plants manufacturing ammonium perchlorate (AP) were located in Henderson, Nevada, a suburb of Las Vegas. The attractions of the site were the cheap electricity from the Hoover Dam and the very dry climate, which makes the handling and storage of the hygroscopic ammonium perchlorate much easier.

Ammonium perchlorate decomposes when heated above 200°C:

$$2\ NH_4ClO_4(s) \xrightarrow{\Delta} N_2(g) + Cl_2(g) + 2\ O_2(s) + 4\ H_2O(g)$$

On 4 May 1988, this decomposition occurred on a massive scale at one of the manufacturing plants. A series of explosions destroyed half of the nation's AP production capacity, as well as causing death, injury, and extensive property damage. Several issues were raised by the accident, such as the feasibility of constructing such plants close to residential areas and the dependence of the space and military rocket programs on only two manufacturing facilities for the nation's entire supply of a crucial chemical compound.

Interhalogen Compounds and Polyhalide Ions

There is an enormous number of combinations of pairs of halogens forming interhalogen compounds and polyhalide ions. The neutral compounds fit the formulas XY, XY_3, XY_5, and XY_7, where X is the halogen of higher atomic mass and Y, that of lower atomic mass. All permutations are known for XY and XY_3, but XY_5 is only known where Y is fluorine. Thus, once again, it is only with fluorine that the highest oxidation states are obtained. The formula XY_7, in which X would have the oxidation state of +7, is found only in IF_7. The common argument for the lack of chlorine and bromine analogs is simply that of size: Only the iodine atom is large enough to accommodate seven fluorine atoms.

All of the interhalogens can be prepared by combination reactions of the constituent elements. For example, mixing chlorine and fluorine in a 1:3 ratio gives chlorine trifluoride, ClF_3:

$$Cl_2(g) + 3\ F_2(g) \rightarrow 2\ ClF_3(g)$$

The simple interhalogens such as chlorine monofluoride and bromine monochloride have colors intermediate between those of the constituents. However, the melting points and boiling points of the interhalogens are slightly higher than the mean values of the constituents because the interhalogen molecules are polar. More important, the chemical reactivity of an interhalogen compound is usually similar to that of the more reactive parent halogen. To chlorinate an element or compound, it is often more convenient to use solid iodine monochloride than chlorine gas, although sometimes the nonhalogen atom in the two products has different oxidation states. This outcome can be illustrated for the chlorination of vanadium:

$$V(s) + 2\ Cl_2(g) \rightarrow VCl_4(l)$$

$$V(s) + 3\ ICl(s) \rightarrow VCl_3(s) + 1\tfrac{1}{2}\ I_2(s)$$

In solution, interhalogen molecules are hydrolyzed to the hydrohalic acid of the more electronegative halogen and the hypohalous acid of the less electronegative halogen. For example

$$BrCl(g) + H_2O(l) \rightarrow HCl(aq) + HBrO(aq)$$

Ruby-red, solid iodine monochloride is used in biochemistry as the Wij reagent for the determination of the number of double carbon-carbon bonds in an oil or fat. When we add the brown solution of the interhalogen to the unsaturated fat, decolorization occurs as the halogens add across the double bond:

$$—C{=}C— + ICl \rightarrow —\underset{\displaystyle I}{\underset{|}{CH}}—\underset{\displaystyle Cl}{\underset{|}{CH}}—$$

When a permanent brown color remains, the reaction has been completed. The results are reported as the iodine number—the volume (milliliters) of a standard iodine monochloride solution needed to react with a fixed mass of fat.

The only interhalogen compound produced on an industrial scale is chlorine trifluoride, a liquid that boils at 11°C. It is a convenient and extremely powerful fluorinating agent, as a result of its high fluorine content and high bond polarity. It is particularly useful in the separation of uranium from most of the fission products in used nuclear fuel. At the reaction temperature of 70°C, uranium forms liquid uranium(VI) fluoride. However, most of the major reactor products, such as plutonium, form solid fluorides:

$$U(s) + 3\ ClF_3(l) \rightarrow UF_6(l) + 3\ ClF(g)$$

$$Pu(s) + 2\ ClF_3(l) \rightarrow PuF_4(s) + 2\ ClF(g)$$

The uranium compound can then be separated from the mixture by distillation.

The interhalogen compounds are of particular interest to inorganic chemists because of their geometries. The shapes of the compounds all follow the VSEPR rules, even iodine heptafluoride, IF_7, which has the rare pentagonal bipyramidal shape of a seven-coordinate species (Figure 16.9).

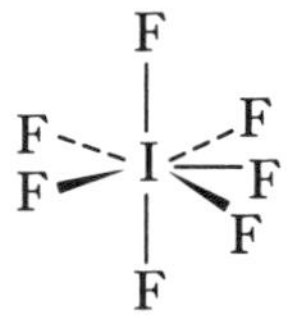

Figure 16.9 The iodine heptafluoride molecule, which has a pentagonal arrangement in the horizontal plane.

The halogens also form polyatomic ions. Iodine is the only halogen to readily form polyhalide anions by itself. The triiodide ion, I_3^-, is important because its formation provides a means of "dissolving" molecular iodine in water by using a solution of the iodide ion:

$$I_2(s) + I^-(aq) \rightleftharpoons I_3^-(aq)$$

The ion is linear and has equal iodine-iodine bond lengths of about 293 pm; these bonds are slightly longer than the single bond in the diiodine molecule (272 pm). There are many other polyiodide ions, including I_5^- and I_7^-, but these are less stable than the triiodide ion.

There also are a wide variety of interhalogen cations and anions, for example, the dichloroiodine ion, ICl_2^+, and the tetrachloroiodate ion, ICl_4^-. These are mainly of interest in terms of the match of their actual shape with that predicted by VSEPR theory (Figure 16.10).

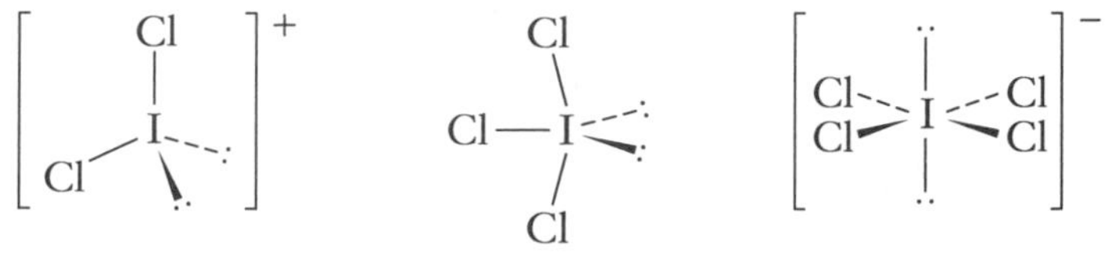

Figure 16.10 (*a*) The dichloroiodine ion, ICl_2^+; (*b*) the iodine trichloride molecule, ICl_3; and (*c*) the tetrachloroiodate ion, ICl_4^-.

Pseudohalogens

Just as the ammonium ion emulates many aspects of the alkali metals, so there are some polyatomic anions that emulate the halide ions. The most common of these are the cyanide ion, CN^-, and the thiocyanate ion, SCN^-. The cyanide ion resembles a halide ion in a remarkable number of ways:

1. It forms a weak acid, hydrocyanic acid, HCN.
2. It can be oxidized to form a colorless poisonous gas, cyanogen, $(CN)_2$.
3. Like the chloride ion, its compounds with silver, lead(II), and mercury(I) ions are insoluble in water.
4. It will form "pseudointerhalogen" compounds, such as BrCN.
5. Like silver chloride, silver cyanide reacts with ammonia to give the diamminesilver(I) cation.
6. It forms numerous complex ions such as $[Cu(CN)_4]^{2-}$, which is similar to its chloride analog, $[CuCl_4]^{2-}$.

Cyanide ion most closely resembles iodide ion. For example, just as iodide is oxidized by copper(II) ion to iodine, so cyanide is oxidized to cyanogen:

$$2\,Cu^{2+}(aq) + 4\,I^-(aq) \rightarrow 2\,CuI(s) + I_2(aq)$$

$$2\,Cu^{2+}(aq) + 4\,CN^-(aq) \rightarrow 2\,CuCN(s) + (CN)_2(aq)$$

Biological Aspects

The halogens are unique in that every stable member of the group has a biological function.

Fluorine

The fluoride ion appears to be an essential trace element, although its function has not been established. At the same time, it is very toxic in high concentrations, because it is a dangerous competitor for the hydroxide ion in enzyme reactions. The main interest has revolved around its use in preventing dental caries. A dentist, Frederick McKay, noticed the remarkable lack of cavities in the population in the Colorado Springs, Colorado, area in 1902. He tracked down the apparent cause as being the higher than average levels of fluoride ion in the drinking water. We now know that a concentration of about 1 ppm is required to convert the softer tooth material hydroxyapatite, $Ca_5(PO_4)_3(OH)$, to the tougher $Ca_5(PO_4)_3F$. A higher concentration of 2 ppm results in a brown mottling of the teeth; and at 50 ppm, toxic effects start to be noticed. Calcium ion appears to inhibit the harmful effects of any excess fluoride, presumably through the formation of insoluble calcium fluoride. Tea leaves contain high levels of fluoride ion, and a heavy tea drinker ingests about 1 mg of fluoride ion per day.

Most of the western and southwestern regions of the United States contain adequate fluoride ion concentrations in the natural water supply. In some parts of Texas, the levels in the natural water supply exceed that recommended. However, in the East and on the Pacific Coast, the levels are below the optimum required for caries prevention.

Chlorine

The chloride ion has a vital role in the ion balance in our bodies. It does not appear to play an active role but simply acts to balance the positive ions of sodium and potassium. However, covalently bonded chlorine is far less benign. Most of the toxic compounds with which we are currently concerned—for example, DDT and PCBs—are chlorine-containing molecules.

Bromine

This element appears to play an important role as a trace element in certain enzymes. Potassium bromide has been used in medicine as a sedative and as an anticonvulsant in the treatment of epilepsy.

Iodine

About 75 percent of the iodine in the human body is found in one location—the thyroid gland. Iodine is utilized in the synthesis of the hormones thyroxine (Figure 16.11) and triiodothyronine. These hormones are essential for growth, for the regulation of neuromuscular functioning, and for the maintenance of male and female reproductive functions. Yet goiter, the disease resulting from a deficiency of the thyroid hormones, is found throughout the world, including a band across the northern United States, much of South America, and Southeast Asia; there are localized areas of deficiency in most other countries of the world. One common cause of the disease is a lack of iodide ion in the diet. To remedy the iodine deficiency, potassium iodide is added to common household salt (iodized salt).

A symptom of goiter, which can have causes other than simple iodide deficiency, is a swollen lower part of the neck. The enlargement is the attempt of the thyroid gland to maximize absorption of iodide in iodine-deficient circumstances. In previous centuries, women with mild goiter were favored as marriage partners because their swollen necks enabled them to display their expensive and ornate necklaces more effectively.

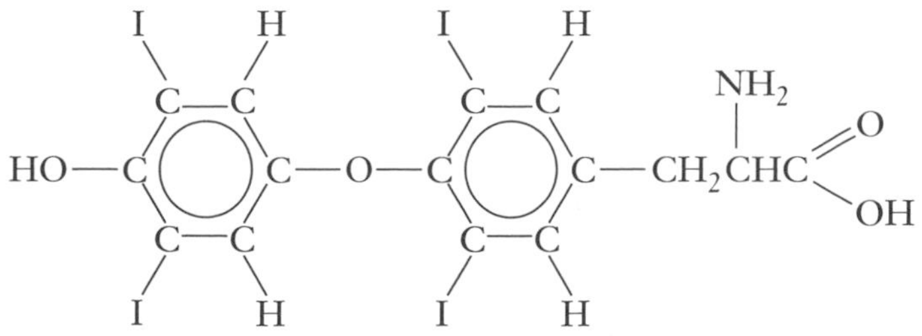

Figure 16.11 The thyroxine molecule.

Exercises

16.1. Write balanced chemical equations for the following chemical reactions:
(a) uranium(IV) oxide with hydrogen fluoride
(b) calcium fluoride with concentrated sulfuric acid
(c) liquid sulfur tetrachloride with water
(d) aqueous dichlorine and hot sodium hydroxide solution
(e) diiodine with difluorine in a 1:5 mole ratio
(f) bromine trichloride and water

16.2. Write balanced chemical equations for the following chemical reactions:
(a) lead metal with excess dichlorine
(b) magnesium metal with dilute hydrochloric acid
(c) hypochlorite ion with sulfur dioxide gas
(d) mild heating of potassium chlorate
(e) solid iodine monobromide with water
(f) phosphorus and iodine monochloride

16.3. Summarize the unique features of fluorine chemistry.

16.4. Suggest an explanation for why difluorine is so reactive toward other nonmetals.

16.5. Use the formation of solid iodine heptafluoride to indicate why entropy cannot be a driving force in the reactivity of fluorine.

16.6. Why can't difluorine be produced electrolytically from an aqueous solution of sodium fluoride by a process similar to that used to produce dichlorine from sodium chloride solution?

16.7. In the Frost diagram for chlorine, the Cl_2/Cl^- lines are identical for acidic and basic solutions. Explain why.

16.8. Why, in the Frost diagram (Figure 16.1), is the acid species of chloric acid written as ClO_3^-, whereas that of chlorous acid is written $HClO_2$?

16.9. Suggest a reason why hydrofluoric acid is a weak acid, whereas the binary acids of the other halogens are all strong acids.

16.10. Explain why hydrofluoric acid ionizes to a greater extent as the solution becomes more concentrated.

16.11. If annual hydrogen fluoride production is 1.2×10^6 tonnes per year, calculate the mass of calcium sulfate produced per annum by this process.

16.12. Why would you expect the hydrogen difluoride ion to form a solid compound with potassium ion?

16.13. Deduce the oxidation number for oxygen in hypofluorous acid, HOF.

16.14. Why is hydrochloric acid used as a common laboratory acid in preference to nitric acid?

16.15. Suggest how you would prepare (a) chromium(III) chloride, $CrCl_3$, from chromium metal; and (b) chromium(II) chloride, $CrCl_2$, from chromium metal.

16.16. Suggest how you would prepare (a) selenium tetrachloride, $SeCl_4$, from selenium; (b) diselenium dichloride, Se_2Cl_2, from selenium.

16.17. Explain why iron(III) iodide is not a stable compound.

16.18. Describe the tests used to identify each of the halide ions.

16.19. Calculate the enthalpy of reaction of ammonium perchlorate with aluminum metal. Apart from the exothermicity of the reaction, what other factors would make it a good propellant mixture?

16.20. Construct an electron-dot formula for the triiodide ion. Hence deduce the shape of the ion.

16.21. The concentration of hydrogen sulfide in a gas supply can be measured by passing a measured volume of gas over solid diiodine pentoxide. The hydrogen sulfide reacts with the diiodine pentoxide to give sulfur dioxide, diiodine, and water. The diiodine can then be titrated with thiosulfate ion and the hydrogen sulfide concentration calculated. Write chemical equations corresponding to the two reactions.

16.22. Carbon tetrachloride has a melting point of –23°C; carbon tetrabromide, +92°C; and carbon tetraiodide, +171°C. Provide an explanation for this trend. Estimate the melting point of carbon tetrafluoride.

16.23. The highest fluoride of sulfur is sulfur hexafluoride. Suggest why sulfur hexaiodide does not exist.

16.24. Construct electron-dot structures of chlorine dioxide that have zero, one, and two double bonds (one of each) and decide which would be preferred on the basis of formal charge assignments.

16.25. Another compound of chlorine and oxygen, Cl_2O_4, is more accurately represented as chlorine perchlorate, $ClOClO_3$. Draw the electron-dot structure of this compound and determine the oxidation number of each chlorine in the compound.

16.26. Describe the uses of (a) sodium hypochlorite; (b) chlorine dioxide; (c) ammonium perchlorate; (d) iodine monochloride.

16.27. Iodine pentafluoride undergoes self-ionization. Deduce the formulas of the cation and anion formed in the equilibrium and write a balanced equation for the equilibrium. Construct electron-dot diagrams for the molecule and the two ions. Which ion is the Lewis acid and which the Lewis base? Explain your reasoning.

16.28. Predict some physical and chemical properties of astatine as an element.

16.29. Explain why the cyanide ion is often considered as a pseudohalogen.

16.30. The thiocyanate ion, SCN^-, is linear. Construct reasonable electron-dot representations of this ion by assigning formal charges. The carbon-nitrogen bond length is known to be close to that of a triple bond. What does this tell you about the relative importance of each representation?

16.31. How does fluoride ion affect the composition of teeth?

CHAPTER

17

The Group 18 Elements: The Noble Gases

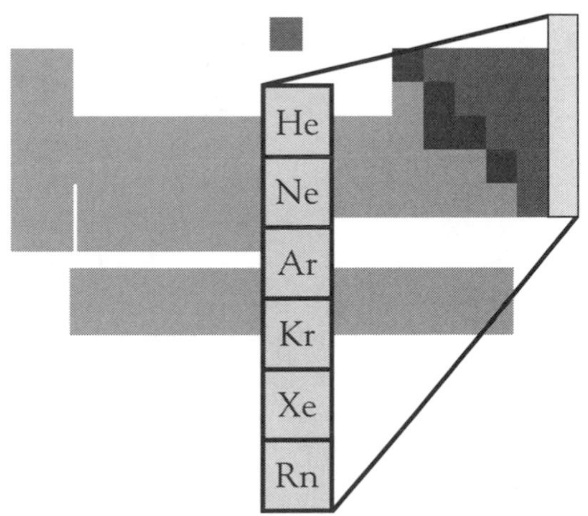

The noble gases make up the least reactive group of elements in the periodic table. In fact, xenon is the only noble gas to form a wide range of compounds. It is doubtful that stable chemical compounds will ever be made of helium or neon.

Although it had been noted as early as 1785 that there was something else in air besides oxygen and nitrogen, it was not until 100 years later that Sir William Ramsay showed that this other gas produced a previously unknown spectrum when an electric discharge was passed through it. Because every element has a unique spectrum, the gas producing the new spectrum had to be a new element. He named it argon, from the Greek word for lazy, because of its unreactive nature; and he suggested that it might be the first member of a new group in the periodic table.

In fact, one other element in this group—helium—had already been discovered, but not on Earth. Observations of the spectrum of the sun had shown some lines that did not belong to any element known at that time. The

Table 17.1 Melting and boiling points of the noble gases

Noble gas	Melting point (°C)	Boiling point (°C)	Number of electrons
He	—	–269	2
Ne	–249	–245	10
Ar	–189	–186	18
Kr	–157	–152	36
Xe	–112	–109	54
Rn	–71	–62	86

new element was named hel-ium, the first part of the name indicating that it was first discovered in the sun (Greek, *helios*) and the ending indicating that it was expected to be a metal. The element was first isolated on Earth in 1894 from uranium ores; and a few years later, it was realized that the helium is produced during the radioactive decay of uranium and its daughter elements. In 1926 it was suggested that the name of the element be changed to helion, to indicate that it was not a metal, but by that time the former name was too well established.

Every one of the noble gases was first identified by its unique emission spectrum. Hence it was really physical chemists rather than inorganic chemists who founded the study of this group of elements.

Group Trends

All the elements in Group 18 are colorless, odorless, monatomic gases at room temperature. They neither burn nor support combustion; in fact, they make up the least reactive group in the periodic table. The very low melting and boiling points of the noble gases indicate that the dispersion forces holding the atoms together in the solid and liquid phases are very weak. The trend in the melting and boiling points, shown in Table 17.1, corresponds to the increasing number of electrons and, hence, greater polarizability.

Table 17.2 Densities of the noble gases (at SATP)

Noble gas	Density ($g{\cdot}L^{-1}$)	Molar mass ($g{\cdot}mol^{-1}$)
He	0.2	4
Ne	1.0	20
Ar	1.9	39
Kr	4.1	84
Xe	6.4	131
Rn	10.6	222

Because the elements are all monatomic gases, there is a well-behaved trend in densities at the same pressure and temperature. The trend is a simple reflection of the increase in molar mass (Table 17.2). Air has a density of about 2.8 $g{\cdot}L^{-1}$; so, relative to air, helium has an extremely low density. Conversely, radon is among the densest of gases at SATP.

To date, chemical compounds have been isolated only for the three heaviest members of the group: krypton, xenon, and radon. Few compounds of krypton are known, whereas xenon has an extensive chemistry. The study of radon chemistry is very difficult because all the isotopes of radon are highly radioactive.

Unique Features of Helium

When helium is cooled to as close to absolute zero as we can reach, it is still a liquid. In fact, at that low temperature, a pressure of about 2.5 MPa is required to cause it to solidify. However, liquid helium is an amazing substance. When it condenses at 4.2 K, it behaves like an ordinary liquid (referred to as helium I); but when cooled below 2.2 K, the properties of the liquid (now helium II) are dramatically different. For example, helium II is an incredibly good thermal conductor, 10^6 times greater than helium I and much better than even silver, the best metallic conductor at room temperature. Even more amazing, its viscosity drops to close to 0. When helium II is placed in an open container, it literally "climbs the walls" and runs out over the edges. These and many other bizarre phenomena exhibited by helium II are best interpreted in terms of quantum behavior in the lowest possible energy states of the element. A full discussion is in the realm of quantum physics.

Uses of the Noble Gases

All the stable noble gases are found in the atmosphere, although only argon is present in a high proportion (Table 17.3). Helium is found in high concentrations in some underground natural gas deposits, where it has been accumulating from the decay of radioactive elements in Earth's crust. Gas reservoirs in the southwestern United States are among the largest in the

Table 17.3 Abundance of the noble gases in the atmosphere

Noble gas	Abundance (% by volume)
He	0.00052
Ne	0.0015
Ar	0.93
Kr	0.00011
Xe	0.0000087
Rn	Trace

world, and the United States is the world's largest producer of helium. In fact, when the deposits were first discovered, the price of helium gas dropped from $88·L^{-1} to 0.05¢·L^{-1}.

Because it is the gas with the second lowest density (dihydrogen having the lowest), helium is used to fill balloons. Dihydrogen would provide more "lift," but its flammability is a major disadvantage. Almost everyone has heard of the Hindenberg disaster, the burning of a transatlantic airship. Yet few are aware that the airship was designed to use helium. When the National Socialist party came to power in Germany in the 1930s, the U.S. government placed an embargo on helium shipments to Germany, fearing that the gas would be used for military purposes. Thus, when the airship was completed, dihydrogen had to be used. Today, the public thinks of airships solely in their advertising role. However, they have also been used as long-endurance flying radar posts by the U.S. Coast Guard to identify illegal drug-carrying flights. And an airship has been used to study the upper canopy of the rain forest in the Amazon basin, a vital task that would be very difficult to do by any other means.

Helium is used in deep-sea diving gas mixtures as a replacement for the more blood-soluble nitrogen gas in air. The velocity of sound is much greater in low-density helium than in air. As many people are aware, this gives breathers of helium "Mickey Mouse" voices. It should be added that the combination of dry helium gas and the higher frequency of vibrations in the larynx can cause voice damage to those who frequently indulge in the gas for fun.

Of great scientific importance, liquid helium is the only safe means of cooling scientific apparatus to very low temperatures. Many pieces of equipment use superconducting magnets to obtain very high magnetic fields, but at present the coils only become superconducting at temperatures close to 0 K.

All the other gases are obtained as by-products of the production of dioxygen from air. Some argon also is obtained from industrial ammonia synthesis, where it accumulates during the recycling of the unused atmospheric gases. Argon production is quite large, approaching 10^6 tonnes per year. Its major use is as an inert atmosphere for high-temperature metallurgical processes. Argon and helium are both used as an inert atmosphere in welding; neon, argon, krypton, and xenon are used to provide different colors in "neon" lights. The denser noble gases, particularly argon, have been used to fill the air space between the glass layers of thermal insulating windows. This use is based on the low thermal conductivities of these gases; for example, that of argon at 0°C is 0.017 $J{\cdot}s^{-1}{\cdot}m^{-1}{\cdot}K^{-1}$. Dry air at the same temperature has a thermal conductivity of 0.024 $J{\cdot}s^{-1}{\cdot}m^{-1}{\cdot}K^{-1}$.

The high abundance of argon in the atmosphere is a result of the radioactive decay of potassium-40, the naturally occurring radioactive isotope of potassium. This isotope captures a core electron to form argon-40:

$$^{40}_{19}K + ^{0}_{-1}e \rightarrow ^{40}_{18}Ar$$

Clathrates

Until the synthesis of the first compound of the noble gases, their only known chemical behavior was the formation of hydrates. For example, when xenon dissolves in water under pressure and the solution is cooled below 0°C, crystals with the approximate composition of $Xe{\cdot}6H_2O$ are formed. Warming the crystals causes immediate release of the gas. There is no chemical interaction between the noble gas and the water molecule; the gas atoms are simply locked into cavities in the hydrogen-bonded ice structure. A substance in which molecules or atoms are trapped within the crystalline framework of other molecules is called a *clathrate*. The name is derived from the Latin word *clathratus*, which means "enclosed behind bars."

A Brief History of Noble Gas Compounds

The story of the discovery of the noble gas compounds has become part of the folklore of inorganic chemistry. Unfortunately, like most folklore, the "true" story has been buried by myth. It was in 1924 that the German chemist von Antropoff made the suggestion that is obvious to us today: Because they have eight electrons in their valence level, the noble gases could form compounds with up to eight covalent bonds. Following that, in 1933 the American chemist Linus Pauling predicted the formulas of some possible noble gas compounds, such as oxides and fluorides. Two chemists at Caltech, Don Yost and Albert Kaye, set out to make compounds of xenon and fluorine. At the time, they thought they had been unsuccessful, but there is evidence that they did, in fact, make the first noble gas compound.

It was only after Yost and Kaye's admitted failure that the myth of the inertness of the noble gases spread. The "full octet" was claimed to be the reason, even though every inorganic chemist knew that many compounds involving nonmetals beyond the second period violated this "rule." So things remained, with this dogma being accepted by generation after generation of chemistry students, until the upsurge in interest in inorganic chemistry in the 1960s. It was Neil Bartlett, working at the University of British Columbia, who then approached the problem from a different direction in 1962.

Bartlett had been working with platinum(VI) fluoride, which he found was such a strong oxidizing agent that it oxidized dioxygen gas to form the compound $O_2^+PtF_6^-$. While teaching a first-year chemistry class, he noticed that the first ionization energy of xenon was almost identical to that of the dioxygen molecule. Despite the skepticism of his colleagues and students, he managed to synthesize a yellow compound that he claimed was $Xe^+PtF_6^-$. This reaction was the first proven formation of a compound of a noble gas. However, the compound did not have this simplistic formula, and it is now believed to have been a mixture of compounds that contained the XeF^+ ion. Unknown to Bartlett, Rudolf Hoppe in Germany had for some years been working with enthalpy cycles, and he had come to the conclusion that xenon fluorides should exist, on thermodynamic grounds. By passing an electric discharge through a mixture of xenon and difluorine, he was able to prepare xenon difluoride. Unfortunately for Hoppe, this discovery came a few weeks after Bartlett's discovery.

Since then, the field of noble gas chemistry has blossomed. In spite of much work, only a few compounds of krypton and radon have been isolated. Xenon is the sole noble gas to form a rich diversity of compounds, and then only with electronegative elements, such as fluorine, oxygen, and nitrogen.

Xenon Fluorides

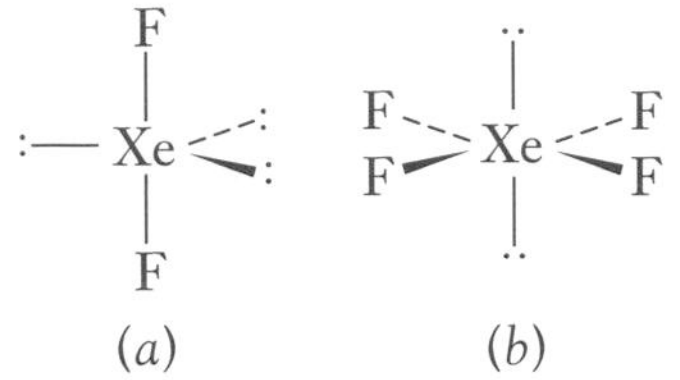

Figure 17.1 (*a*) Xenon difluoride; (*b*) xenon tetrafluoride.

Xenon forms three fluorides:

$$Xe(g) + F_2(g) \rightarrow XeF_2(s)$$

$$Xe(g) + 2\ F_2(g) \rightarrow XeF_4(s)$$

$$Xe(g) + 3\ F_2(g) \rightarrow XeF_6(s)$$

The product depends on the mole ratios of the reactants and on the exact reaction conditions of temperature and pressure, although very high partial pressures of difluorine are needed to form the xenon hexafluoride.

All three xenon fluorides are white solids and are stable with respect to dissociation into elements at ordinary temperatures; that is, they have negative free energies of formation at 25°C. As noted earlier, it is not necessary to invoke any novel concepts to explain the bonding; in fact, the three compounds are isoelectronic with those of well-established iodine polyfluoride anions. Table 17.4 shows the formulas of the compounds and the number of electron pairs around the central atom.

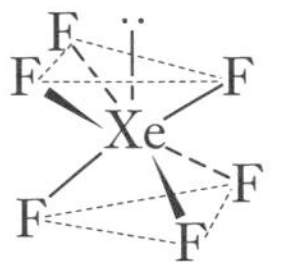

Figure 17.2 Probable capped octahedral structure of xenon hexafluoride in the gas phase.

The shapes of xenon difluoride and tetrafluoride are exactly those predicted from simple VSEPR theory (see Figure 17.1). Xenon hexafluoride, with six bonding pairs and one lone pair around the xenon atom, might be expected to adopt some form of pentagonal bipyramid like iodine heptafluoride. As discussed in Chapter 3, there are three possible arrangements: pentagonal bipyramid, capped trigonal prism, and capped octahedron. The structural studies of xenon hexafluoride in the gas phase indicate that it adopts the capped octahedral arrangement (Figure 17.2).

What is the driving force in the formation of the xenon fluorides? Let us take xenon tetrafluoride as an example. If we look at the equation for the formation of the compound from its elements, we see that the entropy change must be negative, considering that one mole of solid is being formed from three moles of gas:

$$Xe(g) + 2\ F_2(g) \rightarrow XeF_4(s)$$

The negative free energy must therefore result from a negative enthalpy change—an exothermic reaction. Figure 17.3 shows the enthalpy cycle for the formation of this compound from its elements. In the cycle, two moles of difluorine are dissociated into atoms, then four moles of xenon-fluorine

Table 17.4 Isoelectronic xenon halides and iodine polyfluoride anions

Number of electron pairs	Xenon halides	Iodine polyfluoride anions
5	XeF_2	IF_2^-
6	XeF_4	IF_4^-
7	XeF_6	IF_6^-

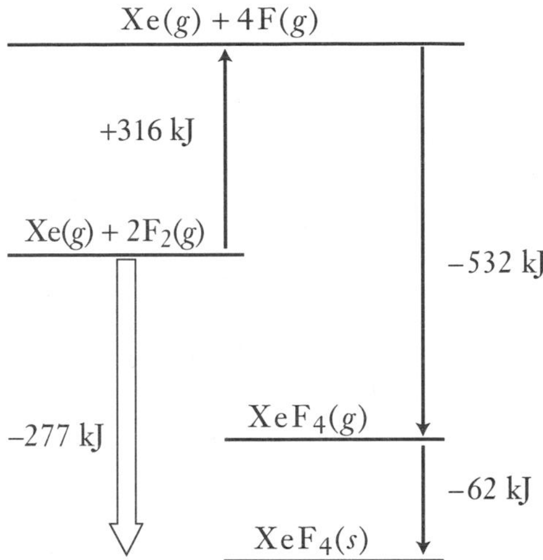

Figure 17.3 Enthalpy cycle for the formation of xenon tetrafluoride.

bonds are formed, followed by solidification of the product. The stability of this compound clearly depends on the moderately high Xe–F bond energy and the low dissociation energy of the fluorine molecule.

All the fluorides hydrolyze in water; for example, xenon difluoride is reduced to xenon gas:

$$2\ XeF_2(s) + 2\ H_2O(l) \rightarrow 2\ Xe(g) + O_2(g) + 4\ HF(l)$$

Xenon hexafluoride is first hydrolyzed to xenon oxide tetrafluoride, $XeOF_4$, which in turn is hydrolyzed to xenon trioxide:

$$XeF_6(s) + H_2O(l) \rightarrow XeOF_4(l) + 2\ HF(l)$$

$$XeOF_4(l) + 2\ H_2O(l) \rightarrow XeO_3(s) + 4\ HF(l)$$

The fluorides are strong fluorinating agents. For example, xenon difluoride can be used to fluorinate double bonds in organic compounds. It is a very "clean" fluorinating agent, in that the inert xenon gas can be easily separated from the required product:

$$XeF_2(s) + CH_2{=}CH_2(g) \rightarrow CH_2FCH_2F(g) + Xe(g)$$

Furthermore, a fluoride in which the other element is in its highest possible oxidation state can be produced by using xenon fluorides as reagents. Thus xenon tetrafluoride will oxidize sulfur tetrafluoride to sulfur hexafluoride:

$$XeF_4(s) + 2\ SF_4(g) \rightarrow 2\ SF_6(g) + Xe(g)$$

Xenon Oxides

Xenon forms two common oxides, xenon trioxide and xenon tetroxide. As mentioned in Chapter 15, oxygen usually "brings out" a higher oxidation number of an element than does fluorine; and with xenon, this is certainly the case.

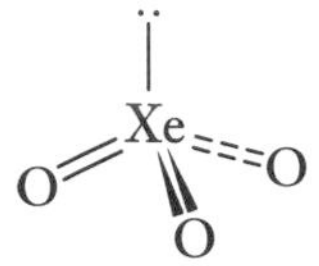

Xenon trioxide is a colorless, deliquescent solid that is quite explosive. The oxide is an extremely strong oxidizing agent, although its reactions are often kinetically slow. Because of its lone pair, it is a trigonal pyramidal molecule, as predicted by VSEPR theory (Figure 17.4). The bond length indicates that, as discussed before for silicon, phosphorus, and sulfur, there is multiple bonding resulting from the overlap of filled oxygen $2p$ orbitals and empty xenon $5d$ orbitals.

Figure 17.4 The xenon trioxide molecule.

Xenon trioxide reacts with dilute base to give the hydrogen xenate ion, $HXeO_4^-$. However, this ion is not stable, and disproportionation to xenon gas and the perxenate ion, XeO_6^{4-}, occurs:

$$XeO_3(s) + OH^-(aq) \rightarrow HXeO_4^-(aq)$$

$$2\ HXeO_4^-(aq) + 2\ OH^-(aq) \rightarrow XeO_6^{4-}(aq) + Xe(g) + O_2(g) + H_2O(l)$$

Alkali and alkaline earth metal salts of the perxenate ion can be crystallized; they are all colorless, stable solids. In the perxenate ion, the xenon is surrounded by the six oxygen atoms in an octahedral arrangement. Perxenates are among the most powerful oxidizing agents known, which is not really surprising considering that the xenon is in the formal oxidation state of +8. For example, they rapidly oxidize manganese(II) ion to permanganate, themselves being reduced to the hydrogen xenate ion:

$$XeO_6^{4-}(aq) + 5\ H^+(aq) + 2\ e^- \rightarrow HXeO_4^-(aq) + 2\ H_2O(l)$$

$$4\ H_2O(l) + Mn^{2+}(aq) \rightarrow MnO_4^-(aq) + 8\ H^+(aq) + 5\ e^-$$

Xenon tetroxide is prepared by adding concentrated sulfuric acid to solid barium perxenate:

$$Ba_2XeO_6(s) + 2\ H_2SO_4(aq) \rightarrow 2\ BaSO_4(s) + XeO_4(g) + 2\ H_2O(l)$$

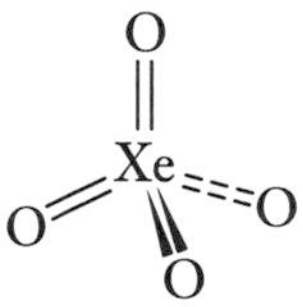

This oxide, also with xenon in the oxidation state of +8, is an explosive gas. Its structure has been shown to be tetrahedral (Figure 17.5). This geometry is expected from VSEPR arguments. Furthermore, the fact that it is a gas at room temperature suggests that it is probably a nonpolar molecule.

Figure 17.5 The xenon tetroxide molecule.

Biological Aspects

None of the noble gases has any positive biological functions. Radon, however, has recently been in the news because it accumulates inside buildings. The radiation it releases as it decays may be a significant health hazard. Radon isotopes are produced during the decay of uranium and thorium. Only one isotope, radon-222, has a half-life long enough (3.8 days) to cause major problems, and this particular isotope is produced in the decay of uranium-238. This process is happening continuously in the rocks and soils, and the radon produced normally escapes into the atmosphere. However, the radon formed beneath dwellings permeates through cracks in concrete floors and basement walls, a process that is enhanced when the pressure inside the house is lower than the external value. This pressure differential occurs when ventilation fans, clothes dryers, and other mechanical devices pump air out of the house. Furthermore, our concern about saving energy has prompted us to build houses that are more airtight, thereby preventing exchange of radon-rich interior air with exterior fresh air.

It is not actually the radon that is the problem but the solid radioactive isotopes produced by its subsequent decay. These solid particles attach themselves to lung tissue, subsequently irradiating it with α and β particles, disrupting the cells, and even initiating lung cancer. Awareness of the problem arose from an incident at the Limerick Nuclear Generating Station, Pennsylvania. When leaving such an installation, workers have to pass through a radiation detector to ensure they have not become contaminated with radioactive materials (this process is shown vividly in the movie *Silkwood*). By accident, one of the workers, Stanley Watras, entered the plant through the detector, setting it off. Investigators were puzzled until they checked his house, which showed very high levels of radiation to which he and his family were constantly being exposed. The radiation was a result of enormous levels of radon and its decay products leaking into the house from a vein of uranium-bearing ore that lay under the house.

There is clear evidence that exposure to high levels of radon does increase the probability of lung cancer. The concentration of radon at which significant hazard exists is still under debate. Certainly, cigarette smoking presents a far greater hazard than radon exposure for the average person. However, investigators have discovered houses where the radiation levels are about 100 times greater than normal levels. Usually these houses are built over geological deposits that produce high levels of radon. It is possible to have a home tested for radon by a certified technician or testing process, but it is always advisable to have a properly ventilated house, to prevent not only possible radon accumulation but also, more generally, to flush out continuously all the air pollutants that are present in most modern well-sealed houses and offices.

Exercises

17.1. Write balanced chemical equations for the following chemical reactions:
(a) xenon with difluorine in a 1:2 mole ratio
(b) xenon tetrafluoride with phosphorus trifluoride

17.2. Write balanced chemical equations for the following chemical reactions:
(a) xenon difluoride with water
(b) solid barium perxenate with sulfuric acid

17.3. Describe the trends in the physical properties of the noble gases.

17.4. Why is argon (thermal conductivity 0.017 $J{\cdot}s^{-1}{\cdot}m^{-1}{\cdot}K^{-1}$ at 0°C) more commonly used as a thermal insulation layer in glass windows than xenon (thermal conductivity 0.005 $J{\cdot}s^{-1}{\cdot}m^{-1}{\cdot}K^{-1}$ at 0°C)?

17.5. What are the unusual features of helium?

17.6. Why would we expect noble gas compounds to exist?

17.7. A bright green ion, Xe_2^+, has been identified. Suggest the bond order for this ion, showing your reasoning.

17.8. Bartlett's noble gas compound is now known to contain the XeF^+ ion. Construct the electron-dot formula for this ion. By comparison with interhalogen chemistry, would this ion be predicted to exist?

17.9. What are the key thermodynamic factors in the formation of xenon-fluorine compounds?

17.10. For the formation of xenon tetrafluoride, $\Delta G_f^\circ = -121.3\ kJ{\cdot}mol^{-1}$ and $\Delta H_f^\circ = -261.5\ kJ{\cdot}mol^{-1}$. Determine the value for the standard entropy of formation of this compound. Why do you expect the sign of the entropy change to be negative?

17.11. Estimate the enthalpy of formation of xenon tetrachloride from the following data: bond energy (Cl–Cl) = 242 $kJ{\cdot}mol^{-1}$; bond energy (Xe–Cl) [estimated]

= 86 kJ·mol^{-1}; enthalpy of sublimation of solid xenon tetrachloride [estimated] = 60 kJ·mol^{-1}.

17.12. One of the few krypton compounds known is krypton difluoride, KrF_2. Calculate the enthalpy of formation of this compound from tabulated data. The Kr–F bond energy is 50 kJ·mol^{-1}.

17.13. Construct an electron-dot structure for $XeOF_4$ with a xenon-oxygen (a) single bond; (b) double bond. Decide which is more significant on the basis of formal charge.

17.14. Determine the shapes of the following ions: (a) XeF_3^+; (b) XeF_5^+; (c) XeO_6^{4-}.

17.15. Determine the oxidation number of xenon in each of the compounds in Exercise 17.14.

17.16. It is possible to prepare a series of compounds of formula $MXeF_7$, where M is an alkali metal ion. Which alkali metal ion should be used in order to prepare the most stable compound?

17.17. Explain why xenon forms compounds with oxygen in the +8 oxidation state but with fluorine only up to an oxidation state of +6.

17.18. Briefly discuss why radon is a health hazard.

CHAPTER

18

Introduction to Transition Metal Complexes

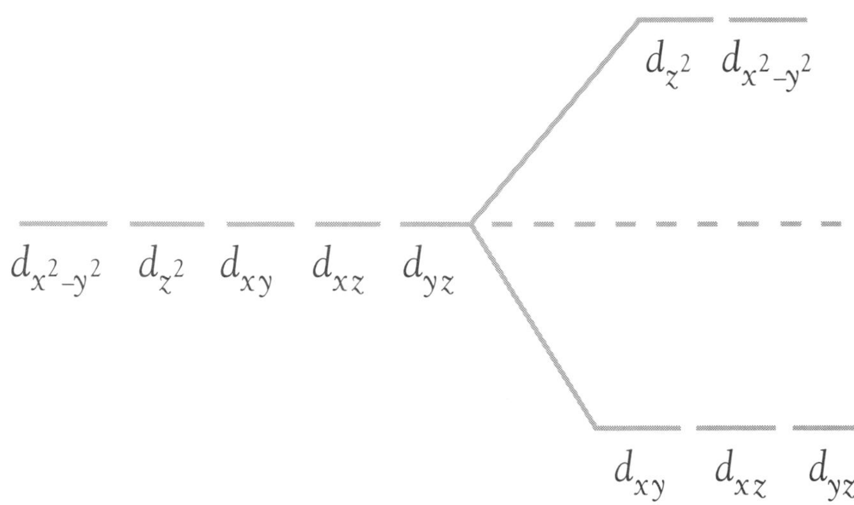

It is the enormity of the number of compounds that is the obvious feature of the transition metals. And the variety of colors is very striking in comparison to the predominant white of the compounds of the main group metals. We will see the ways in which transition metals can form this variety of compounds, introduce the naming system used specifically for them, and discuss the theories of bonding that are used to explain this diversity.

The compounds of the transition metals have always held a special interest for inorganic chemists. Whereas the compounds of the main group metals are almost always white, the transition metal compounds come in every color of the rainbow. Chemists were fascinated by the fact that it was sometimes possible to make compounds of the same formula but in different colors. For example, chromium(III) chloride hexahydrate, $CrCl_3 \cdot 6\ H_2O$, can be synthesized in purple, pale green, and dark green forms.

The initial explanation for this multitude of compounds was that, like organic compounds, the components of the transition metal compounds were strung out in chains. It was the Swiss chemist Alfred Werner who, during a restless night in 1893, devised the concept that transition metal compounds consisted of the metal ion surrounded by the other ions and molecules. This novel theory was accepted in Germany, but it received a hostile reception in the English-speaking world. Over the next eight years, Werner and his students prepared several series of transition metal compounds, searching for proof of his theory. As more and more evidence accumulated, the opposition disintegrated, and he was awarded the Nobel prize for chemistry in 1913 in recognition of his contribution. Although Werner deserved the credit for devising this theory, we must always keep in mind that the toil at the research bench was done mainly by his research students. In particular, one of the most crucial pieces of evidence was established by a young English student, Edith Humphrey.

Transition Metals

Chemists generally restrict the term *transition metal* to some, but not all, of the *d*-block elements (Figure 18.1). The generally accepted definition of a transition metal is an element that has at least one simple ion with an incomplete outer set of *d* electrons. This definition excludes the Group 12 elements—zinc, cadmium, and mercury—because these metals always maintain a d^{10} electron configuration. These elements are the subject of Chapter 20. The Group 3 elements—scandium, yttrium, and lanthanum—also can be excluded because they almost always exhibit the +3 oxidation state, which has a d^0 electron configuration. These metals more closely resemble those of the 4*f*-block elements in their chemistry and hence are discussed together in Chapter 21. Also excluded are the postactinoid metals. These actually are transition metals, but because they are all short-lived radioactive elements, it is common to discuss them with the actinoid metals. Thus these too are described in Chapter 21.

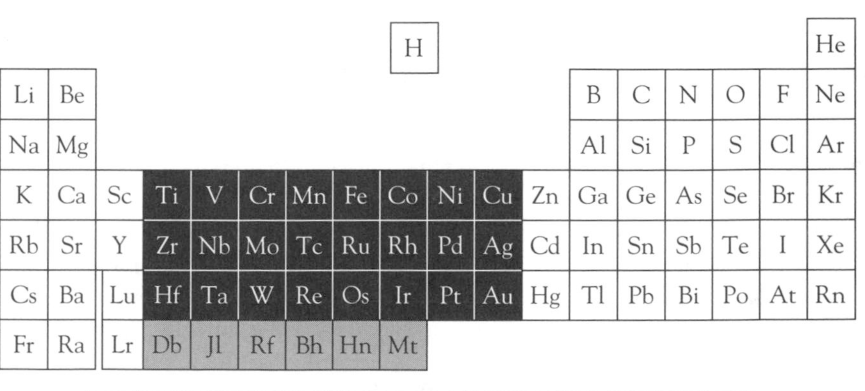

Figure 18.1 A periodic table with the elements usually defined as transition metals shown in black and postactinoid metals shaded.

Table 18.1 Formulas and structures of a series of platinum(II) complexes

Composition	Number of ions	Modern formulation
$PtCl_2 \cdot 4\,NH_3$	3	$[Pt(NH_3)_4]^{2+}\ 2Cl^-$
$PtCl_2 \cdot 3\,NH_3$	2	$[PtCl(NH_3)_3]^+\ Cl^-$
$PtCl_2 \cdot 2\,NH_3$	0	$[PtCl_2(NH_3)_2]$ (two forms)
$KPtCl_3 \cdot NH_3$	2	$K^+\ [PtCl_3(NH_3)]^-$
K_2PtCl_4	3	$2K^+\ [PtCl_4]^{2-}$

Transition Metal Complexes

We rarely encounter a "naked" transition metal ion, because the ion is usually covalently bonded to other ions or molecules. These groupings are called metal *complexes*, and it is the number and diversity of the metal complexes that provide the wealth of transition metal chemistry.

It was Alfred Werner's proposal that metal ions had not only a particular value of charge but also some characteristic "combining power." That is, there was a specific number of molecules or ions with which a transition metal would combine. We now refer to this number (or numbers) as the *coordination number(s)* of the element, and it is usually four or six. The molecules or ions that are covalently bonded to the central metal ion are called *ligands*.

One of the best illustrations of the concept is shown by the series of compounds that can be prepared from platinum(II) and ammonia, chloride ions, and potassium ions. These compounds are shown in Table 18.1. The key to understanding this multiplicity of compounds was provided by measurements of the electrical conductivity of their solutions. Thus the presence of three ions in solution in the first case can only be explained if the two chloride ions are not covalently bonded to the platinum. In the second complex, the presence of two ions shows that only one chloride ion is ionic and that the other must be part of the coordination sphere of the platinum. Similar arguments can be made for the other compounds.

The bonding theories will be discussed shortly, but for the moment we can consider complex formation to be the result of coordinate covalent bond formation, the metal ion acting as a Lewis acid and the ligands as Lewis bases.

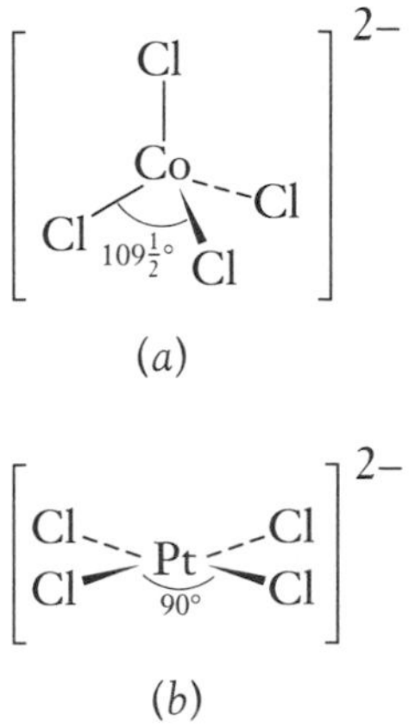

Stereochemistries

Transition metal complexes have a wide range of shapes. With four ligands, there are two alternatives: tetrahedron and square plane. Tetrahedrons are more common in Period 4 transition metals, and square planar complexes are more prevalent among the Periods 5 and 6 transition metal series. Figure 18.2a shows the tetrahedral geometry of the tetrachlorocobaltate(II) ion, $[CoCl_4]^{2-}$, and Figure 18.2b shows the square planar configuration of the tetrachloroplatinate(II) ion, $[PtCl_4]^{2-}$.

There are few simple complexes with five ligands, but it is interesting to find that, like the four-ligand situation, these complexes have two

Figure 18.2 (*a*) The tetrahedral tetrachlorocobaltate(II) ion and (*b*) the square planar tetrachloroplatinate(II) ion.

stereochemistries: trigonal bipyramid, like the main group compounds, and square-based pyramid (Figure 18.3). The energy difference between these two configurations must be very small, because the pentachlorocuprate(II) ion, $[CuCl_5]^{3-}$, adopts both structures in the solid phase, the preference depending on the identity of the cation.

The most common number of simple ligands is six, and almost all of these complexes adopt the octahedral arrangement. This configuration is shown in Figure 18.4 for the hexafluorocobaltate(IV) ion, $[CoF_6]^{2-}$. Recall that cobalt compounds usually have cobalt oxidation states of +2 and +3; thus, once again, it is fluoride that has to be used to attain the higher oxidation state of +4.

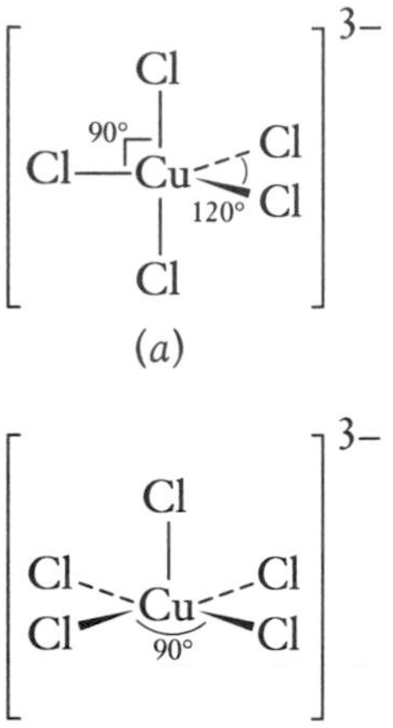

Figure 18.3 The two stereochemical arrangements of the pentachlorocuprate(II) ion: (*a*) trigonal bipyramid and (*b*) square-based pyramid.

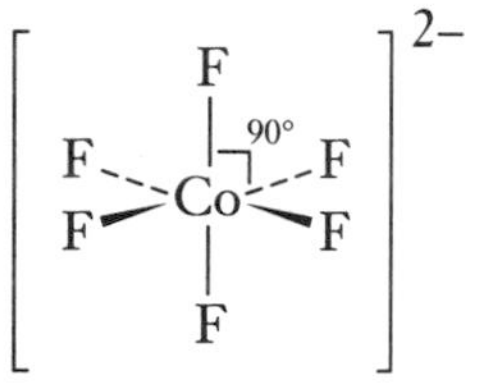

Figure 18.4 The octahedral hexafluorocobaltate(IV) ion.

Ligands

As mentioned earlier, the atoms, molecules, or ions attached to the metal ion are called ligands. For most ligands, such as water or chloride ion, each occupies one coordination site. These species are known as *monodentate* ligands (from the Greek word meaning "one-toothed"). There are several molecules and ions that take up two bonding sites; common examples are the 1,2-diaminoethane molecule, $H_2NCH_2CH_2NH_2$, and the oxalate ion, $^-O_2CCO_2^-$. Such groups are called *bidentate* ligands (Figure 18.5). More complex ligands can be synthesized and will bond to three, four, five, and even six coordination sites. These species are called tridentate, tetradentate, pentadentate, and hexadentate ligands, respectively. All ligands that form more than one attachment to a metal ion are called *chelating* ligands (from the Greek *chelos*, meaning "clawlike").

Ligands and Oxidation States of Transition Metals

Another feature common to transition metals is their wide range of oxidation states. The preferred oxidation state is very dependent on the nature of the ligand; that is, various types of ligands stabilize low, normal, or high oxidation states.

Ligands that tend to stabilize low oxidation states. The two common ligands that particularly favor metals in low oxidation states are the carbon monoxide molecule and the isoelectronic cyanide ion. For example, iron has an oxidation number of 0 in $Fe(CO)_5$.

Ligands that tend to stabilize "normal" oxidation states. Most common ligands, such as water, ammonia, and halide ions, fall in this category. For example, iron exhibits its common oxidation states of +2 and +3 with water:

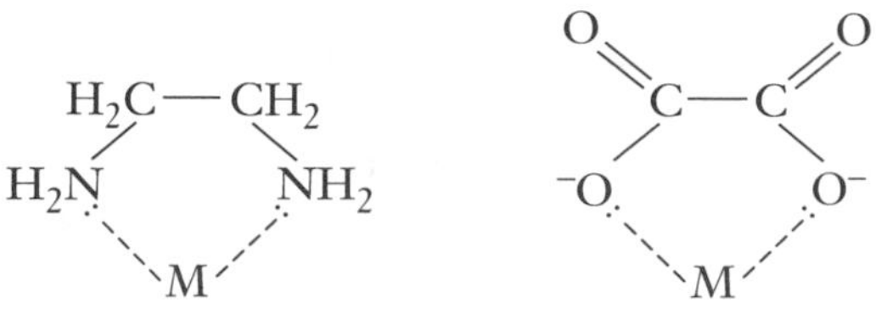

Figure 18.5 (*a*) The 1,2-diaminoethane molecule, $H_2NCH_2CH_2NH_2$, and (*b*) the oxalate ion, $^-O_2CCO_2^-$. The atoms that coordinate to the metal have dashed lines to a metal ion, M, showing how the bonding will occur.

$[Fe(OH_2)_6]^{2+}$ and $[Fe(OH_2)_6]^{3+}$. There are many cyanide complexes in normal oxidation states as well. This is not unexpected, for the ion is a pseudohalide ion (as discussed in Chapter 16) and hence capable of behaving like a halide ion.

Ligands that tend to stabilize high oxidation states. Like nonmetals, transition metals only adopt high oxidation numbers when complexed with fluoride and oxide ions. We have already mentioned the hexafluorocobaltate(IV) ion, $[CoF_6]^{2-}$, as one example. In the tetraoxoferrate(VI) ion, $[FeO_4]^{2-}$, the oxide ions stabilize the abnormal +6 oxidation state of iron!

Isomerism in Transition Metal Complexes

As mentioned in Table 18.1, the compound $Pt(NH_3)_2Cl_2$ exists as two forms (isomers). These isomers are members of one of two common classes of isomers in organic chemistry: stereoisomers and structural isomers. For stereoisomers, the bonds to the metal ion are identical, whereas the bonds of structural isomers are different. These categories can be further subdivided, as shown in Figure 18.6.

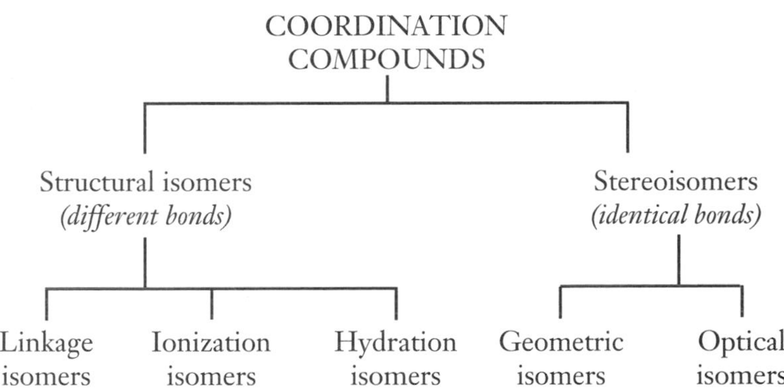

Figure 18.6 Isomer types.

Stereoisomerism

The two types of inorganic stereoisomers, geometric and optical, are parallel to those found in organic chemistry, except that in inorganic chemistry, optical isomerism is most common for a metal ion in an octahedral environment rather than for the tetrahedral environment of organic carbon compounds.

Geometric isomerism. Inorganic geometric isomers are analogous to organic geometric isomers that contain carbon-carbon double bonds. Geometric isomers must have two different ligands, A and B, attached to the same metal, M. For square planar compounds, geometric isomerism occurs in compounds of the form MA_2B_2. The term *cis-* is used for the isomer in which ligands of one kind are neighbors, and *trans-* is used to identify the isomer in which ligands of one kind are opposite each other (Figure 18.7).

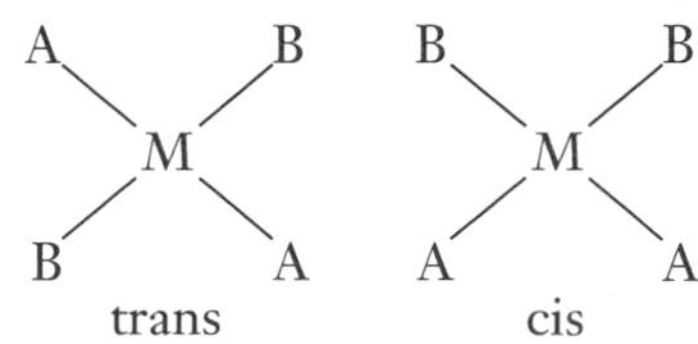

Figure 18.7 The geometric isomers of a square planar MA_2B_2 arrangement.

In the earlier discussion of platinum(II) complexes (Table 18.1), we noted that there are two chemically different forms of $[PtCl_2(NH_3)_2]$. This observation indicates the probable geometry, even though there are two

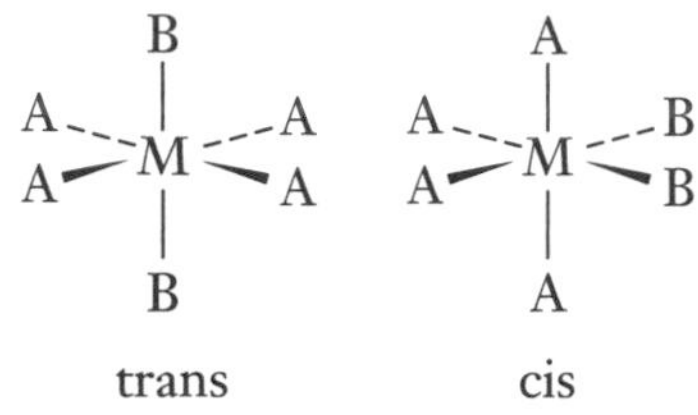

Figure 18.8 The geometric isomers of an octahedral MA_4B_2 arrangement.

possibilities: tetrahedral and square planar. If the isomers were tetrahedral, all of the bond directions would be equivalent and only one form of the compound would be possible. However, as Figure 18.7 shows, the square planar arrangement allows two geometric isomers.

There are two formulas of octahedral compounds for which geometric isomers are possible. Compounds with the formula MA_4B_2 can have the two B ligands on opposite sides or as neighbors. Hence these, too, are known as trans and cis isomers (Figure 18.8). Octahedral compounds with the formula MA_3B_3 also can have isomers (Figure 18.9). If one set of ligands occupies three sites of the horizontal plane (in this case, the A ligands) and the other set, three sites in the vertical plane, then the prefix *mer*- (for meridional) is used. In the other isomer, the A ligands are all clustered together, each being only 90° apart from the other two A ligands; the B ligands are similarly clustered. The prefix *fac*- (for facial) is used to describe this form.

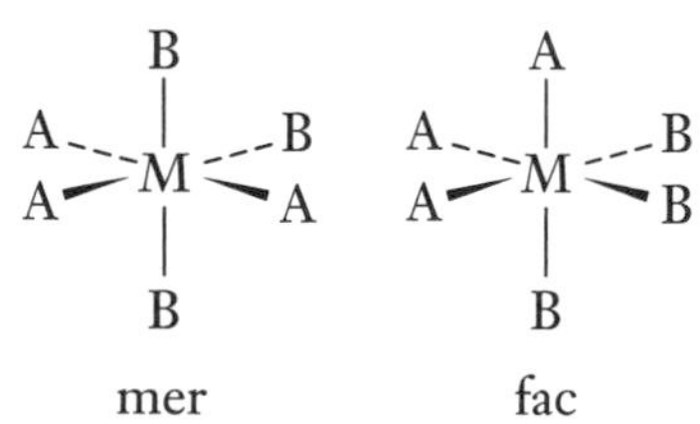

Figure 18.9 The geometric isomers of an octahedral MA_3B_3 arrangement.

Optical isomerism. Again, inorganic optical isomerism is analogous to that of organic chemistry. Optical isomers are pairs of compounds in which one isomer is a nonsuperimposable mirror image of the other. One of the characteristics of optical isomers is that they rotate the plane of polarized light, one isomer rotating the light in one direction and the other isomer in the opposite direction. Compounds that exist as optical isomers are called *chiral compounds*.

This form of isomerism is found most commonly when a metal is surrounded by three bidentate ligands; 1,2-diaminoethane, $H_2NCH_2CH_2NH_2$, is one such ligand. For convenience, we often use the abbreviation *en* for this ligand, from the traditional organic chemistry name of ethylenediamine. Hence the complex ion would have the formula $[M(en)_3]^{n+}$, where $n+$ is the charge of the transition metal ion. The two optical isomers of this complex ion are shown in Figure 18.10, where the 1,2-diaminoethane molecules are depicted schematically as pairs of linked nitrogen atoms.

The metal-ligand covalent bond is comparatively weak; thus complexed 1,2-diaminoethane molecules are continuously exchanging with free 1,2-diaminoethane molecules in solution. Complexes in which rapid ligand exchange occurs are called labile complexes. For labile complexes, it is impossible to physically separate the two optical isomers. The two important exceptions among the Period 4 transition metal series are the cobalt(III) and chromium(III) ions, which retain the ligands for lengthy periods of time, thereby enabling the optical isomers to be crystallized separately. Complexes showing slow exchange are called inert complexes (a reason for the inertness will be discussed later in the chapter). This is a somewhat misleading term, because the slowness of ligand exchange is a kinetic factor, not a thermodynamic one.

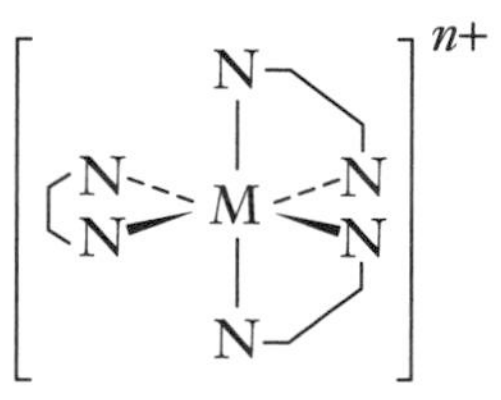

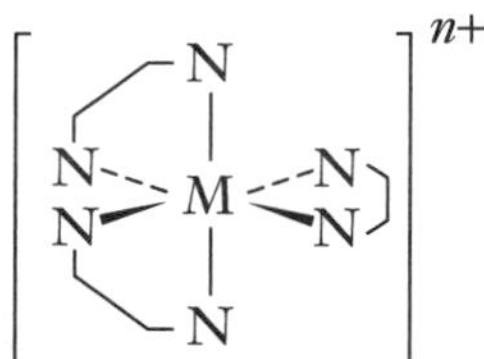

Figure 18.10 The two optical isomers of the $[M(en)_3]^{n+}$ ion. The linked nitrogen atoms represent the 1,2-diaminoethane bidentate ligands.

Structural Isomerism

Structural isomerism is unique to metal complexes. It has three common types: linkage isomerism, ionization isomerism, and hydration isomerism. Ionization and hydration isomerism are sometimes categorized together as coordination-sphere isomerism because in both cases it is the identity of the ligands that differs.

Platinum Complexes and Cancer Treatment

It is a common misconception that scientific research works the same way as technology, where goals are set and the appropriate solutions found. In science, however, there is so much that is not known that we still rely to a large extent on observing the unexpected. It was in 1965 that Barnett Rosenberg of Michigan State University was studying the rate of bacterial growth in the presence of electric fields. Both he and his coresearchers were surprised to find that the bacteria in electric fields were growing without dividing. The group spent a considerable amount of time looking for possible causes of this, such as pH changes, temperature changes, and so on. Having excluded every probable cause, they examined the electrodes that they were using to generate the electric charge. These were made of platinum, a metal that was "well known" to be extremely unreactive.

However, their tests showed that some of the platinum metal was being oxidized, and it was the oxidation products, the diamminodichloroplatinum(II), $PtCl_2(NH_3)_2$, and diamminotetrachloroplatinum(IV), $PtCl_4(NH_3)_2$, molecules, that were causing the bacterial abnormalities. Further, only the cis geometric isomers were active. This biological activity of platinum compounds was completely unexpected. Because they prevented cell division, the compounds were tested for antitumor activity, and the *cis*-diamminodichloroplatinum(II) compound seemed particularly effective. The compound is now available for cancer treatment under the name cisplatin. It does have side effects; and chemists, now that they are aware of the potential of platinum compounds, are looking for more effective and less toxic analogs. The key to this compound's effectiveness seems to be the ability of the *cis*-$(H_3N)_2Pt$ unit to link DNA units, thereby preventing further DNA synthesis. The trans isomers show no biological activity. Hence these compounds demonstrate the influence of isomerism on the chemical behavior of compounds.

Linkage isomerism. Some ligands can form bonds through more than one atom. The classic example is the nitrite ion, which can form bonds through the nitrogen atom or through one of the oxygen atoms. Thus two separate compounds can have the same formula, $Co(NH_3)_5Cl_2(NO_2)$, for example. One of these, the red form, contains the $[Co(ONO)(NH_3)_5]^{2+}$ ion, in which one of the oxygen atoms of the nitrite ion is bonded to the cobalt(III) ion. The other isomer, the yellow form, contains the $[Co(NO_2)(NH_3)_5]^{2+}$ ion, in which the nitrogen atom is bonded to the cobalt(III) ion.

Ionization isomerism. Ionization isomers give different ions when dissolved in solution. Again, there is a classic example: $Co(NH_3)_5Br(SO_4)$. If barium ion is added to a solution of the red-violet form, a white precipitate of barium sulfate forms. Addition of silver ion has no effect. Hence the complex ion must have the formula $[CoBr(NH_3)_5]^{2+}$, with an ionic sulfate ion. A solution of the red form, however, does not give a precipitate with barium ion; instead, a cream-colored precipitate is formed with silver ion. Hence this complex ion must have the structure of $[CoSO_4(NH_3)_5]^+$, with an ionic bromide ion.

Hydration isomerism. Hydration isomerism is very similar to ionization isomerism, in that the identity of the ligand species is different for the two isomers. In this case, rather than different types of ions, it is the proportion of coordinated water molecules that differs between isomers. It is possible to have a series of complexes in which the proportion of coordinated water molecules differs. The three structural isomers of formula $CrCl_3{\cdot}6H_2O$ provide the best example. In the violet form, the six water molecules are coordinated; hence the formula for this compound is more correctly written as $[Cr(OH_2)_6]^{3+}\ 3Cl^-$. As evidence, all three chloride ions are precipitated from solution by silver ion. In the light green form, one of the chloride ions is not precipitated by silver ion; hence the complex is assigned the structure $[CrCl(OH_2)_5]^{2+}\ 2Cl^-{\cdot}H_2O$ (light green). Finally, only one chloride ion can be precipitated by silver ion from a solution of the dark green form; hence this compound must have the structure $[CrCl_2(OH_2)_4]^+\ Cl^-{\cdot}2H_2O$.

Naming Transition Metal Complexes

Because of the complexity of the transition metal complexes, the simple system of inorganic nomenclature proved unworkable. As a result, special rules for naming transition metal complexes were devised:

1. Nonionic species are written as one word; ionic species are written as two words with the cation first.

2. The central metal atom is identified by name, which is followed by the formal oxidation number in Roman numerals in parenthesis, such as (IV) for a +4 state and (–II) for a –2 state. If the complex is an anion, the ending *-ate* adds to the metal name or replaces any *-ium*, *-en*, or *-ese* ending. Thus we have cobaltate and nickelate, but chromate and tungstate (not chromiumate or tungstenate). For a few metals, the anion name is derived from the old Latin name of the element: ferrate (iron), argentate (silver), cuprate (copper), and aurate (gold).

3. The ligands are written as prefixes of the metal name. Neutral ligands are given the same name as the parent molecule, whereas the names of negative ligands are given the ending *-o* instead of *-e*. Thus sulfate becomes sulfato and nitrite becomes nitrito. Anions with *-ide* endings have them completely replaced by *-o*. Hence chloride ion becomes chloro; iodide, iodo; cyanide, cyano; and hydroxide, hydroxo. There are three special names: Coordinated water is commonly named *aqua*; ammonia, *ammine*; and carbon monoxide, *carbonyl*.

4. Ligands are always placed in alphabetical order (this system will also be used here for the order in the chemical formulas).

5. For multiple ligands, the prefixes *di-*, *tri-*, *tetra-*, *penta-*, and *hexa-* are used for two, three, four, five, and six, respectively.

6. For multiple ligands already containing numerical prefixes (such as 1,2-diaminoethane), the prefixes used are *bis-*, *tris-*, and *tetrakis-* for two, three, and four. This is not a rigid rule. Many chemists use these prefixes for all polysyllabic ligands.

Examples

Let us try naming some of the platinum metal complexes discussed earlier in this chapter. Notice that, in chemical formulas, square brackets, [ ], are used to enclose all units linked together by covalent bonds.

Example 1: $[Pt(NH_3)_4]Cl_2$. Because this compound has separate ions, the name will consist of (at least) two words (rule 1). There are two negative chloride ions outside of the complex, so the complex itself must have the formula $[Pt(NH_3)_4]^{2+}$. The ammonia ligands are neutral; hence the platinum must have an oxidation state of +2. Hence we start with the stem name platinum(II) (rule 2). The ligand is ammonia; hence we use the name ammine (rule 3). But there are four ammonia ligands, so we add a prefix and get tetraammine (rule 5). Finally, the chloride anions must be included. They are free, uncoordinated chloride ions, so they are called chloride, not chloro. We do not identify the number of chloride ions because the oxidation number enables us to deduce it. Hence the full name is tetraammineplatinum(II) chloride.

Example 2: $[PtCl_2(NH_3)_2]$. This is a nonionic species; hence it will have a one-word name (rule 1). Again, to balance the two chloride ions, the platinum is in the +2 oxidation state, so we start with platinum(II) (rule 2). The ligands are named ammine for ammonia and chloro for chloride (rule 3). Alphabetically, ammine comes before chloro (rule 4); thus we have the prefix diamminedichloro (rule 5). The whole name is diamminedichloroplatinum(II). As mentioned earlier, this particular compound is square planar and exists as two geometric isomers. We refer to these isomers as *cis*-diamminedichloroplatinum(II) and *trans*-diamminedichloroplatinum(II).

Example 3: $K_2[PtCl_4]$. Again, two words are needed (rule 1), but in this case, the platinum is in the anion, $[PtCl_4]^{2-}$. The metal is in the +2 oxidation state; thus the anionic name will be platinate(II) (rule 2). There are four chloride ligands, giving the prefix tetrachloro (rules 3 and 5), and the separate potassium cations. The complete name is potassium tetrachloroplatinate(II).

Example 4: $[Co(en)_3]Cl_3$. The complex ion is $[Co(en)_3]^{3+}$. Because (en), $H_2NCH_2CH_2NH_2$, is a neutral ligand, the cobalt must be in a +3 oxidation state. The metal, then, will be cobalt(III). The full name of the ligand is 1,2-diaminoethane and contains a numerical prefix, so we use the alternate prefix set (rule 6) to give tris(1,2-diaminoethane)—parentheses are used to separate the ligand name from the other parts of the name. Finally, we add the chloride anions. The full name is tris(1,2-diaminoethane)cobalt(III) chloride.

Unfortunately, there are a number of transition metal compounds that have well-known common names in addition to their systematic names. For example, a few complexes are identified by the name of their discoverer, such as Zeise compound, Wilkinson catalyst, and Magnus green salt.

The Ewens-Bassett Nomenclature System

The nomenclature system that we have used to this point was first devised by Alfred Stock in 1919, and it is still the one most widely used. As we have seen, the system uses Roman numerals to indicate the oxidation state of the central atom, and from that, the ion charge can be calculated. An alternative system was devised by R. Ewens and H. Bassett in 1949. According to their rules, the ion charge is bracketed in Arabic numerals. It was, in fact, the Ewens-Bassett system that we used in Chapter 10 to distinguish the O_2^- ion from the O_2^{2-} ion, the former being called the dioxide(1–) ion and the latter the dioxide(2–) ion (these are much more useful names than the traditional superoxide and peroxide). The Ewens-Bassett system is particularly useful for systematic naming of polyatomic ions containing one element: for example, the dioxides mentioned above; C_2^{2-}, dicarbide(2–) (instead of carbide or acetylide); O_3^-, trioxide(1–) (not ozonide); and N_3^-, trinitride(1–) (not azide).

Apart from indicating charge rather than oxidation number, the Stock and Ewens-Bassett systems employ the same nomenclature rules. The use of either Roman or Arabic numerals identifies the method used in a particular name; for example, $K_4[Fe(CN)_6]$ is called potassium hexacyanoferrate(II) by the Stock method, because the iron in the complex ion has a formal oxidation state of +2. The Ewens-Bassett name for this compound is potassium hexacyanoferrate(4–), because the complex anion has a 4– charge. For neutral molecules, no number is shown; thus *cis*-$[Pt(NH_3)_2Cl_2]$ would be *cis*-diamminedichloroplatinum rather than the Stock name of *cis*-diamminedichloroplatinum(II). The Ewens-Bassett name, then, can be found just from the charge balance. And this system is also useful when the ion is so complicated that any formal oxidation state is difficult to identify. Conversely, seeing the oxidation state in the name enables us to identify whether the metal is in a typical, high, or low oxidation state. In the following chapters, we will consistently use the Stock system for transition metal complexes.

An Overview of Bonding Theories of Transition Metal Compounds

For many decades, chemists and physicists struggled with possible explanations to account for the large number of transition metal compounds. Such explanations had to account for the variety of colors found among the compounds, the wide range of stereochemistries, and the paramagnetism of many compounds. One of the first approaches was to regard the bonding as that between a Lewis acid (the metal ion) and Lewis bases (the ligands). This model produced the 18-electron rule or the effective atomic number (EAN) rule. As we will see, this model works for many metal compounds in which the metal is in a low oxidation state; but it does not work for most compounds, nor does it explain the color or paramagnetism of many transition metal compounds. Following from this, the American chemist Linus Pauling proposed the valence-bond theory in which he assumed that the bonding of transition metals was similar to that of typical main group elements, assigning different modes of hybridization to the metal ion depending on the known geometry of the compound. This theory did account for the different stereochemistries and formulas, but it too failed to account for colors and unpaired electrons.

Two physicists, Hans Bethe and Johannes Van Vleck, approached the problem from a completely different direction. They assumed that the interaction between metal ion and ligands was totally electrostatic in nature. Known as crystal field theory, it has been remarkably successful in accounting for the properties of transition metal complexes. To include covalent contributions to the bonding, crystal field theory has been modified and given the name ligand field theory. Finally, molecular orbital theory was used to obtain an overall set of energy levels for these compounds. Molecular orbital theory is clearly the most complete picture of the bonding system, but for our discussions, crystal field theory provides an adequate and simple explanation for the properties and behavior of most transition metal compounds.

The 18-Electron Rule

For the main group elements, the octet rule is sometimes used to predict the formulas of covalent compounds. On the basis of that rule, we assumed that the central atom formed bonds such that the total number of electrons about that atom was eight—the maximum occupancy of the *s* and *p* orbitals. The rule is only valid for the nonmetallic Period 2 elements; and even then there are exceptions, as noted in Chapter 3.

The 18-electron rule is based on a similar concept—the central transition metal ion can accommodate electrons in the *d*, *s*, and *p* orbitals, giving a maximum of 18. Thus, to the number of electrons in the outer electron set, a metal can add electron pairs from Lewis bases to bring the total up to 18. This rule also has its great limitations; in this case, it is valid only for metals in very low oxidation states.

The classic examples are the complexes in which carbon monoxide is the ligand. In Chapter 13, for example, we mentioned tetracarbonylnickel(0), $Ni(CO)_4$, a compound used in the purification of nickel metal. Nickel has an electron configuration of $[Ar]4s^23d^8$, a total of 10 outer electrons. According to the electron-dot approach, each carbon monoxide molecule donates the two electrons of the lone pair on the carbon atom. The bonding of four carbon monoxide molecules would provide eight additional electrons, resulting in a total of 18.

$$\begin{aligned} \text{nickel electrons } (4s^23d^8) &= 2 + 8 = 10 \\ \text{carbon monoxide electrons} &= 4 \times 2 = \ \ 8 \\ \text{total} &= 18 \end{aligned}$$

Many, but not all, low oxidation state metal complexes obey the 18-electron rule.

Valence-Bond Theory

In valence-bond theory, we again consider the interaction between the metal ion and its ligands to be that of a Lewis acid with Lewis bases, but the ligand electron pairs are simply placed in the empty higher orbitals of the metal ion. This arrangement is shown for the tetrahedral tetrachloronickelate(II) ion, $[NiCl_4]^{2-}$, in Figure 18.11. The free nickel(II) ion has an electron configuration of $[Ar]3d^8$ with two unpaired electrons. According to the theory, the 4*s* and 4*p* orbitals of the nickel hybridize to form four sp^3 hybrid orbitals, and these are occupied by an electron pair from each chloride ion (the Lewis bases).

This representation accounts for the two unpaired electrons in the complex ion and the tetrahedral shape expected for sp^3 hybridization. However, we can only construct the orbital diagrams once we know from a crystal

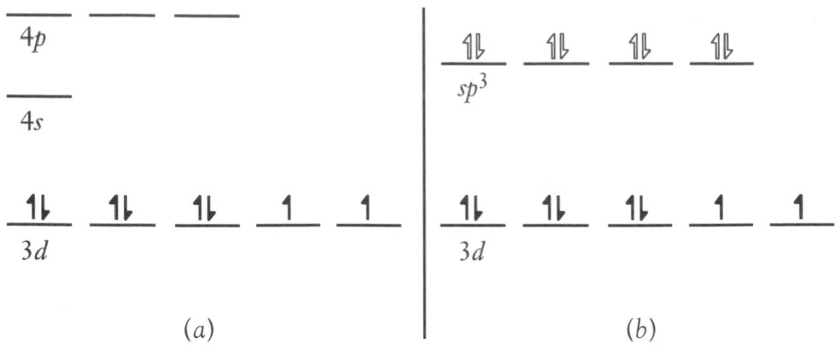

Figure 18.11 (*a*) The electron distribution of the free nickel(II) ion. (*b*) The hybridization and occupancy of the higher energy orbitals by electron pairs (open half-headed arrows) of the chloride ligands.

structure determination and magnetic measurements what the ion's shape and number of unpaired electrons actually are. For chemists, a theory should be predictive, if possible, and valence-bond theory is not. For example, some iron(III) compounds have five unpaired electrons and others have one unpaired electron, but valence-bond theory cannot predict this.

The theory also has some conceptual flaws. In particular, it does not explain why the electron pairs occupy higher orbitals, even though there are vacancies in the 3*d* orbitals. For some Period 4 transition metal complexes, the ligand electron pairs have to be assigned to 4*d* orbitals as well as to 4*s* and 4*p* orbitals, even though there is room in the 3*d* orbitals. In addition, the theory fails to account for the color of the transition metal complexes, one of the most obvious features of these compounds. For these reasons, the valence-bond theory has become little more than a historical footnote.

Crystal Field Theory

As we have seen, the classic valence-bond theory was unable to explain many of the aspects of transition metal complexes. In particular, valence-bond theory did not satisfactorily explain the different numbers of unpaired electrons that we find among the transition metal ions. For example, the hexaaquairon(II) ion, $[Fe(OH_2)_6]^{2+}$, has four unpaired electrons, whereas the hexacyanoferrate(II) ion, $[Fe(CN)_6]^{4-}$, has no unpaired electrons.

A radically different approach was based on an electrostatic model and led to crystal field theory. In spite of its simplistic nature, crystal field theory has proved remarkably useful for explaining the properties of Period 4 transition metal complexes. The theory assumes that the transition metal ion is free and gaseous; that the ligands behave like point charges; and that there are no interactions between metal orbitals and ligand orbitals.

We can consider complex formation as a series of events:

1. The initial approach of the ligand electrons forms a spherical shell around the metal ion. Repulsion between the ligand electrons and the metal ion electrons will cause an increase in energy of the metal ion *d* orbitals.

2. The ligand electrons rearrange so that they are distributed in pairs along the actual bonding directions (such as octahedral or tetrahedral). The mean *d* orbital energies will stay the same, but the orbitals oriented along the bonding directions will increase in energy and those between the bonding directions will decrease in energy. This loss in *d* orbital degeneracy will be the focus of the crystal field theory discussion.

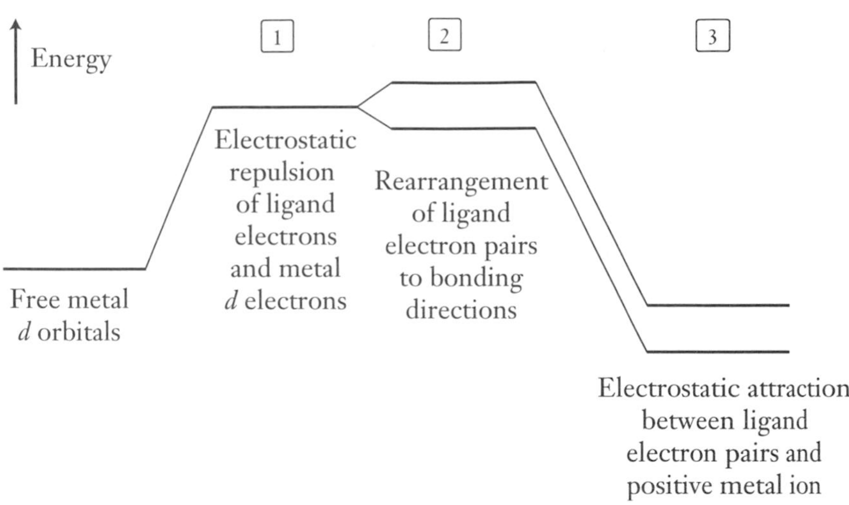

Figure 18.12 The hypothetical steps in complex ion formation according to crystal field theory.

3. Up to this point, complex formation would not be favored, because there has been a net increase in energy as a result of the ligand electron–metal electron repulsion (step 1). Furthermore, the decrease in the number of free species means that complex formation will generally result in a decrease in entropy. However, there will be an attraction between the ligand electrons and the positively charged metal ion that will result in a net decrease in energy. It is this third step that provides the driving force for complex formation.

These three hypothetical steps are summarized in Figure 18.12.

Octahedral Complexes

Although it is the third step that provides the energy for complex formation, it is the second step—the loss of degeneracy of the *d* orbitals—that is crucial for the explanation of the color and magnetic properties of transition metal complexes. Examining the octahedral situation first, we see that the six ligands are located along the Cartesian axes (Figure 18.13). As a result of these negative charges along the Cartesian axes, the energy of the orbitals aligned

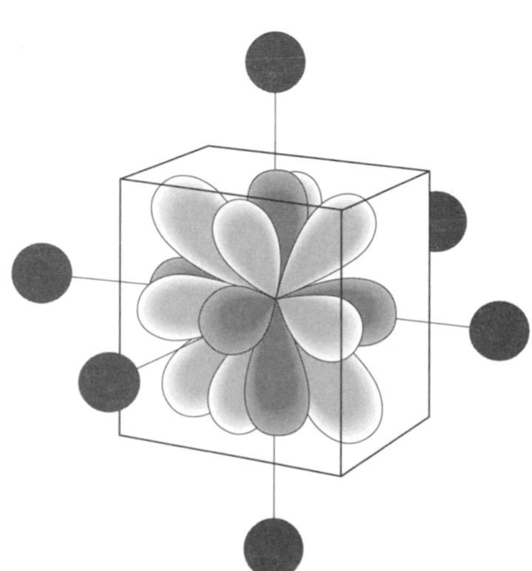

Figure 18.13 The orientation of six ligands with respect to the metal *d* orbitals. (Adapted from J. E. Huheey, E. A. Keiter, and R. L. Keiter, *Inorganic Chemistry*, 4th ed. [New York: HarperCollins, 1993], p. 397.)

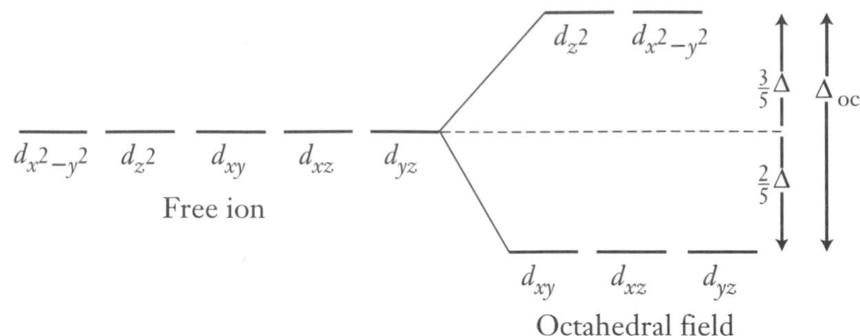

Figure 18.14 The splitting of the *d* orbital energies that occurs when the metal ion is surrounded by an octahedral array of ligands.

along these axes, the $d_{x^2-y^2}$ and d_{z^2} orbitals, will be higher than those of the d_{xy}, d_{xz}, and d_{yz} orbitals. This splitting is represented in Figure 18.14. The energy difference between the two sets of *d* orbitals in the octahedral field is given the symbol Δ_{oct}. The sum of the orbital energies equals the *degenerate energy* (sometimes called the baricenter). Thus the energy of the two higher energy orbitals ($d_{x^2-y^2}$ and d_{z^2}) is $+\frac{3}{5}\Delta_{oct}$, and the energy of the three lower energy orbitals (d_{xy}, d_{xz}, and d_{yz}) is $-\frac{2}{5}\Delta_{oct}$ below the mean.

If we construct energy diagrams for the different numbers of *d* electrons, we see that for the d^1, d^2, and d^3 configurations, the electrons will all fit into the lower energy set (Figure 18.15). This net energy decrease is known as the *crystal field stabilization energy*, or CFSE.

For the d^4 configuration, there are two possibilities: The fourth *d* electron can pair up with an electron in the lower energy level or it can occupy the upper energy level, depending on which situation is more energetically favorable. If the octahedral crystal field splitting, Δ_{oct}, is smaller than the pairing energy, then the fourth electron will occupy the higher orbital. If the pairing energy is less than the crystal field splitting, then it is energetically preferred for the fourth electron to occupy the lower orbital. The two situations are shown in Figure 18.16. The result having the greater number of unpaired electrons is called the *high spin* (or weak field) situation, and that having the lesser number of unpaired electrons is called the *low spin* (or strong field) situation.

Two possible spin conditions exist for each of the d^4, d^5, d^6, and d^7 electron configurations in an octahedral environment. The number of possible unpaired electrons corresponding to each *d* electron configuration is shown in Table 18.2, where h.s. and l.s. indicate high spin and low spin, respectively.

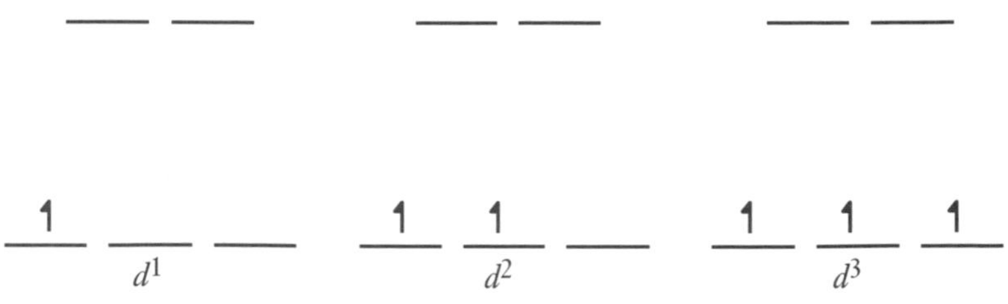

d^1 d^2 d^3

Figure 18.15 The *d* orbital filling for the d^1, d^2, and d^3 configurations.

Table 18.2 The *d* electron configurations and corresponding number of unpaired electrons for an octahedral stereochemistry

Configuration	Number of unpaired electrons	Common examples
d^1	1	Ti^{3+}
d^2	2	V^{3+}
d^3	3	Cr^{3+}
d^4	4 (h.s.), 2 (l.s.)	Mn^{3+}
d^5	5 (h.s.), 1 (l.s.)	Mn^{2+}, Fe^{3+}
d^6	4 (h.s.), 0 (l.s.)	Fe^{2+}, Co^{3+}
d^7	3 (h.s.), 1 (l.s.)	Co^{2+}
d^8	2	Ni^{2+}
d^9	1	Cu^{2+}

The energy level splitting depends on two factors: the oxidation state of the metal and the nature of the ligands. Generally, the higher the oxidation state of the metal, the greater the crystal field splitting, Δ. Thus most cobalt(II) complexes are high spin as a result of the small crystal field splitting, whereas almost all cobalt(III) complexes are low spin as a result of the much larger splitting by the 3+ ion. The common ligands can be ordered on the basis of the effect that they have on the crystal field splitting. This ordered listing is called the *spectrochemical series.* Among the common ligands, the splitting is largest with carbonyl and cyanide and smallest with iodide. The ordering for most metals is

$$I^- < Br^- < Cl^- < F^- < OH^- < OH_2 < NH_3 < en < CN^- < CO$$

Consider the d^6 hexaaquairon(II) ion, $[Fe(OH_2)_6]^{2+}$, which is found to possess four unpaired electrons. The water ligands, being low in the spectrochemical series, produce a small Δ_{oct}; hence the electrons adopt a high spin configuration. Conversely, the hexacyanoferrate(II) ion, $[Fe(CN)_6]^{4-}$, is found to be

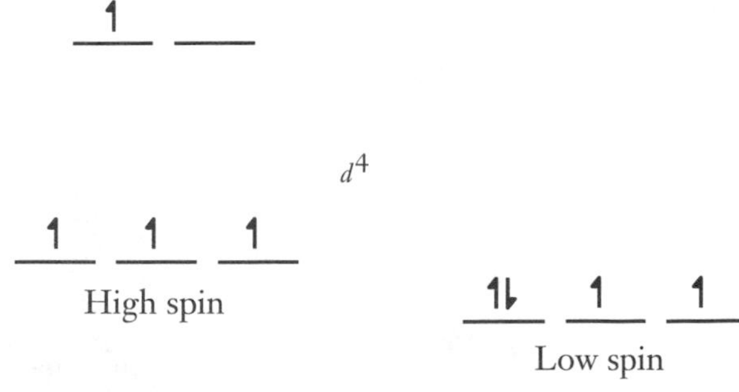

Figure 18.16 The two possible spin situations for the d^4 configuration.

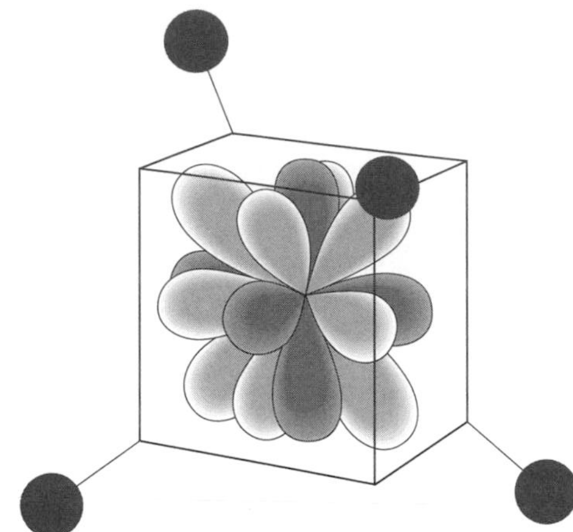

Figure 18.17 The orientation of four ligands with respect to the metal *d* orbitals. (Adapted from J. E. Huheey, E. A. Keiter, and R. L. Keiter, *Inorganic Chemistry*, 4th ed. [New York: HarperCollins, 1993], p. 402.)

diamagnetic (zero unpaired electrons). Cyanide is high in the spectrochemical series and produces a large Δ_{oct}; hence the electrons adopt a low spin configuration.

Tetrahedral Complexes

The second most common stereochemistry is tetrahedral. Figure 18.17 shows the tetrahedral arrangement of four ligands around the metal ion. In this case, it is the d_{xy}, d_{xz}, and d_{yz} orbitals that are more in line with the approaching ligands than the $d_{x^2-y^2}$ and d_{z^2} orbitals. As a result, it is the $d_{x^2-y^2}$ and d_{z^2} orbitals that are lower in energy, and the tetrahedral energy diagram is inverted relative to the octahedral diagram (Figure 18.18).

With only four ligands instead of six, and the ligands not quite pointing directly at the three *d* orbitals, the crystal field splitting is much less than in the octahedral case; in fact, it is about four-ninths of Δ_{oct}. As a result of the small orbital splitting, the tetrahedral complexes are almost all high spin. Tetrahedral geometries are most commonly found for halogen complexes, such as the tetrachlorocobaltate(II) ion, $[CoCl_4]^{2-}$, and for the oxyanions, such as the tetraoxomolybdate(VI) ion, MoO_4^{2-} (commonly called molybdate).

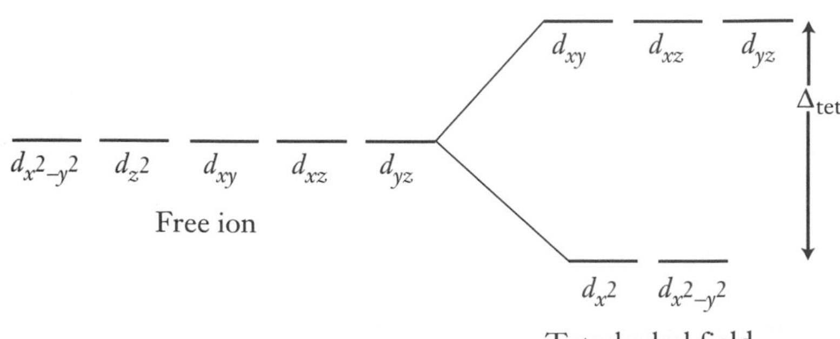

Figure 18.18 The splitting of the *d* orbital energies that occurs when the metal ion is surrounded by a tetrahedral array of ligands.

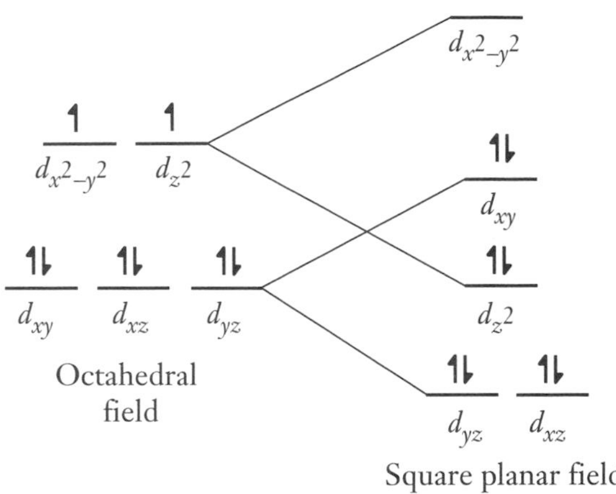

Figure 18.19 The d^8 orbital energy diagram for the square planar environment, as derived from the octahedral diagram.

Square Planar Complexes

For the Period 4 transition metals, it is only nickel that tends to form square planar complexes, such as the tetracyanonickelate(II) ion, $[Ni(CN)_4]^{2-}$. These complexes are diamagnetic. We can develop a crystal field diagram to see why this is so, even though both octahedral and tetrahedral geometries result in two unpaired electrons for the d^8 configuration.

If we start from the octahedral field and withdraw the ligands from the z-axis, the d_{z^2} orbital will no longer feel the electrostatic repulsion from the axial ligands; hence it will drop substantially in energy. The other two orbitals with z-axis components—the d_{xz} and d_{yz}—will also undergo a decrease in energy. Conversely, with the withdrawal of the axial ligands, there will be a greater electrostatic attraction on the ligands in the plane, and these will become closer to the metal ion. As a result, the $d_{x^2-y^2}$ orbital will increase substantially in energy and the d_{xy} orbital will increase to a lesser extent. After we construct the orbital diagram (Figure 18.19), we assume that the splitting of the $d_{x^2-y^2}$ and the d_{xy} orbitals is greater than the pairing energy, thus accounting for the diamagnetism of the ion.

Successes of Crystal Field Theory

A good chemical theory is one that can account for many aspects of physical and chemical behavior. By this standard, crystal field theory is remarkably successful, because it can be used to explain most of the properties that are unique to transition metal ions. Here, we will look at a selection of them.

Magnetic Properties

Any theory of transition metal ions has to account for the paramagnetism of many of the compounds. The degree of paramagnetism is dependent on the identity of the metal, its oxidation state, its stereochemistry, and the nature of the ligand. Crystal field theory explains the paramagnetism very well in terms of the splitting of the d orbital energies, at least for the Period 4 transition metals. For example, we have just seen how crystal field theory can explain

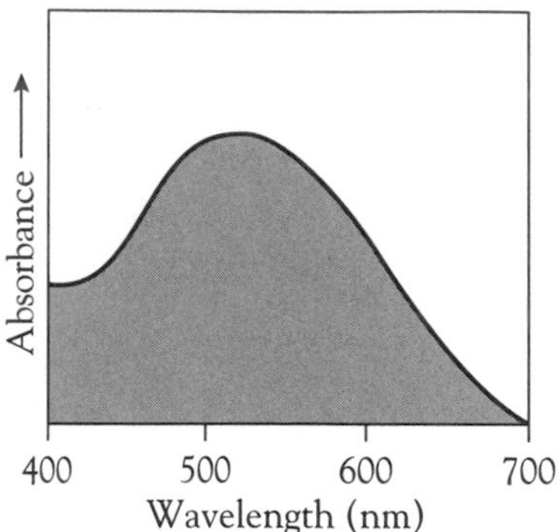

Figure 18.20 The visible absorption spectrum of the hexaaquatitanium(III) ion.

the diamagnetism of square planar nickel(II) ion, which contrasts with the paramagnetism of the tetrahedral and octahedral geometries.

Colors of Transition Metal Complexes

The most striking feature of transition metal complexes is the range of colors that they exhibit. These colors are the result of absorptions in the visible region of the electromagnetic spectrum. For example, Figure 18.20 shows the visible absorption spectrum of the purple hexaaquatitanium(III) ion, $[Ti(OH_2)_6]^{3+}$. This ion absorbs light in the green part of the spectrum, transmitting blue and red light to give the blended purple color.

The titanium(III) ion has a d^1 electron configuration; and with six water molecules as ligands, we can consider the ion to be in an octahedral field. The resulting *d* orbital splitting is shown on the left-hand side of Figure 18.21. An absorption of electromagnetic energy causes the electron to shift to the upper *d* orbital set as shown on the right-hand side of Figure 18.21. The electron subsequently returns to the ground state, and the energy is released as thermal motion rather than as electromagnetic radiation. The absorption maximum is at about 520 nm, which represents an energy difference between the upper and lower *d* orbital sets of about 230 kJ·mol^{-1}. This energy difference represents the value of Δ, the crystal field splitting.

As is apparent from Figure 18.20, the electronic absorption bands are very broad. These bands are broad because the electron transition time is much shorter than the vibrations occurring within the molecule. When the ligands are farther away from the metal than the mean bond length, the field is weaker, the splitting is less, and hence the transition energy is smaller than the "normal" value. Conversely, when the ligands are closer to the metal, the field is stronger, the splitting is greater, and the transition energy is larger than the normal value. We can confirm this explanation by cooling the complex to close to absolute zero, thereby reducing the molecular vibrations. When we do, as predicted, the visible absorption spectrum becomes much narrower.

The hexachlorotitanate(III) ion, $[TiCl_6]^{3-}$, has an orange color as a result of an absorption centered at 770 nm. This value corresponds to a crystal field splitting of about 160 kJ·mol^{-1}. The lower value reflects the weakness of the chloride ion as a ligand relative to the water molecule; that is, chloride is lower than water in the spectrochemical series.

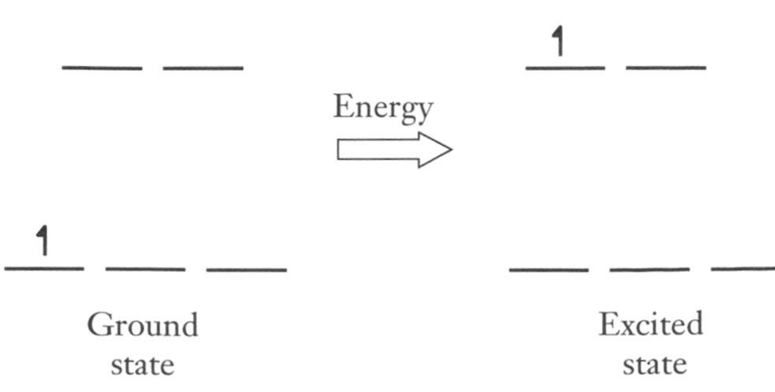

Figure 18.21 The electron transition corresponding to the visible absorption of the titanium(III) ion.

For most branches of chemistry, we measure energy differences in kilojoules per mole, but transition metal chemists usually report crystal field splittings in a frequency unit called wavenumbers. This is simply the reciprocal of the wavelength expressed as centimeters. Thus the units of wavenumbers will be cm^{-1}, called reciprocal centimeters. For example, the crystal field splitting for the hexaaquatitanium(III) ion is usually cited as 19 200 cm^{-1} rather than 230 $kJ{\cdot}mol^{-1}$.

We will discuss the visible absorptions of other electron configurations later.

Hydration Enthalpies

Another of the parameters that can be explained by crystal field theory is the enthalpy of hydration of transition metal ions. This is the energy released when gaseous ions are hydrated, a topic discussed in Chapter 6.

$$M^{n+}(g) + 6\,H_2O(l) \rightarrow [M(OH_2)_6]^{n+}(aq)$$

As the effective nuclear charge of metal ions increases across a period, we expect the electrostatic interaction between the water molecules and the metal ions to increase regularly along the transition metal series. In fact, we find deviations from a linear relationship (Figure 18.22). To explain this observation, we assume that the greater hydration enthalpy is the result of the crystal field stabilization energy, which can be calculated in terms of Δ_{oct}, the crystal field splitting. Recall that for an octahedral field, the d_{xy}, d_{xz}, and d_{yz} orbitals are lowered in energy by $\frac{2}{5}\,\Delta_{oct}$ and the $d_{x^2-y^2}$ and d_{z^2} orbitals are raised in energy by $\frac{3}{5}\,\Delta_{oct}$. Hence for a particular electron configuration, it is possible to calculate the net contribution of the crystal field to the hydration enthalpy. Figure 18.23 illustrates the situation for the d^4 high spin ion. This ion would have a net stabilization energy of

$$-[3(\tfrac{2}{5}\,\Delta_{oct})] + [1(\tfrac{3}{5}\,\Delta_{oct})] = -0.6\,\Delta_{oct}$$

The complete set of crystal field stabilization energies is listed in Table 18.3.

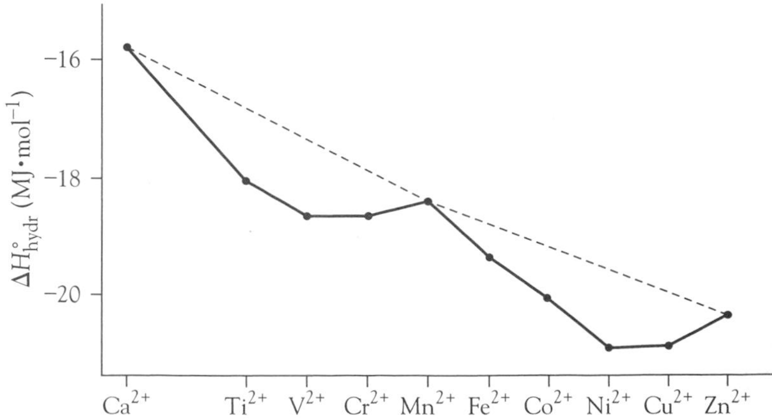

Figure 18.22 Hydration enthalpies of the dipositive ions of the Period 4 transition metals.

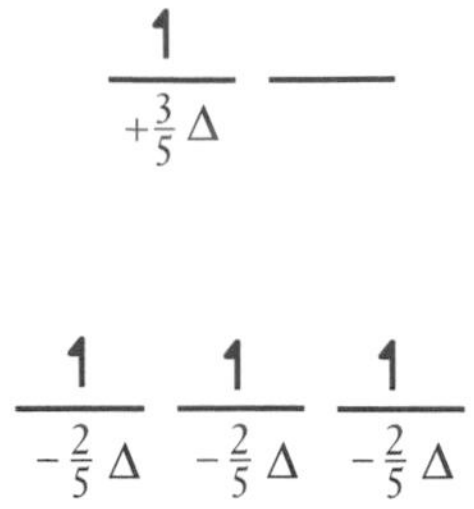

$$\underset{+\frac{3}{5}\Delta}{\underline{1}} \quad \underline{}$$

$$\underset{-\frac{2}{5}\Delta}{\underline{1}} \quad \underset{-\frac{2}{5}\Delta}{\underline{1}} \quad \underset{-\frac{2}{5}\Delta}{\underline{1}}$$

Figure 18.23 The crystal field stabilization energy for the d^4 high spin electron configuration.

Table 18.3 Crystal field stabilization energies (CFSE) for the dipositive, high spin ions of various Period 4 metals

Ion	Configuration	CFSE
Ca^{2+}	d^0	$-0.0\ \Delta_{oct}$
—	d^1	$-0.4\ \Delta_{oct}$
Ti^{2+}	d^2	$-0.8\ \Delta_{oct}$
V^{2+}	d^3	$-1.2\ \Delta_{oct}$
Cr^{2+}	d^4	$-0.6\ \Delta_{oct}$
Mn^{2+}	d^5	$-0.0\ \Delta_{oct}$
Fe^{2+}	d^6	$-0.4\ \Delta_{oct}$
Co^{2+}	d^7	$-0.8\ \Delta_{oct}$
Ni^{2+}	d^8	$-1.2\ \Delta_{oct}$
Cu^{2+}	d^9	$-0.6\ \Delta_{oct}$
Zn^{2+}	d^{10}	$-0.0\ \Delta_{oct}$

These values correspond remarkably well with the deviations of the hydration enthalpies. Of particular note, it is only the d^0, d^5 (high spin), and d^{10} ions that fit the expected near-linear relationship, and these all have zero crystal field stabilization energy.

Spinel Structures

Yet another triumph of crystal field theory is the explanation for the transition metal ion arrangements in the spinel structures that we first met in Chapter 12. The spinel is a mixed oxide of general formula $(M^{2+})(M^{3+})_2(O^{2-})_4$, with the metal ions occupying both octahedral and tetrahedral sites. In a normal spinel, all of the 2+ ions are in the tetrahedral sites and the 3+ ions are in the octahedral sites; whereas in an inverse spinel, the 2+ ions are in the octahedral sites and the 3+ ions fill the tetrahedral sites and the remaining octahedral sites.

The choice of normal spinel or inverse spinel for mixed transition metal oxides is determined usually (but not always) by the option that will give the greater crystal field stabilization energy. This can be illustrated by a pair of oxides that each contains ions of one metal in two different oxidation states: Fe_3O_4, containing Fe^{2+} and Fe^{3+}; and Mn_3O_4, containing Mn^{2+} and Mn^{3+}. The former adopts the inverse spinel structure: $(Fe^{3+})_t(Fe^{2+},Fe^{3+})_oO_4$. All these ions are high spin, so the Fe^{3+} ion (d^5) has a zero CFSE; but the Fe^{2+} ion (d^6) has a nonzero CFSE. Because a crystal field splitting for the tetrahedral geometry is four-ninths that of the equivalent octahedral environment, the CFSE of an octahedrally coordinated ion will be greater than that of a tetrahedrally coordinated ion. This energy difference accounts for the octahedral site preference of the Fe^{2+} ion. Unlike the mixed iron oxide, the mixed

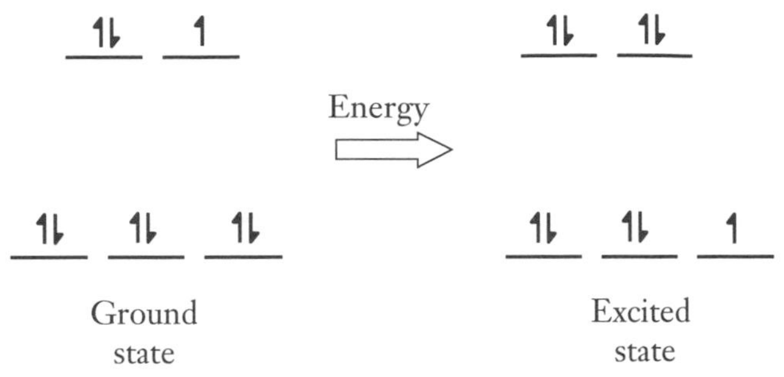

Figure 18.24 The electron transition for the d^9 electron configuration.

manganese oxide has the normal spinel structure: $(Mn^{2+})_t(Mn^{3+})_oO_4$. In this case, it is the Mn^{2+} ion (d^5) that has a zero CFSE and the Mn^{3+} ion (d^4) that has a nonzero CFSE. Hence it is the Mn^{3+} ion that preferentially occupies the octahedral sites.

More on Electronic Spectra

In a previous section, we saw that the single visible absorption of the titanium(III) ion can be explained in terms of a *d* electron transition from the lower level to the upper level in the crystal field. For a copper(II) ion (d^9) in an octahedral environment, a single broad visible absorption is also observed. As in the d^1 situation, the absorption of this ion can be interpreted to mean that one electron is excited to the upper level in the octahedral crystal field (Figure 18.24).

For d^2 ions, we might expect two absorption peaks, corresponding to the excitation of one or both of the electrons. However, a total of three fairly strong absorptions are observed. To explain this, we have to consider interelectronic repulsions. In the ground state, a d^2 ion, such as the hexaaquavanadium(III) ion, has two electrons with parallel spins in any two of the three lower energy orbitals: d_{xy}, d_{xz}, and d_{yz}. When one electron is excited, the resulting combination can have different energies, depending on whether or not the two electrons are occupying overlapping orbitals and therefore repelling each other. For example, an excited configuration of $(d_{xy})^1(d_{z^2})^1$ will be lower in energy because the two electrons occupy very different volumes of space, whereas the $(d_{xy})^1(d_{x^2-y^2})^1$ configuration will be higher in energy because both electrons occupy space in the *x* and *y* planes.

By calculation, it can be shown that the combinations $(d_{xy})^1(d_{z^2})^1$, $(d_{xz})^1(d_{x^2-y^2})^1$, $(d_{yz})^1(d_{x^2-y^2})^1$ all have the same lower energy, and $(d_{xy})^1(d_{x^2-y^2})^1$, $(d_{xz})^1(d_{z^2})^1$, and $(d_{yz})^1(d_{z^2})^1$ all have the same higher energy. This accounts for two of the transitions, and the third transition corresponds to the excitation of both electrons into the upper levels to give the configuration $(d_{x^2-y^2})^1(d_{z^2})^1$. Three possibilities are shown in Figure 18.25.

Sometimes very weak absorptions also appear in the visible spectrum. These correspond to transitions in which an electron has reversed its spin, such as that shown in Figure 18.26. Such transitions, which involve a change in spin state (spin-forbidden transitions), are of low probability and hence have very low intensity in the spectrum.

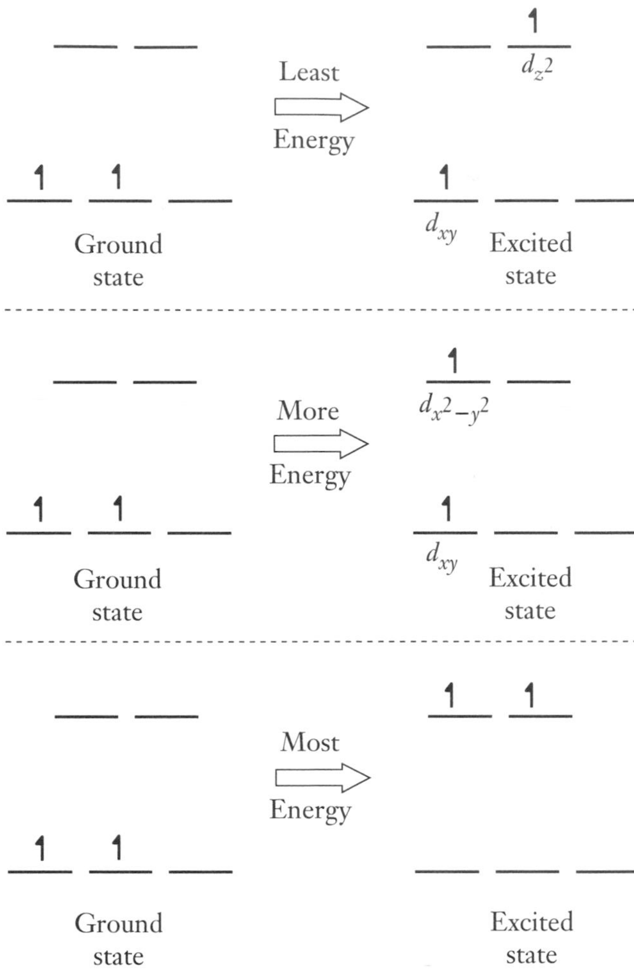

Figure 18.25 Three possible electron transitions for the d^2 electron configuration.

We find that the visible spectra of d^1, d^4(high spin), d^6(high spin), and d^9 can be interpreted in terms of a single transition, whereas d^2, d^3, d^7(high spin), and d^8 spectra can be interpreted in terms of three transitions. The remaining high spin case, d^5, is unique in that the only possible transitions are spin-forbidden. As a result, complexes such as hexaaquamanganese(II) ion and hexaaquairon(III) ion have very pale colors. A more detailed study of transition metal ion spectra is the domain of theoretical inorganic chemistry.

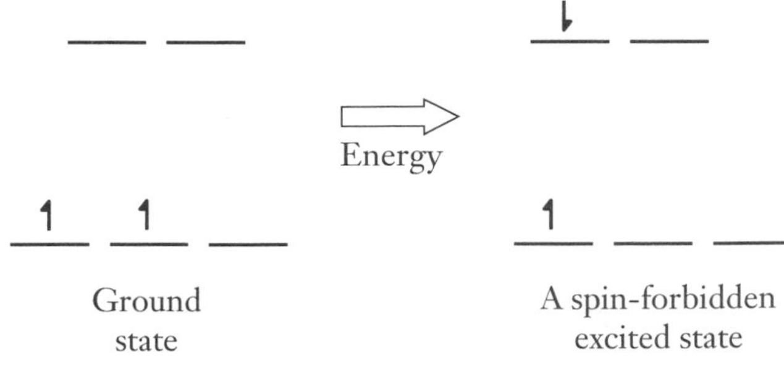

Figure 18.26 A possible spin-forbidden transition for the d^2 electron configuration.

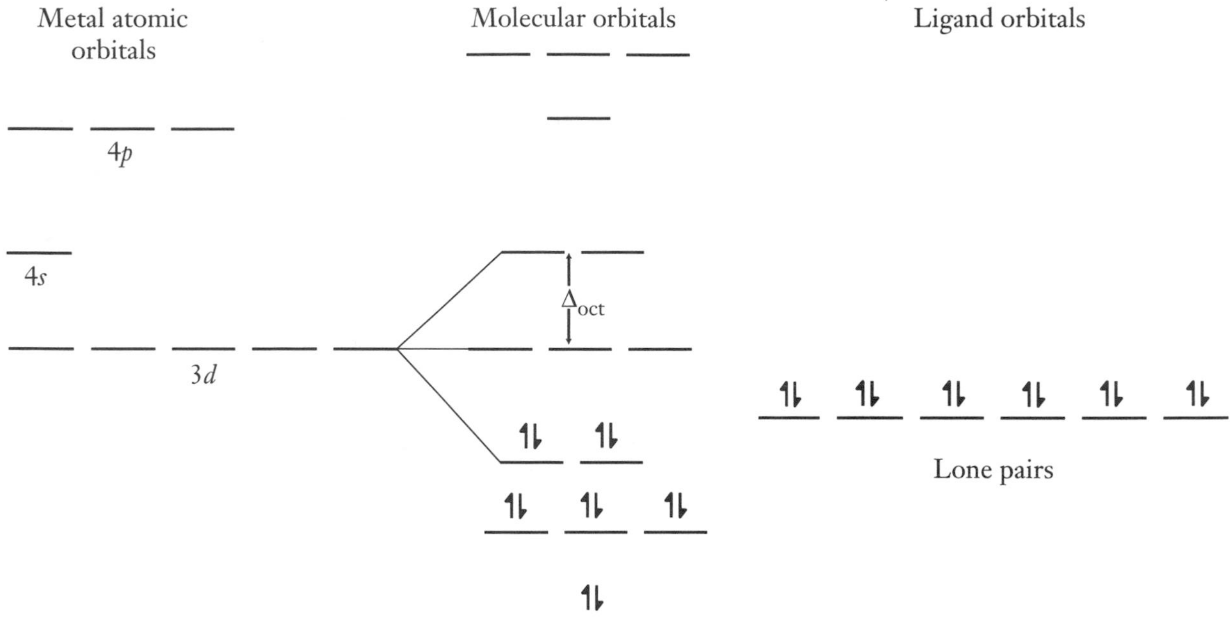

Figure 18.27 A simplified molecular orbital diagram for an octahedral Period 4 transition metal complex. Connecting lines are only shown for molecular orbitals involving 3*d* atomic orbitals.

Molecular Orbital Theory

Crystal field theory works satisfactorily for most transition metal compounds and, fortunately, is quite easy to comprehend. However, it is not a very realistic model of the metal-ligand interactions, and there are some aspects of transition metal chemistry that it cannot explain—for example, why the cyanide ion is such a strong ligand. Crystal field theory assumes that the interaction between metal and ligand is purely electrostatic in nature; but, in fact, there is a tremendous amount of evidence that the metal and ligand orbitals overlap, providing a high degree of covalency to the bonding.

Molecular orbital theory provides a much more elegant picture of the bonding, and it is more consistent with the orbital approach that we have adopted throughout the earlier part of the text. Qualitatively, it is very easy to develop a molecular orbital diagram (Figure 18.27); and, when we do, it becomes apparent why crystal field theory actually works even though its theoretical basis is untenable.

First of all, we see that for Period 4 transition metals, the 4*s*, 4*p*, and some of the 3*d* orbitals are involved in bonding. The three *d* orbitals oriented between the ligand directions, d_{xy}, d_{xz}, and d_{yz}, are not involved. Hence they become nonbonding molecular orbitals. The six ligand orbitals then interact with the 4*s*, 4*p*, $3d_{z^2}$, and $3d_{x^2-y^2}$ orbitals to give six bonding and six antibonding molecular orbital sets. The six pairs of ligand electrons will fill the six bonding orbitals. Hence, from the molecular orbital model, we see that it is the lowering in energy of the ligand electrons that provides the driving force for complex formation.

If we look now at the fate of any *d* electrons from the transition metal (although, of course, to the complex, all electrons are indistinguishable), we see that they will first occupy the three nonbonding molecular orbitals and then start to fill the two lowest energy unoccupied antibonding orbitals, those derived in part from the 3*d* atomic orbitals. Thus our crystal field model works because it is looking specifically at the "frontier" set of molecular orbitals (the HOMOs) that are being filled as we cross the transition metal series.

But molecular orbital theory is able to do much more than just provide a justification of crystal field theory. It enables us to explain those facets of transition metal chemistry that crystal field theory cannot. In particular, we can explain the strength of the metal-carbon bond in cyanide and carbonyl complexes. Up to this point, we have considered metal-ligand bonding to be exclusively σ in nature. This is certainly true for ligands such as water, but it is not true for many other ligands.

Transition metal ions can interact with π-acceptor or π-donor ligands; in the octahedral case, by using the d_{xy}, d_{xz}, and d_{yz} orbitals that lie between the bonding directions. With π-donor ligands, the ions can form a π bond by sharing electron pairs from the filled ligand π system. To provide a favorable situation for accepting electrons, the d_{xy}, d_{xz}, and d_{yz} orbitals themselves should be empty. Hence complexes with ligands containing filled *p* orbitals that can form an appropriately oriented π orbital set (such as fluoride and oxide) tend to stabilize high metal ion oxidation states. Molecular orbital theory shows that from a set of six ligands there are, in fact, only three π ligand molecular orbitals that can interact with the *d* orbitals of the metal. Figure 18.28 shows the interaction of a π-donor with the molecular orbitals of an octahedral d^0 complex. As can be seen, the ligand electron pairs are reduced in energy, whereas any metal *d* electrons are increased in energy. Hence π-donor ligands, such as oxide and fluoride, preferentially form complexes with metal ions in high oxidation states (empty *d* orbital sets). In addition, the decrease in the crystal field splitting with the π-donor ligands explains why such ligands are low in the spectrochemical series.

With π-acceptor ligands, transition metal ions can form a π bond by sharing electron pairs in the d_{xy}, d_{xz}, and d_{yz} orbitals with an empty ligand π orbital set. To provide a favorable situation for donating electrons, the metal ions need filled d_{xy}, d_{xz}, and d_{yz} orbitals. Hence complexes with π-acceptor ligands, such as cyanide and carbonyl, tend to stabilize low metal ion oxidation states. Figure 18.29 shows the interaction of a π-donor ligand with the molecular orbitals of a low spin octahedral d^6 complex. In this case, the driving force is the decrease in the energy of the metal ion electrons. Hence with π-acceptor ligands, completely filled d_{xy}, d_{xz}, and d_{yz} orbitals will lead to an

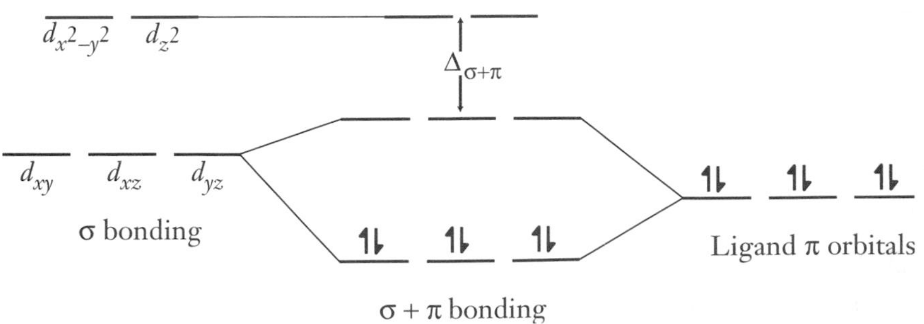

Figure 18.28 The effect on the crystal field splitting, Δ, of interaction of π-donor ligands with a d^0 transition metal ion.

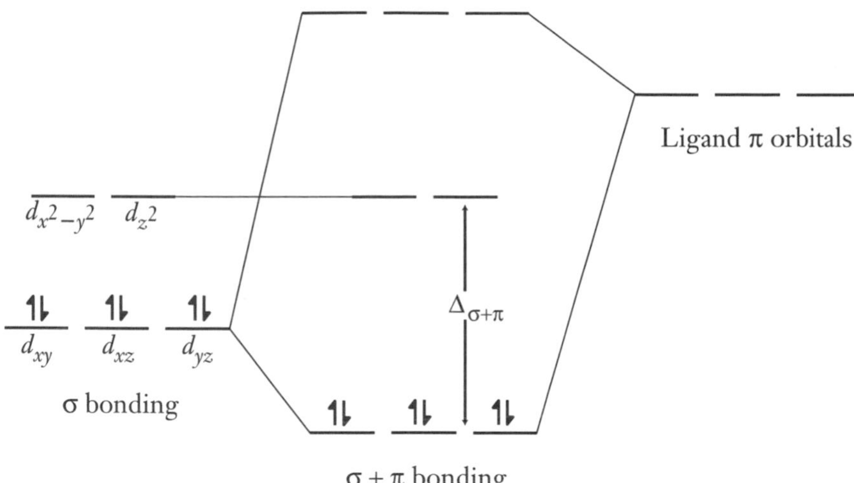

Figure 18.29 The effect on the crystal field splitting, Δ, of interaction of π-acceptor ligands with a d^6 transition metal ion.

optimum bonding situation. That is, π-acceptor ligands preferentially form complexes with metal ions in low oxidation states (filled lower *d* orbital sets). In addition, the increase in the crystal field splitting with the π-acceptor ligands explains why such ligands are at the top of the spectrochemical series.

This result provides an explanation for the 18-electron rule. The optimum octahedral bonding situation will occur when the six ligands each contribute an electron pair to the σ bonding system and the metal atom/ion provides six electrons for the π system. Altogether these species provide a total of 18 electrons. Why, then, do most transition metal complexes fail to follow this pattern? Most values of Δ are sufficiently small that having electrons in the upper *d* orbital set is not a major energy problem. With the very large crystal field splittings that result from π-acceptor ligands, there is a major energy penalty from having the higher set occupied, and, as a result, the 18-electron situation is strongly favored.

Metal Carbonyl Compounds

To conclude this section, we will look at the metal-carbonyl bonding in detail. In Chapter 13, we saw that, for the carbon monoxide molecule, the highest energy occupied molecular orbital (HOMO) is a σ^{NB} orbital essentially derived from the high-energy *2p* carbon atomic orbitals. We assume this orbital resembles a lone pair on the carbon atom. The lowest energy empty molecular orbitals (LUMOs) are the π^*_{2p} antibonding orbitals. Again, the predominant contribution comes from the *2p* atomic orbitals of carbon, so they, too, are focused around the carbon rather than around the oxygen atom. Approximate shapes of these orbitals are shown in Figure 18.30.

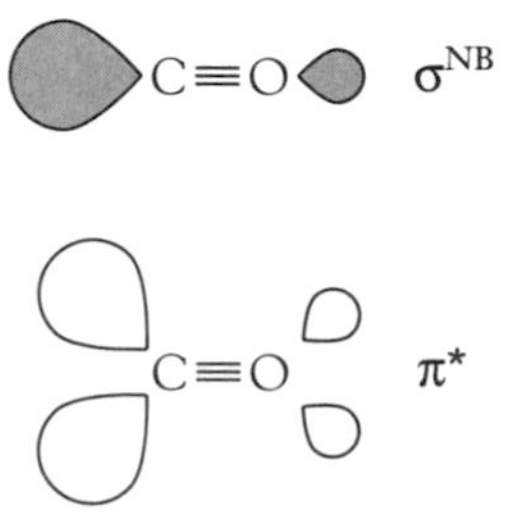

Figure 18.30 The highest energy occupied molecular orbital (HOMO, σ^{NB}), and the lowest energy unoccupied molecular orbital (LUMO, π*) of carbon monoxide.

We picture an overlap of the end of the HOMO of the carbon monoxide with an empty orbital of the metal; that is, the carbon monoxide is acting as a Lewis base and the metal as a Lewis acid. Simultaneously, there is an overlap of a full *d* orbital of the metal with the LUMO of the carbon monoxide (Figure 18.31). In inorganic terminology, the carbon monoxide is said to be a σ-donor and π-acceptor, and the metal is a σ-acceptor and π-donor. Thus there would be a flow of electrons from the carbon monoxide to the

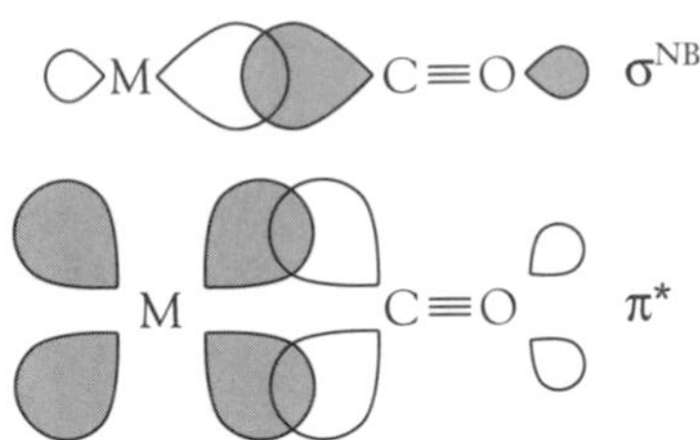

Figure 18.31 The interaction of the HOMO, σ^{NB}, and the LUMO, π^*, of carbon monoxide with the appropriate *d* orbitals of a transition metal ion. The filled orbitals are shaded.

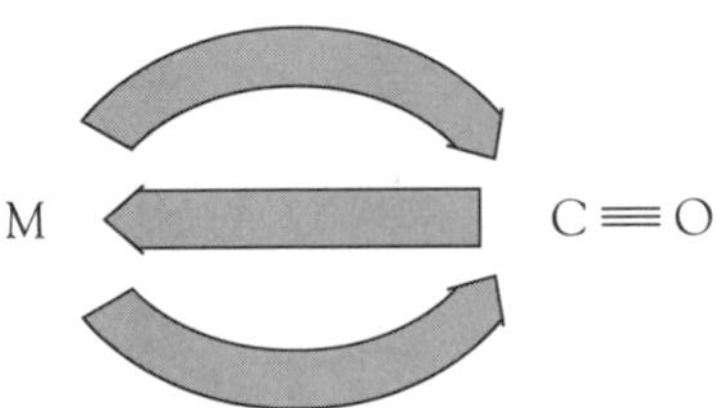

Figure 18.32 A depiction of the synergistic effect in the bonding of carbon monoxide to a low oxidation state metal ion.

metal through the σ system and a flow through the π system in the reverse direction (a "back bond"). This is known as a *synergistic* effect, and it leads to a strong, almost double, covalent bond. The concept is depicted in Figure 18.32.

According to this representation of the bonding, electrons would be "pumped" into the LUMO, a π antibonding orbital of the carbon monoxide. An increased occupancy of antibonding orbitals would lead to a reduction of the bond order below its value of three in the free carbon monoxide molecule. Experimental measurements have shown that, indeed, the carbon-oxygen bond in these carbonyl compounds is longer and weaker than that in carbon monoxide itself. This is good evidence of the validity of our molecular orbital bonding model.

Coordination Equilibria

Many transition metal complexes are synthesized in aqueous solution by displacement of the water ligands. For example, we can produce the hexaamminenickel(II) ion by adding an excess of aqueous ammonia to a solution of the hexaaquanickel(II) ion:

$$[Ni(OH_2)_6]^{2+}(aq) + 6\,NH_3(aq) \rightarrow [Ni(NH_3)_6]^{2+}(aq) + 6\,H_2O(l)$$

Ammonia, which is higher than water in the spectrochemical series, readily replaces the water ligands. In other words, the nickel-ammonia bond is stronger than the nickel-water bond, and the process is exothermic. Hence the reaction is enthalpy-driven.

If we are dealing with a chelating ligand, we find that the equilibrium is driven strongly to the right by entropy factors as well. This situation can be illustrated by the formation of the tris(1,2-diaminoethane)nickel(II) ion from the hexaamminonickel(II) ion:

$$[Ni(OH_2)_6]^{2+}(aq) + 3\,en(aq) \rightarrow [Ni(en)_3]^{2+}(aq) + 6\,H_2O(l)$$

In this case, we have a similar enthalpy increase, but there is also a major entropy increase because the total number of ions and molecules has increased from four to seven. It is the entropy factor that results in such strong complex formation by chelating ligands, behavior that is known as the *chelate effect*.

Thermodynamic versus Kinetic Factors

For a reaction to proceed, there must be a decrease in free energy. However, we must always keep kinetic factors in mind. Most solution reactions of transition metal ions proceed rapidly. For example, addition of a large excess of chloride ion to the pink hexaaquacobalt(II) ion, $[Co(OH_2)_6]^{2+}$, gives the dark blue color of the tetrachlorocobaltate(II) ion, $[CoCl_4]^{2-}$, almost instantaneously:

$$[Co(OH_2)_6]^{2+}(aq) + 4\,Cl^-(aq) \rightleftharpoons [CoCl_4]^{2-}(aq) + 6\,H_2O(l)$$

This reaction is thermodynamically favored and also has a low activation energy. As mentioned earlier, complexes that react quickly (for example, within 1 minute) are said to be labile whereas those that take much longer are called inert.

The two common Period 4 transition metal ions that form inert complexes are chromium(III) and cobalt(III). The former has a d^3 electron configuration and the latter, a low spin d^6 configuration (Figure 18.33). It is the stability of the half-filled and the filled lower set of *d* orbitals that precludes any low-energy pathway of reaction. For example, addition of concentrated acid to the hexaamminecobalt(III) ion should cause a ligand replacement according to free energy calculations. However, such a reaction occurs so slowly that several days must elapse before any color change is noticeable:

$$[Co(NH_3)_6]^{3+}(aq) + 6\ H_3O^+(aq) \rightleftharpoons [Co(OH_2)_6]^{3+}(aq) + 6\ NH_4^+(aq)$$

For this reason, to synthesize specific cobalt(III) or chromium(III) complexes, we usually find a pathway that involves synthesizing the complex of the respective labile 2+ ion and then oxidizing it to the inert 3+ ion.

Cr^{3+}

Co^{3+}

Figure 18.33 Inert complexes are formed by metal ions with the d^3 and low spin d^6 electron configurations.

Biological Aspects

The chelate effect is important in biological complexes. There is a tetradentate ligand of particular importance to biological systems—the porphyrin ring. The basic structure of this complex is shown in Figure 18.34. It is an organic molecule in which alternating double bonds hold the structure in a rigid plane with four nitrogen atoms oriented toward the center. The space in the center is the right size for many metal ions.

In biological systems, the porphyrin ring carries different substituents on the edge of the molecule but the core—a metal ion surrounded by four nitrogen atoms—is consistent throughout the biological world. The hemoglobin molecule, essential to animal life, contains four iron-porphyrin units, whereas the oxygen-binding protein in muscle contains just one unit. Vitamin B_{12} contains a cobalt-porphyrin unit; and plant life depends on the chlorophylls, which contain magnesium-porphyrin units. Thus the porphyrin metal complexes are among the most important in the living world.

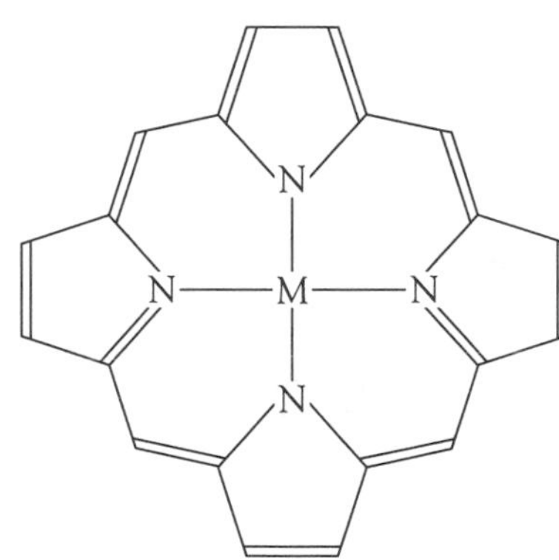

Figure 18.34 The core of metalloporphyrin complexes.

Exercises

18.1. Define the following terms: (a) transition metal; (b) ligand; (c) crystal field splitting.

18.2. Define the following terms: (a) coordination number; (b) chelate; (c) chelate effect.

18.3. Nickel forms two tetracyano complexes, $[Ni(CN)_4]^{2-}$ and $Ni(CN)_4]^{4-}$. Explain why you might expect complexes with nickel in both normal (+2) and low (0) oxidation states.

18.4. Suggest why the fluoride in which chromium has its highest oxidation state is CrF_6 whereas the highest oxidation state chromium assumes in a chloride is $CrCl_4$.

18.5. In addition to the two geometric isomers of $[Pt(NH_3)_2Cl_2]$, there is a third isomer. It has the same empirical formula and square planar geometry, but it is electrically conducting in solution. Write the structure of the compound.

18.6. Identify the probable type of isomerism for (a) $Co(en)_3Cl_3$; (b) $Cr(NH_3)_3Cl_3$.

18.7. Provide systematic names for each of the following compounds: (a) $Fe(CO)_5$; (b) $K_3[CoF_6]$; (c) $[Fe(OH_2)_6]Cl_2$; (d) $[CoCl(NH_3)_5]SO_4$.

18.8. Provide systematic names for each of the following compounds: (a) $(NH_4)_3[CuCl_5]$; (b) $[Co(NH_3)_5(OH_2)]Br_3$; (c) $K_3[Cr(CO)_4]$; (d) $K_2[NiF_6]$; (e) $[Cu(NH_3)_4](ClO_4)_2$.

18.9. Deduce the formula of each of the following transition metal complexes:

(a) hexaamminechromium(III) bromide
(b) aquabis(ethylenediammine)thiocyanato-cobalt(III) nitrate
(c) potassium tetracyanonickelate(II)
(d) tris(1,2-diaminoethane)cobalt(III) iodide

18.10. Deduce the formula of each of the following transition metal complexes:
(a) hexaaquamanganese(II) nitrate
(b) palladium(II) hexafluoropalladate(IV)
(c) tetraaquadichlorochromium(III) chloride dihydrate
(d) potassium octacyanomolybdenate(V)

18.11. Write the names of the following geometric isomers:

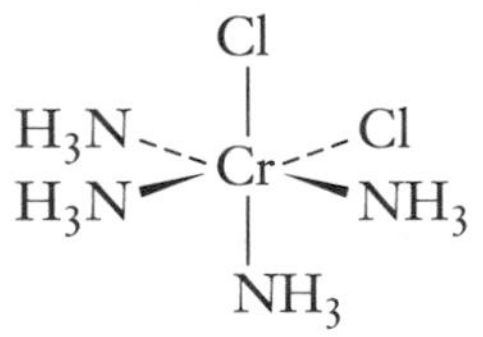

Cl; H_3N, H_3N, Cr, Cl, NH_3; NH_3
(*a*)

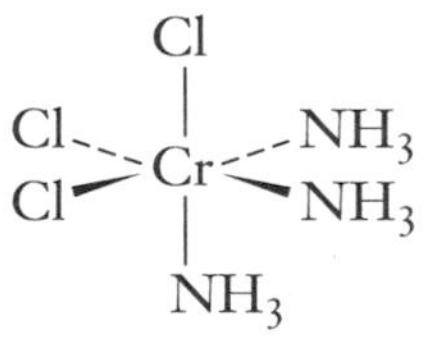

Cl; Cl, Cl, Cr, NH_3, NH_3; NH_3
(*b*)

18.12. For the ion $[Co(en)_2(NO_2)_2]^+$, draw (a) two geometric isomers; (b) a linkage isomer of one of the geometric isomers.

18.13. Which of the following ions obey the 18-electron rule: $[V(CO)_6]$; $[V(CO)_6]^-$; $[Fe(CO)_4]^{2-}$?

18.14. Manganese forms a carbonyl of formula $Mn_2(CO)_{10}$ with a structure of $(OC)_5Mn–Mn(CO)_5$. If the manganese atoms share one electron each, does each manganese atom obey the 18-electron rule? Show your calculations.

18.15. Construct energy level diagrams for both high and low spin situations for the d^6 electron configuration in (a) an octahedral field; (b) a tetrahedral field.

18.16. Which one of the iron(III) complexes, hexacyanoferrate(III) ion or tetrachloroiron(III), is likely to be high spin and which low spin? Give your reason.

18.17. The crystal field splittings, Δ, are listed below for three ammine complexes of cobalt. Explain the differences in values.

Complex	**Δ (cm^{-1})**
$[Co(NH_3)_6]^{3+}$	22 900
$[Co(NH_3)_6]^{2+}$	10 200
$[Co(NH_3)_4]^{2+}$	5 900

18.18. The crystal field splittings, Δ, are listed below for four complexes of chromium(III). Explain the differences in values.

Complex	**Δ (cm^{-1})**
$[CrF_6]^{3-}$	15 000
$[Cr(OH_2)_6]^{3+}$	17 400
$[CrF_6]^{2-}$	22 000
$[Cr(CN)_6]^{3-}$	26 600

18.19. Construct a table of *d* electron configurations and the corresponding number of unpaired electrons for a tetrahedral stereochemistry (similar to Table 18.2).

18.20. Construct a table of crystal field stabilization energies for the dipositive high spin ions of the Period 4 transition metals in a tetrahedral field in terms of Δ_{tet} (similar to Table 18.3).

18.21. Would you predict that $NiFe_2O_4$ would adopt a normal spinel or an inverse spinel structure? Explain your reasoning.

18.22. Would you predict that $NiCr_2O_4$ would adopt a normal spinel or an inverse spinel structure? Explain your reasoning.

18.23. The ligand $H_2NCH_2CH_2NHCH_2CH_2NH_2$, commonly known as det, is a tridentate ligand, coordinating to metal ions through all three nitrogen atoms. Write a balanced chemical equation for the reaction of this ligand with the hexaaquanickel(II) ion and suggest why the formation of this complex would be strongly favored.

CHAPTER

19

Properties of the Transition Metals

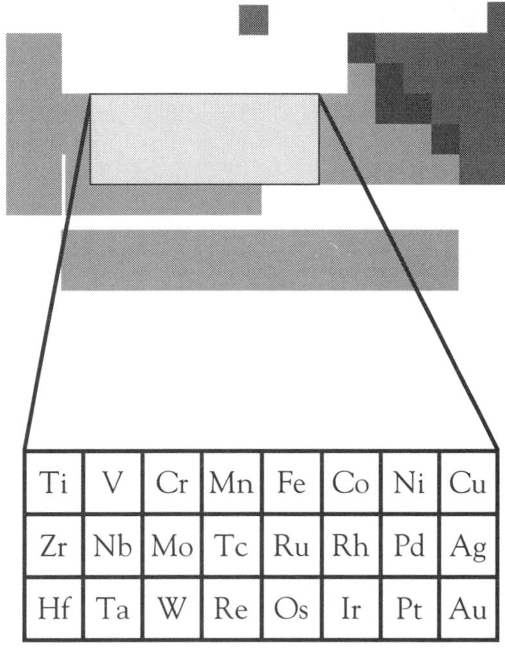

The most striking feature of the transition metals is the variety of colors of their compounds. The large number of compounds is attributable to the many oxidation states that the metals exhibit; and, in most cases, the colors result from electron transitions within the partially filled d orbitals in these species. The Period 4 members of the transition metal series are the most important; hence these are the elements that will be the focus of this chapter.

With its origins linked to the study of minerals, inorganic chemistry was the first branch of chemistry to be pursued rigorously. However, by the first part of the twentieth century, it had fossilized into the memorization of long lists of compounds, their properties, and methods of synthesis. The polymer and pharmaceutical industries became fields of rapid growth, and organic chemistry became the major focus of chemistry.

Ronald Nyholm, an Australian chemist, brought inorganic chemistry to life again. Nyholm was born in 1917 in Broken Hill, Australia, a mining town where the streets have names such as Chloride, Sulphide, Oxide, and Silica. With this childhood environment—and an enthusiastic chemistry teacher in high school—it was natural for him to choose a career in chemistry. Nyholm moved to England to study with some of the great chemists of the time,

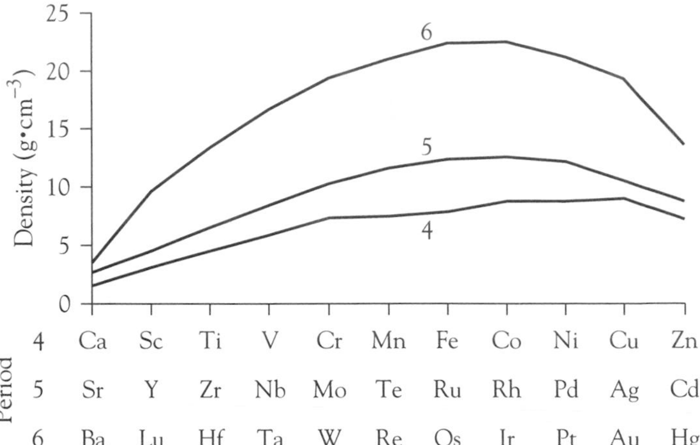

Figure 19.1 Densities of the transition metals in Periods 4, 5, and 6.

where his research opened whole new directions of study by showing that it was the nature of the ligand that determined much of metal ion behavior. For example, using specific ligands, he was able to produce unusual oxidation numbers and coordination numbers in metal complexes. Nyholm, together with Ronald Gillespie, a British chemist, devised the VSEPR method of predicting molecular shape, and he was the first to argue that inorganic chemistry involved the understanding of molecular structure, not just memorization of formulas. Sadly, Nyholm was killed in an automobile accident in 1971, at the peak of his career.

Overview of the Transition Metals

All the transition metals are hard and have very high melting points. In fact, 10 of the metals have melting points above 2000°C and 3 above 3000°C (tantalum, tungsten, and rhenium). The transition metals all have high densities; and the trends in this property are shown in Figure 19.1. The densities increase from the Period 4 elements to the Period 6 elements, with the highest values being those of osmium and iridium (23 $g{\cdot}cm^{-3}$). Chemically, the metals themselves are comparatively unreactive. Only a few of the metals, such as iron, are electropositive enough to react with acids.

Group Trends

For the main group elements, there are clear trends down each group. For the transition metals, the elements of Periods 5 and 6 show very strong similarities in their chemistry within a group. This similarity is to a large extent a result of the filling of the 4*f* orbitals in the elements that lie between these two rows. Electrons in these orbitals are poor shielders of electrons in the outer 6*s* and 5*d* orbitals. With the greater effective nuclear charge, the atomic, covalent, and ionic radii of the Period 6 transition elements are reduced to almost the same as those in Period 5. This effect is illustrated in Table 19.1, where the ionic radii of the Group 2 and Group 5 metals are com-

Table 19.1 Ion sizes of elements in Groups 2 and 5

Group 2 ion	Ion radius (pm)	Group 5 ion	Ion radius (pm)
Ca^{2+}	114	V^{3+}	78
Sr^{2+}	132	Nb^{3+}	86
Ba^{2+}	149	Ta^{3+}	86

pared. The radii of the Group 2 metals increase down the group, whereas the niobium and tantalum ions have identical radii.

There are some superficial similarities in the chemistries of the Periods 5 and 6 elements with those of the Period 4 elements. Thus chromium, molybdenum, and tungsten all form oxides with an oxidation number of +6. However, chromium(VI) oxide, CrO_3, is highly oxidizing, whereas molybdenum(VI) oxide, MoO_3, and tungsten(VI) oxide, WO_3, are the "normal" oxides of these metals.

The limitations of such comparisons are also illustrated by the lower chlorides of chromium and tungsten. Chromium forms a compound, $CrCl_2$ (among others), whereas tungsten forms an apparently analogous compound, WCl_2. The former does contain the chromium(II) ion, but the latter is known to be $[W_6Cl_8]^{4+}{\cdot}4Cl^-$, with the polyatomic cation containing a cluster of tungsten ions at the corners of an octahedron and chloride ions in the middle of the faces. The enthalpy of formation of the theoretical $W^{2+}{\cdot}2Cl^-$ can be calculated as +430 kJ·mol^{-1} (very different from the value of −397 kJ·mol^{-1} for chromium(II) chloride), providing a thermodynamic reason for its nonexistence. The enthalpy difference is mostly due to the very much higher atomization energy of tungsten (837 kJ·mol^{-1}) relative to that of chromium (397 kJ·mol^{-1}). This high atomization energy reflects the strong metal-metal bonding in the Periods 5 and 6 transition metals. As a result, like WCl_2, many compounds of these elements contain groups of metal ions, and these are called *metal cluster compounds.*

The oxidation numbers of the transition metals are higher for the first half of each row than for the later members. The Periods 5 and 6 elements usually have higher common oxidation numbers than does the analogous member of Period 4, as is shown in Table 19.2. Like the main group elements, the highest oxidation number of a transition metal is found in an

Table 19.2 The most common oxidation numbers of the transition metals

Ti	V	Cr	Mn	Fe	Co	Ni	Cu
+4	+3, +4	+3, +6	+2, +3, +7	+2, +3	+2, +3	+2	+1, +2
Zr	Nb	Mo	Tc	Ru	Rh	Pd	Ag
+4	+5	+6	—	+3	+3	+2	+1
Hf	Ta	W	Re	Os	Ir	Pt	Au
+4	+5	+6	+4, +7	+4, +8	+3, +4	+2, +4	+3

oxide. Thus the +8 oxidation number of osmium occurs in osmium(VIII) oxide, OsO_4. Unlike the main group metals, transition metals exhibit almost every possible oxidation number; for example, there are various compounds of manganese in which manganese has every oxidation number from +7 to −1.

One consistent factor found in each transition metal group is the increase in the crystal field splitting, Δ, from Period 4 to Period 6 elements. For example, in the series $[Co(NH_3)_6]^{3+}$, $[Rh(NH_3)_6]^{3+}$, and $[Ir(NH_3)_6]^{3+}$, the Δ_{oct} values are 23 000 cm^{-1}, 34 000 cm^{-1}, and 41 000 cm^{-1}, respectively. Because of the larger crystal field splittings for the Periods 5 and 6 transition metals, almost all compounds of these elements are low spin.

Comparative Stability of Oxidation States of the Period 4 Transition Metals

The transition metals in Period 4 are most common and of greatest industrial importance. Furthermore, the patterns in their properties are the easiest to comprehend. Figure 19.2 summarizes the Frost diagrams for these elements. Titanium metal (oxidation state 0) is strongly reducing, but the elements become less so as we progress along the row. When we reach copper, the metal itself is the most thermodynamically stable oxidation state. As we move across the row, the highest oxidation state becomes less favored; and by chromium, it has become highly oxidizing. The most thermodynamically stable oxidation number is +3 (just) for both titanium and vanadium, whereas +2 is favored by the other elements. For iron, the stabilities of the +3 and +2 oxidation states are very similar. Copper is unique in having a stable +1 oxidation number, but as is apparent from Figure 19.2, it is prone to disproportionation—to the +2 and 0 oxidation states.

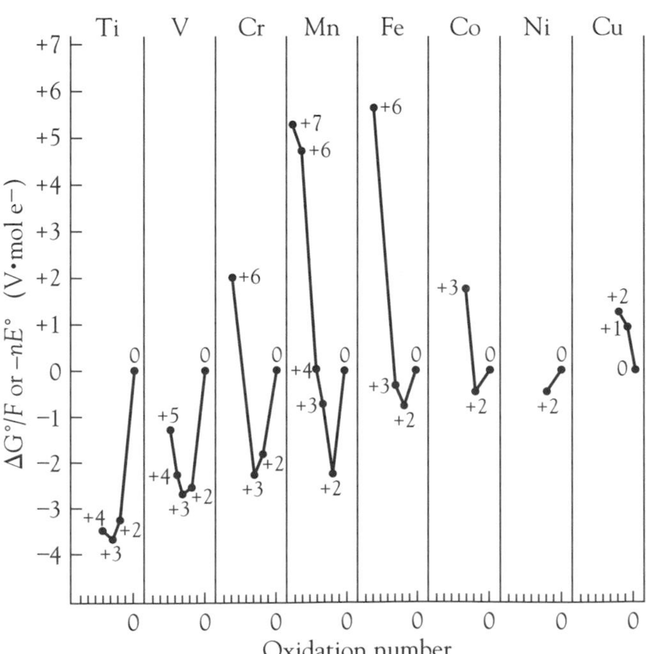

Figure 19.2 Frost diagrams for the Period 4 transition metals under acid conditions.

Group 4: Titanium, Zirconium, and Hafnium

The only widely used element in this group is titanium. Titanium is the ninth most abundant element in Earth's crust, whereas zirconium and hafnium, like most of the Periods 5 and 6 transition metals, are rare.

Titanium

Titanium, a hard, silvery white metal, is the least dense (4.5 $g{\cdot}cm^{-3}$) of the transition metals. This combination of high strength and low density makes it a preferred metal for military aircraft and for nuclear submarines, where cost is less important than performance.

The pure metal is difficult to obtain from the most common titanium compound. Reduction of titanium(IV) oxide, TiO_2, with carbon produces the metal carbide rather than the metal. The only practical route (the Kroll process) involves the initial conversion of the titanium(IV) oxide to titanium(IV) chloride by heating the oxide with carbon and dichlorine:

$$TiO_2(s) + 2\,C(s) + 2\,Cl_2(g) \xrightarrow{\Delta} TiCl_4(g) + 2\,CO(g)$$

The titanium(IV) chloride gas is condensed at 137°C.

We can use an Ellingham diagram for chlorides to examine the possible reduction routes for this compound (Figure 19.3). Carbon is totally unsuitable for reducing the titanium(IV) chloride, because the slope of the line in the Ellingham diagram is the opposite of that required—in other words, the free energy line for the formation of carbon tetrachloride does not cross any metal chloride line. Hydrogen also is unsatisfactory, because it will only reduce titanium(IV) chloride above about 1700°C. The alternative is to find a metal whose metal–metal chloride line lies below that of the titanium–titanium(IV) chloride line. The choice of reactive metal is partially based on cost and partially on the ease of separating the titanium metal from the other metal chloride and from excess metal reactant. Magnesium is usually preferred; and at about 850°C, it will displace the titanium:

$$TiCl_4(g) + 2\,Mg(l) \xrightarrow{\Delta} Ti(s) + 2\,MgCl_2(l)$$

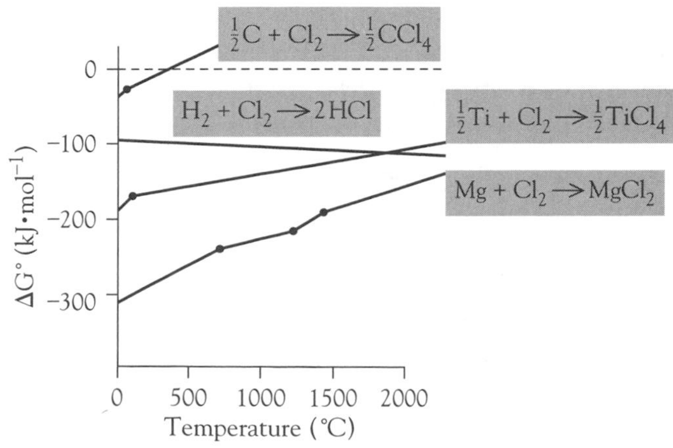

Figure 19.3 Ellingham diagram for various chloride species.

The spongy mass of titanium metal is porous, and the magnesium chloride and excess magnesium metal can be dissolved out by using dilute acid. The titanium metal granules are then fabricated into whatever shape is required.

Titanium(IV) Oxide

Although the production of titanium metal is vital for the defense industry, the enormous quantities of titanium ores mined each year are destined for a more innocuous purpose—as paint pigment. Of the 5 million tonnes of titanium ore produced each year, Canada provides about one-third and Australia about one-fourth. Although the element is often found as the dioxide (mineral name, rutile), it is too impure to be used directly.

The purification process involves the conversion of rutile to the chloride, as in the metal synthesis:

$$TiO_2(s) + 2\ C(s) + 2\ Cl_2(g) \xrightarrow{\Delta} TiCl_4(g) + 2\ CO(g)$$

The chloride is then reacted with dioxygen at about 1200°C to give pure white titanium(IV) oxide:

$$TiCl_4(g) + O_2(g) \xrightarrow{\Delta} TiO_2(s) + 2\ Cl_2(g)$$

The dichlorine is recycled.

Prior to the use of titanium(IV) oxide in paints, the common white pigment was "white lead," $Pb_3(CO_3)_2(OH)_2$. Apart from its toxicity, it discolored in industrial city atmospheres to give black lead(II) sulfide. Titanium(IV) oxide, which is stable to discoloration in polluted air, has now replaced white lead completely. Not only is titanium(IV) oxide of very low toxicity, it has the highest refractive index of any white or colorless inorganic substance—even higher than diamond. As a result of this high light-scattering ability, it covers and hides previous paint layers more effectively. In addition to its use in white paint, it is also added to colored paints to make the colors paler and mask previous colors better.

Zirconium

Although a very rare metal, zirconium is used to make the containers for nuclear fuel because it has a low capture cross section for neutrons—that is, it does not absorb the neutrons that propagate the fission process. Unfortunately, hafnium has a high capture cross section; thus it is crucial to remove hafnium impurities from the chemically similar zirconium. To produce metallic zirconium, the ore baddeleyite (zirconium(IV) oxide), ZrO_2, is processed by a method similar to that for titanium:

$$ZrO_2(s) + 2\ C(s) + 2\ Cl_2(g) \xrightarrow{\Delta} ZrCl_4(g) + 2\ CO(g)$$

At this stage, the 2 percent impurity of hafnium(IV) chloride, $HfCl_4$, can be separated from the zirconium(IV) chloride, $ZrCl_4$, by fractional sublimation. The hafnium compound sublimes at 319°C and the zirconium compound at 331°C. (The closeness in sublimation temperatures shows the great similarity between the two elements.) Then the pure zirconium(IV) chloride is reduced with magnesium metal:

$$ZrCl_4(g) + 2\ Mg(l) \xrightarrow{\Delta} Zr(s) + 2\ MgCl_2(l)$$

In the baddeleyite crystalline form of zirconium(IV) oxide, each zirconium(IV) ion is surrounded by seven oxide ions (Figure 19.4a). Above 2300°C, the compound rearranges to an eight-coordinate fluorite structure (Figure 19.4b), cubic zirconia, which is an excellent diamond substitute in jewelry. Although the refractive index (and hence "sparkle") and hardness of cubic zirconia are less than those of diamond, its melting point of 2700°C makes it more thermally stable than diamond. By a patented process, zirconium(IV) oxide can be produced in a fibrous form. These silky fibers have nearly uniform dimensions: 3 μm in diameter and 2 to 5 cm long. They can be woven into a material that is stable up to 1600°C, making zirconia cloth very useful for high-temperature purposes.

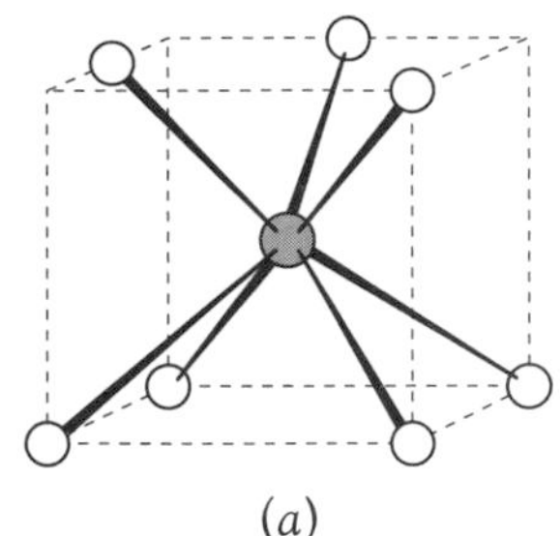

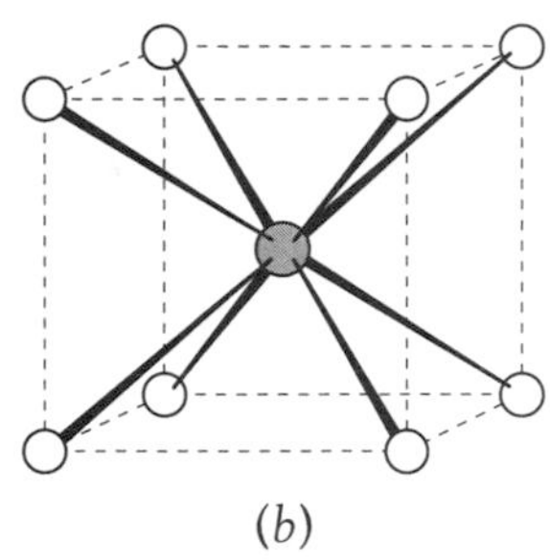

Figure 19.4 The zirconium(IV) oxide ion arrangements in (*a*) baddeleyite and (*b*) cubic zirconia.

Group 5: Vanadium, Niobium, and Tantalum

None of the Group 5 metals has any great usefulness, although vanadium is used for vanadium steels, a particularly hard alloy that is used for knife blades and various workshop tools.

Oxidation States of Vanadium

The simple redox chemistry of vanadium is particularly interesting to inorganic chemists because vanadium readily exists in four different oxidation states: +5, +4, +3, and +2, corresponding to the d^0, d^1, d^2, and d^3 electron configurations. With an oxidation number of +5 for vanadium, the colorless vanadate ion, $[VO_4]^{3-}$, exists in very basic solution; under neutral conditions, conjugate acids such as the pale yellow dihydrogen vanadate ion, $[H_2VO_4]^-$, are formed.

A reducing agent, such as zinc metal in acid solution, can be used to reduce the vanadium(V) to give the characteristically colored ions of vanadium in lower oxidation states:

$$Zn(s) \rightarrow Zn^{2+}(aq) + 2\ e^-$$

Initial reduction of the dihydrogen vanadate ion by zinc metal in acid solution (or by a weak reducing agent, such as sulfur dioxide) gives the deep blue vanadyl ion, VO^{2+} (with +4 oxidation number):

$$[H_2VO_4]^-(aq) + 4\ H^+(aq) + e^- \rightarrow VO^{2+}(aq) + 3\ H_2O(l)$$

This ion is written more precisely as the $[VO(OH_2)_5]^{2+}$, because five water molecules occupy the other coordination sites.

As reduction continues, the bright blue color of the vanadyl ion is replaced by that of the green hexaaquavanadium(III) ion, $[V(OH_2)_6]^{3+}$ (or $V^{3+}(aq)$, for simplicity):

$$VO^{2+}(aq) + 2\ H^+(aq) + e^- \rightarrow V^{3+}(aq) + H_2O(l)$$

Provided air is excluded, further reduction results in the formation of the lavender hexaaquavanadium(II) ion, $[V(OH_2)_6]^{2+}$:

$$[V(OH_2)_6]^{3+}(aq) + e^- \rightarrow [V(OH_2)_6]^{2+}(aq)$$

As soon as this solution is exposed to air, it reoxidizes to the vanadium(III) ion.

Biological Aspects

Vanadium is not widely used in nature. Yet it does appear to be vital to one of the simplest groups of marine organisms, the tunicates, or sea squirts. These organisms belong somewhere between invertebrates and vertebrates. One family of tunicates utilizes very high levels of vanadium in its blood plasma for oxygen transport. Why the tunicates should have picked such a unique element for a biochemical pathway is still unclear. The element also appears to be used by a very different organism, the poisonous mushroom *Amanita muscaria*. Here, too, the reason for this element being utilized is not well understood.

Group 6: Chromium, Molybdenum, and Tungsten

All the stable Group 6 metals are utilized in the manufacture of metal alloys for specialized uses. In addition, chromium provides a shiny protective coating for iron and steel surfaces. Chromium metal is not inert in itself; instead, it has a very thin tough oxide coating that confers the protection. Tungsten is used for the filaments of the traditional electric light bulb, for it has the highest melting point of any metal (3420°C). As a result of the high melting point, the vapor pressure of the hot metal is low and the filament is long lasting. However, as can be seen from the slow darkening of the glass envelope, the metal does, in time, sublime from the filament, thereby weakening and ultimately breaking it.

For molybdenum and tungsten, the oxidation number of +6 is thermodynamically preferred. However, for chromium, the +6 state is highly oxidizing; the oxidation number of +3 is most stable.

Chromates and Dichromates

In spite of their thermodynamic instability, kinetic factors enable several chromium(VI) compounds to exist. The most important of these are the chromates and dichromates. The yellow chromate ion, $[CrO_4]^{2-}$, can only exist in solution under alkaline conditions, and the orange dichromate ion, $[Cr_2O_7]^{2-}$, only under acid conditions, because of the equilibrium

$$2\,[CrO_4]^{2-}(aq) + 2\,H^+(aq) \rightleftharpoons [Cr_2O_7]^{2-}(aq) + H_2O(l)$$

The chromate ion is the conjugate base of the hydrogen chromate ion, $[HCrO_4]^-$; thus a chromate ion solution is always basic because of the equilibrium

$$[CrO_4]^{2-}(aq) + H_2O(l) \rightleftharpoons [HCrO_4]^-(aq) + OH^-(aq)$$

Many chromates are insoluble, and they are often yellow if the cation is colorless, such as lead(II) chromate, $PbCrO_4$. The high insolubility of lead(II) chromate and its high refractive index (hence high opacity) have resulted in its use for yellow highway markings.

In both the chromate and dichromate ions, chromium has an oxidation state of +6; hence the metal has a d^0 electron configuration. Without *d* electrons, we might expect these, and all d^0 configurations, to be colorless. This is obviously not the case. The color comes from an electron transition from the ligand to the metal, a process known as *charge transfer*. That is, an electron is excited from a filled ligand π orbital into the empty metal ion d orbitals. We can depict the process simply as

$$Cr^{6+}—O^{2-} \rightarrow Cr^{5+}—O^{-}$$

Such transitions require considerable energy; hence the absorption is usually centered in the ultraviolet part of the spectrum, with just the edge of the absorption in the visible region. Charge transfer is particularly evident when the metal is in high oxidation states, such as chromates and dichromates.

Silver(I) chromate, Ag_2CrO_4, has a unique brick-red color, making it a useful compound in the analysis of silver ion. One route is a precipitation titration (the Mohr method), in which silver ion is added to chloride ion to give a white precipitate of silver chloride:

$$Ag^+(aq) + Cl^-(aq) \rightarrow AgCl(s)$$

In the presence of chromate ion (usually about 0.01 mol·L^{-1}), the brick-red, slightly more soluble silver chromate will form as soon as the chloride ion is completely consumed, the color change indicating that the equivalence point has been reached (actually, slightly exceeded):

$$2\ Ag^+(aq) + [CrO_4]^{2-}(aq) \rightarrow Ag_2CrO_4(s)$$

The dichromate ion has a structure involving a bridging oxygen atom (Figure 19.5). This ion is a strong oxidizing agent, although the carcinogenic nature of the chromium(VI) ion means that it should be treated with respect, particularly the powdered solid, which can be absorbed through the lungs. The orange dichromate ion is a good oxidizing agent and is reduced to the green hexaaquachromium(III) ion, $[Cr(OH_2)_6]^{3+}$, in the redox reaction

$$[Cr_2O_7]^{2-}(aq) + 14\ H^+(aq) + 6\ e^- \rightarrow 2\ Cr^{3+}(aq) + 7\ H_2O(l) \qquad E^\circ = +1.33V$$

This reaction is used in breath analyzers for the detection of excessive alcohol intake. The ethanol in the breath is bubbled through an acidic solution of dichromate, the color change being detected quantitatively. In the reaction, the ethanol is oxidized to ethanoic (acetic) acid:

$$CH_3CH_2OH(aq) + H_2O(l) \rightarrow CH_3CO_2H(aq) + 4\ H^+(aq) + 4\ e^-$$

The oxidation of organic compounds with dichromate ion is a common reaction in organic chemistry. Sodium dichromate is preferred because it has a higher solubility than potassium dichromate.

For quantitative analysis, sodium dichromate cannot be used as a primary standard because of its deliquescence. However, potassium dichromate is an ideal primary standard because it does not hydrate and because it can be obtained in high purity by recrystallization—its solubility in water increases rapidly with increasing temperature. One application is the determination of iron(II) ion concentrations in an acidic solution. In this titrimetric procedure, the dichromate is reduced to chromium(III) ion and the iron(II) ion is oxidized to iron(III) ion:

$$Fe^{2+}(aq) \rightarrow Fe^{3+}(aq) + e^-$$

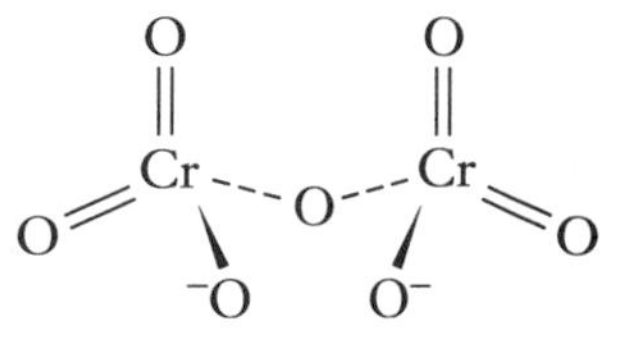

O O
Cr O Cr
O O
$^-$O O$^-$

Figure 19.5 The dichromate ion, $[Cr_2O_7]^{2-}$.

The characteristic color change of orange to green as the dichromate is reduced to the chromium(III) ion is not sensitive enough; thus an indicator, barium diphenylamine sulfonate, has to be used. This indicator is less readily oxidized than iron(II) ions, but it is oxidized to give a blue color once all the iron(II) ions have been converted to the iron(III) state. Because free iron(III) ions affect the indicator and thus give rise to an inaccurate end point, some phosphoric acid is added before starting the titration. This reagent gives a stable iron(III) phosphate complex.

Ammonium dichromate, $(NH_4)_2Cr_2O_7$, is often used in "volcano" demonstrations. If a red-hot wire is touched to a pile of ammonium dichromate, the exothermic decomposition is initiated, emitting sparks and water vapor in a spectacular way. However, this is not a safe demonstration because a dust containing carcinogenic chromium(VI) compounds is usually released. The reaction is nonstoichiometric, producing chromium(III) oxide, water vapor, nitrogen gas, and some ammonia gas. It is commonly represented as

$$(NH_4)_2Cr_2O_7(s) \rightarrow Cr_2O_3(s) + N_2(g) + 4\,H_2O(g)$$

The industrial production of dichromate provides some interesting chemistry. The starting material is a mixed oxide, iron(II) chromium(III) oxide, $FeCr_2O_4$ (commonly called iron chromite), an ore found in enormous quantities in South Africa. The powdered ore is heated to about 1000°C with sodium carbonate in air, thereby causing the chromium(III) to be oxidized to chromium(VI):

$$4\,FeCr_2O_4(s) + 8\,Na_2CO_3(s) + 7\,O_2(g) \xrightarrow{\Delta} 8\,Na_2CrO_4(s) + 2\,Fe_2O_3(s) + 8\,CO_2(g)$$

Addition of water dissolves the sodium chromate, a process called leaching; it leaves the insoluble iron(III) oxide. To obtain sodium dichromate, the Le Châtelier principle is applied. The following equilibrium lies to the left under normal conditions, but under high pressures of carbon dioxide (obtained from the previous reaction), the yield of sodium dichromate is high:

$$2\,Na_2CrO_4(aq) + 2\,CO_2(aq) + H_2O(l) \rightleftharpoons Na_2Cr_2O_7(aq) + 2\,NaHCO_3(s)$$

In fact, the aqueous carbon dioxide is really employed as a low-cost way of decreasing the pH to favor the dichromate ion in the chromate-dichromate equilibrium. It can be seen that the mole ratio of carbon dioxide to chromate produced in the previous step is exactly the same as that employed in this step. The slightly soluble sodium hydrogen carbonate has to be filtered off under pressure to prevent the equilibrium from shifting to the left. The sodium hydrogen carbonate is then reacted with an equimolar proportion of sodium hydroxide to obtain the sodium carbonate that can be reused in the first step. Thus the ore and sodium hydroxide are the only bulk chemicals used in the process.

Chromium(VI) Oxide

The oxide of chromium in which chromium assumes its highest oxidation state is chromium(VI) oxide, CrO_3. It is a red crystalline solid that is prepared by adding concentrated sulfuric acid to a cold concentrated solution of potassium dichromate. The synthesis can be viewed as an initial formation of chromic acid followed by a decomposition to the acidic oxide and water:

$$K_2Cr_2O_7(aq) + H_2SO_4(aq) + H_2O(l) \rightarrow K_2SO_4(aq) + 2\ \text{“}H_2CrO_4(aq)\text{”}$$

$$\text{“}H_2CrO_4(aq)\text{”} \rightarrow CrO_3(s) + H_2O(l)$$

Chromium(VI) oxide is an acidic oxide, as are most metal oxides in which the metal has a very high oxidation number. It is very soluble in water, forming "chromic acid," which is in fact a mixture of species. The strongly oxidizing (and acidic) nature of the solution results in its occasional use as a final resort for cleaning laboratory glassware. However, the hazard of the solution itself (carcinogenic and very acidic) and the potential danger from exothermic redox reactions with glassware contaminants make it a very unwise choice.

The oxide is also strongly oxidizing. For example, ethanol ignites on contact; it is oxidized to a mixture of ethanal, CH_3CHO, and ethanoic acid, CH_3CO_2H, and the chromium(VI) oxide is reduced to chromium(III) oxide.

Chromyl Chloride

Chromyl chloride, a red, oily liquid of formula CrO_2Cl_2, is of interest only as a definitive means of identifying chloride ion if a halide ion is known to be present. When concentrated sulfuric acid is added to a mixture of solid potassium dichromate and an ionic chloride, such as sodium chloride, a dark red liquid is formed:

$$6\ H_2SO_4(l) + K_2Cr_2O_7(s) + 4\ NaCl(s)$$
$$\rightarrow 2\ CrO_2Cl_2(l) + 2\ KHSO_4(s) + 4\ NaHSO_4(s) + 3\ H_2O(l)$$

When heated gently and very cautiously, a deep red, toxic vapor is produced. This gas can be collected and condensed to a dark red, covalent liquid, chromyl chloride. If this liquid is added to a basic solution, it immediately hydrolyzes to the yellow chromate ion:

$$CrO_2Cl_2(l) + 4\ OH^-(aq) \rightarrow CrO_4^{2-}(aq) + 2\ Cl^-(aq) + 2\ H_2O(l)$$

Because bromides and iodides do not form analogous chromyl compounds, this test is specific for chloride ions.

The molecule itself has a tetrahedral arrangement around the central chromium atom, and there is appreciable double bond character in the Cr–O bonds (Figure 19.6).

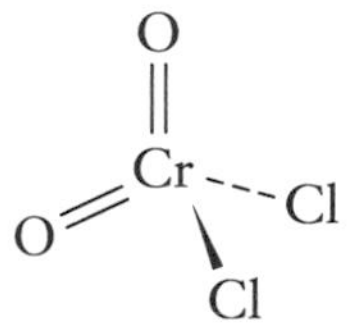

Figure 19.6 The chromyl chloride molecule, CrO_2Cl_2.

Similarities Between Chromium(VI) and Sulfur(VI) Compounds

As we have seen, transition metals can form covalent compounds in which the metal has a very high oxidation number. These compounds often reveal analogous behaviors between the members of the corresponding Groups x and $x + 10$ in the periodic table, particularly in the cases of chromium(VI) (Group

Table 19.3 Formulas of common chromium(VI) and sulfur(VI) species

Formula	Systematic name	Formula	Systematic name
CrO_3	Chromium(VI) oxide	SO_3	Sulfur trioxide
CrO_2Cl_2	Chromyl chloride	SO_2Cl_2	Sulfuryl chloride
CrO_4^{2-}	Chromate ion	SO_4^{2-}	Sulfate ion
$Cr_2O_7^{2-}$	Dichromate ion	$S_2O_7^{2-}$	Pyrosulfate ion

6) and sulfur(VI) (Group 16). The similarity is very apparent in the formulas of their compounds (Table 19.3).

There also are a few similarities in their chemistries. For example, each metal chromate is usually isomorphous with the corresponding sulfate; in other words, potassium chromate and potassium sulfate form identically shaped crystals. However, there are major chemical differences; in particular, chromates and dichromates are strongly oxidizing and colored, whereas the sulfates and pyrosulfates are nonoxidizing and white.

Chromium(III) Oxide

The green, powdery compound, chromium(III) oxide, Cr_2O_3, is a basic oxide, as expected from the lower oxidation number of the metal. Just as lead(II) chromate (chrome yellow) is an important yellow pigment, so chromium(III) oxide is a common green pigment. It is chromium(III) oxide that has, since 1862, provided the pigment for the green U.S. currency ("greenbacks"). Because the pigment is a mineral rather than an organic dye, the green will not fade, nor is it affected by acids, bases, or oxidizing or reducing agents. To prepare a pure pigment, sodium dichromate is reduced; the common reducing agent is sulfur at high temperatures:

$$Na_2Cr_2O_7(s) + S(l) \xrightarrow{\Delta} Cr_2O_3(s) + Na_2SO_4(s)$$

The sodium sulfate is washed away, leaving pure chromium(III) oxide.

Chromium(III) Chloride

Anhydrous chromium(III) chloride, $CrCl_3$, is a reddish violet solid obtained when chlorine is passed over strongly heated chromium metal:

$$2\ Cr(s) + 3\ Cl_2(g) \xrightarrow{\Delta} 2\ CrCl_3(s)$$

When crystallized from aqueous solution, a dark green hexahydrate is obtained. If a solution of this hydrated chromium(III) chloride is treated with a solution of silver nitrate, only one-third of the chloride precipitates out as silver chloride, that is, only one of the chlorides is present as the free ion. This result indicates that the formula of this compound is $[Cr(OH_2)_4Cl_2]^+ Cl^-{\cdot}2H_2O$. As mentioned in Chapter 18, there are actually three hydration isomers of this compound: violet, $[Cr(OH_2)_6]^{3+}\ 3Cl^-$; light green, $[CrCl(OH_2)_5]^{2+}\ 2Cl^-{\cdot}H_2O$; and dark green, $[Cr(OH_2)_4Cl_2]^+\ Cl^-{\cdot}2H_2O$.

Molybdenum(IV) Sulfide

Molybdenum(IV) sulfide is the only industrially important compound of molybdenum. It is the common ore of the metal, and nearly half of the world's supply is in the United States. The purified black molybdenum(IV) sulfide, MoS_2, has a layer structure that resembles graphite. This property has led to its use as a lubricant, both alone and as a slurry mixed with hydrocarbon oils.

Biological Aspects

Although chromium(VI) is carcinogenic when ingested or absorbed through the skin, we require small quantities of chromium(III) in our diet. Insulin and the chromium(III) ion regulate blood glucose levels. A deficiency of chromium(III) or an inability to utilize the chromium ion can lead to diabetes.

However, molybdenum is the most biologically important member of the group. It is the heaviest (highest atomic number) element to have a wide range of functions in living organisms. At the present time, over a dozen known enzymes rely on molybdenum, which is usually absorbed as the molybdate ion, $[MoO_4]^{2-}$. The most crucial molybdenum enzyme (which contains iron as well) is nitrogenase. This enzyme occurs in bacteria that reduce the "inert" dinitrogen of the atmosphere to the ammonium ion, used in protein synthesis by plants. Some of these bacteria have a symbiotic relationship with the leguminous plants (pea and bean family), forming nodules on the roots of the plants. These bacteria process about 2×10^8 tonnes of nitrogen per year in the soils of this planet! Another molybdenum-containing enzyme is sulfite oxidase, which oxidizes harmful sulfite ion to the innocuous sulfate ion in our livers.

Why is a metal as rare as molybdenum so biologically important? There are a number of possible reasons: The molybdate ion has a high aqueous solubility at near-neutral pH values, making it easily transportable by biological fluids. The ion has a negative charge, making it more suitable for different environments than are the cations of the Period 4 transition metals. The element has a wide range of oxidation states (+4, +5, and +6) whose redox potentials overlap with those of biological systems. Being a soft Lewis acid, molybdenum(VI) shows a strong bonding preference for the soft Lewis base, sulfide, another important component of these enzymes. Finally, molybdenum is about eighteenth in the order of abundances of metals in seawater, and much of the choice of elements for biochemical processes was probably determined when the only life on this planet was in the sea.

Group 7: Manganese, Technetium, and Rhenium

Manganese is important as an additive in specific types of steel. Rhenium has little practical use, but technetium, all of whose isotopes are radioactive, has medical uses in radiotherapy.

Oxidation States of Manganese

Manganese readily forms compounds over a range of oxidation states that is wider than that of any other common metal. Figure 19.7 shows the relative

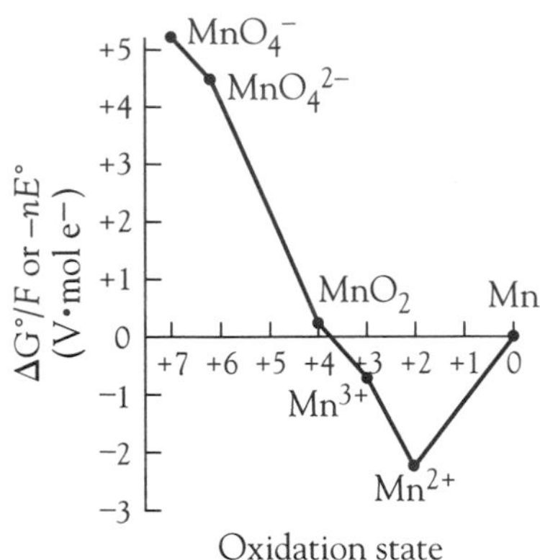

Figure 19.7 Frost (oxidation state) diagram for manganese in acidic solution.

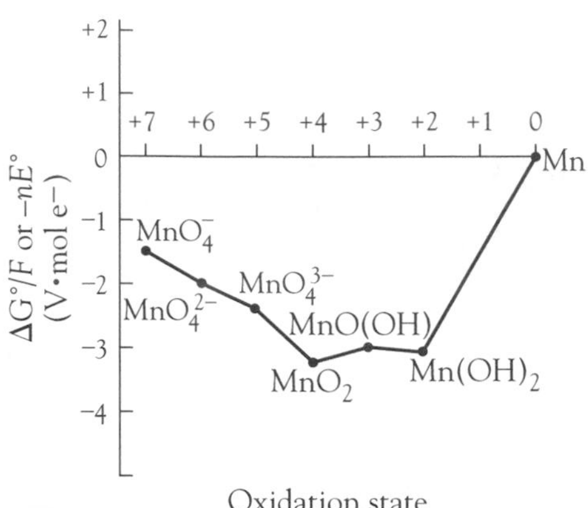

Figure 19.8 Frost (oxidation state) diagram for manganese in basic solution.

stabilities for the oxidation states of manganese in acid solution. From this diagram, we can see that the permanganate ion, $[MnO_4]^-$, or tetraoxomanganate(VI), is very strongly oxidizing in acid solution. The deep green manganate ion, $[MnO_4]^{2-}$, or tetraoxomanganate(V), is also strongly oxidizing, but it disproportionates readily to the permanganate ion and manganese(IV) oxide; thus it is of little importance. Manganese(IV) oxide is oxidizing with respect to the most stable manganese species, the manganese(II) ion. In acid solution, the manganese(III) ion disproportionates, and it also is of little interest. Finally, the metal itself is reducing.

In basic solution, we find a different situation, as can be seen in Figure 19.8. The differences can be summarized as follows:

1. For a particular oxidation state, many of the compounds are unique. Manganese, like most metals, at high pH forms insoluble hydroxides (and oxide hydroxides) in which the metal has low oxidation states.

2. The higher oxidation states are not strongly oxidizing, as they are in acid solution. This difference can be explained simply in terms of reductions that involve hydrogen ion concentration, for these reactions will be strongly pH dependent. In the Frost diagram for acid solution, the concentration of hydrogen ion is 1 mol·L^{-1}; in the diagram for basic solution, the hydrogen ion concentration is 10^{-14} mol·L^{-1} (that is, 1 mol·L^{-1} concentration of hydroxide ion). Using the Nernst equation, we can show that this change in ion concentration will have a major effect on the standard reduction potential.

3. Oxidation states that are very unstable in acid can exist in basic solution (and vice versa). Thus the bright blue manganite ion, $[MnO_4]^{3-}$, can be formed in basic solution.

4. In basic solution, the most thermodynamically stable species is manganese(IV) oxide, although the manganese(III) oxide hydroxide, MnO(OH), and manganese(II) hydroxide also are both moderately stable. In fact, above pH 14, manganese(III) oxide hydroxide is thermodynamically more stable than manganese(II) hydroxide.

Potassium Permanganate

Potassium permanganate, $KMnO_4$, a black solid, is the best known compound containing manganese with an oxidation number of +7. Like chromium(VI) compounds, the color in this d^0 ion is derived from charge transfer electron transitions. It dissolves in water to give a deep purple solution. The permanganate ion is an extremely powerful oxidizing agent; and under acid conditions, it is reduced to the colorless manganese(II) ion:

$$[MnO_4]^-(aq) + 8\ H^+(aq) + 5\ e^- \rightarrow Mn^{2+}(aq) + 4\ H_2O(l) \qquad E° = +1.51\ V$$

It will oxidize concentrated hydrochloric acid to chlorine, and this is one way of producing chlorine gas in the laboratory:

$$2\ HCl(aq) \rightarrow Cl_2(g) + 2\ H^+(aq) + 2\ e^-$$

Potassium permanganate is an important reagent in redox titrations. Unlike potassium dichromate, it is not a suitable primary standard, because its purity cannot be guaranteed. Samples of the substance contain some manganese(IV) oxide, and aqueous solutions slowly deposit brown manganese(IV) oxide on standing. Its precise concentration is determined by titration against a standard solution of oxalic acid. Potassium permanganate solution is run into the oxalic acid solution from a buret, and the purple color disappears as the (almost) colorless manganese(II) ions and carbon dioxide are formed. The permanganate acts as its own indicator, for the slightest excess of permanganate gives a pink tint to the solution:

$$[MnO_4]^-(aq) + 8\ H^+(aq) + 5\ e^- \rightarrow Mn^{2+}(aq) + 4\ H_2O(l)$$

$$[C_2O_4]^{2-}(aq) \rightarrow 2\ CO_2(g) + 2\ e^-$$

This particular reaction has a high activation energy. To provide a reasonable reaction rate, the oxalate solution is initially warmed. However, once some manganese(II) ion is produced, it acts as its own catalyst, and the reaction occurs more quickly as the titration proceeds.

A standardized solution of potassium permanganate can be used for the quantitative determination of iron in samples such as mineral ores or foodstuffs. The iron is converted to iron(II) ion, which is then titrated with standardized permanganate ion solution, again using the permanganate ion as reagent and indicator:

$$[MnO_4]^-(aq) + 8\ H^+(aq) + 5\ e^- \rightarrow Mn^{2+}(aq) + 4\ H_2O(l)$$

$$Fe^{2+}(aq) \rightarrow Fe^{3+}(aq) + e^-$$

One of the few oxidizing agents even more powerful than permanganate is the bismuthate ion, $[BiO_3]^-$. A test for manganese(II) ion is the addition of sodium bismuthate to a sample under cold, acidic conditions. The purple permanganate ion is produced, thereby indicating the presence of manganese:

$$Mn^{2+}(aq) + 4\ H_2O(l) \rightarrow [MnO_4]^-(aq) + 8\ H^+(aq) + 5\ e^-$$

$$[BiO_3]^-(aq) + 6\ H^+(aq) + 2\ e^- \rightarrow Bi^{3+}(aq) + 3\ H_2O(l)$$

In organic chemistry, potassium permanganate is usually used under basic conditions. In base, purple permanganate ion is first reduced to green manganate ion and then to solid brown-black manganese(IV) oxide:

$$[MnO_4]^-(aq) + e^- \rightarrow [MnO_4]^{2-}(aq)$$

$$[MnO_4]^{2-}(aq) + 2\ H_2O(l) + 2\ e^- \rightarrow MnO_2(s) + 4\ OH^-(aq)$$

As an example, the permanganate ion can be used for oxidizing alkenes to diols:

$$-\overset{|}{\underset{|}{C}}{=}\overset{|}{\underset{|}{C}}- + 2\ OH^-(aq) \rightarrow -\overset{HO}{\underset{|}{\overset{|}{C}}}-\overset{OH}{\underset{|}{\overset{|}{C}}}- + 2\ e^-$$

Manganese(VII) Oxide

Manganese(VII) oxide, a dark liquid, is a strongly oxidizing covalent compound. It decomposes explosively to the more stable manganese(IV) oxide:

$$2\ Mn_2O_7(l) \rightarrow 4\ MnO_2(s) + 3\ O_2(g)$$

Similarities Between Manganese(VII) and Chlorine(VII) Compounds

Just as chromium(VI) resembles sulfur(VI), so manganese(VII) resembles chlorine(VII). In fact, there is a closer resemblance between the elements of Groups 7 and 17 than between the elements of Groups 6 and 16. For example, both permanganate ion, $[MnO_4]^-$, and perchlorate ion, $[ClO_4]^-$, are strongly oxidizing. Also, the corresponding permanganate and perchlorate salts are isomorphous. In addition, the two oxides, manganese(VII) oxide, Mn_2O_7, and dichlorine heptoxide, Cl_2O_7, are explosive, covalent liquids. The one major difference is that the manganese compounds are strongly colored and the chlorine compounds are not.

Potassium Manganate

Potassium manganate, a green solid, is about the only common compound of manganese(VI). It is only stable in the solid phase or in extremely basic conditions. When it is dissolved in water, it disproportionates, as predicted by the Frost diagram:

$$3\ [MnO_4]^{2-}(aq) + 2\ H_2O(l) \rightarrow 2\ [MnO_4]^-(aq) + MnO_2(s) + 4\ OH^-(aq)$$

Manganese(IV) Oxide

The only compound of manganese(IV) of any importance is the dioxide, MnO_2, which occurs naturally as the ore pyrolusite. Manganese(IV) oxide is a black insoluble solid and is considered to have an essentially ionic structure. The compound is a strong oxidizing agent; it releases chlorine from concentrated hydrochloric acid and is, at the same time, reduced to manganese(II) chloride:

$$MnO_2(s) + 4\ HCl(aq) \rightarrow MnCl_2(aq) + Cl_2(g) + 2\ H_2O(l)$$

Manganese(II) Compounds

The most thermodynamically stable oxidation number of manganese is +2. Manganese in this oxidation state exists as a very pale pink ion, $[Mn(OH_2)_6]^{2+}$, a species present in all the common salts of manganese, such as nitrate, chloride, and sulfate. The very pale color of this ion contrasts with the strong colors of most other transition metal ions. The reason for the virtual absence of color can be deduced from our earlier remarks on the cause of color in transition metal compounds. The wavelengths absorbed correspond to the energy needed to raise a *d* electron from its ground state to an excited state. However, in the high spin manganese(II) ion, each orbital already contains one electron. The only way an electron can absorb energy in the visible spectrum is by inverting its spin and pairing with another electron during the excitation (Figure 19.9). This process (a spin-forbidden transition) has a very low probability; hence little visible light is absorbed by the manganese(II) ion.

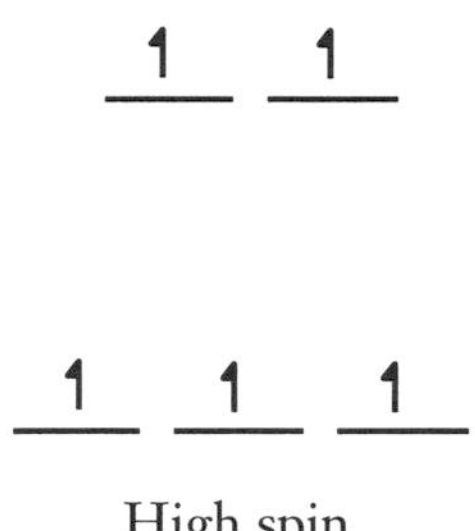

Figure 19.9 The *d* orbital occupancy for the d^5 electron configuration.

When base is added to a solution containing manganese(II) ion, the white manganese(II) hydroxide is formed:

$$Mn^{2+}(aq) + 2\ OH^-(aq) \rightarrow Mn(OH)_2(s)$$

However, the manganese(III) state is favored under basic conditions, and the manganese(II) hydroxide oxidizes in the air to a brown hydrated manganese(III) oxide, MnO(OH):

$$Mn(OH)_2(s) + OH^-(aq) \rightarrow MnO(OH)(s) + H_2O(l) + e^-$$

$$\tfrac{1}{2}\ O_2(g) + H_2O(l) + 2\ e^- \rightarrow 2\ OH^-(aq)$$

Biological Aspects

Manganese is a crucial element in a number of plant and animal enzymes. In mammals, it is used in the liver enzyme arginase, which converts nitrogen-containing wastes to the excretable compound urea. There is a group of enzymes in plants, the phosphotransferases, that incorporate manganese. Like most transition metals, the biological role of manganese seems to be as a redox reagent, cycling between the +2 and +4 oxidation states.

Iron

Iron is believed to be the major component of Earth's core. This metal is also the most important material in our civilization. It does not hold this place because it is the "best" metal; after all, it corrodes much more easily than many other metals. Its overwhelming dominance in our society comes from a variety of factors.

Mining the Seafloor

We usually think of minerals as coming from mines bored into Earth's crust, yet there is increasing interest in mining ore deposits from the seafloor. In 1873 the Challenger expedition to the Pacific Ocean first dredged up mineral nodules from the bottom of the sea. We now know that nodules are widespread over the ocean floors. Generally, manganese and iron each make up between 15 and 20 percent of the content of these nodules; smaller concentrations of titanium, nickel, copper, and cobalt are also present. However, the composition varies from site to site, some nodule beds containing up to 35 percent manganese.

The question of how such nodules formed puzzled chemists for a long time. It was the Swedish chemist I. G. Sillén who proposed that the oceans be considered as a giant chemical reaction vessel. As the metal ions accumulate in the seas from land runoff and undersea volcanic vents, the products of their reactions with anions in the seawater exceed the solubility product. The compounds then start to crystallize out very slowly over thousands and perhaps millions of years in the form of these nodules.

Because the nodules are such concentrated ores, there is much interest in using them for their metal content, particularly by the United States, which has to import much of the manganese, cobalt, and nickel it uses. A number of mining techniques are being developed to remove up to 200 tonnes of nodules per hour. However, there are two concerns, the first of which relates to the life on the seafloor; such large-scale excavations could have a major effect on bottom ecosystems. Furthermore, there is the question of ownership. Should the nodules be the property of whichever company and/or country that can mine them first, or, being in international waters, should they be the collective property of the world? Both of these issues need to be discussed and solved in the very near future.

1. Iron is the second most abundant metal in the Earth's crust, and concentrated deposits of iron ore are found in many localities, thus making it easy to mine.
2. The common ore can be easily and cheaply processed thermochemically to obtain the metal.
3. The metal is malleable and ductile; many metals are relatively brittle.
4. The melting point (1535°C) is low enough that the liquid phase can be handled without great difficulty.
5. By the addition of small quantities of other elements, alloys that have exactly the required combinations of strength, hardness, or ductility for very specific uses can be formed.

The one debatable factor is iron's chemical reactivity. This is considerably less than that of the alkali and alkaline earth elements but is not as low as that of many other transition metals. Its relatively easy oxidation is a major disadvantage—consider all the rusting automobiles, bridges, and other iron and steel structures, appliances, tools, and toys. At the same time, it does mean that our discarded metal objects will crumble to rust rather than remain an environmental blight forever.

Production of Iron

The most common sources of iron are the two oxides: iron(III) oxide, Fe_2O_3, and iron(II) iron(III) oxide, Fe_3O_4. These have the mineral names of hematite and magnetite, respectively. The extraction of iron is carried out in a blast furnace (Figure 19.10), which can be between 25 and 60 m in height and up to 14 m in diameter. The furnace itself is constructed of steel and has a lining of a heat- and corrosion-resistant material. The lining used to be brick, but it is now highly specialized ceramic materials. In fact, half of the high-temperature ceramics used today are produced for iron and steel smelter linings. The main ceramic material used for the lining is aluminum oxide (commonly called corundum), although the lining of the lower parts of the furnace consist of ceramic oxides of formula $Al_xCr_{2-x}O_3$. In these oxides, the chromium(III) ion has replaced some of the aluminum ions. These mixed metal oxide ceramics are more chemical- and temperature-resistant than the pure oxide ceramics.

A mixture of iron ore, limestone, and coke in the correct proportions is fed into the top of the blast furnace through a cone and hopper arrangement

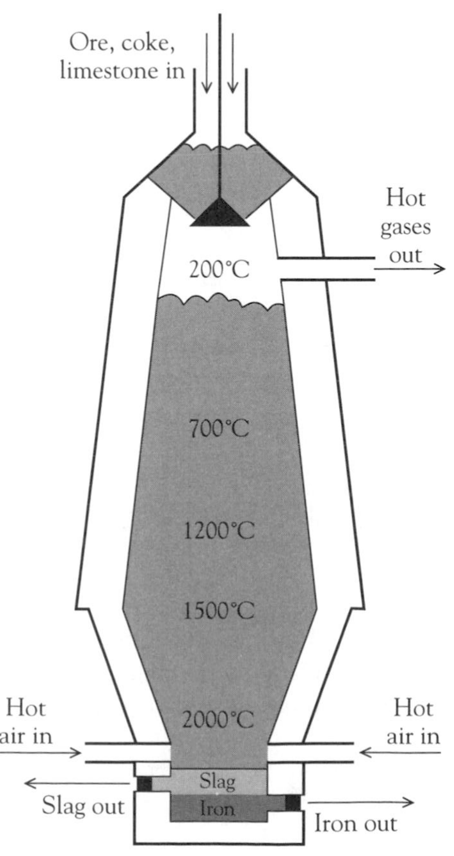

Figure 19.10 A blast furnace.

to prevent escape of the gases. Air, preheated to 600°C by combustion of the exhaust gases, is injected into the lower part of the furnace. The gases move up the furnace while the solids descend as the products are drawn off from the bottom. The heat is generated by the reaction of the dioxygen in the air with the carbon (coke):

$$2\ C(s) + O_2(g) \rightarrow 2\ CO(g)$$

It is the hot carbon monoxide (initially at about 2000°C) that is the reducing agent for the iron ore.

At the top of the furnace, the temperature ranges from 200° to 700°C, a temperature sufficient to reduce iron(III) oxide to iron(II) iron(III) oxide, Fe_3O_4:

$$3\ Fe_2O_3(s) + CO(g) \rightarrow 2\ Fe_3O_4(s) + CO_2(g)$$

Lower in the furnace, at about 850°C, the iron(II) iron(III) oxide is reduced to iron(II) oxide,

$$Fe_3O_4(s) + CO(g) \rightarrow 3\ FeO(s) + CO_2(g)$$

and the temperature is high enough to also decompose the calcium carbonate (limestone) to calcium oxide and carbon dioxide:

$$CaCO_3(s) \rightarrow CaO(s) + CO_2(g)$$

As the mixture descends into the hotter regions (between 850° and 1200°C), the iron(II) oxide is reduced to iron metal and the carbon dioxide formed is re-reduced to carbon monoxide by the coke:

$$FeO(s) + CO(g) \rightarrow Fe(s) + CO_2(g)$$

$$C(s) + CO_2(g) \rightarrow 2\ CO(g)$$

Lower still, at temperatures between 1200° and 1500°C, the iron melts and sinks to the bottom of the furnace, and the calcium oxide reacts with the silicon dioxide impurities in the iron ore to give calcium silicate, commonly called slag:

$$CaO(s) + SiO_2(g) \rightarrow CaSiO_3(l)$$

The blast furnace is provided with two tapholes that are plugged with clay, the lower one for the denser iron metal and the upper one for the less dense slag. These plugs are periodically removed, releasing a stream of molten iron through the lower taphole and liquid slag through the upper. Blast furnaces are run 24 hours a day; and depending on its size, a furnace can produce from 1000 to 10 000 tonnes of iron every 24 hours.

The molten metal is usually conveyed directly in the liquid form to steelmaking plants. The slag can be cooled to the solid phase, ground, and used in concrete manufacture, or, while liquid, mixed with air and cooled into a "woolly" material that can be used for thermal insulation. The hot gases emerging from the top of the furnace contain appreciable amounts of carbon monoxide, and these are burnt to preheat the air for the furnace:

$$2\ CO(g) + O_2(g) \rightarrow 2\ CO_2(g)$$

The iron produced contains a wide range of impurities, such as silicon, sulfur, phosphorus, carbon, and oxygen. The carbon, which can be present in as great a proportion as 4.5 percent, is a particular contributor to the brittleness of the material. Iron is rarely used in pure form; more commonly, we require carefully controlled levels of impurities to provide exactly the properties required. One method for controlling content is the basic oxygen process. A schematic diagram of a typical furnace is shown in Figure 19.11.

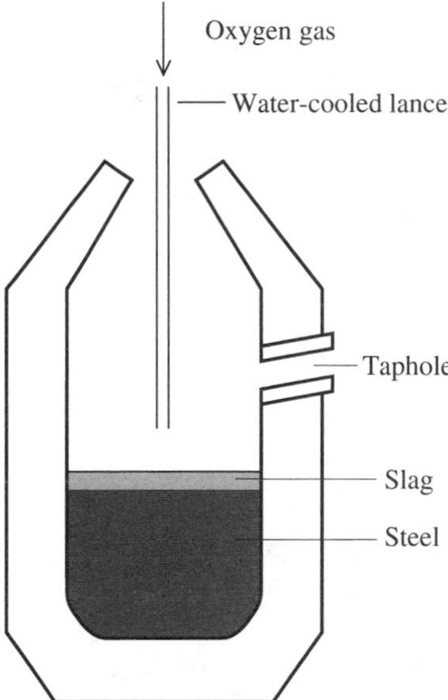

Figure 19.11 A furnace used for the basic oxygen process.

Unlike the blast furnace, this process is not continuous. The converter is filled with about 60 tonnes of molten iron at a temperature of about 1200°C. A blast of oxygen diluted with carbon dioxide is blown through the converter. Oxygen is used instead of air because the nitrogen in air would react with the iron at these temperatures to form a brittle metal nitride. Oxygen reacts with the impurities and raises the temperature in the furnace to about 1900°C, and the diluent carbon dioxide prevents the temperature from increasing excessively. In addition, cold scrap metal is usually added to keep the temperature down.

In the basic oxygen process, carbon is oxidized to carbon monoxide, which burns at the top of the converter to carbon dioxide. The silicon, an impurity, is oxidized to silicon dioxide, which then reacts with the oxides of other elements to form a slag. The furnace also is lined with limestone (calcium carbonate), which reacts with acidic phosphorus-containing impurities. After several minutes, the flame at the top of the converter sinks, indicating that all the carbon has been removed. The slag is poured off, and any required trace elements are added to the molten iron. For normal steel, between 0.1 and 1.5 percent carbon is required. The carbon reacts with the iron to form iron carbide, Fe_3C, commonly called cementite. This compound forms separate small crystals among the crystals of iron. The ductility of the iron is reduced and the hardness increased by the presence of this impurity.

The properties of iron can be altered to suit our needs by adding controlled proportions of other elements. Examples of various iron alloys are given in Table 19.4.

Iron(VI) Compounds

Beyond manganese, the Period 4 transition metals do not form compounds in which they have a d^0 electron configuration. In fact, a compound with a

Table 19.4 Important alloys of iron

Name	Approximate composition	Properties and uses
Stainless steel	73% Fe, 18% Cr, 8% Ni	Corrosion-resistant (tableware, cookware)
Tungsten steel	94% Fe, 5% W	Hard (high-speed cutting tools)
Manganese steel	86% Fe, 13% Mn	Tough (rock drill bits)
Permalloy	78% Ni, 21% Fe	Magnetic (electromagnets)

metal in an oxidation state higher than +3 is very difficult to prepare, and such compounds are only stable in the solid phase. The ferrate ion, $[FeO_4]^{2-}$, in which iron has an oxidation number of +6, is one of these rare compounds. This purple, tetrahedral ion can be stabilized by forming an insoluble ionic compound such as the red-purple solid barium ferrate, $BaFeO_4$. This mixed metal compound is a very powerful oxidizing agent.

Iron(III) Compounds

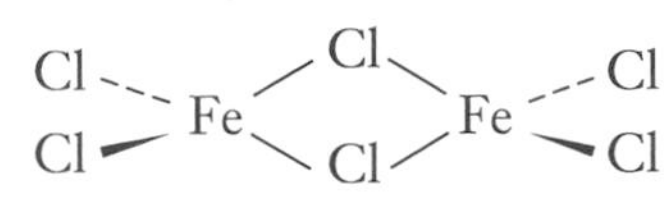

Figure 19.12 The diiron hexachloride molecule, Fe_2Cl_6.

The iron(III) ion itself is small and sufficiently polarizing that its anhydrous compounds exhibit covalent character. For example, iron(III) chloride is a red-black, covalent solid with a network covalent structure. When heated to the gas phase, it exists as the dimeric species Fe_2Cl_6 shown in Figure 19.12. Iron(III) chloride can be made by heating iron in the presence of dichlorine:

$$2\ Fe(s) + 3\ Cl_2(g) \rightarrow 2\ FeCl_3(s)$$

The bromide is similar to the chloride, but the iodide cannot be isolated because the iodide ion reduces iron(III) to iron(II):

$$2\ Fe^{3+}(aq) + 2\ I^-(aq) \rightarrow 2\ Fe^{2+}(aq) + I_2(aq)$$

Anhydrous iron(III) chloride reacts exothermically with water, producing hydrogen chloride gas:

$$FeCl_3(s) + 3\ H_2O(l) \rightarrow Fe(OH)_3(s) + 3\ HCl(g)$$

This reaction contrasts with that of the golden yellow, ionic, hydrated salt, $FeCl_3{\cdot}6H_2O$, which simply dissolves in water to give the hexahydrate ion in solution.

The hexaaquairon(III) ion, $[Fe(OH_2)_6]^{3+}$, is very pale purple, a color that can be seen in the solid iron(III) nitrate nonahydrate. Like the manganese(II) ion, the iron(III) ion is a high spin d^5 species. Lacking any spin-allowed electron transitions, its color is very pale relative to that of other transition metal ions. The yellow color of the chloride compound is due to the presence of ions such as $[Fe(OH_2)_5Cl]^{2+}$ in which a charge transfer transition can occur:

$$Fe^{3+}—Cl^- \rightarrow Fe^{2+}—Cl^0$$

All the iron(III) salts dissolve in water to give an acidic solution, a characteristic of high charge density, hydrated cations. In such circumstances, the coordinated water molecules become sufficiently polarized that other water molecules can function as bases and abstract protons. The iron(III) ion behaves as follows:

$$[Fe(OH_2)_6]^{3+}(aq) + H_2O(l) \rightleftharpoons H_3O^+(aq) + [Fe(OH_2)_5(OH)]^{2+}(aq)$$

$$[Fe(OH_2)_5(OH)]^{2+}(aq) + H_2O(l) \rightleftharpoons H_3O^+(aq) + [Fe(OH_2)_4(OH)_2]^+(aq)$$

and so on. The equilibria are pH dependent; thus addition of hydronium ion will give the almost colorless hexaaquairon(III) ion. Conversely, addition of hydroxide ion gives an increasingly yellow solution, followed by precipitation of a rust-colored gelatinous (jellylike) precipitate of iron(III) oxide hydroxide, FeO(OH):

$Fe^{3+}(aq) + 3\ OH^{-}(aq) \rightarrow FeO(OH)(s) + H_2O(l)$

The pH and E dependence of the iron species are shown in Figure 19.13. For simplicity, the aqueous cations are simply shown as $Fe^{3+}(aq)$ and $Fe^{2+}(aq)$, respectively, although, as we have just seen, there is a whole range of different hydrated iron(III) ions that depend on the pH. The iron(III) ion is only thermodynamically preferred under oxidizing conditions (very positive E) and low pH. The iron(III) oxide hydroxide, however, predominates over much of the basic range. It is the iron(II) ion that is preferred over most of the E range and under acid conditions, whereas the iron(II) hydroxide, $Fe(OH)_2$, is only stable at high pH and strongly reducing conditions (very negative E).

The actual oxidation potential of iron(II) to iron(III) is very dependent on the ligands. For example, the hexacyanoferrate(II) ion, $[Fe(CN)_6]^{4-}$, is much more easily oxidized than the hexaaquairon(II) ion, $[Fe(OH_2)_6]^{2+}$:

$$[Fe(OH_2)_6]^{2+}(aq) \rightarrow [Fe(OH_2)_6]^{3+}(aq) + e^- \qquad E^\circ = -0.77\ V$$

$$[Fe(CN)_6]^{4-}(aq) \rightarrow [Fe(CN)_6]^{3-}(aq) + e^- \qquad E^\circ = -0.36\ V$$

This might seem surprising, considering that cyanide, as a π-acceptor ligand, stabilizes low oxidation states, not high ones, and that, in fact, the iron-carbon bond is stronger in the iron(II) ion than in the iron(III) ion. But there is a thermodynamic aspect to the cyanide equilibrium: The aqueous $[Fe(CN)_6]^{4-}$ is of such high charge density that there is a strongly organized sphere of water molecules around it. Such an ion has a very negative entropy of hydration; but oxidation decreases the charge density, thereby reducing the organization of the hydration sphere and increasing the entropy. The oxidation, then, is entropy-driven.

Although iron(III) species usually adopt an octahedral stereochemistry, the yellow tetrachloroferrate(III) ion, $[FeCl_4]^-$, is tetrahedral. This ion is easily formed by adding hydrochloric acid to a solution of the hexaaquairon(II) ion:

$$[Fe(OH_2)_6]^{3+}(aq) + 4\ Cl^-(aq) \rightleftharpoons [FeCl_4]^-(aq) + 6\ H_2O(l)$$

A specific test for the presence of the iron(III) ion is the addition of a solution of hexacyanoferrate(II) ion, $[Fe(CN)_6]^{4-}$, to give a dark blue precipitate of iron(III) hexacyanoferrate(II), $Fe_4[Fe(CN)_6]_3$:

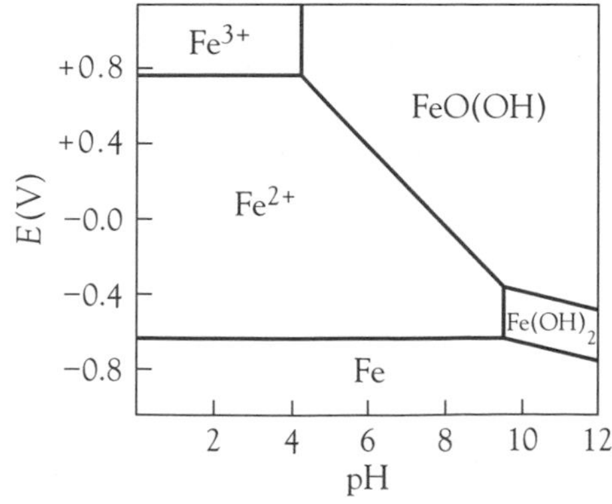

Figure 19.13 Simplified Pourbaix diagram for iron species.

$$4\,Fe^{3+}(aq) + 3\,[Fe(CN)_6]^{4-}(aq) \rightarrow Fe_4[Fe(CN)_6]_3(s)$$

In this compound, commonly called Prussian blue, the crystal lattice contains alternating iron(III) and iron(II) ions. The intense blue of this compound has led to its use in blue inks and paints, including its use as the blue pigment for architectural and engineering blueprints.

The most sensitive test for the iron(III) ion is the addition of potassium thiocyanate solution. The appearance of the intense red color of the pentaaquathiocyanoiron(III) ion, $[Fe(SCN)(OH_2)_5]^{2+}$, indicates the presence of iron(III):

$$[Fe(OH_2)_6]^{3+}(aq) + SCN^-(aq) \rightarrow [Fe(SCN)(OH_2)_5]^{2+}(aq) + H_2O(l)$$

This test for iron(III) ion has to be used cautiously, for a solution of iron(II) ion usually contains enough iron(III) impurity to give some color.

A unique reaction of iron(III) ion is that with an ice-cold solution of thiosulfate. Mixing these two nearly colorless solutions gives the dark violet bis(thiosulfato)ferrate(III) ion:

$$Fe^{3+}(aq) + 2\,[S_2O_3]^{2-}(aq) \rightarrow [Fe(S_2O_3)_2]^-(aq)$$

When the solution is warmed to room temperature, the iron(III) is reduced to iron(II) and the thiosulfate is oxidized to the tetrathionate ion, $[S_4O_6]^{2-}$:

$$[Fe(S_2O_3)_2]^-(aq) + Fe^{3+}(aq) \rightarrow 2\,Fe^{2+}(aq) + [S_4O_6]^{2-}(aq)$$

Similarities Between Iron(III) and Aluminum Ions

Iron(III) and aluminum ions have the same charge and similar sizes (and hence similar charge densities), so their chemistries have some common features. For example, in the vapor phase, both of these ions form covalent chlorides of the form M_2Cl_6. These (anhydrous) chlorides can be used as Friedel-Crafts catalysts in organic chemistry, where they function by the formation of the $[MCl_4]^-$ ion. In addition, the hexaaqua ions of both metals are very strongly acidic, another result of their high charge densities.

Iron forms a series of compounds parallel to the alums. One such pair of parallel compounds are the ammonium salts, $NH_4Al(SO_4)_2{\cdot}12H_2O$ and $NH_4Fe(SO_4)_2{\cdot}12H_2O$. We can account for the similarities between the iron(III) and aluminum ions by noting that the electron configuration of the aqueous iron(III) ion is high spin d^5. Hence there is no crystal field stabilization energy, and the iron(III) ion will behave like a main group metal ion.

There are, however, some significant differences. Iron(III), like other transition metal ions, forms colored compounds (usually as a result of charge transfer transitions), whereas those of aluminum are white. Also, the oxides behave differently: Aluminum oxide is an amphoteric oxide,whereas iron(III) oxide is a basic oxide. This difference is utilized in the separation of pure aluminum oxide from the iron-containing bauxite in the production of aluminum (Chapter 12). The amphoteric aluminum oxide reacts with hydroxide ion to give the soluble tetrahydroxoaluminate ion, $[Al(OH)_4]^-$, whereas the basic iron(III) oxide remains in the solid phase:

$$Al_2O_3(s) + 2\,OH^-(aq) + 3\,H_2O(l) \rightarrow 2\,[Al(OH)_4]^-(aq)$$

Iron(II) Compounds

Anhydrous iron(II) chloride, $FeCl_2$, can be made by passing a stream of dry hydrogen chloride over the heated metal; the hydrogen produced acts as a reducing agent, preventing iron(III) chloride from being formed:

$$Fe(s) + 2\ HCl(g) \rightarrow FeCl_2(s) + H_2(g)$$

The pale green hexaaquairon(II) chloride, $Fe(OH_2)_6Cl_2$, can be prepared by reacting hydrochloric acid with iron metal. Both anhydrous and hydrated forms of iron(II) chloride are ionic.

All the common hydrated iron(II) salts contain the pale green $[Fe(OH_2)_6]^{2+}$ ions, although partial oxidation to yellow or brown iron(III) compounds is quite common. In addition, crystals of the simple salts, such as iron(II) sulfate heptahydrate, $FeSO_4{\cdot}7H_2O$, tend to lose some of the water molecules, a process known as *efflorescence*. In the solid phase, the double salt, ammonium iron(II) sulfate hexahydrate, $(NH_4)_2Fe(SO_4)_2{\cdot}6H_2O$ (or more correctly, ammonium hexaaquairon(II) sulfate), shows the greatest lattice stability. Commonly known as Mohr's salt, it neither effloresces nor oxidizes when exposed to the atmosphere. For this reason, it is used as a primary standard for redox titrations, especially for standardizing potassium permanganate solutions. The tris(1,2-diaminoethane)iron(II) sulfate, $Fe(en)_3SO_4$, is also used as a redox standard.

In the presence of nitrogen monoxide, one molecule of water is displaced from the hexaaquairon(II) ion and replaced by the nitrogen monoxide to give the pentaaquanitrosyliron(II) ion, $[Fe(NO)(OH_2)_5]^{2+}$:

$$NO(aq) + [Fe(OH_2)_6]^{2+}(aq) \rightarrow [Fe(NO)(OH_2)_5]^{2+}(aq) + H_2O(l)$$

This complex is dark brown, and the above reaction is the basis of the "brown ring" test for ionic nitrates (the nitrate having been reduced to nitrogen monoxide by a reducing agent).

Addition of hydroxide ion to iron(II) gives an initial precipitate of green, gelatinous iron(II) hydroxide:

$$Fe^{2+}(aq) + 2\ OH^-(aq) \rightarrow Fe(OH)_2(s)$$

However, as Figure 19.13 shows, except for strongly reducing conditions (or the absence of air), it is the hydrated iron(III) oxide that is thermodynamically stable in basic solution over most of the E range. Thus the green color is replaced by the yellow-brown of the hydrated iron(III) oxide as the oxidation proceeds.

Just as iron(III) ion can be detected with the hexacyanoferrate(II) ion, $[Fe(CN)_6]^{4-}$, so can iron(II) ion be detected with the hexacyanoferrate(III) ion, $[Fe(CN)_6]^{3-}$, to give the same product of Prussian blue (formerly called Turnbull's blue when it was thought to be a different product):

$$3\ Fe^{2+}(aq) + 4\ [Fe(CN)_6]^{3-}(aq) \rightarrow Fe_4[Fe(CN)_6]_3(s) + 6\ CN^-(aq)$$

The Rusting Process

It is a common experiment in junior high science to show that the oxidation of iron (commonly called rusting) requires the presence of both dioxygen and water. By use of an indicator, it can also be shown that around parts of an iron surface, the pH rises. From these observations, the electrochemistry of the rusting process can be determined. This process is really a reflection of the Nernst equation, which states that potential is dependent on concentration—in this case, the concentration of dissolved dioxygen. At the point on the iron surface that has a higher concentration of dioxygen, the element is reduced to hydroxide ion:

$$O_2(aq) + 2\ H_2O(l) + 4\ e^- \rightarrow 4\ OH^-(aq)$$

The bulk iron acts like a wire connected to a battery, conveying electrons from another point on the surface that has a lower oxygen concentration, a point at which the iron metal is oxidized to iron(II) ions:

$$Fe(s) \rightarrow Fe^{2+}(aq) + 2\ e^-$$

These two ions diffuse through the solution, and where they meet, insoluble iron(II) hydroxide is formed:

$$Fe^{2+}(aq) + 2\ OH^-(aq) \rightarrow Fe(OH)_2(s)$$

Like hydrated manganese(III) oxide, the iron(III) oxide hydroxide (rust) is thermodynamically preferred to iron(II) hydroxide in basic solution:

$$Fe(OH)_2(s) + OH^-(aq) \rightarrow FeO(OH)(s) + H_2O(l) + 2\ e^-$$

$$\tfrac{1}{2}\ O_2(aq) + H_2O(l) + 2\ e^- \rightarrow 2\ OH^-(aq)$$

Iron Oxides

There are three common oxides of iron: iron(II) oxide, FeO; iron(III) oxide, Fe_2O_3; and iron(II) iron(III) oxide, Fe_3O_4. Black iron(II) oxide is actually a nonstoichiometric compound, always being slightly deficient in iron(II) ions. The most accurate formulation is $Fe_{0.95}O$. The oxide is basic, dissolving in acids to give the aqueous iron(II) ion:

$$FeO(s) + 2\ H^+(aq) \rightarrow Fe^{2+}(aq) + H_2O(l)$$

Iron(III) oxide, or hematite, is found in large underground deposits. The oldest beds of iron(III) oxide have been dated at about 2 billion years old. Because iron(III) oxide can only form in an oxidizing atmosphere, our current dioxygen-rich atmosphere must have appeared at that time. The appearance of dioxygen, in turn, indicates that photosynthesis (and life itself) became widespread about 2 billion years ago. The oxide can be made in the laboratory by heating the iron(III) oxide hydroxide, which was generated by adding hydroxide ion to iron(III) ion. The product formed by this route, α-Fe_2O_3, consists of a hexagonal close-packed array of oxide ions with the iron(III) ions in two-thirds of the octahedral holes. A different structural form, γ-Fe_2O_3, is produced by oxidizing iron(II) iron(III) oxide. In this form, the iron(III) oxide has a cubic close-packed array of oxide ions with the iron(III) ions distributed randomly among the tetrahedral and octahedral holes.

The third common oxide of iron contains iron in both the +2 and +3 oxidation states, and we have mentioned this compound, $(Fe^{2+})(Fe^{3+})_2O_4$, previously in the context of normal and inverse spinels (Chapter 18). This compound is found naturally as magnetite or lodestone, and a piece of this magnetic compound, suspended by a thread, was used as a primitive compass.

The iron oxides are in great demand as paint pigments. Historically, colors such as yellow ochre, Persian red, and umber (brown) were obtained from iron ore deposits containing certain particle sizes of iron oxides, often with consistent levels of specific impurities. Most yellow, red, and black paints are still made from iron oxides, but they are industrially synthesized to give precise compositions and particle sizes to ensure the production of consistent colors. Of special note is the organic process for the manufacture of aniline, $C_6H_5NH_2$, an important organic chemical. This process used to produce waste iron oxides, but a modification of the synthesis has enabled the composition and particle size of the iron oxide to be adjusted. As a result, the oxide by-product can be marketed as a pigment rather than dumped:

$$4\,C_6H_5NO_2(l) + 9\,Fe(s) + 4\,H_2O(l) \xrightarrow{FeCl_2} 4\,C_6H_5NH_2(l) + 3\,Fe_3O_4(s)$$

If one were to consider which single chemical compound has most revolutionized our modern lives, it would be γ-iron(III) oxide. It is this form of iron(III) oxide that has precisely the magnetic characteristics needed for audio- and videotapes and for the surfaces of the hard and floppy disks in computers. Thus if any compound could be said to be indispensable, this would be it. It is somewhat ironic that iron(III) oxide, which is present on Earth in such vast amounts, should be so valuable in an ultrapure state of precise particle size range, the essential form for magnetic recording purposes.

Ferrites

It is not just iron oxides that are important magnetic materials. There are several mixed metal oxides, one metal being iron, that have valuable properties. These magnetoceramic materials are called ferrites. There are two classes of ferrites, the "soft" ferrites and the "hard" ferrites. These terms do not refer to their physical hardness but to their magnetic properties.

The soft ferrites can be magnetized rapidly and efficiently by an electromagnet, but they lose their magnetism as soon as the current is discontinued. Such properties are essential for the record-erase heads in videotape and audiotape systems and computer drive heads. The compounds have the formula MFe_2O_4, where M is a dipositive metal ion such as Mn^{2+}, Ni^{2+}, Co^{2+}, or Mg^{2+}, and the iron is in the form of Fe^{3+}. These compounds also have spinel structures.

The hard ferrites retain their magnetic properties constantly, that is, they are permanent magnets. These materials are used in DC motors, alternators, and other electrical devices. The general formula of these compounds is $MFe_{12}O_{19}$, where again the iron is Fe^{3+}; the two preferred dipositive metal ions are Ba^{2+} and Sr^{2+}. The hard ferrites adopt a more complex structure than the soft ferrites. The use of both ferrites is not particularly large in terms of mass, but in terms of value, annual world sales are several billion dollars.

Biological Aspects

The biological roles of iron are so numerous that whole books have been written on the subject. Here we focus on three specific types of iron-containing macromolecules: hemoglobin, ferritin, and the ferredoxins.

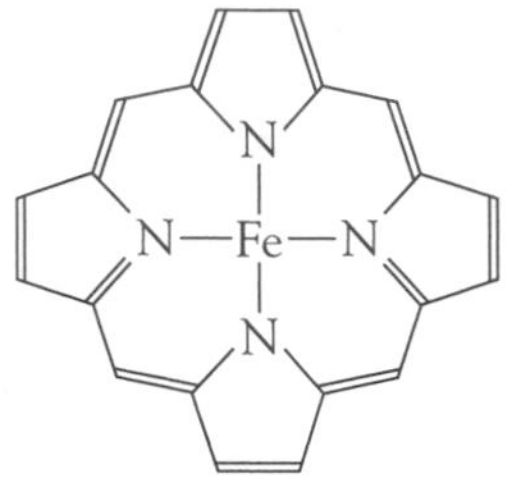

Figure 19.14 The simplified structure of a iron-porphyrin complex.

In hemoglobin, iron has an oxidation state of +2. There are four iron ions in each hemoglobin molecule, each iron ion being surrounded by a porphyrin unit (see Figure 19.14). Each hemoglobin molecule reacts with four molecules of dioxygen to form oxyhemoglobin. The bonding to the dioxygen molecules is weak enough that, upon reaching the site of oxygen utilization such as the muscles, the oxygen can be released. Carbon monoxide is extremely toxic to mammals, because the carbonyl ligand bonds very strongly to the iron of the hemoglobin, thus preventing it from carrying dioxygen molecules.

In oxyhemoglobin, the iron(II) is in the diamagnetic, low spin state. It is just the right radius (75 pm) to fit in the plane of the porphyrin ring. Once the dioxygen is lost, iron in the deoxyhemoglobin molecule shifts below the plane of the porphyrin ring and away from the vacant coordination site, because it has become a larger (radius 92 pm), paramagnetic, high spin iron(II) ion. Throughout the cycle, the iron stays in the iron(II) state, merely alternating between high and low spin forms. It is only when exposed to air that the red iron(II)-containing hemoglobin is oxidized to the brown iron(III) species, an irreversible reaction.

Both plants and animals need to store iron for future use. To accomplish this, members of an amazing protein family, the ferritins, are utilized. They consist of a shell of linked amino acids (peptides) surrounding a core of an iron(III) oxohydroxophosphate. This core is a cluster of iron(III) ions, oxide ions, hydroxide ions, and phosphate ions. It is very large, containing up to 4500 iron ions. With its hydrophilic coating, this large aggregate is water-soluble, concentrating in the spleen, liver, and bone marrow.

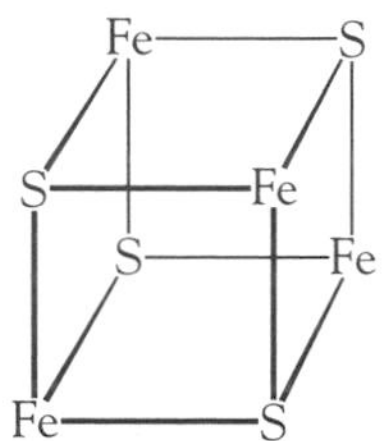

Figure 19.15 The Fe_4S_4 core of a four-iron ferredoxin protein.

Plants and bacteria use a family of iron(III)-sulfur structures as the core of their redox proteins, the ferredoxins. These proteins contain covalently bonded iron and sulfur, and they act as excellent electron transfer agents. Most interesting are the Fe_4S_4 cores, where the iron and sulfur atoms occupy the alternating corners of a cubic arrangement (Figure 19.15).

Cobalt

Cobalt is a bluish white, hard metal; and, like iron, cobalt is a magnetic (ferromagnetic) material. The element is quite unreactive chemically. The most common oxidation numbers of cobalt are +2 and +3, the former being the "normal" state for simple cobalt compounds. For cobalt, the +3 state is more oxidizing than the +3 state of iron.

Cobalt(III) Compounds

All the cobalt(III) complexes are octahedral in shape; and, like chromium(III), the complexes are very kinetically inert, meaning that we can separate different optical isomers, where they are feasible. Typical examples of cobalt(III) complexes are the hexaamminecobalt(III) ion, $[Co(NH_3)_6]^{3+}$, and the hexacyanocobaltate(III) ion, $[Co(CN)_6]^{3-}$.

An unusual complex ion is the hexanitrocobaltate(III) ion, $[Co(NO_2)_6]^{3-}$, which is usually synthesized as the sodium salt, $Na_3[Co(NO_2)_6]$. As would be expected for an alkali metal salt, the compound is water-soluble. However, the potassium salt is quite insoluble (as are the rubidium and cesium salts), the reason relating to relative ion sizes. The potassium ion is much closer in size to the polyatomic anion; hence it has a higher lattice energy and a lower solubility. This is one of the few precipitation reactions that can be used for the potassium ion:

$$3\ K^+(aq) + [Co(NO_2)_6]^{3-}(aq) \rightarrow K_3[Co(NO_2)_6](s)$$

As noted for the iron ions, altering the ligands produces dramatic changes in $E°$ values, which in turn affects the stabilities of the various oxidation states. For example,

$$[Co(OH_2)_6]^{3+}(aq) + e^- \rightarrow [Co(OH_2)_6]^{2+}(aq) \qquad E° = +1.82\ V$$

$$[Co(NH_3)_6]^{3+}(aq) + e^- \rightarrow [Co(NH_3)_6]^{2+}(aq) \qquad E° = +0.10\ V$$

The value for the oxidation of the hexaamminecobalt(II) ion of +0.10 V is less positive than the value +1.23 V for the reduction of oxygen:

$$\tfrac{1}{2}\ O_2(g) + 2\ H^+(aq) + 2\ e^- \rightarrow H_2O(l) \qquad E° = +1.23\ V$$

Hence oxygen should be potentially capable of oxidizing $[Co(NH_3)_6]^{2+}$ to $[Co(NH_3)_6]^{3+}$ in solution; and this is, in fact, found to be the case in the presence of charcoal as a catalyst:

$$2\ [Co(NH_3)_6]^{2+}(aq) + \tfrac{1}{2}\ O_2(g) + H_2O(l) \rightarrow 2\ [Co(NH_3)_6]^{3+}(aq) + 2\ OH^-(aq)$$

Complex ion formation is a general method used to stabilize oxidation states of elements that would otherwise be unstable.

By comparing the electron configurations in an octahedral field, we can see why ligands causing a larger crystal field splitting would enable cobalt(II) to be readily oxidized. For cobalt(II), nearly all the complexes are high spin, whereas cobalt(III), with its higher charge, is almost always low spin. Hence the oxidation results in a much greater crystal field stabilization energy (Figure 19.16). The higher the ligands in the spectrochemical series, the greater the Δ_{oct} value and the greater the CFSE increase obtained by oxidation.

Cobalt(III) parallels chromium(III) by forming a diverse range of complexes, many of which exhibit geometric and optical isomerism. Table 19.5 shows an interesting series of complexes derived from cobalt(III) chloride and ammonia. It is very easy to determine the formulation of these compounds.

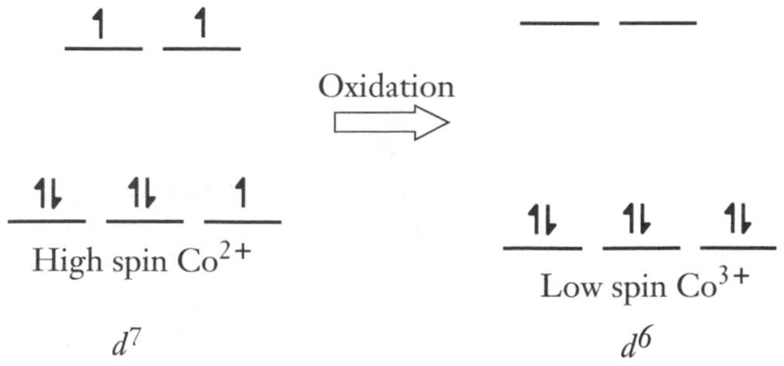

Figure 19.16 Comparison of cobalt(II) and cobalt(III) crystal field stabilization energies.

Table 19.5 Complexes derived from cobalt(III) chloride and ammonia

Formula	Color
$[Co(NH_3)_6]^{3+}\ 3Cl^-$	Orange-yellow
$[Co(NH_3)_5Cl]^{2+}\ 2Cl^-$	Purple
$[Co(NH_3)_4Cl_2]^+\ Cl^-$ (cis)	Violet
$[Co(NH_3)_4Cl_2]^+\ Cl^-$ (trans)	Green

The number of ions can be identified from conductivity measurements, and the free chloride ion can be precipitated with silver ion.

Cobalt(II) Compounds

In solution, cobalt(II) salts are pink, the color being due to the presence of the hexaaquacobalt(II) ion, $[Co(OH_2)_6]^{2+}$. When a solution of a cobalt(II) salt is treated with concentrated hydrochloric acid, the color changes to deep blue. This color is the result of the formation of the tetrahedral tetrachlorocobaltate(II) ion, $[CoCl_4]^{2-}$:

$$[Co(OH_2)_6]^{2+}(aq) + 4\ Cl^-(aq) \rightleftharpoons [CoCl_4]^{2-}(aq) + 6\ H_2O(l)$$

This color change is characteristic for the cobalt(II) ion.

Addition of hydroxide ion to the aqueous cobalt(II) ion results in the formation of cobalt(II) hydroxide, which precipitates first in a blue form and then changes to a pink form on standing:

$$Co^{2+}(aq) + 2\ OH^-(aq) \rightarrow Co(OH)_2(s)$$

The cobalt(II) hydroxide is slowly oxidized by the dioxygen in the air to a cobalt(III) oxide hydroxide, CoO(OH):

$$Co(OH)_2(s) + OH^-(aq) \rightarrow CoO(OH)(s) + H_2O(l) + e^-$$

Cobalt is an amphoteric metal. When concentrated hydroxide ion is added to cobalt(II) hydroxide, a deep blue solution of the tetrahydroxocobaltate(II) ion, $[Co(OH)_4]^{2-}$, is formed:

$$Co(OH)_2(s) + 2\ OH^-(aq) \rightarrow [Co(OH)_4]^{2-}(aq)$$

Biological Aspects

Cobalt is yet another essential element. Of particular importance, vitamin B_{12} has cobalt(III) at the core of the molecule, surrounded by a ring structure similar to the porphyrin ring. Injections of this vitamin are used in the treatment of pernicious anemia. Certain anaerobic bacteria use a related molecule, methylcobalamin, in a cycle to produce methane. Unfortunately, this same biochemical cycle converts elemental mercury and insoluble inorganic mercury compounds in mercury-polluted waters to soluble, highly toxic methyl mercury(II), $[HgCH_3]^+$, and dimethyl mercury(II), $Hg(CH_3)_2$.

Cobalt is also involved in some enzyme functioning. A deficiency disease among sheep in Florida, Australia, Britain, and New Zealand was traced to a lack of cobalt in the soil. To remedy this, sheep are given a pellet of cobalt metal in their food, some of which remains in their digestive system for life.

Nickel

Nickel is a silvery white metal that is quite unreactive. In fact, nickel plating is sometimes used to protect iron. The only common oxidation number for nickel is +2. Most nickel complexes have an octahedral geometry, but some tetrahedral and square-planar complexes are known. Square-planar geometry is otherwise exceedingly rare for the compounds of Period 4 transition metal.

Extraction of Nickel

Although the extraction of nickel from its compounds is complex, the last two steps—the conversion of nickel(II) oxide to pure nickel metal—are of particular interest. As we saw from the Ellingham diagram for oxides (see Figure 9.8), the line of the oxidation of carbon has a steep negative slope, crossing most of the metal oxides. The line for nickel(II) oxide is crossed at an obtainable temperature. Thus the reduction can be performed by the inexpensive thermal smelting route rather than having to use a more expensive electrochemical process:

$$NiO(s) + CO(g) \xrightarrow{\Delta} Ni(s) + CO_2(g)$$

However, this nickel is impure. To extract the nickel from the other metals, such as cobalt and iron, there are two alternatives. One is an electrolytic process whereby impure nickel is cast into anodes and, by using solutions of nickel sulfate and chloride as electrolytes, 99.9 percent pure nickel is deposited at the cathode. The other process utilizes a reversible chemical reaction known as the Mond process. In this reaction, nickel metal reacts at about 60°C with carbon monoxide gas to form a colorless gas, tetracarbonylnickel(0), $Ni(CO)_4$ (b.p. 43°C):

$$Ni(s) + 4\ CO(g) \rightarrow Ni(CO)_4(g)$$

The highly toxic compound can be piped off, for it is the only metal to form a volatile carbonyl compound so easily. Heating the gas to 200°C shifts the equilibrium in the opposite direction, depositing 99.95 percent pure nickel metal:

$$Ni(CO)_4(g) \xrightarrow{\Delta} Ni(s) + 4\ CO(g)$$

The carbon monoxide can then be reused.

Nickel(II) Compounds

The hexaaquanickel(II) ion is a pale green color. Addition of ammonia gives the blue hexaamminenickel(II) ion:

$$[Ni(OH_2)_6]^{2+}(aq) + 6\ NH_3(aq) \rightarrow [Ni(NH_3)_6]^{2+}(aq) + 6\ H_2O(l)$$

Nickel(II) hydroxide can be precipitated as a green gelatinous solid by adding sodium hydroxide solution to a solution of a nickel(II) salt:

$$Ni^{2+}(aq) + 2\ OH^-(aq) \rightarrow Ni(OH)_2(s)$$

Like cobalt(II), the only common complexes having tetrahedral geometry are the halides, such as the blue tetrachloronickelate(II) ion. This complex is formed by adding concentrated hydrochloric acid to aqueous nickel(II) ion:

$$[Ni(OH_2)_6]^{2+}(aq) + 4\ Cl^-(aq) \rightleftharpoons [NiCl_4]^{2-}(aq) + 6\ H_2O(l)$$

In addition to octahedral and tetrahedral complexes, nickel forms a few square-planar complexes. One such complex is the yellow tetracyanonickelate(II) ion, $[Ni(CN)_4]^{2-}$, and another is bis(dimethylglyoximato)nickel(II), $[Ni(C_4N_2O_2H_7)_2]$, which precipitates as a red solid when dimethylglyoxime is added to a solution of a nickel salt made just alkaline by the addition of ammonia. The formation of this characteristic red complex is used as a test for nickel(II) ions. Abbreviating dimethylglyoxime ($C_4N_2O_2H_8$), a bidentate ligand, as DMGH, we write the equation for its formation as

$$Ni^{2+}(aq) + 2\ DMGH(aq) + 2\ OH^-(aq) \rightarrow Ni(DMG)_2(s) + 2\ H_2O(l)$$

Octahedral versus Tetrahedral Stereochemistry

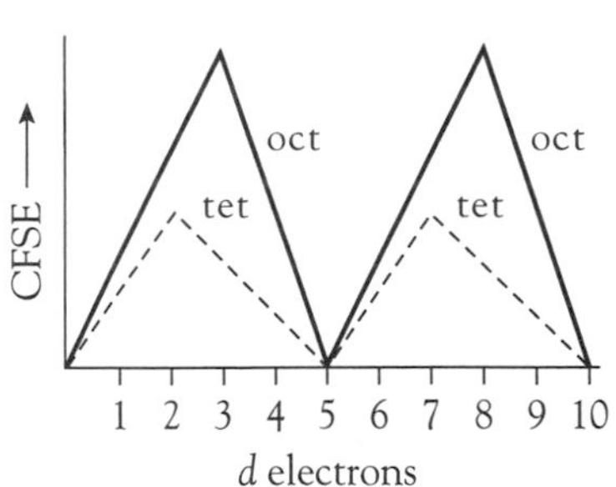

Figure 19.17 Variation of CFSE for octahedral and tetrahedral environments for M^{2+} ions with high spin electron configuration.

Cobalt(II) readily forms tetrahedral complexes, but as we have just seen, nickel(II) complexes are usually octahedral; however, a few are square planar, and a very few are tetrahedral. Several factors determine the choice of stereochemistry, but one in particular is the crystal field stabilization energy (CFSE). We can calculate this for each configuration and plot the values as a function of the number of *d* orbital electrons. Because Δ_{tet} is four-ninths that of Δ_{oct}, the CFSE for a tetrahedral environment will always be less than that of the isoelectronic octahedral environment. Figure 19.17 shows the variation of CFSE with high spin electron configurations. The differences between octahedral and tetrahedral CFSE energies are greatest for the high spin d^3 and d^8 cases, and these are the electron configurations for which we find the fewest tetrahedral complexes.

Nevertheless, there are a few tetrahedral nickel(II) complexes. These are formed with large, negatively charged, weak field ligands (that is, ligands low in the spectrochemical series). With large size and negative charge, these complexes will have considerable electrostatic repulsion between the neighboring ligands, so four ligands will be preferred over six. Thus the tetrahalonickelate(II) ions, $[NiX_4]^{2-}$, where X is chloride, bromide, or iodide, are the most common examples. Even then, to crystallize the tetrahedral ion, it is necessary to use a large cation; otherwise, the nickel ion will acquire other ligands (such as water molecules) to attain an octahedral environment.

Biological Aspects

The biochemistry of nickel is the most poorly understood of all those of the Period 4 transition elements. It is known to be present in some enzyme systems in the form of porphyrin-type complexes. Also, certain tropical trees concentrate nickel to such an extent that it makes up about 15 percent of their dry mass.

The Platinum Metals

The concept of the periodic table is based on the similarity of properties among (vertical) groups of elements. This intragroup similarity is generally true even for the main group metals and nonmetals. For the transition metals, we find that the Period 5 and Period 6 members are much more closely related to each other than to the Period 4 member of a particular group. However, Groups 8, 9, and 10 show a different pattern. The members of Period 4—iron, cobalt, and nickel—have considerable (horizontal) similarity: They are all ferromagnetic; they have almost identical melting points (between 1535°C and 1455°C); and +2 is their most stable oxidation number.

The Periods 5 and 6 members of Groups 8, 9, and 10—ruthenium, rhodium, palladium, osmium, iridium, platinum—form an interesting (horizontal and vertical) block among themselves. They are often called the platinum group metals. These elements are all extremely rare, unreactive, silvery, lustrous metals that are found together in nature, the annual total production being about 300 tonnes. Some of that quantity is used to produce jewelry and bullion coinage, but most is used in the chemical industry as catalysts for various reactions. These catalytic processes include nitric acid production, organic hydrogenation reactions, petroleum re-forming, and the automobile exhaust catalytic converter. In some cases, the catalysis results from the unique range of values and energies of the oxidation states of these metals.

The densities of these metals show a strong horizontal relationships: The Period 5 metals have densities of about 12 $g{\cdot}cm^{-3}$, whereas those of the Period 6 metals are about 21 $g{\cdot}cm^{-3}$. The melting points of the platinum metals are also high, with values ranging from 1500° to 3000°C.

Although all the platinum elements have chemical similarities, such as an enormous range of compounds with π-acceptor ligands (because they are electron-rich metals), there are significant differences among the three vertical pairs. For example, only ruthenium and osmium form stable compounds with very high oxidation states; in fact, osmium(VIII) oxide, OsO_4, is a useful oxidizing agent in organic chemistry. Both rhodium and iridium readily form complexes with the +3 oxidation state, which, like the cobalt analogs, are kinetically inert. Many rhodium(I), iridium(I), palladium(II), and platinum(II) complexes are known, and these all have the square-planar stereochemistry that we discussed for the d^8 nickel(II) complexes.

Group 11: Copper, Silver, and Gold

Copper, silver, and gold are sometimes called the coinage metals, because historically they were the three metals used for this purpose. The reasons for this were fourfold: They are readily obtainable in the metallic state; they are malleable, so disks of the metal can be stamped with a design; they are quite unreactive chemically; and, in the case of silver and gold, the rarity of the metals meant that the coins had the intrinsic value of the metal itself (these days, our coins are merely tokens with little actual value). Copper and gold are the two yellow metals, although a thin coating of copper(I) oxide, Cu_2O, often makes copper look reddish.

All three metals exhibit an oxidation number of +1. This group and the alkali metals are the only metals that normally do so. For copper, the oxidation number of +2 is more common than that of +1, whereas, for gold, the +3

Table 19.6 Comparison of the alkali metals and the coinage metals

Property	Alkali metals	Coinage metals
Common oxidation numbers	Always +1	Silver always +1; copper and gold rarely +1
Chemical reactivity	Very high; increasing down the group	Very low; decreasing down the group
Density	Very low; increasing down the group (0.5 to 1.9 $g{\cdot}cm^{-3}$)	High; increasing down the group (9 to 19 $g{\cdot}cm^{-3}$)
Melting points	Very low; decreasing down the group (181° to 29°C)	High; all about 1000°C

state is thermodynamically preferred. The metals are not easily oxidized, as can be seen from the following (positive) reduction potentials:

$$Cu^{2+}(aq) + 2\ e^- \rightarrow Cu(s) \qquad E° = +0.34\ V$$

$$Ag^{+}(aq) + e^- \rightarrow Ag(s) \qquad E° = +0.80\ V$$

$$Au^{3+}(aq) + 3\ e^- \rightarrow Au(s) \qquad E° = +1.68\ V$$

The copper group and alkali metal group were once considered to be related, particularly because they are the only common metals to exhibit an oxidation number of +1. However, we can see from Table 19.6 that oxidation number is about all that the two groups have in common.

Extraction of Copper

Although copper does not occur abundantly in nature, many copper-containing ores are known. The most common ore is copper(I) iron(III) sulfide, $CuFeS_2$, a metallic-looking solid that has the two mineralogical names of chalcopyrites and copper pyrites. A rarer mineral, $CuAl_6(PO_4)_4(OH)_8{\cdot}4H_2O$, is the valued blue gemstone turquoise.

The extraction of copper from the sulfide can be accomplished by using either a thermal process (pyrometallurgy) or an aqueous process (hydrometallurgy). For the pyrometallurgical process, the concentrated ore is heated (a process called roasting) in a limited supply of air. This reaction decomposes the mixed sulfide, to give iron(III) oxide and copper(I) sulfide:

$$4\ CuFeS_2(s) + 9\ O_2(g) \rightarrow 2\ Cu_2S(l) + 6\ SO_2(g) + 2\ Fe_2O_3(s)$$

Sand is added to the molten mixture, converting the iron(III) oxide into a slag of iron(III) silicate:

$$Fe_2O_3(s) + 3\ SiO_2(s) \rightarrow Fe_2(SiO_3)_3(l)$$

This liquid floats on the surface and can be poured off. Air is again added, causing copper(I) sulfide to be oxidized to copper(I) oxide:

$$2\ Cu_2S(l) + 3\ O_2(g) \rightarrow 2\ Cu_2O(s) + 2\ SO_2(g)$$

The air supply is discontinued after about two-thirds of the copper(I) sulfide has been oxidized. The mixture of copper(I) oxide and copper(I) sulfide then undergoes an unusual redox reaction to give impure copper metal:

$$Cu_2S(l) + 2\ Cu_2O(s) \rightarrow 6\ Cu(l) + SO_2(g)$$

The pyrometallurgical process has a number of advantages: Its chemistry and technology are well known; there are many existing copper smelters; and it is a fast process. The process also has disadvantages: The ore must be fairly concentrated; there is a large energy requirement for the smelting process; and there are large emissions of sulfur dioxide.

Most metals are extracted by pyrometallurgical processes, using high temperatures and a reducing agent such as carbon monoxide. However, as mentioned, pyrometallurgy requires high energy input, and the wastes are often major air and land pollutants. Hydrometallurgy—the extraction of metals by using solution processes—had been known for centuries but did not become widely used until the twentieth century and then only for specific metals, such as silver and gold. This method has many advantages: Its byproducts are usually less of an environmental problem than the flue gases and slag of a smelter; plants can be built on a small scale and then expanded, whereas a smelter needs to be built on a large scale to be economical; high temperatures are not required, so less energy is consumed than by smelting; and hydrometallurgy can process lower grade ores (less metal content) than can pyrometallurgy.

In general, hydrometallurgical processes consist of three general steps: leaching, concentration, and recovery. The leaching is often accomplished by crushing and heaping the ore, then spraying it with some reagent, such as dilute acid (for copper extraction) or cyanide ion (for silver and gold extraction). Sometimes, instead of chemicals, solutions of the bacterium *Thiobacillus ferrooxidans* are used (this process is actually biohydrometallurgical). This bacterium oxidizes the sulfide in insoluble metal sulfides to a soluble sulfate. The dilute metal ion solution is then removed and concentrated by a variety of means. Finally, the metal itself is produced by chemical precipitation by using a single replacement reaction or by an electrochemical process.

In the specific hydrometallurgical process for copper, copper pyrites are air oxidized in acid suspension to give a solution of copper(II) sulfate:

$$2\ CuFeS_2(s) + H_2SO_4(aq) + 4\ O_2(g) \rightarrow 2\ CuSO_4(aq) + 3\ S(s) + Fe_2O_3(s) + H_2O(l)$$

Thus, in this method, sulfur is released in the forms of sulfate ion solution and solid elemental sulfur, rather than as the sulfur dioxide produced in the pyrometallurgical method. The copper metal is then obtained by electrolysis, and the oxygen gas formed can be utilized in the first step of the process:

$$2\ H_2O(l) \rightarrow O_2(g) + 4\ H^+(aq) + 4\ e^-$$

$$Cu^{2+}(aq) + 2\ e^- \rightarrow Cu(s)$$

Table 19.7 Important alloys of copper

Alloy	Approximate composition	Properties
Brass	77% Cu, 23% Zn	Harder than copper
Bronze	80% Cu, 10% Sn, 10% Zn	Harder than brass
Nickel coins	75% Ni, 25% Cu	Corrosion resistant
Sterling silver	92.5% Ag, 7.5% Cu	More durable than pure silver

Copper is refined electrolytically to give a product about 99.95 percent pure. This impure copper (formerly the cathode) is now made the anode of an electrolytic cell that contains pure strips of copper as the cathode and an electrolyte of copper(II) sulfate solution. During electrolysis, copper is transferred from the anode to the cathode; an anode sludge containing silver and gold is produced during this process, thus helping to make the process economically feasible:

$$Cu(s) \rightarrow Cu^{2+}(aq) + 2\ e^-$$

$$Cu^{2+}(aq) + 2\ e^- \rightarrow Cu(s)$$

Because there is no net electrochemical reaction in this purification step, the voltage required is minimal (about 0.2 V), and the power consumption is very small. Of course, the environmentally preferred route for our copper needs is the recycling of previously used copper.

Pure copper has the highest thermal conductivity of all metals. For this reason, copper is used in premium cookware so that the heat is distributed rapidly throughout the walls of the container. An alternative approach is to apply a copper coating to the base of cookware made from other materials. Copper is second only to silver as an electrical conductor; hence electrical wiring represents a major use of this metal. Copper is comparatively expensive for a common metal. To make penny coins of copper would now cost more than 1¢, so more recent coins have an outer layer of copper over a core of the lower priced zinc.

Copper is a soft metal and it is often used as alloys—brass (for plumbing fixtures) and bronze (for statues). Also, it is often a minor component in nickel and silver alloys. Some of the common alloy compositions are shown in Table 19.7.

Copper(II) Compounds

Although copper forms compounds in both the +1 and +2 oxidation states, it is the +2 state that dominates the aqueous chemistry of copper. In aqueous solution, almost all copper(II) salts are blue, the color being due to the presence of the hexaaquacopper(II) ion, $[Cu(OH_2)_6]^{2+}$. The common exception is copper(II) chloride. A concentrated aqueous solution of this compound is green, the color caused by the presence of complex ions such as the nearly planar tetrachlorocuprate(II) ion, $[CuCl_4]^{2-}$. When diluted, the color of the solution changes to blue. These color transformations are due to the successive replacement of chloride ions in the complexes by water molecules, the final color being that of the hexaaquacopper(II) ion. The overall process can be summarized as

$$[CuCl_4]^{2-}(aq) + 6\ H_2O(l) \rightleftharpoons [Cu(OH_2)_6]^{2+}(aq) + 4\ Cl^-(aq)$$

If a solution of ammonia is added to a copper(II) ion solution, the blue color of the hexaaquacopper(II) ion is replaced by the deep blue of the square-planar tetraamminecopper(II) ion, $[Cu(NH_3)_4]^{2+}$:

$$[Cu(OH_2)_6]^{2+}(aq) + 4\ NH_3(aq) \rightarrow [Cu(NH_3)_4]^{2+}(aq) + 6\ H_2O(l)$$

Addition of hydroxide ion to a copper(II) ion solution causes the precipitation of the copper(II) hydroxide, a blue-green gelatinous solid:

$$Cu^{2+}(aq) + 2\ OH^-(aq) \rightarrow Cu(OH)_2(s)$$

However, warming the solution causes the hydroxide to decompose to the black copper(II) oxide and water:

$$Cu(OH)_2(s) \rightarrow CuO(s) + H_2O(l)$$

Copper(II) hydroxide is insoluble in dilute base, but it dissolves in concentrated hydroxide solution to give the deep blue tetrahydroxocuprate(II) ion, $[Cu(OH)_4]^{2-}$:

$$Cu(OH)_2(s) + 2\ OH^-(aq) \rightarrow [Cu(OH)_4]^{2-}(aq)$$

Copper(II) hydroxide also dissolves in an aqueous solution of ammonia to give the tetraamminecopper(II) ion:

$$Cu(OH)_2(s) + 4\ NH_3(aq) \rightleftharpoons [Cu(NH_3)_4]^{2+}(aq) + 2\ OH^-(aq)$$

For most ligands, the copper(II) oxidation state is the more thermodynamically stable, although reducing ligands, such as iodide, will reduce copper(II) ions to the copper(I) state:

$$2\ Cu^{2+}(aq) + 4\ I^-(aq) \rightarrow 2\ CuI(s) + I_2(aq)$$

Jahn-Teller Effect

Copper(II) often forms complexes that are square planar. When copper(II) compounds with six occupied ligand sites are synthesized, it is usually found that the two axial ligands are more distant from the metal than those in the equatorial plane.

This preference for a distortion from a true octahedron can be explained simply in terms of the *d* orbital splittings. In an octahedral d^9 electron configuration, a slight energy benefit can be obtained by a splitting of the $d_{x^2-y^2}$ and the d_{z^2} energies, one increasing in energy and the other decreasing in energy by the same amount. The electron pair will occupy the lower orbital, and the single electron will occupy the higher energy orbital. Thus two

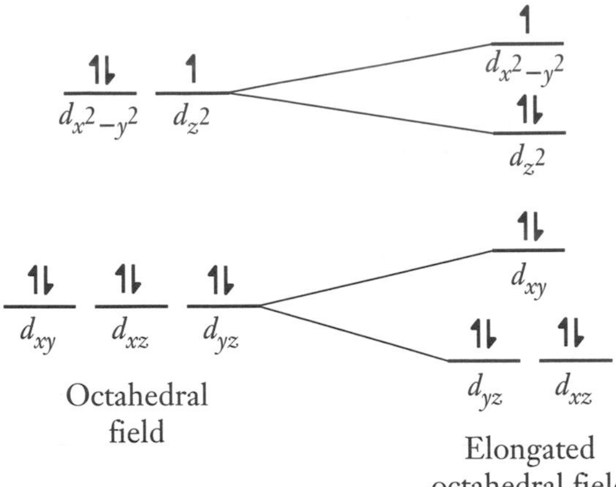

Figure 19.18 The splitting of the *d* orbital energies as a result of the Jahn-Teller effect.

electrons will have lower energies and only one will have higher energy. This splitting can be accomplished by lengthening the axial bond and weakening the electron-electron repulsion along the *z*-axis (Figure 19.18). The phenomenon of octahedral distortion is known as the *Jahn-Teller effect.* This effect can occur with other electron configurations, but it has been studied most in copper(II) compounds. A continuation of the distortion leads to the square planar situation that is found for some d^8 ions (see Figure 18.19).

Copper(I) Compounds

Even though copper is a relatively unreactive metal, it is attacked by concentrated acids. In particular, copper reacts with boiling hydrochloric acid to give a colorless solution and hydrogen gas. This reaction is particularly surprising because hydrochloric acid is not a strong oxidizing acid like nitric acid. The copper(I) ion formed in the oxidation is rapidly complexed by the chloride ion to produce the colorless dichlorocuprate(I) ion, $[CuCl_2]^-$. It is this second equilibrium step that lies far to the right and "drives" the first step:

$$2\,Cu(s) + 2\,H^+(aq) \rightleftharpoons 2\,Cu^+(aq) + H_2(g)$$

$$Cu^+(aq) + 2\,Cl^-(aq) \rightleftharpoons [CuCl_2]^-(aq)$$

When the solution is poured into air-free distilled water, copper(I) chloride precipitates as a white solid:

$$[CuCl_2]^-(aq) \rightarrow CuCl(s) + Cl^-(aq)$$

It must be rapidly washed, dried, and sealed in the absence of air, because a combination of air and moisture oxidizes it to copper(II) compounds.

In organic chemistry, the dichlorocuprate ion is used for converting benzene diazonium chloride into chlorobenzene (the Sandmeyer reaction):

$$[C_6H_5N_2]^+Cl^-(aq) \xrightarrow{[CuCl_2]^-} C_6H_5Cl(l) + N_2(g)$$

Generally, copper(I) compounds are colorless or white, because the ion has a d^{10} electron configuration. That is, with a filled set of *d* orbitals, there can be no *d* electron transitions to cause visible light absorption.

In aqueous solution, the hydrated copper(I) ion is unstable and disproportionates into the copper(II) ion and copper, as the Frost diagram in Figure 19.2 predicts:

$$2\ Cu^{+}(aq) \rightleftharpoons Cu^{2+}(aq) + Cu(s)$$

Silver

Silver is found mostly as the free element and as silver(I) sulfide, Ag_2S. Significant amounts of silver are also obtained during the extraction of lead from its ores and the electrolytic refining of copper. One method of extraction of the metal involves the treatment of pulverized silver(I) sulfide with an aerated solution of sodium cyanide, a process that extracts the silver as the dicyanoargentate(I) ion, $[Ag(CN)_2]^-$:

$$2\ Ag_2S(s) + 8\ CN^-(aq) + O_2(g) + 2\ H_2O(l) \rightarrow 4\ [Ag(CN)_2]^-(aq) + 2\ S(s) + 4\ OH^-(aq)$$

Addition of metallic zinc causes a single replacement reaction in which the very stable tetracyanozincate ion, $[Zn(CN)_4]^{2-}$, is formed:

$$2\ [Ag(CN)_2]^-(aq) + Zn(s) \rightarrow 2\ Ag(s) + [Zn(CN)_4]^{2-}(aq)$$

The pure metal is obtained by electrolysis, using an electrolyte of acidified silver nitrate solution with the impure silver as the anode and a pure strip of silver as the cathode:

$$Ag(s) \rightarrow Ag^{+}(aq) + e^- \qquad \text{(anode)}$$

$$Ag^{+}(aq) + e^- \rightarrow Ag(s) \qquad \text{(cathode)}$$

Silver Compounds

In almost all the simple compounds of silver, the metal has a +1 oxidation number; and the Ag^+ ion is the only water-stable ion of the element. Hence it is common to substitute "silver" for "silver(I)."

The most important compound of silver is white silver nitrate. The only highly water-soluble salt of silver, it gives the colorless, hydrated silver ion when dissolved in water. Silver nitrate is used industrially for the preparation of other silver compounds, particularly the silver halides that are used in photography. In the laboratory, a standard solution of silver nitrate is used to test for the presence of chloride, bromide, and iodide ion. In qualitative analysis, the halide can be identified by color:

$$Ag^{+}(aq) + Cl^-(aq) \rightarrow AgCl(s) \qquad \text{(white)}$$

$$Ag^{+}(aq) + Br^-(aq) \rightarrow AgBr(s) \qquad \text{(cream)}$$

$$Ag^{+}(aq) + I^-(aq) \rightarrow AgI(s) \qquad \text{(yellow)}$$

Because the intensity of the color depends on particle size, it can be difficult to differentiate chloride and bromide or bromide and iodide. Hence there is a secondary confirmatory test. This test involves addition of dilute ammonia solution. Silver chloride reacts with dilute ammonia solution to give the diamminesilver(I) ion:

$$AgCl(s) + 2\ NH_3(aq) \rightarrow [Ag(NH_3)_2]^+(aq) + Cl^-(aq)$$

Silver bromide is only slightly soluble, and silver iodide is insoluble in dilute ammonia. However, silver bromide will react with concentrated ammonia:

$$AgBr(s) + 2\ NH_3(aq) \rightarrow [Ag(NH_3)_2]^+(aq) + Br^-(aq)$$

To understand this difference in behavior, we must compare the equation for the precipitation reaction (where X is any of the halides) with the equation for the complexation reaction:

$$Ag^+(aq) + X^-(aq) \rightleftharpoons AgX(s)$$

$$Ag^+(aq) + 2\ NH_3(aq) \rightleftharpoons [Ag(NH_3)_2]^+(aq)$$

There are two competing equilibria for the silver ion. In qualitative terms, it is the one with the larger equilibrium constant that will predominate. Hence, in the case of very insoluble silver iodide, it is the precipitation equilibrium that will dominate. Conversely, the more soluble silver chloride will result in a silver ion concentration high enough to drive the complexation reaction:

$$Ag^+(aq) + Cl^-(aq) \leftarrow AgCl(s)$$

$$\downarrow$$

$$Ag^+(aq) + 2\ NH_3(aq) \rightarrow [Ag(NH_3)_2]^+(aq)$$

The quantitative estimation of chloride, bromide, and iodide ions can be accomplished gravimetrically, by weighing the silver halide produced, or titrimetrically, by using an indicator such as potassium chromate (the Mohr method) that we discussed earlier in the context of the chromate ion.

The insolubility of the silver chloride, bromide, and iodide was explained in terms of covalent character in Chapter 5. The silver fluoride, AgF, is a white, water-soluble solid and is considered to be ionic in both solid and aqueous solution.

Silver chloride, bromide, and iodide are sensitive to light, and the ready reduction of the silver ion results in a darkening of the solid. (This is why silver compounds and their solutions are stored in dark bottles.)

$$Ag^+(s) + e^- \rightarrow Ag(s)$$

This reaction is the key to the photographic process. In black-and-white photography, it is simply the impact of light on sensitized silver halide microcrystals that initiates the production of the negative image. For color photography, the film is composed of layers, with organic dyes acting as color filters for the silver bromide. Thus only light from a particular spectral region will activate the silver bromide in a specific layer. The photographic process was briefly described in Chapter 15.

Although nearly all simple silver compounds exhibit the +1 oxidation state, there are exceptions. For example, silver metal can be oxidized to black AgO, which is actually silver(I) silver(III) oxide, $Ag^+Ag^{3+}(O^{2-})_2$. This compound reacts with perchloric acid to give the paramagnetic tetraaquasilver(II) ion, $[Ag(OH_2)_4]^{2+}$. The reaction is the reverse of a disproportionation, and the highly oxidizing perchlorate ion stabilizes the +2 oxidation state of the silver:

$$AgO(s) + 2\ H^+(aq) \rightarrow Ag^{2+}(aq) + H_2O(l)$$

Gold

With its very high reduction potential, this element is always found in nature as the free metal. For extraction from ores, the same cyanide process is used as is used for silver metal. Gold does form a variety of complexes but few simple inorganic compounds. Gold(I) oxide, Au_2O, is one of the few stable gold compounds in which the metal has an oxidation number of +1. Like copper, this oxidation state is only stable in solid compounds, because solutions of all gold(I) salts disproportionate to gold metal and gold(III) ions:

$$3\ Au^+(aq) \rightarrow 2\ Au(s) + Au^{3+}(aq)$$

One of the most common compounds of gold is gold(III) chloride, $AuCl_3$; this can be prepared simply by reacting the two elements together:

$$2\ Au(s) + 3\ Cl_2(g) \rightarrow 2\ AuCl_3(s)$$

Dissolving gold(III) chloride in concentrated hydrochloric acid gives the tetrachloroaurate(III) ion, $[AuCl_4]^-$, an ion that is one of the components in "liquid gold," a mixture of gold species in solution that will deposit a film of gold metal when heated.

Biological Aspects

Copper is the third most biologically important transition metal after iron and zinc. About 5 mg are required in the daily human diet. A deficiency of this element renders the body unable to use iron stored in the liver. There are numerous copper proteins throughout the living world, the most intriguing being the hemocyanins. These molecules are common oxygen carriers in the invertebrate world: Crabs, lobsters, octopi, scorpions, and snails have a bright blue blood.

At the same time, an excess of copper is highly poisonous, particularly to fish. This is why copper coins should never be thrown into fish pools for "good luck." Humans usually excrete any excess, but a biochemical (genetic) defect can result in copper accumulation in the liver, kidneys, and brain. This illness, Wilson's disease, can be treated by administering chelating agents, which complex the metal ion and allow it to be excreted harmlessly.

Both silver and gold have specific medical applications. Silver ion is a bactericide, and dilute solutions of silver nitrate are placed in the eyes of newborn babies to prevent infection. Gold compounds, such as the drug auranofin, are used in the treatment of arthritis.

Exercises

19.1. Write balanced equations for
(a) the reaction between titanium(IV) chloride and oxygen gas
(b) sodium dichromate with sulfur at high temperature
(c) the warming of copper(II) hydroxide
(d) the reaction between the dicyanoargentate ion and zinc metal
(e) the reaction between gold and dichlorine

19.2. Write balanced equations for
(a) the vanadyl ion with zinc metal in acidic solution (two equations)

(b) the oxidation of chromium(III) ion to dichromate ion by ferrate ion, which itself is reduced to iron(III) ion in acid solution (write initially as two half-equations)

(c) the warming of copper(II) hydroxide

(d) the reaction between copper(II) ion and iodide ion

(e) the reaction between gold and dichlorine

19.3. Discuss briefly how the stability of the oxidation states of the Period 4 transition metals changes along the row.

19.4. Identify uses for (a) titanium(IV) oxide; (b) chromium(III) oxide; (c) molybdenum(IV) sulfide; (d) silver nitrate.

19.5. What evidence do you have that titanium(IV) chloride is a covalent compound? Suggest why this is to be expected.

19.6. The equation for the first step in the industrial extraction of titanium is

$$TiO_2(s) + 2\ C(s) + 2\ Cl_2(g) \rightarrow TiCl_4(g) + 2\ CO(g)$$

Which element is being oxidized and which is being reduced in this process?

19.7. Write half-equations for the reduction of permanganate ion in (a) acidic solution; (b) basic solution.

19.8. Aluminum is the most abundant metal in Earth's crust. Discuss the reasons why it is iron, not aluminum, that is the most important metal.

19.9. Compare and contrast the chemistry of (a) manganese(VII) and chlorine(VII); (b) iron(III) and aluminum; (c) silver(I) and rubidium.

19.10. Contrast how iron(II) chloride and iron(III) chloride are synthesized.

19.11. In the purification of nickel metal, tetracarbonylnickel(0) is formed from nickel at a lower temperature, while the compound decomposes at a higher temperature. Qualitatively discuss this equilibrium in terms of the thermodynamic factors, enthalpy and entropy.

19.12. Identify the elements that are called (a) the coinage metals; (b) the noble metals.

19.13. Identify each metal from the following tests and write balanced equations for each reaction:

(a) Addition of chloride ion to a pink aqueous solution of this cation gives a deep blue solution.

(b) Concentrated hydrochloric acid reacts with this metal to give a colorless solution; when diluted, a white precipitate is formed.

(c) Addition of acid to this yellow anion results in an orange solution.

19.14. Identify each metal from the following tests and write balanced equations for each reaction:

(a) Addition of excess chloride ion to a pale violet solution of this cation gives a yellow solution.

(b) Addition of ammonia to this pale blue cation gives a deep blue solution.

(c) Addition of thiocyanate solution to this almost colorless cation gives a deep red color.

19.15. Addition of a solution of this halide ion to a silver ion solution gives a whitish precipitate that is insoluble in dilute ammonia solution but soluble in a concentrated ammonia solution. Identify the halide ion and write a balanced chemical equation for each step.

19.16. A colorless anion solution is added to a cold pale yellow cation solution. A violet solution is formed that becomes colorless when warmed to room temperature. Identify the ions and write a balanced chemical equation for each step.

19.17. You wish to prepare a tetrahedral complex of vanadium(II). Suggest the best choice of a ligand (two reasons).

19.18. The highest oxidation state for nickel in a simple compound is found in the hexafluoronickelate(IV) ion, $[NiF_6]^{2-}$.

(a) Why would you expect fluorine to be the ligand?

(b) Would you expect the complex to be high spin or low spin? Give your reasoning.

19.19. Suggest why aluminum oxide is amphoteric and iron(III) oxide is basic.

19.20. There is only one simple anion of cobalt(III) that is high spin. Identify the likely ligand and write the formula of this octahedral ion.

19.21. Suggest why copper(I) chloride is insoluble in water.

19.22. Suggest why silver bromide and iodide are colored, even though both silver and halide ions are colorless.

19.23. Use the 18-electron rule to predict the formula of the complex anion formed between copper(I) ion and cyanide ion.

19.24. Taking the Jahn-Teller effect into account, how many absorptions would you expect from *d* electron transitions for the octahedral copper(II) ion? (In fact, the energies are so close that we usually see one very broad absorption in the visible spectrum.)

19.25. Identify which transition metal(s) is(are) involved in each of the following biochemical molecules: (a) hemocyanin; (b) ferredoxin; (c) nitrogenase; (d) vitamin B_{12}.

CHAPTER

The Group 12 Elements

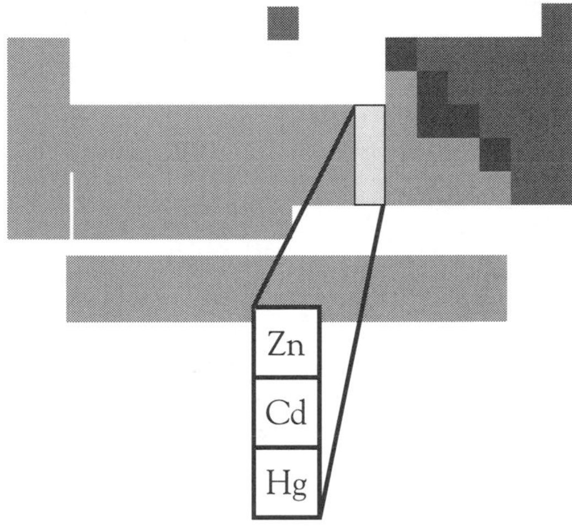

The Group 12 elements, although at the end of the transition metal series, behave like main group metals. In fact, they have many similarities to the Group 2 metals. Zinc is the most commonly encountered member of the group, both chemically and biochemically.

Mercury, the metal that flows like water, has been a source of fascination for millennia. Accounts of the metal are found in ancient Chinese and Indian writings, and Egyptian specimens date to about 1500 B.C. From 200 B.C., a mine in Spain supplied mercury (as mercury(II) sulfide) to the Roman Empire. One of the most feared punishments for Roman convicts was a sentence to the mercury mine, for working in this mercury vapor–rich atmosphere almost guaranteed an agonizing death within months. The same mine continues in production to the present day. Not until 1665 was it made illegal to work in the mine more than eight days a month and more than six hours a day—although this concern for workers' health was more related to production efficiency than to the welfare of the workers themselves.

Medieval alchemists used mercury in their attempts to turn common metals into gold. However, the surge in demand for mercury came when, in about 1570, it was realized that mercury could be used to extract silver from silver-containing ores. The solution of silver in mercury was separated from the solid residue and then heated strongly—the mercury vaporized and dissipated into the atmosphere. This process was obviously dangerous for the workers, and we are still living with the resulting mercury pollution 300 years later. It has been estimated that in the Americas alone about 250 000 tonnes of mercury have been released into the environment by precious metal extraction, and the fate of most of this mercury is unknown. Even today, this primitive and environmentally dangerous method is being used, this time to extract gold from gold deposits in the Amazon basin.

Group Trends

This group of silvery metals superficially appears to belong to the transition metals, but in fact the chemistry of these elements is distinctly different. For example, the melting points of zinc and cadmium are 419° and 321°C, respectively, much lower than the typical values of the transition metals, which are close to 1000°C. The liquid phase of mercury at room temperature can be best explained in terms of relativistic electron effects—namely, that the contraction of the outer orbitals makes the element behave more like a "noble liquid."

The Group 12 elements have filled *d* orbitals in all their compounds, so they are better considered as main group metals. Consistent with this assignment, most of the compounds of the Group 12 metal are white, except when the anion is colored. Zinc and cadmium are very similar in their chemical behavior, having an oxidation number of +2 in all their simple compounds. Mercury exhibits oxidation numbers of +1 and +2, although the Hg^{+} ion itself does not exist. Instead, a Hg_2^{2+} ion is formed. The only real similarity between the Group 12 elements and the transition metals is complex formation, particularly with ligands such as ammonia, cyanide ions, and halide ions. All of the metals, but especially mercury, tend to form covalent rather than ionic compounds.

Zinc and Cadmium

These two soft metals are chemically reactive. For example, zinc reacts with dilute acids to give the zinc ion:

$$Zn(s) + 2\,H^{+}(aq) \rightarrow Zn^{2+}(aq) + H_2(g)$$

The metal also burns when heated gently in chlorine gas:

$$Zn(s) + Cl_2(g) \rightarrow ZnCl_2(s)$$

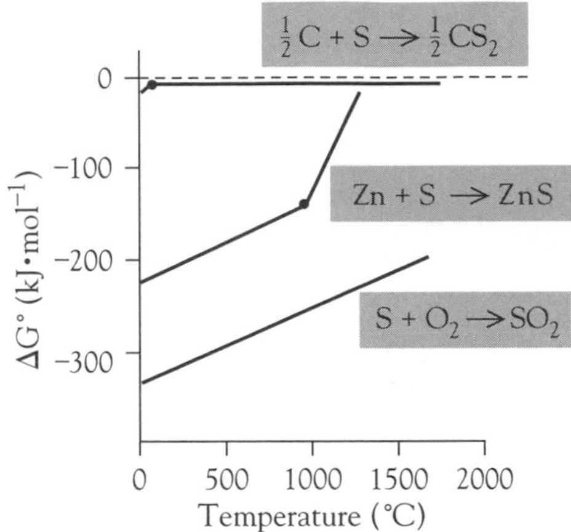

Figure 20.1 Ellingham diagram for selected sulfur compounds.

Extraction of Zinc

The principal source of zinc is zinc sulfide, ZnS, an ore that is called zinc blende and occurs in Australia, Canada, and the United States. Extraction of the metal is not simple, as the Ellingham diagram shows (Figure 20.1).

For oxides, the carbon-oxygen line has a low free energy that becomes even more negative with increasing temperature. For the carbon-sulfur line, the free energy change and the slope are both close to zero. The difference between the lines is, in part, accounted for by the very different enthalpy of formation values, which reflect the fact that the C=S bond is much weaker than the C=O bond. As a result, carbon reduction of sulfides is impractical.

$$C(s) + \tfrac{1}{4}\,S_8(s) \rightarrow CS_2(l) \qquad \Delta H^\circ = +117\ \text{kJ·mol}^{-1}$$

$$C(s) + O_2(g) \rightarrow CO_2(g) \qquad \Delta H^\circ = -394\ \text{kJ·mol}^{-1}$$

However, the diagram also shows that the sulfur-oxygen line is below that of the zinc–zinc sulfide line, a relation indicating that we can use the oxidation of the sulfide ion by dioxygen as a source of free energy to convert zinc sulfide to zinc oxide. Thus the first step in the extraction of zinc is "roasting" the zinc sulfide in air at about 800°C, converting it to the oxide:

$$2\ ZnS(s) + 3\ O_2(g) \xrightarrow{\Delta} 2\ ZnO(s) + 2\ SO_2(g)$$

It is then possible to use coke to reduce the metal oxide to the metal, as is apparent from the oxide Ellingham diagram (Figure 20.2):

$$ZnO(s) + C(g) \xrightarrow{\Delta} Zn(g) + CO(g)$$

Unlike the smelting of other metals, the two lines do not cross until zinc is in the gas phase. In fact, a temperature of about 1400°C is used. At these temperatures, the zinc is readily reoxidized, for example, by any carbon dioxide formed:

$$Zn(g) + CO_2(g) \xrightarrow{\Delta} ZnO(s) + CO(g)$$

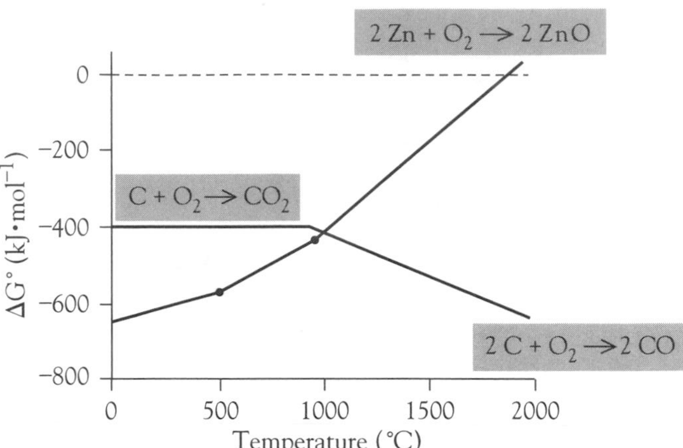

Figure 20.2 Ellingham diagram for the oxides of zinc and carbon.

To prevent this reaction, an excess of carbon is used, so that any carbon dioxide is reduced to carbon monoxide:

$$CO_2(g) + C(s) \xrightarrow{\Delta} 2\,CO(g)$$

In addition, the hot gas produced is rapidly cooled by spraying it with molten lead. The two metals are then easily separated, because the liquid metals are immiscible. The zinc (density 7 g·cm^{-3}) floats on the lead (density 11 g·cm^{-3}), and the lead is recycled.

Zinc is mainly used as a anticorrosion coating for iron. This process is called galvanizing, a term that recognizes the electrochemical nature of the process. Actually, the metal is not quite as reactive as one would expect. This results from the formation of a protective layer in damp air. Initially this is the oxide, but over a period of time the basic carbonate, $Zn_2(OH)_2CO_3$, is formed. The advantage of zinc plating is that the zinc will be oxidized in preference to the iron, even when some of the iron is exposed. This is a result of the more negative reduction potential of the zinc than of the iron, the zinc acting as a sacrificial anode:

$$Zn(s) \rightarrow Zn^{2+}(aq) + 2\,e^- \qquad E^\circ = +0.76\text{ V}$$

$$Fe^{2+}(aq) + 2\,e^- \rightarrow Fe(s) \qquad E^\circ = -0.44\text{ V}$$

Zinc Salts

Most zinc salts are soluble in water, and these solutions contain the colorless hexaaquazinc(II) ion, $[Zn(OH_2)_6]^{2+}$. The solid salts are often hydrated; for example, the nitrate is a hexahydrate and the sulfate a heptahydrate, just like those of magnesium and cobalt(II). The structure of the sulfate heptahydrates is $[M(OH_2)_6]^{2+}$ $[SO_4{\cdot}H_2O]^{2-}$.

The zinc ion has a d^{10} electron configuration, so there is no crystal field stabilization energy. Hence it is often the anion size and charge that determine whether the zinc ion adopts octahedral or tetrahedral stereochemistry. Solutions of zinc salts are acidic as the result of a multistep hydrolysis similar to that of aluminum or iron(III):

The Preservation of Books

Most inexpensive paper, such as newsprint, discolors and rots from reactions that produce acid within the fibers of the paper. For archivists, the rotting and decay of rare books, manuscripts, and old newspapers is a worrisome problem. In recent years, there have been many attempts to find a means of preserving large archives by a low-cost route that will not damage the paper or the ink.

The most promising solution utilizes the first synthesized organometallic compound, a compound containing metal carbon bonds. This compound is diethylzinc, $Zn(C_2H_5)_2$. It was synthesized by Edward Frankland in 1849. In the preservation process used by the Library of Congress, up to 9000 books are placed in a chamber. The air is pumped out and the chamber refilled with pure nitrogen gas under low pressure. It is essential to remove all dioxygen from the chamber because diethylzinc is highly flammable:

$$Zn(C_2H_5)_2(g) + 7\,O_2(g) \rightarrow ZnO(s) + 4\,CO_2(g) + 5\,H_2O(l)$$

Next, diethylzinc vapor is pumped into the chamber, permeating the pages of the books. There it reacts with any hydrogen ions to give zinc ions and ethane gas:

$$Zn(C_2H_5)_2(g) + 2\,H^+(aq) \rightarrow Zn^{2+}(aq) + 2\,C_2H_6(g)$$

The compound also reacts with any moisture in the paper to form zinc oxide:

$$Zn(C_2H_5)_2(g) + H_2O(l) \rightarrow ZnO(s) + 2\,C_2H_6(g)$$

The zinc oxide, being a basic oxide, serves as a reserve of alkalinity in case any more acid is produced by continued rotting of the paper.

The excess diethylzinc and the ethane formed during the reaction are pumped out, and the chamber is flushed with dinitrogen and then air, after which the books can be removed. This procedure takes from three to five days for each batch of books, a slow process but one that results in the survival of many precious documents.

$$[Zn(OH_2)_6]^{2+}(aq) \rightleftharpoons H_3O^+(aq) + [Zn(OH)(OH_2)_5]^+(aq) + H_2O(l)$$

Addition of hydroxide ion causes precipitation of white, gelatinous zinc hydroxide, $Zn(OH)_2$:

$$Zn^{2+}(aq) + 2\,OH^-(aq) \rightarrow Zn(OH)_2(s)$$

With excess hydroxide ion, the soluble tetrahydroxozincate(II) ion, $[Zn(OH)_4]^{2-}$, is formed:

$$Zn(OH)_2(s) + 2\,OH^-(aq) \rightarrow [Zn(OH)_4]^{2-}(aq)$$

The precipitate will also react with ammonia to give a solution of the tetraamminezinc(II) ion, $[Zn(NH_3)_4]^{2+}$:

$$Zn(OH)_2(s) + 4\ NH_3(aq) \rightarrow [Zn(NH_3)_4]^{2+}(aq) + 2\ OH^-(aq)$$

Zinc chloride is one of the most commonly used zinc compounds. It is obtainable as the dihydrate, $Zn(OH_2)_2Cl_2$, and as sticks of the anhydrous zinc chloride. The latter is very deliquescent and extremely soluble in water. It is also soluble in organic solvents such as ethanol and acetone, and this property indicates the covalent nature of its bonding. Zinc chloride is used as a flux in soldering and as a timber preservative. Both uses depend on the ability of the compound to function as a Lewis acid. In soldering, the oxide film on the surfaces to be joined must be removed, otherwise the solder will not bond to these surfaces. Above 275°C, the zinc chloride melts and removes the oxide film by forming covalently bonded complexes with the oxygen ions. The solder can then adhere to the molecular–clean metal surface. When it is applied to timber, zinc chloride forms covalent bonds with the oxygen atoms in the cellulose molecules. As a result, the timber is coated with a layer of zinc chloride, a substance toxic to living organisms.

Zinc Oxide

Zinc oxide can be obtained by burning the metal in air or by the thermal decomposition of the carbonate:

$$2\ Zn(s) + O_2(g) \rightarrow 2\ ZnO(s)$$

$$ZnCO_3(s) \xrightarrow{\Delta} ZnO(s) + CO_2(g)$$

Zinc oxide is a white solid. It is assumed to contain network covalent bonding with a diamond structure. In the crystal, each zinc atom is surrounded tetrahedrally by four oxygen atoms, and each oxygen atom is likewise surrounded by four zinc atoms. Unlike other white metal oxides, zinc oxide develops a yellow color when heated. The reversible change in color that depends on temperature is known as *thermochromism*. In this case, the color change results from the loss of some oxygen from the lattice, thus leaving it with an excess negative charge. The excess negative charge (electrons) can be moved through the lattice by applying a potential difference; thus this oxide is a semiconductor. Zinc oxide returns to its former color when cooled, because the oxygen that was lost during heating returns to the crystal lattice.

Zinc oxide is the most important compound of zinc. It is used as a white pigment, as a filler in rubber, and as a component in various glazes, enamels, and antiseptic ointments. In combination with chromium(III) oxide, it is used as a catalyst in the manufacture of methanol from synthesis gas.

Comparison of Zinc and Magnesium

There are major similarities between the Group 2 and Group 12 elements. This is, in part, a reflection of the similarity in their electron configuration: ns^2 in Group 2 and $(n-1)d^{10}ns^2$ in Group 12. It is the ns^2 electrons that are lost in both cases during ionization. Table 20.1 compares the chemistry of zinc and magnesium.

Table 20.1 Comparison of the properties of zinc and magnesium

Property	Zinc	Magnesium
Ionic radius	74 pm	72 pm
Oxidation state	+2	+2
Ion color	Colorless	Colorless
Hydrated ion	$[Zn(OH_2)_6]^{2+}$	$[Mg(OH_2)_6]^{2+}$
Soluble salts	Chloride, sulfate	Chloride, sulfate
Insoluble salt	Carbonate	Carbonate
Chloride	Covalent, hygroscopic	Covalent, hygroscopic
Hydroxide	Amphoteric	Basic

Zinc also resembles aluminum because both are amphoteric metals:

$$Zn(s) + 2\,H^+(aq) + 4\,H_2O(l) \rightarrow [Zn(OH_2)_4]^{2+}(aq) + H_2(g)$$

$$Zn(s) + 2\,OH^-(aq) + 2\,H_2O(l) \rightarrow [Zn(OH)_4]^{2-}(aq) + H_2(g)$$

Also, like aluminum, the zinc cation is a strong Lewis acid that hydrolyzes in water to give an acidic solution, as discussed earlier.

Cadmium Sulfide

The only commercially important compound of cadmium is cadmium sulfide, CdS. Whereas zinc sulfide has the typical white color of Group 12 compounds, cadmium sulfide is an intense yellow. As a result, the compound is used as a pigment. Cadmium sulfide is prepared in the laboratory and industry by the same route, the addition of sulfide ion to cadmium ion:

$$Cd^{2+}(aq) + S^{2-}(aq) \rightarrow CdS(s)$$

Even though cadmium compounds are highly toxic, cadmium sulfide is so insoluble that it presents little hazard.

Mercury

With the weakest metallic bonding of all, mercury is the only liquid metal at 20°C. Mercury's weak bond also results in a high vapor pressure at room temperature. Because the toxic metal vapor can be absorbed through the lungs, spilled mercury globules from broken mercury thermometers are a major hazard in the traditional chemistry laboratory. Mercury is a very dense liquid ($13.5\ g{\cdot}cm^{-3}$), and it freezes at –39°C and boils at 357°C.

Extraction of Mercury

The only mercury ore is mercury(II) sulfide, HgS, the mineral cinnabar. The deposits in Spain and Italy account for about three-quarters of the world's supply of the metal. Many mercury ores contain considerably less than 1 percent of the sulfide, which accounts for the high price of the metal. Mercury is readily extracted from the sulfide ore by heating it in air. Mercury vapor is evolved and is then condensed to the liquid metal:

$$HgS(s) + O_2(g) \xrightarrow{\Delta} Hg(l) + SO_2(g)$$

Mercury is used in thermometers, barometers, electrical switches, and mercury arc lights. Solutions of other metals in mercury are called amalgams. Sodium amalgam and zinc amalgam are used as laboratory reducing agents; and the most common amalgam of all, dental amalgam (which contains mercury mixed with one or more of the metals silver, tin, and copper), is used for filling cavities in back teeth. It is suitable for this purpose for several reasons: It expands slightly as the amalgam forms, thereby anchoring the filling to the surrounding material. It does not fracture easily under the extreme localized pressures exerted by our grinding teeth. And it has a low coefficient of thermal expansion; thus contact with hot substances will not cause it to expand and crack the surrounding tooth. In terms of total consumption, the major uses of mercury compounds are in agriculture and in horticulture; for example, organomercury compounds are used as fungicides and as timber preservatives.

Mercury(II) Compounds

Virtually all mercury(II) compounds utilize covalent bonding. Mercury(II) nitrate is one of the few compounds believed to contain the Hg^{2+} ion. It is also one of the few water-soluble mercury compounds.

Mercury(II) chloride can be formed by mixing the two elements:

$$Hg(l) + Cl_2(g) \rightarrow HgCl_2(s)$$

This compound dissolves in warm water but the non-electrically conducting behavior of the solution shows that it is present as $HgCl_2$ molecules, not as ions. Mercury(II) chloride solution is readily reduced to white insoluble mercury(I) chloride and then to black mercury metal by the addition of tin(II) chloride solution. This is a convenient test for the mercury(II) ion:

$$2\ HgCl_2(aq) + SnCl_2(aq) \rightarrow SnCl_4(aq) + Hg_2Cl_2(s)$$

$$Hg_2Cl_2(s) + SnCl_2(aq) \rightarrow SnCl_4(aq) + 2\ Hg(l)$$

Red mercury(II) oxide forms when mercury is heated for a long time in air at about 350°C:

$$2\ Hg(l) + O_2(g) \xrightarrow{\Delta} 2\ HgO(s)$$

Mercury(II) oxide is thermally unstable and decomposes into mercury and dioxygen when heated more strongly. This decomposition is an interesting demonstration because mercury(II) oxide, a red powder, "disappears" as silvery globules of metallic mercury form on the cooler parts of the container. However, the reaction is very hazardous because a significant portion of the mercury metal escapes as vapor into the laboratory. The experiment is of historical interest, for it was the method used by Joseph Priestley to obtain the first sample of pure oxygen gas:

$$2\ HgO(s) \xrightarrow{\Delta} 2\ Hg(l) + O_2(g)$$

Mercury(I) Compounds

An interesting feature of mercury is its ability to form the $[Hg–Hg]^{2+}$ ion in which the two mercury atoms are united by a single covalent bond. In fact, there are no known compounds containing the simple mercury(I) ion.

Mercury(I) chloride, Hg_2Cl_2, and mercury(I) nitrate, $Hg_2(NO_3)_2$, are known; but compounds with other common anions, such as sulfide, have never been synthesized. To understand the reason for this, we must look at the disproportionation equilibrium,

$$Hg_2^{2+}(aq) \rightleftharpoons Hg(l) + Hg^{2+}(aq)$$

which has an equilibrium constant, K_{dis}, of about 6×10^{-3} at 25°C. The low value for the equilibrium constant implies that under normal conditions, there is little tendency for the mercury(I) ion to disproportionate into the mercury(II) ion and mercury. However, anions such as sulfide form highly insoluble compounds with mercury(II) ion:

$$Hg^{2+}(aq) + S^{2-}(aq) \rightarrow HgS(s)$$

This precipitation "drives" the disproportionation equilibrium to the right. As a result, the overall reaction of mercury(I) ion with sulfide ion becomes

$$Hg_2^{2+}(aq) + S^{2-}(aq) \rightarrow Hg(l) + HgS(s)$$

Batteries

One common thread among the Group 12 elements is their use in batteries, each metal being important in a different type of cell. It is unfortunate that some of the best battery materials—mercury, cadmium, and lead (discussed in Chapter 13)—are among the most toxic elements, hence presenting major disposal problems.

The Alkaline Battery

This cell has become the most popular household battery. The cell consists of a zinc case (the anode) with a central rod as cathode. This rod consists of a compressed mixture of graphite (a good electrical conductor) and manganese(IV) oxide. The electrolyte is a potassium hydroxide solution. In the cell reaction, the zinc is oxidized to zinc hydroxide while the manganese(IV) oxide is reduced to the manganese(III) oxide hydroxide, MnO(OH):

$$Zn(s) + 2\ OH^-(aq) \rightarrow Zn(OH)_2(s) + 2\ e^-$$

$$2\ MnO_2(s) + 2\ H_2O(l) + 2\ e^- \rightarrow 2\ MnO(OH)(s) + 2\ OH^-(aq)$$

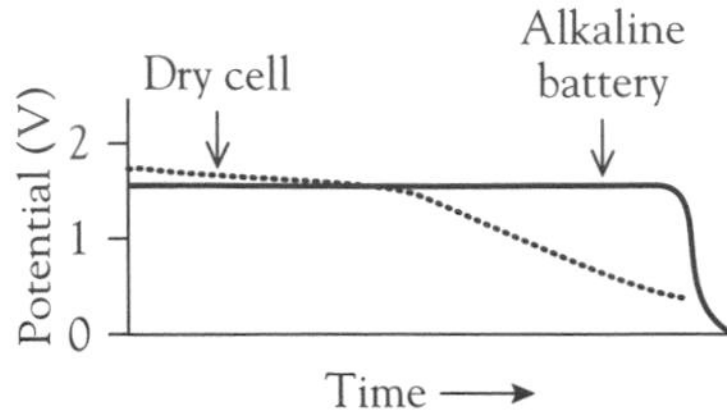

Figure 20.3 Decrease in voltage with time for an alkaline battery and a "dry cell."

In the overall process, one mole of hydroxide ion is consumed at the anode and one mole of hydroxide ion is produced at the cathode. As a result of the constancy of hydroxide ion concentration, the cell potential remains constant, a useful advantage over the old "dry cell" in which the deliverable voltage drops over the lifetime of the battery (Figure 20.3).

The Mercury Cell

For very compact power needs, such as hearing aids, the mercury cell is often used. In this cell, again zinc is the anode, but mercury(II) oxide (mixed with conducting graphite) is the cathode. The zinc is oxidized to zinc hydroxide while the mercury(II) oxide is reduced to mercury metal:

$$Zn(s) + 2\ OH^-(aq) \rightarrow Zn(OH)_2(s) + 2\ e^-$$

$$HgO(s) + H_2O(l) + 2\ e^- \rightarrow Hg(l) + 2\ OH^-(aq)$$

Again the electrolyte concentration remains constant, thus providing a steady cell potential.

The NiCad Battery

Unlike the previous two cells, the NiCad battery can be recharged. In the discharge cycle, cadmium is oxidized to cadmium hydroxide while nickel is reduced from the unusual oxidation state of +3 in nickel(III) oxide hydroxide, NiO(OH), to the more normal +2 state, as nickel(II) hydroxide. Once more the electrolyte is hydroxide ion:

$$Cd(s) + 2\ OH^-(aq) \rightarrow Cd(OH)_2(s) + 2\ e^-$$

$$2\ NiO(OH)(s) + 2\ H_2O(l) + 2\ e^- \rightarrow 2\ Ni(OH)_2(s) + 2\ OH^-(aq)$$

In the charging process, the reverse reactions occur. There are two major reasons for using a basic reaction medium: The nickel(III) state is only stable in base; and the insolubility of the hydroxides means that the metal ions will not migrate far from the metal surface, thus allowing the reverse reactions to happen readily at the same site. This battery, which is commonly used in rechargeable flashlights, portable home appliances, and inexpensive portable computers has one disadvantage—a charging "memory." This unusual phenomenon means that if the NiCad battery is only partially discharged, then recharged, it "remembers" the previous level of discharge and will only discharge to this level again. Thus it is important to discharge the cell completely each time before recharging.

Biological Aspects

This group contains one essential element, zinc, and two very toxic elements.

The Essentiality of Zinc

Among trace essential elements, zinc is second only to iron in importance. By the mid-1980s, over 200 zinc enzymes had been identified, and the figure must be much larger now. Zinc enzymes that perform almost every possible type of enzyme function are known, but the most common function is hydrolysis; the zinc-containing hydrolases are enzymes that catalyze the hydrolysis of P–O–P, P–O–C, and C–O–C bonds.

The question arises as to what makes zinc such a useful ion, considering that it cannot serve a redox function. The answer lies in its Lewis acid properties; that is, the bonds it forms are polarized. Hence they are very reactive. But this is not the only answer, for there are even better Lewis acids. Zinc, unlike many other metals, prefers tetrahedral geometries, a common feature of the metal site in zinc enzymes. Furthermore, zinc has a d^{10} electron configuration, so there is no crystal field stabilization energy associated with exact geometries as there are with the transition metals. Hence, the environment around the zinc can be distorted from the exact tetrahedral to allow for the precise bond angles needed for its function without an energy penalty.

The Toxicity of Cadmium

Cadmium is a toxic element that is present in our foodstuffs and is normally ingested at levels that are close to the maximum safe level. The kidney is the organ most susceptible to cadmium; about 200 ppm cause severe damage. Cigarette smokers absorb significant levels of cadmium from tobacco smoke.

Exposure to cadmium from industrial sources is a major concern. In particular, the nickel-cadmium battery is becoming a major waste-disposal problem. Many battery companies now accept return of defunct NiCad batteries so that the cadmium metal can be safely recycled. Cadmium poisonings in Japan have resulted from cadmium-contaminated water produced by mining operations. The ensuing painful bone degenerative disease has been called *itai-itai*.

The Many Hazards of Mercury

As mentioned earlier, mercury is hazardous because of its relatively high vapor pressure. The mercury vapor is absorbed in the lungs, dissolves in the blood, and is then carried to the brain, where irreversible damage to the central nervous system results. The metal is also slightly water soluble, again a result of its very weak metallic bonding. The escape of mercury metal from leaking chlor-alkali electrolysis plants into nearby rivers has led to major pollution problems in North America.

Inorganic compounds of mercury are usually less of a problem because they are not very soluble. A note of historical interest: At one time, mercury ion solutions were used in the treatment of animal furs for hat manufacture. Workers in the industry were prone to mercury poisoning, and the symptoms of the disease were the model for the Mad Hatter in the book *Alice in Wonderland*.

The organomercury compounds pose the greatest danger. These compounds, such as methylmercury, $HgCH_3$, and dimethylmercury, $Hg(CH_3)_2$, are readily absorbed and are retained by the body much more strongly than the simple mercury compounds. The symptoms of methylmercury poisoning were first established in Japan, where a chemical plant had been pumping mercury wastes into Minamata Bay, a rich fishing area. Inorganic mercury compounds were converted by bacteria in the marine environment to organomercury compounds. These compounds were absorbed by the fatty tissues of fish, and the mercury-laden fish were consumed by the unsuspecting local inhabitants. The unique symptoms of this horrible poisoning have been named Minamata disease. Another major hazard are the organomercury fungicides. In one particularly tragic case, farm families in Iraq were sent mercury fungicide–treated grain, some of which they use for bread making, being unaware of the toxicity. As a result, 450 people died, and over 6500 became ill.

Exercises

20.1. Write balanced chemical equations for the following chemical reactions:
(a) zinc metal with liquid bromine
(b) the effect of heat on solid zinc carbonate

20.2. Write balanced chemical equations for the following chemical reactions:
(a) aqueous zinc ion with ammonia solution
(b) heating mercury(II) sulfide in air

20.3. Suggest a two-step reaction sequence to prepare zinc carbonate from zinc metal.

20.4. Explain briefly the reasons for considering the Group 12 elements separately from the transition metals.

20.5. Compare and contrast the properties of (a) zinc and magnesium; (b) zinc and aluminum.

20.6. Normally, metals in the same group have fairly similar chemical properties. Contrast and compare the chemistry of zinc and mercury by this criterion.

20.7. In the industrial extraction of zinc, molten lead is used to cool the zinc vapor until it liquefies. Molten zinc and molten lead do not mix; thus they can be easily separated. Suggest why the two metals do not mix to any significant extent.

20.8. You are an artist and you wish to make your "cadmium yellow" paint paler. Why is it not a good idea to mix in some "white lead," $Pb_3(CO_3)_2(OH)_2$, to accomplish this?

20.9. Both cadmium ion, Cd^{2+}, and sulfide ion, S^{2-}, are colorless. Suggest an explanation for the color of cadmium sulfide.

20.10. Mercury(I) selenide is unknown. Suggest an explanation.

20.11. Mercury(II) iodide is insoluble in water. However, it will dissolve in a solution of potassium iodide to give a dinegative anionic species. Suggest a formula for this ion.

20.12. Write the two half-equations for the charging process of the NiCad battery.

20.13. Cadmium-coated paper clips were once common. Suggest why they were used and why their use was discontinued.

CHAPTER

21

The Rare Earth and Actinoid Elements

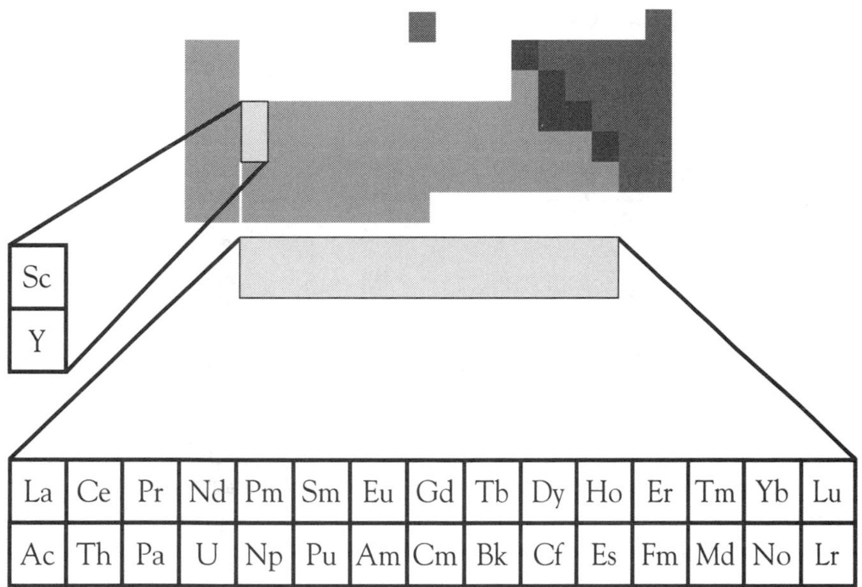

The elements in these groups are rarely discussed in chemistry courses, yet their properties are very interesting. The postactinoid elements, which are synthesized in nuclear reactions like those of many actinoids, are also considered here even though they are members of the Period 7 transition metals.

The discoveries of the lanthanoid and actinoid elements have had a major effect on our knowledge of chemistry. The discovery of a collection of metallic elements with atomic masses in the range of 140 to 175 $g{\cdot}mol^{-1}$ was a major concern of chemists at the beginning of the twentieth century because the original Mendeleev periodic table could not accommodate them. The English chemist Sir William Crookes summed up the situation in 1902:

> The rare earths perplex us in our researches, baffle us in our speculations, and haunt us in our very dreams. They stretch like an unknown sea before us, mocking, mystifying, and murmuring strange revelations and possibilities.

								H									He
Li	Be											B	C	N	O	F	Ne
Na	Mg											Al	Si	P	S	Cl	Ar
K	Ca	Sc	Ti	V	Cr	Mn	Fe	Co	Ni	Cu	Zn	Ga	Ge	As	Se	Br	Kr
Rb	Sr	Y	Zr	Nb	Mo	Tc	Ru	Rh	Pd	Ag	Cd	In	Sn	Sb	Te	I	Xe
Cs	Ba	La	Hf	Ta	W	Re	Os	Ir	Pt	Au	Hg	Tl	Pb	Bi	Po	At	Rn
Fr	Ra	Ac	Th	Pa	U	Np	Pu										
		Ce	Pr	Nd	Pm	Sm	Eu	Gd	Tb	Dy	Ho	Er	Tm	Yb	Lu		

Figure 21.1 The periodic table in 1943.

The final solution was to place this "orphan" set of 14 elements below the body of the periodic table. It was only with the development of the models of electronic structure that it became apparent that these elements corresponded to the filling of the 4*f* orbital set.

By the 1940s, most elements of the modern periodic table up to element 92 were known. However, elements 90 to 92 (thorium, protactinium, and uranium) were considered to be transition metals. During the 1940s, two new chemical elements were synthesized in nuclear reactors: neptunium and plutonium. These elements also were considered to be members of the Period 7 transition series (Figure 21.1). However, neptunium and plutonium had little in common with their vertical neighbors, rhenium and osmium. Instead, they were more chemically similar to their horizontal neighbors, uranium, protactinium, and thorium. Glenn Seaborg first proposed a revised design for the periodic table that would include a whole new series of elements (Figure 21.2). Seaborg showed his revised periodic table to two prominent inorganic chemists of the time. They warned him against publishing it, because they

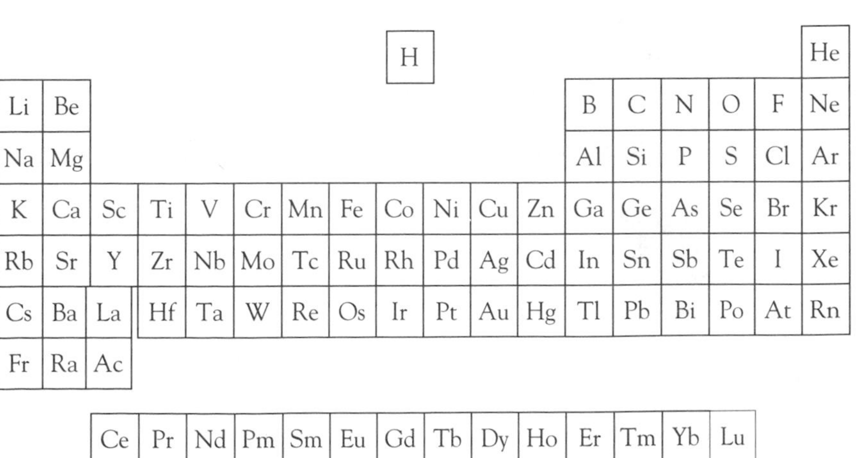

Figure 21.2 Seaborg's 1944 design for the periodic table.

believed tampering with the established periodic table would destroy his professional reputation. As Seaborg later remarked, "I didn't have any scientific reputation, so I published it anyway." Now, of course, we assume that the elements 89 to 102 correspond to the filling of the $5f$ orbitals and do indeed fit in Seaborg's proposed order.

Properties of the Rare Earth Elements

The first problem with the group of elements numbered 57 to 70 is the terminology. The elements lanthanum to ytterbium are known as the lanthanoids, and these elements correspond to the filling of the $4f$ orbital set. However, three other elements are often considered as part of the same set: the Group 3 elements scandium, yttrium, and lutetium. To refer collectively to the lanthanoid and Group 3 elements, the term rare earth metals can be used. However, the term *rare earth* is itself misleading, because many of these elements are quite common. For example, cerium is as abundant as copper.

There is also dissension among chemists as to which group of elements actually constitutes the lanthanoids. Some claim cerium to lutetium while others argue lanthanum to ytterbium. The problem becomes apparent when we look at the electron configurations (Table 21.1). Although most conventional designs of the periodic table show lutetium as a lanthanoid, its electron configuration as an element actually fits the pattern for the third transition series: [Xe]$6s^24f^{14}5d^n$ (where n is 1 in this case). However, because all 15 elements from lanthanum to lutetium share common chemical features, it makes

Table 21.1 Electron configurations of elements 57–71

Element	Atom configuration	3+ ion configuration
Lanthanum	[Xe] $6s^24f^05d^1$	[Xe] $4f^0$
Cerium	[Xe] $6s^24f^15d^1$	[Xe] $4f^1$
Praseodymium	[Xe] $6s^24f^3$	[Xe] $4f^2$
Neodymium	[Xe] $6s^24f^4$	[Xe] $4f^3$
Promethium	[Xe] $6s^24f^5$	[Xe] $4f^4$
Samarium	[Xe] $6s^24f^6$	[Xe] $4f^5$
Europium	[Xe] $6s^24f^7$	[Xe] $4f^6$
Gadolinium	[Xe] $6s^24f^75d^1$	[Xe] $4f^7$
Terbium	[Xe] $6s^24f^9$	[Xe] $4f^8$
Dysprosium	[Xe] $6s^24f^{10}$	[Xe] $4f^9$
Holmium	[Xe] $6s^24f^{11}$	[Xe] $4f^{10}$
Erbium	[Xe] $6s^24f^{12}$	[Xe] $4f^{11}$
Thulium	[Xe] $6s^24f^{13}$	[Xe] $4f^{12}$
Ytterbium	[Xe] $6s^24f^{14}$	[Xe] $4f^{13}$
Lutetium	[Xe] $6s^24f^{14}5d^1$	[Xe] $4f^{14}$

more sense to consider them together. For example, the only common ion for each of these elements has the charge 3+, and the electron configurations for this ion form a simple sequence of 4*f* orbital filling from 0 to 14.

The metals themselves are all soft and moderately dense (about 7 g·cm^{-3}); they have melting points near 1000°C and boiling points near 3000°C. Chemically, the metals are about as reactive as the alkaline earths. For example, they all react with water to give the metal hydroxide and hydrogen gas:

$$2\ M(s) + 6\ H_2O(l) \rightarrow 2\ M(OH)_3(s) + 3\ H_2(g)$$

The similarity among these elements comes in part from the lack of involvement of the 4*f* electrons in bonding. Thus the progressive filling of these orbitals along the row has no effect on the chemistry of the elements. As mentioned earlier, the common oxidation state of all the elements is +3; for example, they all form oxides of the type M_2O_3, where M is the metal ion.

The ionic radii of the 3+ ions decrease smoothly from 117 pm for lanthanum to 100 pm for lutetium. Because the *f* orbitals do not shield the outer 5*s* and 5*p* electrons effectively, the increase in nuclear charge causes the ions to decrease in size. Such large ions have a high coordination number. For example, the hydrated lanthanum ion is a nonahydrate, $[La(OH_2)_9]^{3+}$.

Only cerium has a second common oxidation state, which is +4 and corresponds to the noble gas core configuration of [Xe]. This oxidation state is strongly oxidizing; and cerium(IV) compounds can be used in redox titrations, often in the form of ammonium hexanitratocerate(IV), $(NH_4)_2[Ce(NO_3)_6]$:

$$Ce^{4+}(aq) + e^- \rightarrow Ce^{3+}(aq) \qquad E° = +1.44\ V$$

Because of its usefulness for redox reactions, this is the only rare earth compound that is commonly found in chemistry laboratories.

Tripositive cations of many of the lanthanoids are colored, commonly green, pink, and yellow. These colors are the result of electron transitions among the *f* orbitals. Unlike the spectra of transition metal ions, the spectra of the lanthanoids do not show any significant variations for the different ligands. Furthermore, the absorptions are at very precise wavelengths, unlike the broad absorbance bands of the transition metal ions. The mixed oxides of neodymium and praseodymium absorb much of the yellow range; and this pinkish tan mixture is sometimes used as a filter in sunglasses, because the eye is most sensitive to the yellow part of the spectrum.

As mentioned earlier, scandium and yttrium are often included in discussions of the lanthanoids. Both these elements are soft, reactive metals that also exhibit the +3 oxidation state. Yttrium is found in the same ores that contain lanthanoids. The first discovery of a rare earth mineral was near the town of Ytterby in Sweden, as names of several of these elements testify: yttrium, terbium, erbium, and ytterbium. Both scandium and yttrium differ from their transition metal neighbors in that their only oxidation state is a d^0 electron configuration. Hence they do not exhibit the range of oxidation states that is characteristic of the transition metals as a whole.

The rare earth metals have few uses, and their annual production amounts to about 20 000 tonnes. Most of the metals are used as additives in special purpose steels. However, there is another use that places lanthanoid compounds in almost every household—the phosphors inside color television tubes. The impact of electrons on certain mixed lanthanoid compounds results in the emission of visible light over a small wavelength range. Thus the inside surface of a television tube (and computer color monitor) is coated with tiny patches of three different lanthanoid compositions to give the three colors that make up the color image. For example, a mixed oxide of europium and yttrium, $(Eu,Y)_2O_3$, releases an intense red color when bombarded by high-energy electrons.

In 1911 it was found that cooling certain metals to close to absolute zero caused them to lose all electrical resistance and become superconductors. Later it was found that many of the superconducting materials repelled a magnetic field—the Meissner effect. The first compounds to show superconductivity were discovered in the 1950s, but they had to be cooled to a temperature close to absolute zero before superconductivity was attained. The breakthrough came in Switzerland during 1985, when George Bedornz and Alex Müller prepared an oxide containing lanthanum, barium, and copper(II) ions. This compound becomes superconducting at 35 K, and Bedornz and Müller were awarded the Nobel prize in physics for their work. One year later, Paul Chu at the University of Houston and Maw-Kuen Wu at the University of Alabama, Huntsville, synthesized a compound, $YBa_2Cu_3O_7$, that is superconducting above 77 K. Since then, other mixed metal oxides that are superconducting at a comparatively high temperature have been prepared.

Like many superconducting compounds, $YBa_2Cu_3O_7$ has a perovskite-derived structure. Figure 21.3 shows the structure of perovskite itself, $CaTiO_3$; it has the large calcium ion in the center and the small titanium(IV) ions at the unit cell corners. To construct the superconductor, we stack three perovskite unit cells and replace the titanium(IV) ions by copper(II) ions, the top and bottom calcium ions by barium ions, and the central barium ion by a yttrium ion (Figure 21.4a). This gives a formula of "$YBa_2Cu_3O_9$," which is erroneous, because there is an excess of anions. Thus to arrive at the actual

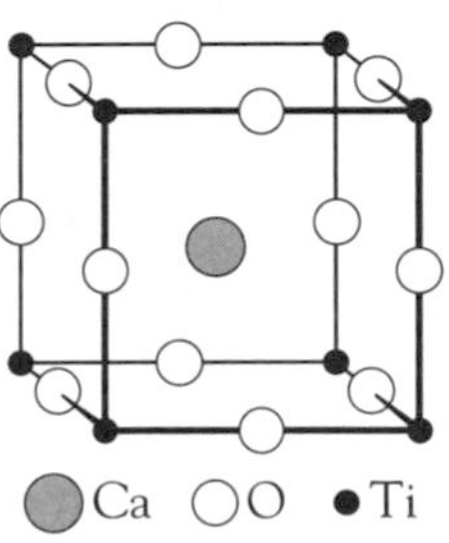

Figure 21.3 The structure of perovskite, $CaTiO_3$.

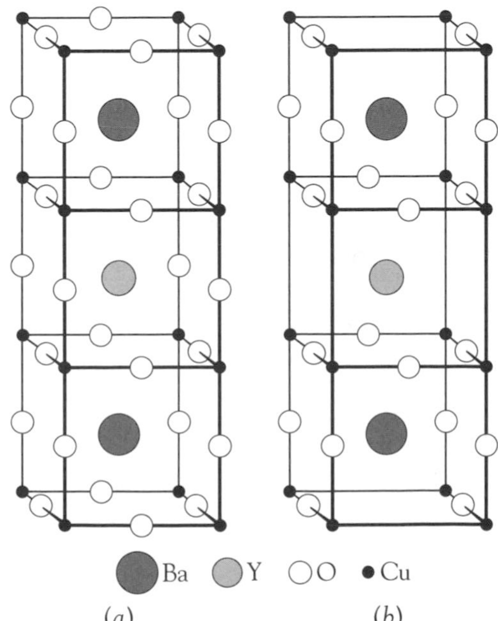

Figure 21.4 (*a*) The perovskite-like structure of the theoretical "$YBa_2Cu_3O_9$." (*b*) The related structure of the superconducting $YBa_2Cu_3O_7$.

formula of $YBa_2Cu_3O_7$, eight edge-shared oxygen ions are removed (a total of two complete ions) to give the structure shown in Figure 21.4b.

Most superconducting compounds have three common features: Their structures are related to the perovskite crystal lattice; they always contain slightly fewer oxygen atoms than the stoichiometry requires; and they usually contain copper as one of the metal ions. Why is superconductivity of such interest? There are many ways in which high-temperature superconductors would change our lives. For example, a tremendous proportion of the generated electricity is lost in the transmission cables as a result of the resistance of the wires and the ensuing conversion of electrical energy to heat energy. Superconducting wires would economize our energy use.

Properties of the Actinoids

Earlier we considered lanthanum as a member of the lanthanoid series. Similarly, we will consider actinium as an actinoid. Again, physical and chemical similarities provide the basis for this inclusion.

The actinoids are all radioactive. The half-lives of the isotopes of both thorium and uranium are long enough to allow appreciable quantities of these elements to exist in the rocks on Earth. The isotope half-lives of the longest lived isotopes of these elements are shown in Table 21.2. The values show that there is a dramatic reduction in an isotope's half-life as atomic number increases.

It is obviously the long-lived elements that have been studied in the most detail (thorium, protactinium, uranium, neptunium, plutonium, and ameri-

Table 21.2 Half-lives of the longest lived isotope of each actinoid

Element	Half-life
Actinium	22 years
Thorium	1×10^{10} years
Protactinium	3×10^{4} years
Uranium	4.5×10^{9} years
Neptunium	2×10^{6} years
Plutonium	8×10^{7} years
Americium	8×10^{3} years
Curium	18 years
Berkelium	3×10^{2} days
Californium	360 years
Einsteinium	250 days
Fermium	4.5 days
Mendelevium	1.5 hours
Nobelium	3 seconds

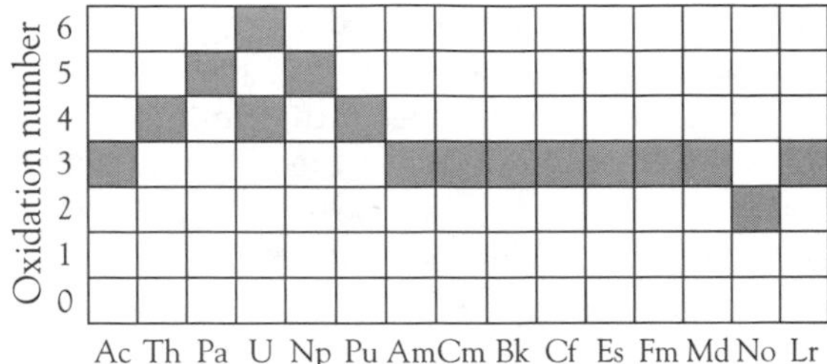

Figure 21.5 The most common oxidation numbers of the actinoids.

cium). These metals are dense (about 15–20 g·cm^{-3}) and have high melting points (about 1000°C) and high boiling points (about 3000°C). The actinoids are not as reactive as the lanthanoids; for example, they react with hot, but not cold, water to give the hydroxide and hydrogen gas. They also differ from the lanthanoids in that they exhibit a range of oxidation numbers in their compounds. The most common oxidation numbers of the actinoids are shown in Figure 21.5.

The pattern of the highest common oxidation states of the early actinoids reflects the loss of all outer electrons, and it is more parallel to that of the transition metals than to that of the lanthanoids (see Table 21.3). For example, uranium has the electron configuration of [Rn]$7s^25f^36d^1$. Thus the formation of the common oxidation state of +6 corresponds to an electron configuration of [Rn]. Like the lanthanoids, formation of the 3+ ion corresponds to the loss of the *s* and *d* electrons before those of the *f* orbitals.

Table 21.3 Electron configurations of elements 89–103

Element	Atom configuration	3+ ion configuration
Actinium	[Rn] $7s^25f^06d^1$	[Rn] $5f^0$
Thorium	[Rn] $7s^25f^06d^2$	[Rn] $5f^1$
Protoactinium	[Rn] $7s^25f^26d^1$	[Rn] $5f^2$
Uranium	[Rn] $7s^25f^3\,6d^1$	[Rn] $5f^3$
Neptunium	[Rn] $7s^25f^46d^1$	[Rn] $5f^4$
Plutonium	[Rn] $7s^25f^6$	[Rn] $5f^5$
Americium	[Rn] $7s^25f^7$	[Rn] $5f^6$
Curium	[Rn] $7s^25f^76d^1$	[Rn] $5f^7$
Berkelium	[Rn] $7s^25f^9$	[Rn] $5f^8$
Californium	[Rn] $7s^25f^{10}$	[Rn] $5f^9$
Einsteinium	[Rn] $7s^25f^{11}$	[Rn] $5f^{10}$
Fermium	[Rn] $7s^25f^{12}$	[Rn] $5f^{11}$
Mendelevium	[Rn] $7s^25f^{13}$	[Rn] $5f^{12}$
Nobelium	[Rn] $7s^25f^{14}$	[Rn] $5f^{13}$
Lawrencium	[Rn] $7s^25f^{14}6d^1$	[Rn] $5f^{14}$

The ready loss of the 5*f* electrons by the early actinoids indicates that these electrons are much closer in energy to the 7*s* and 6*d* electrons than the 4*f* electrons are to the 6*s* and 5*d* electrons in the lanthanoids. An explanation for this difference can be found in terms of the relativistic effect that we discussed in the context of the so-called inert-pair effect. As a result of the relativistic increase in the mass of the 7*s* electrons, the 7*s* orbital undergoes a contraction. Because the electrons in the 5*f* and 6*d* orbitals are partially shielded from the nuclear attraction by the 7*s* electrons, these orbitals expand. As a result, all three orbital sets have very similar energies. In fact, even the middle actinoids very often exhibit the +4 oxidation state, in which a second 5*f* electron must have been lost.

It is important to realize that the half-lives of the natural thorium and uranium isotopes are so long that the radiation from these elements and their compounds is quite negligible. Hence we find these elements in everyday use. For example, thorium(IV) oxide, ThO_2, mixed with 1 percent cerium(IV) oxide converts heat energy from burning natural gas or propane to an intense light. Before the incandescent light bulb, a gauze (gas mantle) of this mixed oxide was placed around a gas flame to provide the major source of indoor lighting. Even today, there is a significant demand for these mantles in camping lights. Thorium(IV) oxide ceramic is also used for high-temperature reaction crucibles because it will withstand temperatures up to 3300°C.

The only actinoid element found in almost every home is americium-241. Because it has such a short half-life, americium-241 does not occur naturally, so it is obtained from nuclear reactor wastes. This isotope is at the heart of all common smoke detectors. It functions by ionizing the air in a sensing chamber, causing an electric current to flow. Smoke particles block the flow of ions, and the drop in current initiates the alarm. Of increasing concern is the disposal of defunct smoke detectors—particularly in areas where incineration is used for garbage disposal. It is preferable to contact the manufacturer to obtain an address to which the old unit can be shipped and the americum-241 recycled.

Extraction of Uranium

Uranium is the one actinoid in large demand—because of its use in nuclear reactors. Uranium is found in ore deposits around the world. Furthermore, seawater contains about 3 ppb, which does not appear to be much; but the total amount in all the oceans is about 5×10^9 tonnes. At present, the cheapest extraction method uses mined uranium(IV) oxide, UO_2, commonly called pitchblende. The shafts of uranium mines must be ventilated with massive volumes of fresh air to prevent the levels of radon in the mine atmosphere, released by radioactive decay of the uranium, from exceeding safe values.

Like most metal extractions, a number of routes are used. The following method has the most interesting chemistry. The ore containing the uranium(IV) oxide is first treated with an oxidizing agent, such as the iron(III) ion, to produce uranium(VI) oxide, UO_3:

$$UO_2(s) + H_2O(l) \rightarrow UO_3(s) + 2\ H^+(aq) + 2\ e^-$$

$$Fe^{3+}(aq) + e^- \rightarrow Fe^{2+}(aq)$$

Addition of sulfuric acid produces a solution of uranyl sulfate, which contains the uranyl cation, UO_2^{2+}:

$$UO_3(s) + H_2SO_4(aq) \rightarrow UO_2SO_4(aq) + H_2O(l)$$

After removal of impurities, ammonia is added to the solution to give a bright yellow precipitate of ammonium diuranate, $(NH_4)_2U_2O_7$:

$$2\ UO_2SO_4(aq) + 6\ NH_3(aq) + 3\ H_2O(l) \rightarrow (NH_4)_2U_2O_7(s) + 2\ (NH_4)_2SO_4(aq)$$

This precipitate, often called "yellow cake," is the common marketable form of uranium.

For use in most types of nuclear reactors and for bomb manufacture, the two common isotopes of uranium, U-235 and U-238, must be separated. This is usually accomplished by allowing gaseous uranium(VI) fluoride to diffuse through a membrane; the lower mass molecules containing U-235 generally pass through more quickly. Again, there are several ways to manufacture this compound. One route is to heat the yellow cake to give the mixed oxide, uranium(IV) uranium(VI) oxide, U_3O_8:

$$9\ (NH_4)_2U_2O_7(s) \xrightarrow{\Delta} 6\ U_3O_8(s) + 14\ NH_3(g) + 15\ H_2O(g) + N_2(g)$$

The mixed uranium oxide is then reduced with hydrogen to uranium(IV) oxide:

$$U_3O_8(s) + 2\ H_2(g) \rightarrow 3\ UO_2(s) + 2\ H_2O(g)$$

The uranium(IV) oxide is treated with hydrogen fluoride to give uranium(IV) fluoride, UF_4:

$$UO_2(s) + 4\ HF(g) \rightarrow UF_4(s) + 2\ H_2O(l)$$

Finally, the solid green uranium(IV) fluoride is oxidized to the required gaseous uranium(VI) fluoride, UF_6, by using difluorine synthesized at the site:

$$UF_4(s) + F_2(g) \rightarrow UF_6(g)$$

The low boiling point of uranium(VI) fluoride is crucial to the purification of uranium and its isotopic separation. If we compare uranium(IV) fluoride and uranium(VI) fluoride, the contrast in physical properties becomes apparent. For example, uranium(IV) fluoride, UF_4, melts at 960°C, whereas uranium(VI) fluoride, UF_6, sublimes at 56°C. The difference can be interpreted in terms of charge densities—uranium(IV) ion, 140 $C{\cdot}mm^{-3}$, and uranium(VI) ion, 348 $C{\cdot}mm^{-3}$. Thus the latter (theoretical) 6+ ion would be sufficiently polarizing to cause covalent behavior.

The Postactinoid Elements

Even though the postactinoid elements are transition metals, it is more instructive to consider them in this chapter because they can only be synthesized in nuclear reactions, like most of the actinoid elements. In fact, the

A Natural Fission Reactor

The atomic mass values that we use for elements assume that the isotope ratio is always constant. But this is not always true. For example, it was variations in the atomic mass values for lead—different values for lead from different sources—that first led Sir Frederick Soddy, a British chemist, to deduce the existence of isotopes. More recently, in a sample of uranium ore, only 0.296 percent of the uranium was found to be uranium-235, much less than the "normal" value of 0.720 percent.

This discrepancy might seem to be of little interest, but it brought scientists from around the world to the mine site at Okla in Gabon (in western Africa). We know that the U-235 isotope spontanously fissions to give energy and various fission products. When nuclear chemists and physicists examined the chemical composition of the ore, they found 15 common fission products. In other words, at some time in the past, Okla had been the site of a nuclear reaction.

The existence of this buried nuclear reaction was not a sign of some visitors from outer space nor of some previous civilization. Instead, it was a result of the early uranium composition on this planet. Uranium-235 has a much shorter half-life than U-238; hence the proportion of U-235 is steadily decreasing. About 2 billion years ago, when the Okla nuclear reaction occurred (an event that lasted between 2×10^5 and 1×10^6 years), there was about 3 percent U-235 in the Okla rocks. Rainwater is believed to have leached the uranium salts into pockets, where the uranium was concentrated enough to initiate the fission chain reaction. Equally important, the water acted as a moderator, slowing the emitted neutrons so that they could fission a neighboring nucleus and continue the chain reaction. The discovery of the ancient reaction was an interesting event for scientists, even if it did not have tabloid newspaper appeal.

short-lived elements with atomic numbers greater than 100—the later actinoids and the postactinoids—are sometimes called the *transfermium elements*. Up to now, six postactinoid elements have been definitely synthesized (but claims of the synthesis of element 110 have been made). However, their short half-lives have made it very difficult to study their chemistry. Indeed, the only known isotope of element 109 has a half-life of 5×10^{-6} seconds.

A major controversy arose over the naming of these elements. The choice of an element name belongs to its discoverer. In the case of the postactinoids, three nuclear facilities have competed for the first synthesis of each of these elements: Berkeley, California; Dubna, Russia; and Darmstadt, Germany. Unfortunately, it was sometimes difficult to identify which country had the most valid claim. For example, the first discovery of element 104 was claimed by both the Berkeley and Dubna groups, the former naming the element rutherfordium, and the latter, kurchatovium. While the conflicting claims were being settled, the International Union of Pure and Applied Chemistry (IUPAC) devised a hybrid Latin-Greek numerical method of providing provisional names and symbols for all newly discovered elements. In a

Table 21.4 Names of the postactinoid elements

Element number	Former IUPAC name	New IUPAC name
104	Unnilquadium (Unq)	Dubnium (Db)
105	Unnilpentium (Unp)	Joliotium (Jl)
106	Unnilhexium (Unh)	Rutherfordium (Rf)
107	Unnilseptium (Uns)	Bohrium (Bh)
108	Unniloctium (Uno)	Hahnium (Hn)
109	Unnilennium (Une)	Meitnerium (Mt)

recent report of the IUPAC committee, the conflicting claims have been adjudicated and the successful claimants have provided names and symbols for their discoveries (Table 21.4). The only rejected name was that of seaborgium (the U.S. preference for element 106) because the IUPAC committee refused to allow the name of a living person to be used.

Little is known about the chemistry of these elements. Dubnium forms a chloride, $DbCl_4$, which seems to be analogous with hafnium(IV) chloride, $HfCl_4$; and joliotium forms a chloride, $JlCl_5$, similar to tantalum(V) chloride, $TaCl_5$.

Exercises

21.1. Write balanced chemical equations for the following chemical reactions:
(a) europium with water
(b) uranium(VI) oxide with sulfuric acid

21.2. Although +3 is the common oxidation state of the rare earth elements, europium and ytterbium can form an ion with a 2+ charge. Suggest an explanation for this. What other oxidation state might terbium adopt?

21.3. The europium ion is almost identical in size to the strontium ion. Which simple europium salts would you expect to be water-soluble and which -insoluble?

21.4. Discuss the reasons for and against including scandium and yttrium with the lanthanoids.

21.5. A solution of cerium(IV) ion is acidic. Write a chemical equation to account for this.

21.6. Lanthanum only forms a trifluoride, whereas cerium forms both a trifluoride and tetrafluoride. Identify the reason for the difference on the basis of Born-Haber cycle calculations for each of the four possibilities (LaF_3, LaF_4, CeF_3, and CeF_4). Assume the MX_3 lattice energy to be –5000 $kJ{\cdot}mol^{-1}$ and the MX_4 lattice energy to be –8400 $kJ{\cdot}mol^{-1}$. Obtain the other necessary values from data tables.

21.7. Explain why the longest lived isotopes of actinium and protactinium have much shorter half-lives than those of thorium and uranium.

21.8. Explain why nobelium is the only actinoid for which the +2 oxidation state is most common.

21.9. There were convincing chemical reasons why the early actinoids seemed to fit with the transition metals. Suggest one of the most important reasons and, in particular, mention the diuranate ion in your discussion.

21.10. Use the accompanying Frost diagram to comment on the redox chemistry of uranium.

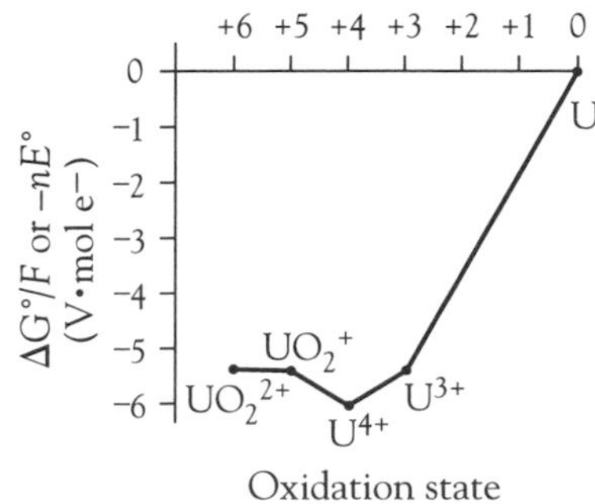

21.11. Suggest the probable formula of the oxide of rutherfordium, should it be synthesized. Explain your reasoning.

APPENDIX

1 Study Questions

1. Identify each of the following ions and write net ionic equations for each reaction.

(a) A colorless cation that gives a white precipitate with chloride ion and a red-brown precipitate with chromate ion.

(b) A pale pink cation that gives a deep blue color with chloride ion. The cation gives a blue solid with hydroxide ion.

(c) A colorless anion that gives a yellowish precipitate with silver ion. Addition of aqueous chlorine to the anion produces a deep brown color that extracts into organic solvents as a purple color.

(d) A yellow anion that gives a yellow precipitate with barium ion. Addition of acid to the anion causes a color change to orange. The orange anion is reduced by sulfur dioxide to give a green cation; the other product is a colorless anion that gives a white precipitate with barium ion.

(e) A pale blue cation that reacts with zinc metal to give a red-brown solid. Addition of the pale blue cation to excess ammonia gives a deep blue color.

(f) A colorless anion that gives a colorless, odorless gas when treated with an acid. The gas gives a white precipitate when bubbled through a solution of calcium hydroxide.

2. Identify each of the following reactants, writing balanced chemical equations for each reaction.

(a) A metal (A) reacts with water to give a colorless solution of compound (B) and a colorless gas (C). Common dilute diprotic acid (D) is added to (B), forming a dense white precipitate (E).

(b) A solution of (F) slowly decomposes to give a liquid (G) and a colorless gas (H). Gas (H) reacts with colorless gas (C) to give liquid (G).

(c) Under certain conditions, colorless acidic gas (I) will react with gas (H) to give a white solid (I). Addition of (G) to (I) gives a solution of acid (D).

(d) Metal (A) burns in excess gas (H) to give compound (J). Compound (J) dissolves in water to produce a solution of (B) and (F).

3. A gas (A) was bubbled into a solution of a common monopositive hydroxide (B) to give a solution of the salt (C). The cation of (B) gives a precipitate with the tetraphenylborate ion. Heating yellow solid (D) with a solution of (C) and evaporating the water gives crystals containing anion (E). Addition of iodine to a solution of anion (E) gives iodide ion and a solution of anion (F). Addition of hydrogen ion to a solution of anion (E) initially produces acid (G), which decomposes to form solid (D) and gas (A). Identify (A) through (G), writing balanced equations for each step.

4. A red substance (A), when heated in the absence of air, vaporized and recondensed to give a yellow waxy substance (B). (A) did not react with air at room temperature, but (B) burned spontaneously to give clouds of a white solid (C). (C) dissolved exothermically in water to give a solution containing a triprotic acid (D).

(B) reacted with a limited amount of chlorine to give a colorless fuming liquid (E), which in turn reacted further with chlorine to give a white solid (F). (F) gave a mixture of (D) and hydrochloric acid when treated with water. When water was added to (E), a diprotic acid (G) and hydrochloric acid were produced.

Identify substances (A) to (G) and write equations for all reactions.

5. Dilute hydrochloric acid was added to a metallic-looking compound (A). A colorless gas (B) with a characteristic odor was formed together with a pale green solution of the cation (C).

The gas (B) was burned in air to give another colorless gas (D) that turned yellow dichromate paper green. Mixing (B) and (D) gave a yellow solid element (E). Depending on the mole ratios, (E) reacted with chlorine gas to give two possible chlorides, (F) and (G).

Addition of ammonia to a sample of the green cation solution (C) gave a pale blue complex ion (H). Addition of hydroxide ion to another sample of the green solution gave a green gelatinous precipitate (I). Addition of zinc metal to a third sample of the green solution gave a metal (J) that on drying could be reacted with carbon monoxide to give a compound (K) with a low boiling point.

Identify each of the substances and write balanced chemical equations for each reaction.

6. A compound (A) of a dipositive metal ion is dissolved in water to give a colorless solution. Hydroxide ion is added to the solution. A gelatinous white precipitate (B) initially forms, but in excess hydroxide ion, the precipitate redissolves to give a colorless solution of complex ion (C). Addition of concentrated ammonia solution to the precipitate (B) gives a colorless solution of complex ion (D). Addition of sulfide ion to a solution of compound (A) gives a highly insoluble white precipitate (E).

Addition of silver ion to a solution of compound (A) results in a yellow precipitate (F). Addition of aqueous bromine to a solution of (A) gives a black solid (G), which can be extracted into an organic solvent and gives a purple solution. The solid (G) reacts with thiosulfate ion to give a colorless solution containing ions (H) and (I), the latter being an oxyanion.

Identify (A) to (I) and write balanced equations for each reaction.

7. When a very pale pink salt (A) is heated strongly, a brown-black solid (B) is produced; a deep brown gas (C) is the only other product. Addition of concentrated hydrochloric acid to (B) gives a colorless solution of salt (D), a pale green gas (E), and water. When the pale green gas is bubbled into a solution of sodium bromide, the solution turns brown. The yellow-brown substance (F) can be extracted into dichloromethane and other low-polarity solvents.

The brown solid (B) can also be produced when a deep purple solution of the anion (G) is reacted in basic solution with a reducing agent, such as hydrogen peroxide. The other product is a gas (H), which will relight a glowing splint. The anion of compound (A) does not form any insoluble salts, whereas the gas (C) is in equilibrium with colorless gas (I), the latter being favored at low temperatures.

Identify (A) through (I), writing balanced equations for each step.

APPENDIX 2

Charge Density of Selected Ions

Charge densities ($C{\cdot}mm^{-3}$) are calculated according to the formula

$$\frac{ne}{(4/3)\pi r^3}$$

where the ionic radii r are the Shannon-Prewitt values in millimeters (*Acta Cryst.*, 1976, A32, 751), e is the electron charge (1.60×10^{-19} C), and n represents the ion charge. The radii used are the values for six-coordinate ions except where noted by (T) for four-coordinate tetrahedral ions; (HS) and (LS) designate the high spin and low spin radii for the transition metal ions.

Cation	Charge density
Ac^{3+}	57
Ag^{+}	15
Ag^{2+}	60
Ag^{3+}	163
Al^{3+}	770 (T)
Al^{3+}	364
Am^{3+}	82
As^{3+}	307
As^{5+}	884
At^{7+}	609
Au^{+}	11
Au^{3+}	118
B^{3+}	7334 (T)
B^{3+}	1663
Ba^{2+}	23
Be^{2+}	1108 (T)
Bi^{3+}	72
Bi^{5+}	262
Bk^{3+}	86
Br^{7+}	1796
C^{4+}	6265 (T)
Ca^{2+}	52
Cd^{2+}	59
Ce^{3+}	75
Ce^{4+}	148
Cf^{3+}	88
Cl^{7+}	3880
Cm^{3+}	84
Co^{2+}	155 (LS)
Co^{2+}	108 (HS)
Co^{3+}	349 (LS)
Co^{3+}	272 (HS)
Co^{4+}	508 (HS)
Cr^{2+}	116 (LS)
Cr^{2+}	92 (HS)
Cr^{3+}	261
Cr^{4+}	465
Cr^{5+}	764
Cr^{6+}	1175
Cs^{+}	6
Cu^{+}	51
Cu^{2+}	116
Dy^{2+}	43
Dy^{3+}	99
Er^{3+}	105
Eu^{2+}	34
Eu^{3+}	88
F^{7+}	25 110
Fe^{2+}	181 (LS)
Fe^{2+}	98 (HS)
Fe^{3+}	349 (LS)
Fe^{3+}	232 (HS)
Fe^{6+}	3864
F^{4+}	5
Ga^{3+}	261
Gd^{3+}	91
Ge^{2+}	116
Ge^{4+}	508
Hf^{4+}	409
Hg^{+}	16
Hg^{2+}	49
Ho^{3+}	102
I^{7+}	889
In^{3+}	138
Ir^{3+}	208
Ir^{5+}	534
K^{+}	11
La^{3+}	72
Li^{+}	98 (T)
Li^{+}	52
Lu^{3+}	115
Mg^{2+}	120
Mn^{2+}	144 (LS)
Mn^{2+}	84 (HS)
Mn^{3+}	307 (LS)
Mn^{3+}	232 (HS)
Mn^{4+}	508
Mn^{7+}	1238
Mo^{3+}	200
Mo^{6+}	589
NH_4^{+}	11
Na^{+}	24
Nb^{3+}	180
Nb^{5+}	402
Nd^{3+}	82
Ni^{2+}	134
No^{2+}	40
Np^{5+}	271
Os^{4+}	335
Os^{6+}	698
Os^{8+}	2053
P^{3+}	587
P^{5+}	1358
Pa^{5+}	245
Pb^{2+}	32
Pb^{4+}	196
Pd^{2+}	76
Pd^{4+}	348
Pm^{3+}	84
Po^{4+}	121
Po^{6+}	431
Pr^{3+}	79
Pr^{4+}	157
Pt^{2+}	92
Pt^{4+}	335
Pu^{4+}	153
Ra^{2+}	18
Rb^{+}	8

Cation	Charge density
Re^{7+}	889
Rh^{3+}	224
Ru^{3+}	208
S^{4+}	1152
S^{6+}	2883
Sb^{3+}	157
Sb^{5+}	471
Sc^{3+}	163
Se^{4+}	583
Se^{6+}	1305
Si^{4+}	970
Sm^{3+}	86
Sn^{2+}	54
Sn^{4+}	267

Cation	Charge density
Sr^{2+}	33
Ta^{3+}	180
Ta^{5+}	402
Tb^{3+}	96
Tc^{4+}	310
Tc^{7+}	780
Te^{4+}	112
Te^{6+}	668
Th^{4+}	121
Ti^{2+}	76
Ti^{3+}	216
Ti^{4+}	362
Ti^{+}	9
Tl^{3+}	105

Cation	Charge density
Tm^{2+}	48
Tm^{3+}	108
U^{4+}	140
U^{6+}	348
V^{2+}	95
V^{3+}	241
V^{4+}	409
V^{5+}	607
W^{4+}	298
W^{6+}	566
Y^{3+}	102
Yb^{3+}	111
Zn^{2+}	112
Zr^{4+}	240

Anion	Charge density
As^{3-}	12
Br^{-}	6
CN^{-}	7
CO_3^{2-}	17
Cl^{-}	8
ClO_4^{-}	3
F^{-}	24

Anion	Charge density
I^{-}	4
MnO_4^{-}	4
N^{3-}	50
N_3^{-}	6
NO_3^{-}	9
O^{2-}	40
O_2^{-}	13

Anion	Charge density
O_2^{2-}	19
OH^{-}	23
P^{3-}	14
S^{2-}	16
SO_4^{2-}	5
Se^{2-}	12
Te^{2-}	9

APPENDIX

3 Selected Bond Energies

Hydrogen

H–H	432	H–S	363
H–B	389	H–F	565
H–C	411	H–Cl	428
H–N	386	H–Br	362
H–O	459	H–I	295

Group 13

B–C	372	B–F	613
B–O	536	B–Cl	456
		B–I	377

All values are in units of $kJ \cdot mol^{-1}$.

Group 14

C–C	346	C–O	358
C=C	602	C=O	799
C≡C	835	C≡O	1072
C–N	305	C–F	485
C=N	615	C–Cl	327
C≡N	887	C–Br	285
C–P	264	C–I	213
Si–Si	222	Si–Cl	381
Si–O	452	Si–Br	310
Si–F	565	Si–I	234

Group 15

N–N	247	N–O	201
N=N	418	N=O	607
N≡N	942	N–F	283
		N–Cl	313
P–F	490	P–Br	264
P–Cl	326	P–I	184

Group 16

O–O	207	O–Cl	218
O=O	494	O–Br	201
O–F	190	O–I	201

Group 17

F–F	155	F–I	278
F–Cl	249	F–Xe	130
F–Br	249		
Cl–Cl	240	Cl–I	208
Cl–Br	216		
Br–Br	190	Br–I	175
I–I	149		

APPENDIX

4 Electron Configurations of the Elements

Element	1s
Hydrogen	1
Helium	2

Element	1*s*	2*s*	2*p*
Lithium	2	1	
Beryllium	2	2	
Boron	2	2	1
Carbon	2	2	2
Nitrogen	2	2	3
Oxygen	2	2	4
Fluorine	2	2	5
Neon	2	2	6

Element	1*s*	2*s*	2*p*	3*s*	3*p*
Sodium	2	2	6	1	
Magnesium	2	2	6	2	
Aluminum	2	2	6	2	1
Silicon	2	2	6	2	2
Phosphorus	2	2	6	2	3
Sulfur	2	2	6	2	4
Chlorine	2	2	6	2	5
Argon	2	2	6	2	6

Element	1*s*	2*s*	2*p*	3*s*	3*p*	3*d*	4*s*	4*p*
Potassium	2	2	6	2	6		1	
Calcium	2	2	6	2	6		2	
Scandium	2	2	6	2	6	1	2	
Titanium	2	2	6	2	6	2	2	
Vanadium	2	2	6	2	6	3	2	
Chromium	2	2	6	2	6	5	1	
Manganese	2	2	6	2	6	5	2	
Iron	2	2	6	2	6	6	2	
Cobalt	2	2	6	2	6	7	2	
Nickel	2	2	6	2	6	8	2	
Copper	2	2	6	2	6	10	1	
Zinc	2	2	6	2	6	10	2	
Gallium	2	2	6	2	6	10	2	1
Germanium	2	2	6	2	6	10	2	2
Arsenic	2	2	6	2	6	10	2	3
Selenium	2	2	6	2	6	10	2	4
Bromine	2	2	6	2	6	10	2	5
Krypton	2	2	6	2	6	10	2	6

Element	1*s*	2*s*	2*p*	3*s*	3*p*	3*d*	4*s*	4*p*	4*d*	4*f*	5*s*	5*p*
Rubidium	2	2	6	2	6	10	2	6			1	
Strontium	2	2	6	2	6	10	2	6			2	
Yttrium	2	2	6	2	6	10	2	6	1		2	
Zirconium	2	2	6	2	6	10	2	6	2		2	
Niobium	2	2	6	2	6	10	2	6	4		1	
Molybdenum	2	2	6	2	6	10	2	6	5		1	
Technetium	2	2	6	2	6	10	2	6	5		2	
Ruthenium	2	2	6	2	6	10	2	6	7		1	
Rhodium	2	2	6	2	6	10	2	6	8		1	
Palladium	2	2	6	2	6	10	2	6	10		0	
Silver	2	2	6	2	6	10	2	6	10		1	
Cadmium	2	2	6	2	6	10	2	6	10		2	
Indium	2	2	6	2	6	10	2	6	10		2	1
Tin	2	2	6	2	6	10	2	6	10		2	2
Antimony	2	2	6	2	6	10	2	6	10		2	3
Tellurium	2	2	6	2	6	10	2	6	10		2	4

Element	**1*s***	**2*s***	**2*p***	**3*s***	**3*p***	**3*d***	**4*s***	**4*p***	**4*d***	**4*f***	**5*s***	**5*p***
Iodine	2	2	6	2	6	10	2	6	10		2	5
Xenon	2	2	6	2	6	10	2	6	10		2	6

Element	**1*s***	**2*s***	**2*p***	**3*s***	**3*p***	**3*d***	**4*s***	**4*p***	**4*d***	**4*f***	**5*s***	**5*p***	**5*d***	**5*f***	**6*s***	**6*p***
Cesium	2	2	6	2	6	10	2	6	10		2	6			1	
Barium	2	2	6	2	6	10	2	6	10		2	6			2	
Lanthanum	2	2	6	2	6	10	2	6	10		2	6	1		2	
Cerium	2	2	6	2	6	10	2	6	10	1	2	6	1		2	
Praseodymium	2	2	6	2	6	10	2	6	10	3	2	6			2	
Neodymium	2	2	6	2	6	10	2	6	10	4	2	6			2	
Promethium	2	2	6	2	6	10	2	6	10	5	2	6			2	
Samarium	2	2	6	2	6	10	2	6	10	6	2	6			2	
Europium	2	2	6	2	6	10	2	6	10	7	2	6			2	
Gadolinium	2	2	6	2	6	10	2	6	10	7	2	6	1		2	
Terbium	2	2	6	2	6	10	2	6	10	9	2	6			2	
Dysprosium	2	2	6	2	6	10	2	6	10	10	2	6			2	
Holmium	2	2	6	2	6	10	2	6	10	11	2	6			2	
Erbium	2	2	6	2	6	10	2	6	10	12	2	6			2	
Thulium	2	2	6	2	6	10	2	6	10	13	2	6			2	
Ytterbium	2	2	6	2	6	10	2	6	10	14	2	6			2	
Lutetium	2	2	6	2	6	10	2	6	10	14	2	6	1		2	
Hafnium	2	2	6	2	6	10	2	6	10	14	2	6	2		2	
Tantalum	2	2	6	2	6	10	2	6	10	14	2	6	3		2	
Tungsten	2	2	6	2	6	10	2	6	10	14	2	6	4		2	
Rhenium	2	2	6	2	6	10	2	6	10	14	2	6	5		2	
Osmium	2	2	6	2	6	10	2	6	10	14	2	6	6		2	
Iridium	2	2	6	2	6	10	2	6	10	14	2	6	7		2	
Platinum	2	2	6	2	6	10	2	6	10	14	2	6	9		1	
Gold	2	2	6	2	6	10	2	6	10	14	2	6	10		1	
Mercury	2	2	6	2	6	10	2	6	10	14	2	6	10		2	
Thallium	2	2	6	2	6	10	2	6	10	14	2	6	10		2	1
Lead	2	2	6	2	6	10	2	6	10	14	2	6	10		2	2
Bismuth	2	2	6	2	6	10	2	6	10	14	2	6	10		2	3
Polonium	2	2	6	2	6	10	2	6	10	14	2	6	10		2	4
Astatine	2	2	6	2	6	10	2	6	10	14	2	6	10		2	5
Radon	2	2	6	2	6	10	2	6	10	14	2	6	10		2	6

Element	1*s*	2*s*	2*p*	3*s*	3*p*	3*d*	4*s*	4*p*	4*d*	4*f*	5*s*	5*p*	5*d*	5*f*	6*s*	6*p*	6*d*	7*s*
Francium	2	2	6	2	6	10	2	6	10	14	2	6	10		2	6		1
Radium	2	2	6	2	6	10	2	6	10	14	2	6	10		2	6		2
Actinium	2	2	6	2	6	10	2	6	10	14	2	6	10		2	6	1	2
Thorium	2	2	6	2	6	10	2	6	10	14	2	6	10		2	6	2	2
Protactinium	2	2	6	2	6	10	2	6	10	14	2	6	10	2	2	6	1	2
Uranium	2	2	6	2	6	10	2	6	10	14	2	6	10	3	2	6	1	2
Neptunium	2	2	6	2	6	10	2	6	10	14	2	6	10	4	2	6	1	2
Plutonium	2	2	6	2	6	10	2	6	10	14	2	6	10	6	2	6		2
Americium	2	2	6	2	6	10	2	6	10	14	2	6	10	7	2	6		2
Curium	2	2	6	2	6	10	2	6	10	14	2	6	10	7	2	6	1	2
Berkelium	2	2	6	2	6	10	2	6	10	14	2	6	10	9	2	6		2
Californium	2	2	6	2	6	10	2	6	10	14	2	6	10	10	2	6		2
Einsteinium	2	2	6	2	6	10	2	6	10	14	2	6	10	11	2	6		2
Fermium	2	2	6	2	6	10	2	6	10	14	2	6	10	12	2	6		2
Mendelevium	2	2	6	2	6	10	2	6	10	14	2	6	10	13	2	6		2
Nobelium	2	2	6	2	6	10	2	6	10	14	2	6	10	14	2	6		2
Lawrencium	2	2	6	2	6	10	2	6	10	14	2	6	10	14	2	6	1	2

Index